BIOLOGIA
CENTRALI-AMERICANA

Reptilia and Batrachia

Yours very truly
A Günther

ALBERT C. L. G. GÜNTHER

BIOLOGIA CENTRALI-AMERICANA

REPTILIA AND BATRACHIA

With Introductions by Hobart M. Smith and A. E. Gunther

Published in Cooperation With
COMITÉ HERPETOLÓGICO NACIONAL (MÉXICO)

1 9 8 7

SOCIETY FOR THE STUDY OF AMPHIBIANS AND REPTILES

FRONTISPIECE AND TITLEPAGE

The frontispiece depicts Albert Günther reading correspondence at his writing table at home, 1900 (see page LVIII); from a watercolor drawing by Miss Lucy Gee (later Mrs. H. Coxeter), courtesy National Portrait Gallery, London. Signature from a memorandum dated October 1881, courtesy British Museum (Nat. Hist.), London.

The drawing on the titlepage, which is also used as a headpiece in the introductory sections of the book, is adapted from a flat pottery stamp found in Veracruz. The motif is an entwined pair of "coatl" (rattlesnakes). The original stamp is in the collection of the National Museum of Anthropology of Mexico (courtesy Jorge Encisco, *Design Motifs of Ancient Mexico*. New York, Dover, 1953).

This book is issued to commemorate the Society's first meeting outside the United States of America, held in Veracruz, México, August 1987. Publication of this book also marks the 25th Anniversary of the initiation of the series *Facsimile Reprints in Herpetology*. During this period 61 separate items have been issued totalling more than 7500 pages. A complete list is printed at the end of this volume.

FACSIMILE REPRINTS IN HERPETOLOGY

KRAIG ADLER, *Editor*
Cornell University
Section of Neurobiology and Behavior
Seeley G. Mudd Hall
Ithaca, New York 14853-2702, U.S.A.

Books in the *Facsimile Reprints in Herpetology* series can be purchased from the Publications Secretary, Douglas H. Taylor, Department of Zoology, Miami University, Oxford, Ohio 45056, U.S.A. A list of all Society publications, including those of The Ohio Herpetological Society and the *Catalogue of American Amphibians and Reptiles*, is printed at the end of this book; additional copies are available from Dr. Taylor. *Facsimile Reprints* are published irregularly and ordered by separate subscription, although Society members receive a special prepublication discount. Persons having suggestions for future *Facsimile Reprints* should contact the Editor.

Members of the Society receive a quarterly technical journal (*Journal of Herpetology*) and a quarterly newsletter (*Herpetological Review*). Currently, dues are $20.00 for students, $27.00 for all others, world-wide; institutional subscriptions are $50.00. All memberships and subscriptions outside the U.S.A. require either a $5.00 surcharge to defray the *additional* surface mailing costs in excess of domestic rates or $25.00 for airmail delivery. Society members receive special prepublication discounts on *Herpetological Circulars*, *Facsimile Reprints*, and on books in the *Contributions to Herpetology* series. The *Catalogue* is available by separate subscription. Apply to the Society's Treasurer, Henri C. Seibert, Department of Zoology, Ohio University, Athens, Ohio 45701, U.S.A. Payments from overseas should be made in U.S.A. funds, by International Money Order, or may be charged to MasterCard or VISA (include account number and expiration date of credit card).

Library of Congress Catalog Number: 87-60187. ISBN: 0-916984-17-6.
Production specifications are given on the last page of this book.

TABLE OF CONTENTS

PATRONS

The Society gratefully acknowledges support from these individuals and organizations whose aid has made the publication of this book possible

BENEFACTOR

ERNEST A. LINER, Houma, Louisiana

FRIENDS OF THE SOCIETY

Individuals

KRAIG ADLER, Ithaca, New York
ROBERT D. ALDRIDGE, Saint Louis, Missouri
RONALD ALTIG, Mississippi State, Mississippi
ROBIN ANDREWS, Blacksburg, Virginia
RALPH W. AXTELL, East Alton, Illinois
JOSEPH R. BAILEY, Durham, North Carolina
ELIZABETH BECHTEL, Valdosta, Georgia
H. BERNARD BECHTEL, M.D., Valdosta, Georgia
JOHN L. BEHLER, Bronx, New York
ELINOR S. BENES, Sacramento, California
JAMES P. BOGART, Guelph, Ontario
CHARLES M. BOGERT, Santa Fe, New Mexico
RONALD A. BRANDON, Carbondale, Illinois
EDMUND D. BRODIE, JR., Arlington, Texas
E. E. BROWN, Davidson, North Carolina
R. BRUCE BURY, Fort Collins, Colorado
JANALEE P. CALDWELL, Los Angeles, California
JONATHAN A. CAMPBELL, Arlington, Texas
DAVID C. CANNATELLA, Berkeley, California
CHARLES CONGDEN CARPENTER, Norman, Oklahoma
ELLEN J. CENSKY, Pittsburgh, Pennsylvania
FRANK A. CLINGENPEEL, Etowah, Tennessee
CHARLES J. COLE, New York, New York
JOSEPH & SUZANNE COLLINS, Lawrence, Kansas
ROGER CONANT, Tucson, Arizona
JOSEPH F. COPP, La Jolla, California
RONALD I. CROMBIE, Washington, District of Columbia
MARTHA L. CRUMP, Gainesville, Florida
WILLIAM G. DEGENHARDT, Albuquerque, New Mexico
JOSEPH R. DINARDO, JR., Levittown, Pennsylvania
MICHAEL DLOOGATCH, Chicago, Illinois
C. KENNETH DODD, JR., Gainesville, Florida
WILLIAM E. DUELLMAN, Lawrence, Kansas
HAROLD DUNDEE, New Orleans, Louisiana
DONALD G. DUNLAP, Vermillion, South Dakota
ARTHUR C. ECHTERNACHT, Knoxville, Tennessee
COLIN R. EDMONDSON, Dunsford, Exeter
MICHAEL A. EWERT, Bloomington, Indiana
HENRY S. FITCH, Lawrence, Kansas

OSCAR FLORES-VILLELA, México, Distrito Federal
WAYNE FRAIR, Briarcliff, New York
MARA GRADY FREED, Houston, Texas
PAUL S. FREED, Houston, Texas
THOMAS H. FRITTS, Albuquerque, New Mexico
FREDRIC L. FRYE, Davis, California
CARL GANS, Ann Arbor, Michigan
ITZCHAK GILBOA, New York, New York
JAMES C. GILLINGHAM, Mount Pleasant, Michigan
JAMES L. GLENN, Salt Lake City, Utah
GEORGE GORMAN, Berkeley, California
N. B. GREEN, Huntington, West Virginia
PAUL GRITIS, Bethlehem, Pennsylvania
ARNOLD B. GROBMAN, Saint Louis, Missouri
LOUIS J. GUILLETTE, Gainesville, Florida
DONALD E. HAHN, Cottonwood, Arizona
STEPHEN HAMMACK, Dallas, Texas
JAMES HANKEN, Boulder, Colorado
JAMES H. HARDING, Birmingham, Michigan
DAVID L. HARDY, M.D., Tucson, Arizona
FRED S. HENDRICKS, Bryan, Texas
MARVIN M. HENSLEY, Williamston, Michigan
LLOYD W. HILL, Winter Haven, Florida
MEGAN HOPE (In honor of), Garden City, Kansas
THOMAS A. HUFF, Picton, Ontario
ARTHUR C. HULSE, Kittanning, Pennsylvania
DAVID JARDINE, Cincinnati, Ohio
THOMAS A. JENSSEN, Blacksburg, Virginia
W. HUBERT KEEN, Cortland, New York
J. P. KENNEDY, Houston, Texas
WILLIAM W. LAMAR, Tyler, Texas
ROBIN LAWSON, Menlo Park, California
LAWRENCE E. LICHT, North York, Ontario
CARL S. LIEB, El Paso, Texas
WILLIAM H. LOERY, Belle Mead, New Jersey
EDMOND V. MALNATE, Philadelphia, Pennsylvania
DALE L. MARCELLINI, Arlington, Virginia
WILLIAM H. MARTIN, Harpers Ferry, West Virginia
C. J. MCCOY, JR., Pittsburgh, Pennsylvania
HUGH K. MCCRYSTAL, Tucson, Arizona
ROY W. MCDIARMID, Washington, District of Columbia
HARRY S. MCDONALD, Nacogdoches, Texas
DONALD A. MERKLE, Farmville, Virginia
MADGE R. MINTON, Indianapolis, Indiana
SHERMAN A. MINTON, M.D., Indianapolis, Indiana
PAUL E. MOLER, Gainesville, Florida
EDWARD O. MOLL, Charleston, Illinois
PATRICK SEARLE MULVANY, Tulsa, Oklahoma
JOHN MURPHY, Plainfield, Illinois
ROBERT W. MURPHY, Toronto, Ontario
KEITH E. NEITMAN, Houston, Texas
CRAIG E. NELSON, Bloomington, Indiana
PETER D. OLDAK, M.D., South Hampton, New Hampshire
REBECCA E. PAPENDICK, Coral Gables, Florida
THEODORE J. PAPENFUSS, Berkeley, California
WILLIAM S. PARKER, Columbus, Mississippi
RAY PAWLEY, Hinsdale, Illinois
ROBERT PRICE, Wantager, New York
ROBERT W. REESE, Austin, Texas
ANDERS G. J. RHODIN, Fitchburg, Massachusetts
CHRISTINA M. RICHARDS, Ann Arbor, Michigan
LAWRENCE P. RICHARDS, Ann Arbor, Michigan
MARTIN J. ROSENBERG, Cleveland, Ohio
JAY M. SAVAGE, Coral Gables, Florida
ALAN H. SAVITZKY, Norfolk, Virginia
NORMAN J. SCOTT, JR., Albuquerque, New Mexico
D. M. SECOY, Regina, Saskatchewan
HENRI C. SEIBERT, Athens, Ohio
H. BRADLEY SHAFFER, Irvine, California
C. ROBERT SHOOP, Kingston, Rhode Island
ROBERT C. SHUPAK, Wyndmoor, Pennsylvania
JACK W. SITES, Orem, Utah
C. DOUGLAS SMITH, Fort Worth, Texas
HOBART M. SMITH, Boulder, Colorado
JAMES R. SPOTILA, Buffalo, New York
GLENN R. STEWART, Pomona, California
MARGARET M. STEWART, Albany, New York
ROBERT M. STORM, Corvallis, Oregon
WILMER W. TANNER, Provo, Utah
DOUGLAS H. TAYLOR, Oxford, Ohio
JOSEPH A. TIHEN, South Bend, Indiana
CAROL R. TOWNSEND, Monroe, New York
LINDA TRUEB, Lawrence, Kansas
ROBERT G. TUCK, JR., El Paso, Texas
R. WAYNE VAN DEVENDER, Boone, North Carolina

BESS S. VAUGHAN (In memoriam), Spartanburg, South Carolina
JAIME VILLA R., Kansas City, Missouri
LAURIE J. VITT, Los Angeles, California
RICHARD C. VOGT, San Andrés Tuxtla, Veracruz
DAVID B. WAKE, Berkeley, California
MARVALEE H. WAKE, Berkeley, California
WARREN F. WALKER, Ossipee, New Hampshire
RICHARD WASSERSUG, Halifax, Nova Scotia
FRANK T. WATROUS, III, Herndon, Virginia
ROBERT G. WEBB, El Paso, Texas
J. KIRWIN WERNER, Marquette, Michigan
DANIEL C. WILHOFT, Newark, New Jersey
THOMAS E. YARBROUGH, Eastaboga, Alabama
RUTH M. ZANTZINGER, Philadelphia, Pennsylvania
ROBERT T. ZAPPALORTI, Beachwood, New Jersey

Organizations

GAINESVILLE HERPETOLOGICAL SOCIETY, Gainesville, Florida
HERPETOLOGICAL EDUCATION FOUNDATION, Arlington, Texas
IOWA HERPETOLOGICAL SOCIETY, Ames, Iowa
LEHIGH VALLEY HERPETOLOGICAL SOCIETY, Sellersville, Pennsylvania
MICHIGAN SOCIETY OF HERPETOLOGISTS, Lansing, Michigan
MINNESOTA HERPETOLOGICAL SOCIETY, Minneapolis, Minnesota
NORTH TEXAS HERPETOLOGICAL SOCIETY, Fort Worth, Texas

EDITOR'S NOTE

This book has resulted from a happy coincidence that occurred at the Society's 1986 Annual Meeting. As long ago as 1967 the Board of Directors had entertained the possibility of holding a meeting in México and this hope was realized at last when the Board accepted a proposal to meet in Veracruz in 1987, which would represent the first SSAR meeting ever held outside the United States of America. The Society wanted to commemorate this important event in its history by issuing a special publication. Günther's classic had been on our list to reprint for some years and, given its geographic coverage, seemed to be the appropriate choice. Later, in discussing the idea with Ernest A. Liner, whose researches in México are well known, he strongly encouraged this choice by issuing a most generous challenge. If we were to publish Günther's book with all of its 12 color plates in full color, then he would offer a sizeable donation to help cover the considerable additional expense of the color plates.

It was decided to accept this challenge and to use it to raise additional funds on a matching basis. Accordingly, an appeal was made to a select group of members who had assisted the Society in its past projects, a group that happens to include most individuals who have published on the herpetology of México and Central America over the past several decades. In addition, the regional herpetological societies in the United States were solicited for support. The response to our appeal was overwhelming and, collectively, these donations, together with Ernie's original gift, have not only permitted us to publish all of the color plates but to offer the book at a lower prepublication price than would otherwise have been possible.

The Society is pleased to acknowledge this generosity on the preceding pages. We thank Ernie Liner, the Friends of the Society, and the several regional organizations for allowing us to issue this book in a format that we trust is fitting for the special occasion of the Veracruz meeting. SSAR is also pleased to acknowledge the cooperation of its sister organization in México, the Comité Herpetológico Nacional, in sponsoring and distributing this book, in particular the aid of Oscar Flores-Villela.

BIBLIOGRAPHIC INFORMATION

Günther's volume on amphibians and reptiles is one of the most important publications on the herpetology of Middle America and it is still in active use, yet it is not widely available. The original edition consisted of probably no more than 300 copies, according to the best estimates of several natural history booksellers, and is regarded by them as one of the scarcest volumes in the entire *Biologia Centrali-Americana* series. As Hobart Smith notes in his introduction, the herpetology volume cost the equivalent of about US$37 in 1902 when the volume had just been completed, a very considerable sum of money at the time. In recent years copies have sold for as much as $1000 on the second-hand market, when they can be found at all.

Format. The *Biologia* was issued in a royal quarto format; that is, it was printed on sheets of paper of so-called "royal" size each folded twice into a signature consisting of eight pages (or four leaves, hence the term "quarto"). The dimensions of an untrimmed copy of the *Biologia* measures approximately 25.5 x 32.0 cm, but the format has been reduced in this reprint for reasons of convenience and economy. Accordingly, the plates have been reduced to 90 percent of their original size; the text, however, is reprinted actual size.

Dates of Publication. Günther's book was issued over the period 1885 to 1902 and at the foot of the first page of each signature the date of publication is given. Since the plates did not always accompany the corresponding text, these cannot be dated so easily. Fortunately, however, such details are recorded on the wrappers in which the original parts were published, and a copy of the *Biologia* with all of the wrappers intact exists in the Zoology Library of the British Museum (Nat. Hist.), London, from which set the following details are provided.

Pages	*Plates*	*Date*	*Part*
1-16	1-15	April 1885	37
17-24	16-18	May 1885	38
25-32	19-21	August 1885	40
33-56	22-25	October 1885	41
-	26-28	January 1890	82
57-80	29-30	February 1890	83
81-88	31-32	April 1893	108
89-96	33	May 1893	109
97-104	34-35	August 1893	110
105-112	36-37	October 1893	112
-	38-40	November 1893	113
113-128	41	February 1894	115
-	42-43	April 1894	116
129-136	44-45	July 1894	117
137-144	46-48	October 1894	118
145-160	49	January 1895	120
161-168	50	March 1895	121
169-176	51-52	May 1895	122
177-184	53-55	July 1895	123
185-192	56-57	August 1895	124
193-195	58-59	October 1895	125
197-212	60-61	February 1900	153
-	62-63	March 1900	154
213-220	64-66	April 1900	155
221-228	-	June 1900	156
229-236	67-68	August 1900	157
237-252	-	February 1901	162
-	69	April 1901	163
253-260	70	May 1901	164
261-268	71-73	June 1901	165
269-292	74	September 1901	166
293-300	-	December 1901	169
301-308	75-76	January 1902	170
309-326	-	May 1902	172
i-xx	-	August 1902	173

These dates are not disputed by information in either of the two zoology indexing periodicals of the era: *Archiv für Naturgeschichte* and *Zoological Record*. Moreover, I have been able further to confirm the accuracy of these dates through my examination of two sets of proofs for Günther's book in the private collections of E. O. Z. Wade and John Greatwood. Neither set is complete but they contain various stages of paged proof for approximately the first 200 pages and 60 plates. Wade's copy also contains 60 original sketches (54 pencil, 6 ink-and-watercolor) of turtles illustrated in plates 1-18. Both copies have Günther's handwritten corrections and the editor's comments and queries to the author, but more important in this context is the fact that the signatures are dated by hand or by an inked stamp by the printer, Taylor and Francis, Ltd. Although some signatures were in proof for long periods (for example, the earliest date on pages 57-64 is 30 September 1885, but this signature was not actually published until February 1890), in no case are the Taylor and Francis dates later than the month printed at the foot of the first page of the published versions. The latest date I can find, relative to the printed date of publication, is for pages 41-48; the printed date on the signature is October 1885 and the handwritten date on the proof is 14 October 1885. Thus, it seems reasonable to accept the printed dates on the signatures as accurate.

Color Plates. Twelve of the 76 lithographed plates are colored in most copies of the book. Four of these are handcolored (numbers 1, 7, 8 and 32). Since each of these four plates is individually watercolored by hand, there are variations from copy to copy in the intensity of coloration but even occasionally in some details of the patterns as well. The other color plates are chromolithographed--that is, printed in colors--but even in this instance there are occasional variations from copy to copy in the color intensity. In addition, in at least *some* copies of the book certain chromolithographed plates (*e.g.*, 62 and 71) have been touched up by hand to add highlights or to intensify colors. Thus, as with all plates partly or wholly colored by hand, not all copies are alike in every particular.

Occasional copies of Günther's book have come on the market with eight plates colored, presumably only those that were color printed.

ACKNOWLEDGMENTS

Hobart M. Smith has provided two introductory essays which will facilitate the use of Günther's book and put it in a proper historical perspective. A. E. Gunther, the author's grandson, has written an enlightening biography with a herpetological emphasis. The Society is greatly indebted to both of them for their important contributions to this reprint edition.

Several people have provided biographical or bibliographical assistance. Sybil Salvin Calverley Rampen, of Oakville, Ontario, great granddaughter of Osbert Salvin, the senior editor of the *Biologia* series, has sent much original information that was previously unpublished. The authorities of the British Museum (Nat. Hist.) have, as always, been particularly helpful; we acknowledge Rex E. R. Banks, Ann Datta, and Dorothy M. Norman. Additional information has been provided by Anita L. Karg (Hunt Institute for Botanical Documentation, Pittsburgh), James R. Dixon (Texas A & M University), Frances Lesser (Academy of Natural Sciences, Philadelphia), R. Fish (Zoological Society of London) and the librarians at Cornell University, the Athenaeum of Philadelphia, Buffalo and Erie County (New York) Public Library, Yale University and Mount Holyoke College.

Michael R. K. Lambert (The British Herpetological Society, London) has graciously checked the set of the *Biologia* at the British Museum (Nat. Hist.) to confirm the dates of publication. I thank E. O. Z. Wade and John Greatwood, both of South London, for allowing me to examine their proof copies of Günther's book.

Finally, I thank the several natural history booksellers who have provided information about the *Biologia* or on Günther: Donald E. Hahn (Cottonwood, Arizona), Howard E. Radclyffe (Bernard Quaritch Ltd., London) and Rudolph Wm. Sabbot (Woodland Hills, California). In particular I acknowledge Howard K. Swann (Wheldon and Wesley Ltd., Codicote) for information and advice, and for his considerable efforts to locate important bibliographic data for the project.

For editorial assistance with one or more of the introductory sections, I thank John S. Applegarth, Ernie Liner, Sybil Rampen, Hobart M. Smith and Howard Swann.

KRAIG ADLER

Ithaca, New York, USA
5 February 1987

A BRIEF HISTORY OF THE *BIOLOGIA CENTRALI-AMERICANA* WITH BIOGRAPHICAL NOTES ON ITS EDITORS

by Hobart M. Smith*

One of the most monumental biological publications of all time is the 67-volume *Biologia Centrali-Americana*, subtitled "Contributions to the Knowledge of the Fauna and Flora of Mexico and Central America." Edited by the eminent British naturalists Osbert Salvin and Frederick DuCane Godman, the series appeared in parts, with 1677 plates (more than 900 colored), at irregular intervals, often monthly, over a period of 37 years, 1879 to 1915. It was actually more nearly a synopsis of the entire biota, as then known, of Middle America -- México (excluding Baja California) through Panamá -- written by numerous specialists, although "certain groups, such as the Crustacea and most of the lower invertebrates, were never treated owing to a lack of material, while a number of groups of insects were omitted because no workers could be found to undertake them . . . " (Selander and Vaurie, 1962: 5).

THE *BIOLOGIA* SERIES

The work was published in parts, 215 in zoology alone, 25 in botany, 17 in archeology, each part on multiple subjects, as manuscripts became available. The parts were eventually to be broken up and their contents recombined to bring all sections of each single-subject volume together. The volume on Reptilia and Batrachia, for example, appeared in 45 sequentially numbered signatures (1-42, a, b, c), all but three of eight pages (four leaves), published in numerous parts between 1885 and 1902. Unfortunately, the printed wrappers of the many parts were as a rule discarded when the subject volumes were collated, and hence few now exist; each part bore its number, month, year, and for each section contained in it, its title, author and pages or plates or both. However, the first page of each signature fortunately bears the month and date of publication, an abbreviation of its volume title, and its own sequential number, hence greatly simplifying the task of

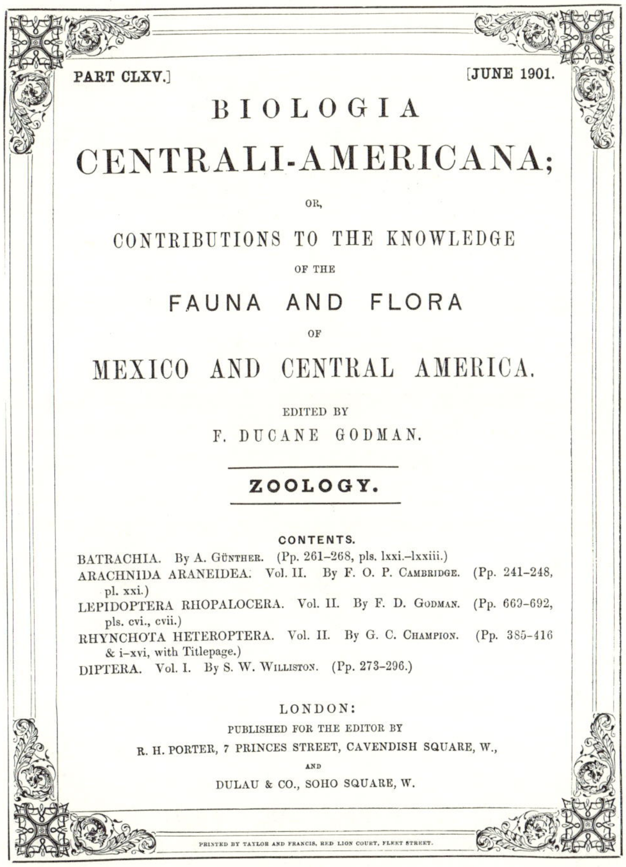
PART CLXV.] [JUNE 1901.

BIOLOGIA

CENTRALI-AMERICANA;

OR,

CONTRIBUTIONS TO THE KNOWLEDGE

OF THE

FAUNA AND FLORA

OF

MEXICO AND CENTRAL AMERICA.

EDITED BY

F. DUCANE GODMAN.

ZOOLOGY.

CONTENTS.

BATRACHIA. By A. Günther. (Pp. 261–268, pls. lxxi.–lxxiii.)
ARACHNIDA ARANEIDEA. Vol. II. By F. O. P. Cambridge. (Pp. 241–248, pl. xxi.)
LEPIDOPTERA RHOPALOCERA. Vol. II. By F. D. Godman. (Pp. 669–692, pls. cvi., cvii.)
RHYNCHOTA HETEROPTERA. Vol. II. By G. C. Champion. (Pp. 385–416 & i–xvi, with Titlepage.)
DIPTERA. Vol. I. By S. W. Williston. (Pp. 273–296.)

LONDON:
PUBLISHED FOR THE EDITOR BY
R. H. PORTER, 7 PRINCES STREET, CAVENDISH SQUARE, W.,
AND
DULAU & CO., SOHO SQUARE, W.

PRINTED BY TAYLOR AND FRANCIS, RED LION COURT, FLEET STREET.

Wrapper from part 165 of the *Biologia*, published in June 1901. This part contains a section of Günther's book--both text and plates--as well as sections of four other volumes (courtesy Kraig Adler).

---

*Department of Environmental, Population and Organismic Biology, University of Colorado, Boulder, Colorado, 80309-0334, U.S.A.

dating any given page. The plates did not always accompany their corresponding text, however. The year of publication of the Reptilia-Batrachia signatures, with their pages and plates, is given by Kraig Adler in the Editor's Note.

Although Godman, in his introduction to the *Biologia* in the introductory volume, stated that the series totalled 63 volumes, Selander and Vaurie (1962: 8) pointed out that "If separate pagination is used as the criterion for a book, the total is actually 67," a figure here accepted. Of those 67, in addition to the introductory volume, "Fifty-two . . . deal with subjects in zoology, five with botany, and nine with archeology. In zoology, one book each is devoted to Mammalia, Reptilia and Batrachia, Pisces, Mollusca, and Chilopoda and Diplopoda. Four books each are concerned with Aves and Arachnida; and the remaining 39, with Insecta . . . insects account for nearly 87 per cent of the 38,637 nominal species of animals treated, and 94 percent of the 19,067 species described as new . . . . The biogeographic and evolutionary aspects of the 'Biologia' were never developed to the extent the editors had hoped, owing in part to the gigantic proportions of the undertaking as well as to the inadequacy of knowledge of the fauna and flora of surrounding regions. Still the 'Biologia' is an incomparable contribution to the study of the natural history of the New World, and it did much to lay the foundation for the study of the biogeography of this area" (Selander and Vaurie, 1962: 5, 8-9).

The core of material on which *Biologia Centrali-Americana* was based was the collections of Salvin and Godman themselves, emphasizing birds and butterflies, respectively, and of their many native and colleagial collectors, spanning half a century or more, from 1857 on. The collections were enormous; some 100,000 butterflies, 200,000 beetles, over 520,000 birds (Sclater, 1919: 330) and untold numbers in other groups (no figures available for amphibians and reptiles) were accumulated. Tens of thousands of specimens were purchased from other private collectors, of all sorts of animals, not only from Middle America but also, for comparative purposes, from North and South America. Godman (1915: 86) puts the total number of animal and plant species treated in the *Biologia* at 50,263, of which 19,263 were described as new.

The enormity of the resource materials on which the work as a whole was based defies comprehension, and obviously involved enormous financial backing in order to pay collectors, purchase other collections, pay for full expenses of the two instigators (who were not content with collecting in the Americas, but diverted themselves with trips also to Africa and Asia), to house the collections, to build their large reference library, and to pay the considerable help they must have needed to curate the material, maintain the library and help with research. In addition there was the major expense of publication of the 67 quarto volumes with their numerous plates, many in color.

Most amazingly, and inconceivably in modern contexts, everything was privately financed, except that publication costs were at least partly covered by subscription, originally at £287 14s (approximately $475 at current exchange rates, but about $1440 then) for the entire series, unbound (Quaritch, 1918: 2). The Reptilia-Batrachia volume alone sold for £7 7s (Quaritch, 1902: 98), but by 1918, during the First World War and when the bookseller Quaritch purchased all remaining stocks of the *Biologia* from Godman, all prices were reduced and the complete herpetology section, unbound, sold for only £5 (Quaritch, 1918: 22). Each part originally was priced at 21s. The collections themselves were private in every sense -- privately obtained, privately housed (in Salvin's home for a time, later in a house they rented) and privately curated -- but eventually everything was donated to the British Museum (Natural History) by the editors, in a series of donations beginning in 1881 and including all of Günther's types (Boulenger, 1906). Obviously Salvin and especially Godman were quite wealthy, and their example of the most extraordinary, scientific dedication and leadership under circumstances that for most others would at best have led only to financial or political exploitation has set a standard that has not otherwise been even distantly approached, and in the world as we now see it probably never again can be. Theirs is a unique achievement before which other publishing ventures pale in comparison, and which can never be adequately honored.

THE EDITORS

That the whole work was finished was due to the tenacity of purpose and administrative skill of Godman and the editorial assistance of George Charles Champion, a beetle specialist, for Salvin, the other editor, died June 1, 1898, at the age of 63, having been born February 25, 1835, in Elmhurst, Finchley, England. From 1898 on, all responsibility for *Biologia Centrali-Americana* and for the joint Salvin-Godman collection fell to Godman. He was then 64, having been born January 15, 1834, in Park Hatch, Surrey, England. Despite advancing years and ill health, he managed to complete the monumental work in all its parts, although the last, introductory volume of 1915 did not do justice to the overall theoretical analysis that had at one time been the editors' hope. The coverage of the work was admittedly not complete, but what was started was brought to a most admirable conclusion. A few years later, on February 19, 1919, Godman died, at the age of 85.

Osbert Salvin was the second of two sons and two daughters of Anne Andrew and Anthony Salvin, an eminent architect (for biography of Anthony Salvin, including his immediate family, see Allibone, 1987). Frederick Godman was the third of seven sons and six daughters of Joseph Godman, a Sussex squire of great wealth. Both attended and graduated from college in Cambridge (Salvin from Trinity Hall, Godman from Trinity College), and there forged a deep, 44-year friendship that dominated their entire lives. Both became active and revered Fellows of the Linnean Society of London, as well as the Royal, Zoological, and Entomological societies, and they shared common interests in natural history, especially insects and birds. They were primary instigators in founding the British Ornithological Union and its journal *The Ibis*, and Salvin held the Strickland Curatorship of Ornithology at the University of Cambridge for nine years (1875-1883), after which he retired to private life on properties inherited from his father, who died in 1880. His last field work abroad was in 1875, but he had enjoyed a full life of collecting and exploring for 18 years, beginning with a birding expedition to Tunisia and Algeria in 1857. Later the same year he visited Central America with George U. Skinner, an orchid and bird collector experienced in the region, spending seven months there. Salvin had been commissioned to examine and report on certain palm tree nuts that were thought to be of possible value for candles; they proved useless, however, so he spent most of his time collecting birds and insects. He returned to Guatemala for a year beginning in the spring of 1859, and then in 1861 both Salvin and Godman made an extensive exploration of Guatemala, visiting Jamaica enroute, Salvin not returning until 1863. That was their last joint trip to the Americas, although they collected together several times in various parts of Europe.

Both men then remained in England for several years, studying their collections. In 1865, Salvin married Caroline Maitland, an accomplished artist who provided the color plates of plants illustrated in the botanical volumes of the *Biologia*. Salvin's last foreign trip, April 1873 to June 1875, accompanied by his wife, was extensive, covering most of Central America (although most time was spent in Guatemala), and even the larger museums of eastern United States. The rest of Salvin's life was centered in Sussex. He died of a "degenerative heart condition" in 1898, brought on, in fact, from the arsenic with which he preserved the thousands of bird skins he put up over the years (Sybil Rampen, pers. comm.).

Osbert Salvin, senior co-editor of the *Biologia*, 1864 (courtesy Sybil Salvin Calverley Rampen).

Godman, on the other hand, continued with expeditions to the Azores in 1865, to the Madeira and Canary islands in 1873, to India and Ceylon (Sri Lanka) in 1886, and to México in 1887-1888. Thereafter he remained in England, working on the *Biologia* with Salvin as long as the latter was able to overcome his deteriorating heart condition. After Salvin's death in 1898, Godman was able to continue work on the *Biologia* only with the vital aid of Champion. Godman apparently never held a salaried position, at least in biology. He was married twice, his first wife dying in 1875, and his second one, Alice Mary Chaplin, to whom he dedicated the introductory volume of the *Biologia*, surviving him with their two daughters.

Both Salvin and Godman were eminently likeable persons, according to all reports (see Anonymous, 1898, 1899, 1908; Poulton, 1919a, 1919b; Sclater, 1919), engendering the unqualified admiration and respect of their associates -- a reaction undoubtedly enhanced by near-total freedom from competitive drives. Salvin's honors began with graduation from Trinity Hall as Senior Optime in the Mathematical Tripos, and ended overtly with election as an Honorary Fellow of the same college in 1897, shortly before his death. He was an accomplished naturalist, as documented by the Royal Society's *Catalogue of Scientific Papers*, which lists 243 articles (not including books or chapters thereof) for him, including 71 by Salvin alone, 60 jointly with Godman, 108 jointly with W. L. Sclater, and four with D. G. Elliot, during the period 1856-1898. Most were on birds, many on insects, and a few early ones on reptiles (Salvin, 1861a, 1861b, 1861c). He edited the third series of *The Ibis*, and jointly with Sclater the fourth series also. Major works included the catalogue of the Strickland collection of birds; the accounts of the Trochilidae and Procellariidae in the *Catalogue of the Birds in the British Museum*; after Lilford's death, the *Coloured Figures of the Birds of the British Islands;* and jointly with Sclater, *Exotic Ornithology* and *Nomenclator Avium Neotropicalium.*

F. DuCane Godman, junior co-editor of the *Biologia*, about 1895 (courtesy Sybil Salvin Calverley Rampen).

As befits a successful collector, Salvin was an exuberant outdoorsman, quite an oarsman, fond of sports and exercise. In addition, he was a skilled mechanic and carpenter, building his own insect cabinets, and constructing at least two steamboats that participated in British regattas and that were sold to ply rivers in India, and a sailboat used to help collecting efforts on the lakes of Guatemala.

Highly compatible though Salvin and Godman were, in scientific interests as well as personality, they nevertheless were quite different. Godman was considerably more indulgently sybaritic although in admirable ways. He was particularly fond of hunting, and horseback riding, annually renting in the 1860's "the Park of the Island of Lewis [for] deerstalking, [becoming] a most accomplished stalker and very deadly rifle-shot. Later on he rented the stalking of a large sheep-farm in West Ross-shire in company with his brother Joseph. . .when Godman rented the Duke of Richmond's forest of Glenavon, he killed in his 70th year eight stags with eight successive shots" (Elwes *in* Sclater, 1919: 332-333). Salmon-fishing was also a favorite sport, for which he rented rivers in both Ireland and Scotland. "As a boy he kept a pack of beagles, and later on a pack of harriers, with which he hunted in the counties of Surrey and Sussex. He was also a constant follower of Lord Leconfield's hounds, and of those of his brother Col. C. B. Godman, for some years Master of the Crawley and Horsham pack.

"Botany and horticulture were always favorite subjects with Godman, and at his country house near Horsham he had formed one of the most beautiful gardens in Sussex, and had one of the best collections in England of rhododendrons, alpine plants, and orchids. He also formed . . . one of the finest collections of Persian and Oriental glazed pottery" (Sclater, 1919: 331). "South Lodge was originally quite a small house, on the south wall of which grew a very fine Camellia, which now covers almost the only part of the house which was left when it was rebuilt. He bought by degrees a good deal of land in the neighbourhood, much of which he farmed himself. I do not think any man ever had a happier life at home, or was more beloved by his numerous relatives, employés, and friends . . . " (Elwes *in* Sclater, 1919: 334).

Although blessed with a rugged constitution, Godman was beset with a number of illnesses and accidents, beginning at the age of 13, when a "low fever" made it impossible for him to work at all. Unable to attend schools, he was tutored for five years, gradually recovering his health sufficiently to tour the Mediterranean, Black Sea and adjacent lands and even "swimming across the Hellespont from Sestos to Abydos" (Sclater, 1919: 326), before entering Trinity College. A hunting accident in 1857, when he was 23, prevented him from joining Salvin on his trip to northern Africa. He suffered an attack of fever while collecting in Guatemala in 1862, collapsed from a heart attack at high altitude in the Himalayas in 1886, and in 1887 contracted phlebitis of the leg, requiring travel abroad for his health (to México and the United States). Yet he recovered well from all such episodes, living into his 86th year, although quite feeble in his last six years.

Travel was one of Godman's chief diversions, obviously. His first extensive foreign expedition came in 1857, after he recovered from his hunting accident, when with his brother Percy he collected birds in northern Norway, hiking across Lapland from the Norwegian sea across Norway and Sweden to the Gulf of Bosnia. Aside from expeditions previously mentioned he travelled to Egypt and South Africa with Mrs. Godman.

With all of these activities demanding so

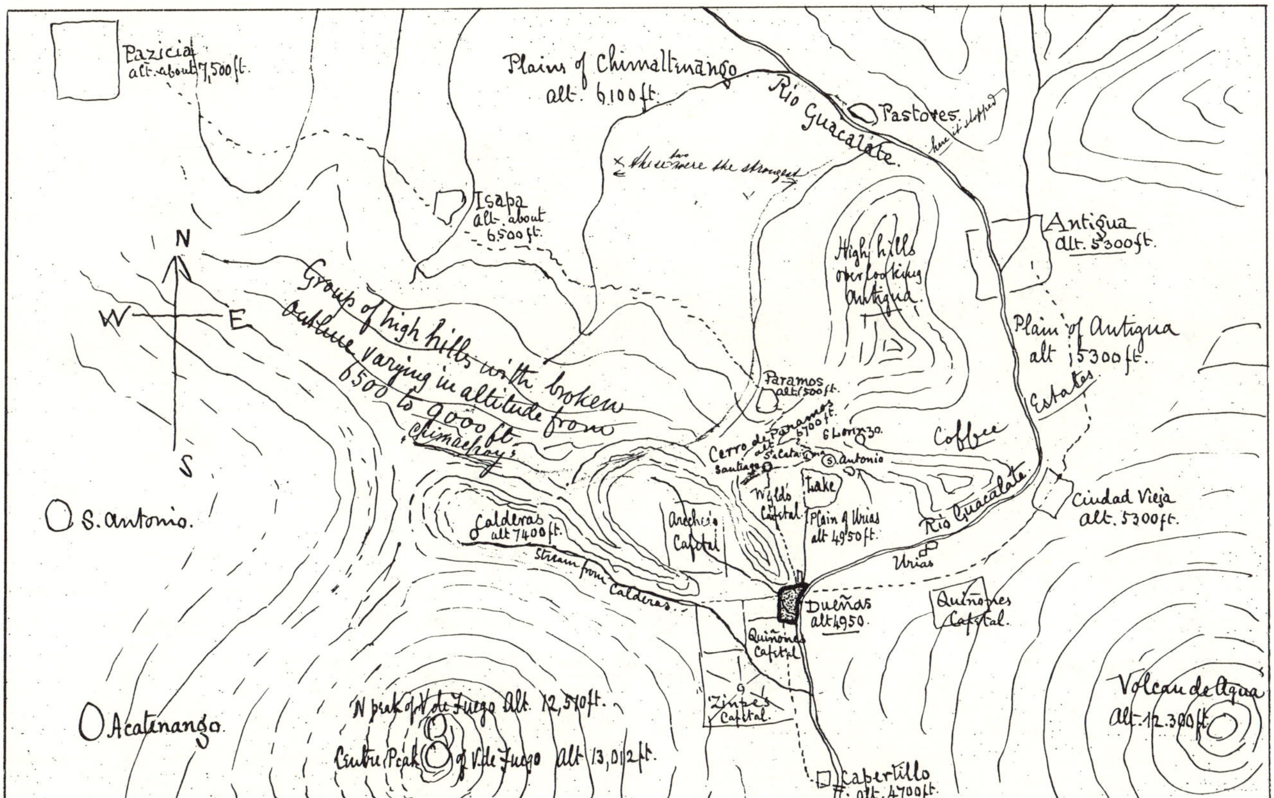

Map of region surrounding the village of Dueñas (near Antigua, southwest of Guatemala City), drawn by Osbert Salvin in 1858 and showing the detail of his field notes. Dueñas was Salvin's main headquarters in Guatemala (courtesy Sybil Salvin Calverley Rampen).

much of his time, it is surprising that Godman managed to be productive at all of scientific publications. Yet he was, of course, a prime motivator as well as contributor to the *Biologia*, and no greater contribution was necessary to leave an indelible imprint upon biology. But he authored another book, *Natural History of the Azores*, analyzing the origin of its biota, and at least 70 articles, one shared with his brother Percy (about the birds they observed and collected on their summer trip to Norway and Sweden), and 60 with Salvin (about half as first author). After 1900 he published but three articles, all by himself, in 1903, 1905 and 1907, all on butterflies. Previous articles were confined to Lepidoptera and birds. His publications, including books, thus spanned 55 years, 1861-1915.

Godman's monumental contributions were recognized by many honors. "He was for many years Vice-President and Member of Council of the Zoological Society, President of the Entomological Society, Gold Medallist of the Linnean Society in 1918, and Trustee of the British Museum; and the University of Oxford conferred on him the honorary degree of D.C.L. He was also a Fellow of the Linnean, Geological and Royal Geographic societies" (Sclater, 1919: 331).

The incredible scientific achievement of the Salvin-Godman team was a cybernetic marvel devolved from diverse talents and resources, pooled in a relentless drive fired by a shared vision of an audaciously ambitious goal. The two men contributed equally, but in different ways and in different roles. Shared vision, however clear, requires a pilot for progress toward it, and in retrospect it appears that Salvin served most regularly in that role, while Godman kept the machinery running smoothly. It was uniquely successful collaboration -- a sobering affirmation of the magnitude of human potential.

MAPS, ITINERARIES AND MANUSCRIPT RESOURCES

Excellent portraits of Salvin and Godman appear in the introductory volume of the *Biologia*, as well as a detailed itinerary (pp. 46-54) and a map showing the travel route in Guatemala, 1879-1883, of one of their most important collectors, G. C. Champion. No maps depicting the travels of either Salvin or Godman, or of their other collectors, have ever been published. Seven other maps in the same volume depict the location of most collecting stations throughout México and Central America, but, as stated by Godman in his

Field party for the *Biologia*, led by A. P. Maudslay, near the Mayan ruins at Quiriguá, Guatemala in 1863. Due to the dense foliage overhead, photography in the tropical forest was difficult and Indians were hired to cut down trees to admit sufficient light (courtesy Sybil Salvin Calverley Rampen).

introduction (p. 45) to the introductory volume, " . . . we are unable to find space . . . to include all the Mexican localities quoted in the Zoology and Botany, some of which, indeed, cannot now be traced." An excellent gazetteer for the volumes on insects in the *Biologia* is available (Selander and Vaurie, 1962) and provides an important source of locality information for all other groups. Webster (1973, 1984) added further information for W. B. Richardson's collections of birds from México, 1897-1898.

Aside from Champion, Godman and Salvin, other especially noteworthy collectors of herpetological material for the *Biologia* include T. Belt, A. Boucard, A. Forrer, G. F. Gaumer, C. T. Höge, H. Lankester, W. B. Richardson, A. Salle, Mr. and Mrs. H. H. Smith, and C. F. Underwood. Most are commemorated by herpetological patronyms.

The British Museum (Natural History) zoology library has a large collection of unpublished material of great historical importance in present contexts. It includes some 1429 letters to both Salvin and Godman, mostly from zoologists, entomologists, botanists and collectors (179 individuals in total), concerned with the *Biologia*; four of Salvin's field notebooks for his Middle American explorations, 1857-1859 and 1873-1874; and 18 of Godman's notebooks, including three of collectors' lists, three about the *Biologia*, and 12 relating to bird collections, 1859-1889. In addition, Sybil Salvin Calverley Rampen of Oakville, Ontario, possesses a collection of original materials, including Salvin's original field notebooks, correspondence concerning the *Biologia* (including letters from Günther) and original drawings and photographs. The correspondence while in the field, between Salvin or his wife and their families, and with Günther, is astonishing in volume and detail, revealing a great warmth of personality, resourcefulness and narrative skill. Salvin's wife, Caroline, made many artistic sketches accompanying her letters. These resources cry out for editorial screening and publication, as irreplaceable historical documents.

ACKNOWLEDGMENTS

I am greatly indebted to Kraig Adler for vital reference material and for the opportunity to participate in the present reprint project, honoring most deservedly not only Albert C. L. G. Günther, as author of the Reptilia and Batrachia volume of the *Biologia Centrali-Americana*, but also the remarkable editors of that series as a whole.

LITERATURE CITED

Allibone, Jill. 1987. Anthony Salvin, Pioneer of Gothic Architecture. Columbia, Univ. Missouri Press, in press.

Anonymous. 1898. Osbert Salvin, F.R.S. Nature, 58: 129.

----------. 1899. Osbert Salvin. . . . Proc. Linnean Soc. London, session 111: 59-61.

----------. 1908. Mr. Osbert Salvin. Ibis, ser. 9, 2 (Jubilee suppl.): 127-128.

Boulenger, G. A. 1906. Reptiles and batrachians, p. 517-531. *In* The History of the Collections Contained in the Natural History Departments of the British Museum. London, Brit. Mus. (Nat. Hist.). Volume 2, iv, 782 pages.

Godman, Frederick DuCane. 1915. Introduction. p. 1-86. *In* Biologia Centrali-Americana, Introductory Volume. London, for the editor by Porter and Dulau. viii, (2), 149, (1) pages, 2 plates, 8 maps.

Poulton, E. B. 1919a. Dr. F. DuCane Godman, F.R.S. Nature, 103: 5-6.

----------. 1919b. Frederick DuCane Godman. Proc. Linnean Soc. London, session 131: 54-55.

Quaritch, Bernard. 1902. Catalogue of works on natural history . . . [no. 215]. London, B. Quaritch. (3), 119, (24) pages.

----------. 1918. Prospectus. Biologia Centrali-Americana. . . . London, B. Quaritch. 47 pages, 1 plate, 1 folding map.

Salvin, Osbert. 1861a. On the reptiles of Guatemala. Proc. Zool. Soc. London, 1860: 451-461, 1 plate.

----------. 1861b. On a collection of reptiles from Guatemala. Proc. Zool. Soc. London, 1861: 227-229.

----------. 1861c. On two new reptiles from Guatemala. Ann. Mag. Nat. Hist., ser. 3, 7: 325-326.

Sclater, William Lutley [inferred]. 1919. Frederick DuCane Godman. Ibis, ser. 11, 1: 326-335, 1 plate.

Selander, Richard B. and Patricia Vaurie. 1962. A gazetteer to accompany the "Insecta" volumes of the "Biologia Centrali-Americana." Amer. Mus. Novit., 2099: 1-70.

Webster, J. D. 1973. Richardson's Zacatecas collection, I. Condor, 75: 239-241.

----------. 1984. Richardson's Mexican collection: birds from Zacatecas and adjoining states. Condor, 86: 204-207.

CURRENT NOMENCLATURE FOR THE NAMES AND MATERIAL CITED IN GÜNTHER'S REPTILIA AND BATRACHIA VOLUME OF THE *BIOLOGIA CENTRALI-AMERICANA*

by Hobart M. Smith*

The accompanying Table 2 lists in the left-hand column all of the names used by Günther, in the same sequence in which they appear in this volume. An asterisk following a name indicates a new taxon. The names used by Günther for orders are centered in the table, in their proper place relative to the species and subspecies names.

The next column of names lists first, directly opposite the Günther name, the current status of the name itself, and if no material or localities were cited that pertain to the current name as now applied, the latter name is placed in parentheses. Beneath that name are listed others currently applied to the material cited (by locality) by Günther. Where two or more non-parenthetic names are listed, their pertinence to the localities cited is indicated, if more than one country is involved, by parenthetic letters, as follows: B, Belize; C, Costa Rica; E, El Salvador; G, Guatemala; H, Honduras; M, México; N, Nicaragua; P, Panamá; SA, South America; US, United States of America; WI, West Indies. Space limitations have prevented a more detailed correlation, as for example with given countries of South America, islands of the West Indies, or states of México or the United States.

In the last two columns the page on which the binomen or trinomen appears, and the plate, if any, on which the taxon is illustrated, are cited.

Current names in this table total 1077; 197 of Günther's names pertain to two or more taxa as now understood, and vary in number from two to 13 (*Tropidonotus ordinatus*; the only others with more than six are *Sceloporus spinosus* with 10, *Geophis chalybaea* 10, *Micrurus fulvius* 10, and *Coronella annulata* 8). The mean number of current names per Günther name is 1.5.

Determination of the identities of all of the material cited, as well as the current status of the name, has occasionally been conjectural, and in some cases incomplete. Unspecified species are referred to as "sp." (plural "spp."), subspecies as "ssp." (plural "sspp."). In a few cases a cited taxon has not yet been named; it is referred to as sp. or subsp. "innominatum" (*Emys ornata*, p. XXVI, and *Chirotes canaliculatus*, p. XXVIII). It has been necessary to propose one new combination (*Sibon pictiventris*). The abbreviation "indet." (indeterminate) is used in reference to taxa of whose identity I am uncertain--a concept quite distinct from that involved by the term "nomen dubium," meaning that the name itself is of uncertain application to any known taxon. A question mark indicates a lesser degree of uncertainty that is thought still to justify a guess of identity.

It was with much reluctance that I resorted to use of the term "sp. indet." or "gen. indet.," but very frequently workers have in the past focussed simply upon correct application of the name used itself, and to a far lesser extent upon the correct identity of all of the material to which the name was applied. In most cases where identity was admitted as indeterminate, examination of the material would be required to establish identity. Undoubtedly I have missed some allocations, as they are too widely scattered in the literature for ready reference, and some are no doubt unfamiliar to me.

*Department of Environmental, Population and Organismic Biology, University of Colorado, Boulder, Colorado, 80309-0334, U.S.A.

It would have been impossible for me to have completed this table as far as it has gone without several indispensable catalogs, and the counsel of several colleagues. Published sources of broadest utility were Duellman's monograph of Middle American hylids (1970); Frost's summary of the amphibians of the world (1985); Peters' catalogs of Neotropical lizards and snakes, with Donoso-Barros (1970) and Orejas-Miranda (1970); Scott, Savage and Robinson's checklist of the Costa Rican herpetofauna (1983), and Savage and Villa's (1986) introduction to the same; Stuart's summary of the herpetofauna of Guatemala (1963); Vanzolini's addenda and corrigenda (1986a, 1986b) to Peters et al. (1970); and Wilson and Meyer's review of the snakes of Honduras (1985). Numerous other works of lesser scope have been consulted, including various reviews of my own of the herpetofauna of México. Specialists consulted are noted under acknowledgments.

As summarized by Godman (1915) in the introductory volume of the *Biologia* series, the "official" count for Günther's volume is 695 species covered, with 58 of them, and three genera, being new; the 76 plates illustrate 138 species. The present survey adds that 163 genera and 32 families, one new, were treated. The family name is recognized as valid at the subfamily level (Lepidophyminae), since it is an older name than the Cricosaurinae Savage, 1963, which it replaces in a classification that places *Cricosaura* and *Lepidophyma* in the same subfamily, separate from the Xantusiinae Baird, 1858. Actually 62 new species-group taxa were proposed (of which 31 are currently accepted as valid at either specific or subspecific levels), including three subspecies (which Godman had not counted) as well as 59 species; Godman missed one in his calculation. Günther also discussed or noted 720 species-group taxa that he accepted as valid; the discrepancy with Godman's figure of 695 (obviously derived from the totals in Günther's introductory tables, pp. viii, x-xvii) is of several sources.

In the first place, Günther's table includes not only the species-group taxa introduced in the text in boldface type, but also quite a few others, their names in italics, discussed or mentioned in the context of related taxa or as incertae sedis. To a certain extent his choice for inclusion of these ancillary taxa today seems somewhat capricious, for others of seemingly equal qualifications were excluded from his table. Furthermore, spellings and generic allocations for a few were different in the table and text (e.g., *Spelerpes rostratum* and *S. sulcatum* of the text, p. 300, are rendered *S. rostratus* and *S. sulcatus* in the table; *Ctenosaura quinquecarinata*, p. 58, as *Ctenosaura quinque-carinata*; *Colobognathus brachycephalus* and *Catostoma brachycephalum*, p. 91, as *Geophis brachycephala*; *Rhabdosoma longiceps*, p. 93, as *Geophis longiceps; Tantilla calamarina*, p. 155, as *Homalocranium calamarinum*).

In the present analysis, all of the taxa listed by Günther in his table are included. To that total of 695 we add 25 names. One is an overlooked main entry, *Spelerpes bocourti*. All subspecies are additional, since Günther did not count them; they are as follows (number of additional entries are given parenthetically, total of 16, since Günther usually did cite the species): *Crocodilus americanus acutus* and *C. a. moreletii* (1); *Cnemidophorus sexlineatus tessellatus* (1); *Sceloporus torquatus mucronatus* and *S. t. cyanogenys* (2); *Phrynosoma orbiculare orbicularis* (sic), *P. o. cortezii*, *P. o. dugesi* and *P. o. boucardii* (3); *Geophis chalybaea quadrangularis* (1); *Streptophorus a. atratus, S. a. labiosus, S. a. drozii, S. a. dorsalis, S. a. punctulatus, S. a. sebae, S. a.* var., and *S. a. maculatus* (7); and *Homalocranium melanocephalum fraseri* and *H. m. pernambucense* (1).

The final source of discrepancy of totals is the group of species and subspecies treated incidentally; many Günther listed in his table, but others were omitted, mostly because they were extralimital, noted only from South America, United States or the West Indies. These are included in the present table, only if Günther seemed not to have rejected them as valid, and if they actually do occur in mainland Middle America. They include a total of eight, as follows: *Basiliscus galeritus, (Ctenosaura) hemilopha, C. erythromelas, Holbrookia lacerata, Dromicus pleii, Herpetodryas bilineatus, H. rappii* and *Bufo melanogaster*.

The numbers of nominal taxa included in the seven major groups of amphibians and reptiles treated by Günther are summarized in Table 1.

It is not surprising that such a massively inclusive work has proved to be indispensable to subsequent workers, and it will no doubt remain so for many years. No work of notable scope on the herpetofauna of Middle America has appeared this century without having to take Günther's account into consideration. Its influence has extended even into South American, North American and West Indian studies.

As all major works do, Günther's monograph bears the indelible stamp of his personality. All of the taxonomic accounts are succinct, many consisting of just a few lines. Clearly Günther possessed an encyclopedic grasp of the American herpetofauna, but he was a driven achiever, impatient with the discipline of thorough presentation of his vast knowledge; disciplined he certainly was, but tuned to a different drummer: to coverage of as many vertebrate subjects as possible. His interests were well-nigh cosmopolitan, and he was driven to exploitation of them all. Thus it was impossible for him, as it would be for anyone, to go into great detail in any one subject; no one work even approached exhaustive presentation of all that he knew about it. Time was too short for it, and besides it wasn't necessary to record details that anyone else, he seemed to think, could determine on his own, if he wished. Nevertheless his boundless energy was enormously productive of many major works on many subjects (see A. E. Gunther's article, and Adler's list of Günther's herpetological publications, in this volume). In addition, his administrative and editorial duties were extensive.

The diversity of Günther's drives partly accounts for the striking disparity between his herpetological volume in the *Biologia* and the corresponding volumes in the concurrent and equally monumental French work, *Mission Scientifique au Mexique et dans l'Amérique Centrale* (Duméril, Bocourt and Mocquard, 1870-1909; Brocchi, 1881-1883), the several volumes of which were reprinted in 1978 (see Literature Cited). Auguste Duméril, also a prolific worker, died in 1870, before he could contribute much to the monograph, but all of the other authors devoted most of their energies to it; the *Mission* was their opus magnum, and no effort was spared to make it the best possible. The illustrations are superb and include a number of critical types in German as well as French museums. The taxonomic accounts in the *Mission* are far more detailed than are those in the *Biologia*, and in that way they are superior. However, their scope was not as broad, since they were based on considerably less material and literature than was available to Günther. Accordingly, these two classic works on the herpetofauna of Middle America, instead of being redundant, are marvelously complementary. Each fills its own distinctive niche.

Acknowledgments. I am much indebted

TABLE 1. Comparison of Total Nominal Taxa Contained in Günther's Major Groups of Amphibians and Reptiles in the *Biologia*.

Numbers of	Chelonia	Crocodilia	Sauria	Ophidia	Salientia	Caudata	Apoda	Total
Species-group taxa	30	3	207	286	164	23	7	720
New species-group taxa*	4(3)	0	9(4)	28(10)	20(13)	0	1(1)	62(31)
Genera	7	2	39	81	26	4	4	163
New genera*	0	0	0	1(1)	2	0	0	3(1)
Current name-entries	38	5	326	458	203	39	8	1077
Names with multiple identifications**	6	1	64	86	33	6	1	197
Supernumerary identifications***	8	2	119	172	41	16	1	359
Families****	1	1	10	9	9	1	1	32
New families	0	0	1	0	0	0	0	1

* Number invalid in parentheses.

** Having more than one name in the "Current Names" column of Table 2 for one name in the column for "Günther's Names."

*** Names in the "Current Names" column of Table 2 in addition to the one nomenclatural equivalent for each Günther name.

**** The "Order Crocodilia" and "Division Batrachia Apoda" are here tabulated as Families also, although Günther listed none for either taxon.

to Ronald A. Brandon, Roger Conant, James R. Dixon, William E. Duellman, Darrel Frost, Carl S. Lieb, Ernest A. Liner, John D. Lynch, Samuel B. McDowell, Janis A. Roze, Jay M. Savage, Paulo E. Vanzolini, David B. Wake, Van Wallach, Ernest E. Williams and Kenneth L. Williams for vital help in specific contexts; no one of them, however, checked all names even in his own specialty. Errors are my responsibility. The account has benefitted immeasurably from Kraig Adler's thorough editing, and from Phyllis O'Connell's inspired word processing.

TABLE 2. Current Equivalents for Günther's Names and Materials Cited Under Them.

Günther's Names (*new)	Current Names	Page	Plate
	CHELONIA		
Cistudo mexicana	*Terrapene carolina mexicana* (Gray, 1849)	1	-
Emys ornata [1§]	*Trachemys scripta ornata* (Gray, 1831)(M)	2	1
	Trachemys scripta venusta (Gray, 1855)(H)		
	Trachemys scripta ssp. innominatum (C, P)[2]		
Emys cataspila[*1]	*Trachemys scripta cataspila* (Günther, 1885)	4	6
Emys salvini[1]	*Trachemys scripta venusta* (Gray, 1855)	4	2-3
Emys mobiliensis[1]	(*Pseudemys concinna hieroglyphica* (Holbrook, 1836))	5	-
	Trachemys scripta elegans (Wied-Neuwied, 1838)(M)[3]		
Emys callirostris[1]	*Trachemys scripta callirostris* (Gray, 1855)	5	-
Emys umbra[1]	*Trachemys scripta grayi* (Bocourt, 1868)	5	4-6
Emys pulcherrima[4]	(*Rhinoclemmys p. pulcherrima* (Gray, 1855))	6	7-8
	Rhinoclemmys pulcherrima rogerbarbouri (Ernst, 1978)		
Emys incisa[4]	*Rhinoclemmys pulcherrima incisa* (Bocourt, 1868)	7	-
Emys gabbi[4]	*Rhinoclemmys annulata* (Gray, 1860)	7	-
Emys funerea[4]	*Rhinoclemmys funerea* (Cope, 1876)	7	-
Emys areolata[4]	*Rhinoclemmys a. areolata* (Duméril & Bibron, 1851)	8	8
Emys rubida[4]	*Rhinoclemmys r. rubida* (Cope, 1870)	8	8
Emys punctularia[4]	*Rhinoclemmys p. punctularia* (Daudin, 1802)(SA)	9	-
	Rhinoclemmys a. areolata (Duméril & Bibron, 1851)(M)		
Dermatemys mawii	*Dermatemys mawii* Gray, 1847	10	-
Chelydra serpentina	*Chelydra s. serpentina* (Linnaeus, 1758)	10	-
Chelydra rossignoni	*Chelydra serpentina rossignonii* (Bocourt, 1868)	10	-
Staurotypus triporcatus	*Staurotypus triporcatus* (Wiegmann, 1828)	11	-
Staurotypus salvini	*Staurotypus salvinii* Gray, 1864	11	-
Claudius angustatus	*Claudius angustatus* (Cope, 1865)	12	-
Cinosternum sonoriense	*Kinosternon s. sonoriense* (LeConte, 1854)	13	-
Cinosternum cruentatum	*Kinosternon scorpioides cruentatum* (Duméril & Bibron, 1851)	13	9-11
Cinosternum triliratum	*Kinosternon scorpioides cruentatum* (Duméril & Bibron, 1851)	14	-
Cinosternum albogulare	*Kinosternon scorpioides albogulare* (Duméril, 1870)	14	-
Cinosternum hirtipes	(*Kinosternon hirtipes* (Wagler, 1830))	15	12-15
	Kinosternon integrum (LeConte,1854)(M)		
	Kinosternon s. sonoriense (LeConte, 1854)(US)		
Cinosternum integrum	*Kinosternon integrum* (LeConte, 1854)	16	0
Cinosternum effeldtii	*Kinosternon acutum* Gray, 1831	16	0
Cinosternum leucostomum	*Kinosternon l. leucostomum* (Duméril & Bibron, 1851) (G, M)	17	16-17
	Kinosternon leucostomum postinguinale (Cope, 1887) (C, P, SA)		

§See endnotes, page XLVII.

Günther's Names (*new)	Current Names	Page	Plate
*Cinosternum brevigulare**	*Kinosternon l. leucostomum* (Duméril & Bibron, 1851)	17	18
*Cinosternum cobanum**	*Kinosternon l. leucostomum* (Duméril & Bibron, 1851)	18	18

CROCODILIA

Günther's Names (*new)	Current Names	Page	Plate
Crocodilus americanus acutus	*Crocodylus acutus* (Cuvier, 1807)	20	-
Crocodilus americanus moreletii	*Crocodylus moreletii* (Duméril & Bocourt, 1851)	21	-
Alligator punctulatus	*Caiman c. crocodilus* (Linnaeus, 1758)(SA)	21	-
	Caiman crocodilus fuscus (Cope, 1868)(G, M)		
	Caiman crocodilus yacare (Daudin, 1802)(SA)		

SAURIA

Günther's Names (*new)	Current Names	Page	Plate
Ameiva surinamensis	(*Ameiva a. ameiva* (Linnaeus, 1758))	22	19
	Ameiva ameiva praesignis (Baird & Girard, 1852)(C, P)[5]		
	Ameiva sp. indet. (N)[6]		
Ameiva gabbiana	*Ameiva quadrilineata* (Hallowell, 1861)	23	-
Ameiva undulata	*Ameiva u. undulata* (Wiegmann, 1834)	23	-
Ameiva festiva	*Ameiva f. festiva* (Lichtenstein, 1856)(P)	24	-
	Ameiva festiva edwardsii Bocourt, 1873 (G)		
	Ameiva festiva occidentalis Taylor, 1956 (C)		
Ameiva quadrilineata	*Ameiva quadrilineata* (Hallowell, 1861)	24	-
Cnemidophorus sex-lineatus	*Cnemidophorus sexlineatus* (Linnaeus, 1766)(US)	25	-
	Cnemidophorus a. angusticeps Cope, 1878 (M)		
	Cnemidophorus angusticeps petenensis Beargie & McCoy, 1964 (G)		
	Cnemidophorus c. communis Cope, 1878 (M)		
	Cnemidophorus costatus mazatlanensis Zweifel, 1959 (M)		
	Cnemidophorus costatus occidentalis Gadow, 1906 (M)		
	Cnemidophorus g. gularis Baird & Girard, 1852 (M)		
Cnemidophorus sex-lineatus tessellatus[7]	*Cnemidophorus tesselatus* (Say, 1823)(US)	26	-
	Cnemidophorus m. marmoratus Baird & Girard, 1852 (US)		
	Cnemidophorus tigris mundus Camp, 1916 (US)		
	Cnemidophorus tigris Baird & Girard, 1852, or *C. marmoratus* Baird & Girard, 1852, subsp. indet. (M)		
Cnemidophorus costatus	*Cnemidophorus c. costatus* Cope, 1878	26	-
Cnemidophorus deppii	*Cnemidophorus d. deppii* Wiegmann, 1834 (E, N?)	27	-
	Cnemidophorus cozumela maslini Fritts, 1969 (G)		
	Cnemidophorus l. lemniscatus (Linnaeus, 1758)(SA)		
	Cnemidophorus l. lineattissimus Cope, 1878 (M)		
Cnemidophorus lineatissimus	*Cnemidophorus l. lineattissimus* Cope, 1878	27	-
	Cnemidophorus lineattissimus duodecemlineatus Lewis, 1956		
*Cnemidophorus mariarum**	*Cnemidophorus communis mariarum* Günther, 1885	28	20
Cnemidophorus guttatus	*Cnemidophorus g. guttatus* Wiegmann, 1834	28	21
	Cnemidophorus guttatus immutabilis Cope, 1878		
Cnemidophorus inornatus	*Cnemidophorus i. inornatus* Baird, 1858	29	-
Gymnophthalmus sumichrasti	*Gymnophthalmus speciosus sumichrasti* (Cope, 1876) (G, M)	29	-
	Gymnophthalmus s. speciosus (Hallowell, 1861)(SA)		
Tretioscincus lævicauda	*Gymnophthalmus s. speciosus* (Hallowell, 1861)	29	-
Argalia pœcilochilus	*Anadia marmorata* (Gray, 1846)	29	-
Anadia metallica	*Anadia m. metallica* (Cope, 1876)	30	-

Günther's Names (*new)	Current Names	Page	Plate
Lepidophyma flavomaculatum	*Lepidophyma f. flavimaculatum* Duméril, 1851 (G)	30	-
	Lepidophyma flavimaculatum obscurum Barbour, 1924 (P)		
	Lepidophyma flavimaculatum smithii Bocourt, 1876 (M)		
Chirotes canaliculatus	*Bipes canaliculatus* Bonnaterre, 1789 (M)	30	-
	Bipes sp. innominatum (US)[8]		
Amphisbœna fuliginosa	*(Amphisbaena f. fuliginosa* Linnaeus, 1758)	31	-
	Amphisbaena fuliginosa varia Laurent, 1768		
Mocoa lateralis	*Scincella laterale* (Say, 1823)(US)	31	-
	Scincella g. gemmingeri (Cope, 1864)(M)		
Mocoa assata	*Sphenomorphus a. assata* (Cope, 1864)(E)	31	-
	Sphenomorphus assata taylori (Oliver, 1937)(N)		
	Sphenomorphus c. cherriei (Cope, 1893)(G)		
Eumeces lynxe	*Eumeces l. lynxe* (Wiegmann, 1828)	32	-
Eumeces brevirostris	*Eumeces b. brevirostris* (Günther, 1860)	32	22
Eumeces bocourti	*Eumeces humilis* (Boulenger, 1887)	32	22
Eumeces callicephalus	*Eumeces callicephalus* Bocourt, 1879	32	-
Eumeces sumichrasti	*Eumeces sumichrasti* (Cope, 1866)	32	-
Eumeces dugesii	*Eumeces dugesi* Thominot, 1883	32	-
Eumeces schwartzii	*Eumeces schwartzei* Fischer, 1884	33	-
Eumeces furcirostris	*Eumeces lynxe furcirostris* Cope, 1885	33	-
Mabuia agilis	*Mabuya agilis* (Raddi, 1823)(SA)	33	-
	Mabuya bistriata (Spix, 1825)(SA)[9]		
	Mabuya unimarginata (Cope, 1862)(G, M, P)		
Mabuia alliacea	*Mabuya unimarginata* (Cope, 1862)	33	-
Anelytropsis papillosus	*Anelytropsis papillosus* Cope, 1885	34	-
Diploglossus bilobatus	*Diploglossus bilobatus* (O'Shaughnessy, 1847)	34	-
Diploglossus steindachneri	*Celestus enneagrammus* (Cope, 1860)(M)	34	22
	Celestus bivittatus (Boulenger, 1894)(G)		
	Celestus rozellae Smith, 1942 (G)		
Diploglossus monotropis	*Diploglossus monotropis* (Kuhl, 1820)	34	23
Ophisaurus ventralis	*Ophisaurus ventralis* (Linnaeus, 1766)(US)	35	-
	Ophisaurus ceroni Holman, 1965 (M)		
Gerrhonotus webbii	*Elgaria multicarinata webbii* (Baird, 1858)	35	-
Gerrhonotus fulvus	*(Barisia moreletii fulva* (Bocourt, 1871))(G)	35	-
	Barisia monticola (Cope, 1877)(C)		
Gerrhonotus deppii	*Abronia deppii* (Wiegmann, 1828)	35	-
Gerrhonotus gramineus	*Abronia graminea* (Cope, 1864)	36	24
*Gerrhonotus oaxacæ**	*Abronia oaxacae* (Günther, 1885)	36	24
Gerrhonotus tæniatus	*Abronia taeniata* (Wiegmann, 1828)	36	-
Gerrhonotus vasconcelosii	*Abronia vasconcelosii* (Bocourt, 1871)	37	-
Gerrhonotus auritus	*Abronia aurita* (Cope, 1869)	37	-
Gerrhonotus fimbriatus	*Abronia aurita* (Cope, 1869)	37	-
Gerrhonotus antauges	*(Barisia antauges* (Cope, 1866))	37	-
	Barisia viridiflava (Bocourt, 1873)		
Gerrhonotus rudicollis	*Barisia rudicollis* (Wiegmann, 1828)	38	-
Gerrhonotus imbricatus	*Barisia i. imbricata* (Wiegmann, 1828)[10]	38	-
Gerrhonotus planifrons	*Barisia imbricata planifrons* (Bocourt, 1878)	38	-
Gerrhonotus cæruleus	*Elgaria c. coerulea* (Wiegmann, 1828)(US)	38	25
	Barisia monticola (Cope, 1877)(C)		
	Elgaria kingii nobilis Baird & Girard, 1852 (US)		
*Gerrhonotus obscurus**	*Barisia viridiflava* (Bocourt, 1873)	40	25

Günther's Names (*new)	Current Names	Page	Plate
Gerrhonotus moreletii	*Barisia m. moreletii* (Bocourt, 1871)(G)	41	24
	Barisia moreletii fulva (Bocourt, 1871)(G)		
	Barisia antauges (Cope, 1866)(M)		
Gerrhonotus liocephalus	*Gerrhonotus l. liocephalus* Wiegmann, 1828 (M)	41	-
	Gerrhonotus liocephalus infernalis Baird, 1858 (US)		
	Gerrhonotus liocephalus ophiurus Cope, 1866 (M)		
Gerrhonotus rhombifer	*Coloptychon rhombifer* (Peters, 1877)	42	-
Gerrhonotus modestus	*Barisia antauges* (Cope, 1866)	42	-
Xenosaurus grandis	*Xenosaurus g. grandis* (Gray, 1856)	43	-
Heloderma horridum	*Heloderma h. horridum* (Wiegmann, 1829)	43	26
Heloderma suspectum	*Heloderma s. suspectum* Cope, 1869	44	-
Anolis sagræ	*Anolis*[11] *sagrei* Duméril & Bibron, 1837	45	-
Anolis beckeri	*Anolis pentaprion beckeri* Boulenger, 1881	45	-
Anolis rodriguezi	*Anolis r. rodriguezii* Bocourt, 1873	45	-
Anolis aureolus	*Anolis r. rodriguezii* Bocourt, 1873	45	-
Anolis limifrons	*Anolis limifrons* Cope, 1862	45	-
Anolis bransfordii	*Anolis limifrons* Cope, 1862	45	-
Anolis trochilus	*Anolis limifrons* Cope, 1862	45	-
Anolis ortonii	*Anolis ortonii* Cope, 1868 (SA)	45	-
	Anolis bouvierii Bocourt, 1873 (G)		
Anolis schiedi	*Anolis schiedii* (Wiegmann, 1834)(M)	46	-
	Anolis cobanensis Stuart, 1942 (G)		
Anolis polylepis	*Anolis polylepis* Peters, 1873	46	-
Anolis tropidolepis	*Anolis tropidolepis* Boulenger, 1885	46	27
Anolis pulchripes	*Anolis limifrons* Cope, 1862	46	-
Anolis baccatus	*Anolis baccatus* Bocourt, 1873	46	-
Anolis guentheri	*Anolis r. rodriguezii* Bocourt, 1873	46	-
Anolis squamulatus	*Anolis squamulatus* Peters, 1863	47	-
Anolis latifrons	*Anolis latifrons* Berthold, 1846	47	-
Anolis microtus	*Anolis microtus* Cope, 1871	47	-
Anolis insignis	*Anolis insignis* Cope, 1871	47	-
Anolis pachypus	*Anolis pachypus* Cope, 1876	47	-
Anolis pentaprion	*Anolis p. pentaprion* Cope, 1862	47	-
Anolis copii	*Anolis b. biporcatus* (Wiegmann, 1834)	47	-
Anolis petersi	*Anolis petersii* Bocourt, 1873	47	-
Anolis heliactin	*Anolis sericeus* Hallowell, 1856	48	-
Anolis ustus	*Anolis sericeus* Hallowell, 1856	48	27
Anolis cymbops	*Anolis cymbops* Cope, 1864	48	-
Anolis concolor	*Anolis concolor* Cope, 1862	48	-
Anolis bourgeaui	*Anolis lemurinus bourgeaei* Bocourt, 1873	48	-
Anolis salvini	*Anolis salvini* Boulenger, 1885	48	27
Anolis læviventris	*Anolis laeviventris* (Wiegmann, 1834)	49	-
Anolis nebulosus	*Anolis nebulosus* (Wiegmann, 1834)	49	-
	Anolis quercorum Fitch, 1978		
	Anolis sp. indet.[11]		
Anolis nebuloides	*Anolis nebuloides* Bocourt, 1873	49	-
	Anolis sp. indet.[11]		
Anolis palpebrosus	*Anolis l. lemurinus* Cope, 1861	49	-
Anolis intermedius	*Anolis laeviventris* (Wiegmann, 1834)	49	-
Anolis sallæi	*Anolis sericeus* Hallowell, 1856	49	27
Anolis binotatus	*Anolis binotatus* Peters, 1863	50	-
Anolis cumingii	*Anolis cumingii* Peters, 1863	50	-

Günther's Names (*new)	Current Names	Page	Plate
Anolis cupreus	*Anolis c. cupreus* Hallowell, 1860 (N)	50	-
	Anolis cupreus hoffmanni Peters, 1863 (C)		
	Anolis cupreus macrophallus Werner, 1917 (G)		
	Anolis dollfusianus Bocourt, 1873		
Anolis crassulus	*Anolis crassulus* Cope, 1864 (G)	50	27
	Anolis haguei Stuart, 1942 (G)		
	Anolis sp. indet. (M)[11]		
Anolis humilis	*Anolis h. humilis* Peters, 1863 (P)	50	-
	Anolis uniformis Cope, 1885 (G)		
Anolis quaggulus	*Anolis h. humilis* Peters, 1863	50	-
Anolis tropidonotus	*Anolis t. tropidonotus* Peters, 1863	51	-
Anolis uniformis	*Anolis uniformis* Cope, 1885	51	-
Anolis metallicus	*Anolis t. tropidonotus* Peters, 1863	51	-
Anolis oxylophus	*Anolis oxylophus* Cope, 1876	51	-
Anolis godmani	*Anolis godmani* Boulenger, 1885 (C, G)[11]	51	27
Anolis radulinus	*Anolis radulinus* Cope, 1862	52	-
Anolis pœcilopus	*Anolis poecilopus* Cope, 1862	52	-
Anolis lemurinus	*Anolis l. lemurinus* Cope, 1861	52	-
Anolis biporcatus	*Anolis b. biporcatus* (Wiegmann, 1834)(G, P, SA)	52	-
	Anolis biporcatus parvauritus Williams, 1966 (SA)		
	Anolis lemurinus bourgeaei Bocourt, 1873 (G)		
	Anolis l. lemurinus Cope, 1861 (P)		
Anolis chrysolepis	*Anolis chrysolepis* Duméril & Bibron, 1857 (SA)	52	-
	Anolis sp. indet. (H)[11]		
	Anolis sp. indet. (M)[11]		
Anolis capito	*Anolis capito* Peters, 1863	52	-
Anolis obtusirostris	*Anolis b. biporcatus* Wiegmann, 1834	52	-
Polychrus gutturosus	*Polychrus gutturosus* Berthold, 1846	53	-
Corythophanes cristatus	*Corytophanes cristatus* (Merrem, 1821)	53	-
Corythophanes percarinatus	*Corytophanes percarinatus* Duméril, 1856	53	-
Corythophanes hernandesii	*Corytophanes hernandesii* (Wiegmann, 1831)	53	-
Læmanctus longipes	*Laemanctus l. longipes* Wiegmann, 1834	54	-
Læmanctus serratus	*Laemanctus s. serratus* Cope, 1864	54	-
	Laemanctus serratus alticoronatus Cope, 1865		
Læmanctus deborrii	*Laemanctus longipes deborrei* Boulenger, 1877	54	-
Læmanctus alticoronatus	*Laemanctus serratus alticoronatus* Cope, 1865	54	-
Basiliscus americanus	*Basiliscus b. basiliscus* (Linnaeus, 1758)	54	-
Basiliscus plumifrons	*Basiliscus plumifrons* Cope, 1876	55	-
Basiliscus vittatus	*Basiliscus vittatus* Wiegmann, 1828	55	28
Basiliscus galeritus	*Basiliscus galeritus* Duméril, 1851	55	-
Enyalius heterolepis	*Enyalioides heterolepis* (Bocourt, 1874)	56	-
Iguana rhinolophus	*Iguana iguana* (Linnaeus, 1758)	56	-
Iguana tuberculata	*Iguana iguana* (Linnaeus, 1758)	56	-
Ctenosaura acanthura	*Ctenosaura acanthura* (Shaw, 1802)(M, US? in error)	56	-
	Ctenosaura pectinata (Wiegmann, 1834)(M)		
	Ctenosaura s. similis (Gray, 1831)(M)		
Ctenosaura completa	*Ctenosaura s. similis* (Gray, 1831)	58	29-30
Ctenosaura quinquecarinata	*Enyaliosaurus quinquecarinatus* (Gray, 1842)(M)	58	-
	Enyaliosaurus palearis (Stejneger, 1899)(H)		
Ctenosaura defensor	*Enyaliosaurus defensor* (Cope, 1866)	58	-
(Ctenosaura) hemilopha	*Ctenosaura h. hemilopha* (Cope, 1863)	59	-
Ctenosaura erythromelas	*Enyaliosaurus defensor* (Cope, 1866)	59	-

Günther's Names (*new)	Current Names	Page	Plate
Crotaphytus wislizenii	*Gambelia w. wislizenii* (Baird & Girard, 1852)	59	-
Crotaphytus collaris	*Crotaphytus c. collaris* (Say, 1823)(US)	59	-
	Crotaphytus collaris baileyi (Stejneger, 1890)(M)		
	Crotaphytus collaris fuscus Ingram & Tanner, 1971 (M)		
	Crotaphytus collaris nebrius Axtell & Montanucci, 1977 (M)		
Holbrookia texana	*Cophosaurus t. texanus* Troschel, 1852	59	-
	Cophosaurus texanus scitulus (Peters, 1951)		
Holbrookia maculata	*Holbrookia m. maculata* Girard,1851 (US)	60	-
	Holbrookia maculata thermophila Barbour, 1921 (M)		
Holbrookia propinqua	*Holbrookia p. propinqua* Baird & Girard, 1852 (US)	60	31
	Holbrookia maculata elegans Bocourt, 1874 (M)		
Holbrookia lacerata	*Holbrookia l. lacerata* Cope, 1880	61	-
Uta lateralis	*Urosaurus ornatus lateralis* (Boulenger, 1883)	61	31
Uta gularis	*Urosaurus ornatus schotti* (Baird, 1858)	61	-
Uta auriculata	*Urosaurus auriculatus* (Cope, 1871)	61	-
Uta irregularis	*Urosaurus irregularis* (Fischer, 1882)	62	-
Uta bicarinata	*Urosaurus b. bicarinatus* (Duméril, 1856)	62	-
	Urosaurus bicarinatus anonymorphus (Mittleman, 1940)		
	Urosaurus bicarinatus tuberculatus (Schmidt, 1921)		
Sceloporus spinosus	*Sceloporus s. spinosus* Wiegmann, 1828	63	-
	Sceloporus c. clarkii Baird & Girard, 1852		
	Sceloporus clarkii boulengeri Stejneger, 1893		
	Sceloporus formosus scitulus Smith, 1942[12]		
	Sceloporus m. magister Hallowell, 1854		
	Sceloporus magister bimaculosus Phelan & Brattstrom, 1955		
	Sceloporus olivaceus Smith, 1934		
	Sceloporus spinosus albiventris Smith, 1939		
	Sceloporus spinosus horridus Wiegmann, 1834		
	Sceloporus spinosus oligoporus Cope, 1864		
Sceloporus viviparus	*Sceloporus f. formosus* Wiegmann, 1834	64	-
	Sceloporus formosus scitulus Smith, 1942		
Sceloporus poinsetti	*Sceloporus p. poinsettii* Baird & Girard, 1854	65	-
	Sceloporus poinsettii macrolepis Smith & Chrapliwy, 1958		
Sceloporus acanthinus	*Sceloporus acanthinus* Bocourt, 1873	65	-
	Sceloporus salvini Günther, 1890		
Sceloporus torquatus	*Sceloporus t. torquatus* Wiegmann, 1828	66	-
	Sceloporus bulleri Boulenger, 1894		
	Sceloporus mucronatus omiltemanus Günther, 1890		
*Sceloporus omiltemanus**	*Sceloporus mucronatus omiltemanus* Günther, 1890	66	32
Sceloporus undulatus	*Sceloporus u. undulatus* (Latreille, 1801)(US)	67	-
	Sceloporus variabilis olloporus Smith, 1937 (G)		
	Sceloporus virgatus Smith, 1938 (M)		
Sceloporus lunæi	*Sceloporus acanthinus lunaei* Bocourt, 1873	67	-
*Sceloporus irazuensis**	*Sceloporus m. malachiticus* Cope, 1864	67	-
*Sceloporus salvini**	*Sceloporus salvini* Günther, 1890 (M)	68	-
	Sceloporus smaragdinus Bocourt, 1873 (G)		
Sceloporus smaragdinus	*Sceloporus smaragdinus* Bocourt, 1873 (G)	68	-
	Sceloporus lundelli gaigeae Smith, 1939 (M)		
	Sceloporus m. malachiticus Cope, 1864 (C)		
Sceloporus yarrovii	*Sceloporus j. jarrovii* Cope, 1875	69	-
	Sceloporus d. dugesii Bocourt, 1873		
Sceloporus consobrinus	(*Sceloporus undulatus consobrinus* Baird & Girard, 1853)	69	-
	Sceloporus subpictus Lynch & Smith, 1965		

Günther's Names (*new)	Current Names	Page	Plate
Sceloporus pyrrhocephalus	*Sceloporus pyrocephalus* Cope, 1864	70	-
Sceloporus chrysostictus	*Sceloporus chrysostictus* Cope, 1867	70	-
Sceloporus grammicus	*Sceloporus g. grammicus* Wiegmann, 1828 (M)[13]	71	-
	Sceloporus m. microlepidotus Wiegmann, 1834 (M)		
	Sceloporus t. taeniocnemis Cope, 1885? (G)		
Sceloporus gratiosus	*Sceloporus g. graciosus* Baird & Girard, 1852 (US)	71	-
	Sceloporus m. megalepidurus Smith, 1934 (M)		
	Sceloporus subpictus Lynch & Smith, 1965 (M)		
Sceloporus ornatus	*Sceloporus o. ornatus* Baird, 1859	72	-
	Sceloporus jarrovii minor Cope, 1885		
*Sceloporus rubriventris**	*Sceloporus g. grammicus* Wiegmann, 1828[13]	72	32
Sceloporus microlepidotus	*Sceloporus m. microlepidotus* Wiegmann, 1834[13]	72	-
	Sceloporus m. disparilis Stejneger, 1916		
	Sceloporus anahuacus Lara-Góngora, 1983		
	Sceloporus palaciosi Lara-Góngora, 1983		
Sceloporus couchii	*Sceloporus couchii* Baird, 1859	73	-
Sceloporus scalaris	*Sceloporus s. scalaris* Wiegmann, 1828	73	-
	Sceloporus cozumelae Jones, 1927		
	Sceloporus scalaris samcolemani Smith & Hall, 1974		
	Sceloporus scalaris slevini Smith, 1937		
	Sceloporus scalaris unicanthalis Smith, 1937		
Sceloporus æneus	*Sceloporus aeneus* Wiegmann, 1828	74	-
	Sceloporus bicanthalis Smith, 1937		
*Sceloporus jalapæ**	*Sceloporus jalapae* Günther, 1890	74	-
*Sceloporus pleurolepis**	*Sceloporus d. dugesii* Bocourt, 1873	74	32
Sceloporus variabilis	*Sceloporus v. variabilis* Wiegmann, 1834 (E, G, M)	75	-
	Sceloporus n. nelsoni Cochran, 1923? (M)		
	Sceloporus ochoterenae Smith, 1934 (M)[12]		
	Sceloporus variabilis marmoratus Hallowell, 1852 (M, US)		
	Sceloporus variabilis smithi Hartweg & Oliver, 1937 (M)		
*Sceloporus teapensis**	*Sceloporus teapensis* Günther, 1890[14]	75	-
Sceloporus fulvus	*Sceloporus squamosus* Bocourt, 1874	76	-
Sceloporus siniferus	*Sceloporus s. siniferus* Cope, 1869	76	-
Sceloporus utiformis	*Sceloporus utiformis* Cope, 1864	76	-
Sceloporus melanogaster	*Sceloporus torquatus melanogaster* Cope, 1885	77	-
Sceloporus ferrari-perezi	*Sceloporus t. torquatus* Wiegmann, 1828	77	-
Sceloporus melanorhinus	*Sceloporus m. melanorhinus* Bocourt, 1876	77	-
Sceloporus torquatus mucronatus	*Sceloporus m. mucronatus* Cope, 1885	77	-
Sceloporus torquatus cyanogenys	*Sceloporus cyanogenys* Cope, 1885	77	-
Sceloporus squamosus	*Sceloporus squamosus* Bocourt, 1874	77	-
Sceloporus serrifer	*Sceloporus s. serrifer* Cope, 1866	77	-
Phrynosoma orbiculare orbicularis	*Phrynosoma o. orbiculare* (Linnaeus, 1789)	78	-
	Phrynosoma orbiculare durangoensis Horowitz, 1955		
Phrynosoma orbiculare cortezii	*Phrynosoma orbiculare cortezii* (Duméril & Bocourt, 1870)	78	-
Phrynosoma orbiculare dugesii	*Phrynosoma orbiculare dugesii* (Duméril & Bocourt, 1870)	78	-
Phrynosoma orbiculare boucardi	*Phrynosoma boucardii* (Duméril & Bocourt, 1870)	78	-
Phrynosoma cornutum	*Phrynosoma cornutum* (Harlan, 1825)	79	-
Phrynosoma asio	*Phrynosoma asio* Cope, 1864	79	-
Phrynosoma modestum	*Phrynosoma modestum* Girard, 1852	79	-
Phrynosoma braconnieri	*Phrynosoma braconnieri* Duméril & Bocourt, 1870	79	-
Phrynosoma taurus	*Phrynosoma taurus* Duméril & Bocourt, 1870	79	-

Günther's Names (*new)	Current Names	Page	Plate
Gonatodes albogularis	*Gonatodes a. albogularis* (Duméril & Bocourt, 1836)(SA)	80	-
	Gonatodes albogularis fuscus (Hallowell, 1865)(E, N, P)		
Gonatodes caudiscutatus	*Gonatodes caudiscutatus* (Günther, 1859)	80	-
Phyllodactylus tuberculosus	*Phyllodactylus t. tuberculosus* Wiegmann, 1835 (E ,G, N)	80	-
	Phyllodactylus h. homolepidurus Smith, 1935 (M)		
	Phyllodactylus l. lanei Smith, 1935 (M)		
	Phyllodactylus lanei rupinus Dixon, 1964 (M)		
	Phyllodactylus palmeus Dixon, 1968 (H)		
	Phyllodactylus xanti nocticolus Dixon, 1964 (US)		
Phyllodactylus ventralis	(*Phyllodactylus ventralis* O'Shaughnessy, 1875)	80	-
	Phyllodactylus t. tuberculosus Wiegmann, 1835 (N)		
Thecadactylus rapicauda	*Thecadactylus rapicaudus* (Houttuyn, 1782)	81	-
Aristelliger præsignis	*Aristelliger p. praesignis* (Hallowell, 1856)(WI)	81	-
	Aristelliger georgeensis (Bocourt, 1870)(B, M, N?)		
Gehyra mutilata	*Gehyra mutilata* (Wiegmann, 1835)	81	-
Sphærodactylus homolepis	*Sphaerodactylus homolepis* Cope, 1886	82	-
Sphærodactylus torquatus	*Sphaerodactylus torquatus* Strauch, 1887[15]	82	-
Sphærodactylus glaucus	*Sphaerodactylus g. glaucus* Cope, 1865 (B, G, M)	82	-
	Sphaerodactylus homolepis Cope, 1886 (C)		
Sphærodactylus lineolatus	*Sphaerodactylus lineolatus* (Lichtenstein, 1856)(P)	82	-
	Sphaerodactylus millepunctatus (Hallowell, 1861)(B, M, N)		
	Sphaerodactylus molei Boettger, 1894 (SA)		
Sphærodactylus casicolus	*Sphaerodactylus lineolatus* (Lichtenstein, 1856)	82	-
Sphærodactylus fantasticus	*Sphaerodactylus f. fantasticus* (Duméril & Bibron, 1836)(WI)	83	-
	Sphaerodactylus spp. indet. (M, SA)[15]		
Sphærodactylus anthracinus	*Sphaerodactylus anthracinus* Cope, 1861 (M)	83	-
	Sphaerodactylus sp. indet. (WI)[15]		
Eublepharis variegatus	*Coleonyx v. variegatus* (Baird, 1858)	83	-
	Coleonyx brevis Stejneger, 1893		
Eublepharis dovii	*Coleonyx mitratus* (Peters, 1863)	83	31
Eublepharis fasciatus	*Coleonyx variegatus fasciatus* (Boulenger, 1885)	84	31
Coleonyx elegans	*Coleonyx e. elegans* Gray, 1845 (B, G, M)	84	-
	Coleonyx mitratus (Peters, 1863)(C)		

OPHIDIA

Günther's Names (*new)	Current Names	Page	Plate
Glauconia albifrons	*Leptotyphlops albifrons* (Wagler, 1824)(SA)[16]	85	-
	Leptotyphlops g. goudotii (Duméril & Bibron, 1844)(C, P, SA)		
	Leptotyphlops goudotii phenops (Cope, 1876)(G, H, M)		
Glauconia myopica	*Leptotyphlops dulcis myopicus* (Garman, 1883)	85	-
Glauconia dulcis	*Leptotyphlops d. dulcis* (Baird & Girard, 1853)(US)	85	-
	Leptotyphlops dulcis dissectus (Cope, 1896)(M)		
Glauconia humilis	(*Leptotyphlops h. humilis* (Baird & Girard, 1853))(US)	86	-
	Leptotyphlops humilis dugesii (Bocourt, 1881)(M)		
	Leptotyphlops humilis tenuiculus (Garman, 1883)(M)		
Typhlops tenuis	*Typhlops tenuis* Salvin, 1860	86	-
Typhlops microstomus	*Typhlops microstomus* Cope, 1866	86	-
Helminthophis frontalis	*Helminthophis frontalis* (Peters, 1860)	86	-
Liotyphlops albirostris	*Liotyphlops albirostris* (Peters, 1857)	87	-
Anomalolepis mexicanus	*Anomalepis mexicanus* Jan, 1861	87	-
Geophis dolichocephala	*Geophis brachycephalus* (Cope, 1871)	87	-

Günther's Names (*new)	Current Names	Page	Plate
Geophis chalybœa	*Geophis chalybeus* (Wagler, 1830)(M)	87	-
	Geophis brachycephalus (Cope, 1871)(C)		
	Geophis championi Boulenger, 1894 (P)		
	Geophis dubius (Peters, 1861)(M)		
	Geophis godmani Boulenger, 1894 (C)		
	Geophis hoffmanni (Peters, 1859)(C)		
	Geophis nasalis (Cope, 1868)(G, M)		
	Geophis petersii Boulenger, 1894 (M)		
	Geophis rhodogaster (Cope, 1868)(G)		
	Geophis sallaei Boulenger, 1894 (M)		
*Geophis chalybœa quadrangularis**	*Geophis brachycephalus* (Cope, 1871)	89	33
Geophis rostralis	*Geophis dubius* (Peters, 1861)	89	-
Geophis semidoliata	*Geophis semidoliatus* (Duméril, Bibron & Duméril, 1854)	90	-
Geophis mœsta	*Geophis brachycephalus* (Cope, 1871)	90	33
(Geophis) brachycephala	*Geophis brachycephalus* (Cope, 1871)	91	-
Geophis dugesii	*Geophis dugesii* Bocourt, 1883	91	-
Geophis bicolor	*Geophis bicolor* Günther, 1868	91	-
Geophis redimita	*Geagras redimitus* Cope, 1876	92	-
*Geophis omiltemana**	*Geophis omiltemanus* Günther, 1893	92	33
Geophis mutitorques	*Geophis mutitorques* (Cope, 1885)	93	-
(Geophis) longiceps	*Geophis mutitorques* (Cope, 1885)	93	-
Geophis latifrontalis	*Geophis latifrontalis* Garman, 1883	93	-
Geophis fasciata	*Sibon f. fischeri* (Boulenger, 1894)	93	34
Geophis psephota	*Ninia psephota* (Cope, 1876)	94	-
Geophis zebrina	*Geophis isthmicus* (Boulenger, 1894)	94	-
Adelphicus quadrivirgatus	*Adelphicos q. quadrivirgatus* Jan, 1862 (M)	94	-
	Adelphicos quadrivirgatus sargii (Fischer, 1885)(G)		
	Adelphicos quadrivirgatus visoninus (Cope, 1866)(B, H)		
Toluca lineata	*Toluca l. lineata* Kennicott, 1859	95	-
Pseudoficimia frontalis	*Pseudoficimia frontalis* (Cope, 1864)	96	-
Sympholis lippiens	*Sympholis l. lippiens* Cope, 1862	96	-
Conopsis nasus	*Conopsis n. nasus* Günther, 1858	97	34
	Conopsis biserialis Taylor & Smith, 1942		
	Toluca conica Taylor & Smith, 1942		
Ficimia olivacea	*Ficimia olivacea* Gray, 1849	98	35
Ficimia publia	*Ficimia publia* Cope, 1866	98	-
Ficimia cana	*Gyalopion canum* Cope, 1860	99	-
*Ficimia quadrangularis**	*Gyalopion quadrangularis* (Günther, 1893)	99	35
Rhinochilus lecontei	*Rhinocheilus l. lecontei* Baird & Girard, 1853 (US)	100	-
	Rhinocheilus lecontei tessellatus Garman, 1883 (M)		
Rhinochilus antonii	*Rhinocheilus lecontei antonii* Dugès, 1886	100	-
Leptocalamus unicolor	*Enulius unicolor* (Fischer, 1882)	100	-
Streptophorus atratus atratus, lansbergii	*Ninia atrata* (Hallowell, 1845)	101	-
Streptophorus atratus labiosus	*Ninia diademata labiosa* (Bocourt, 1883)	101	-
Streptophorus atratus drozii, collaris	*Ninia atrata* (Hallowell, 1845)(H, N)	102	-
	Ninia s. sebae (Duméril, Bibron & Duméril, 1854)(G, M)		
Streptophorus atratus dorsalis	*Ninia s. sebae* (Duméril, Bibron & Duméril, 1854)	102	-
Streptophorus atratus punctulatus	*Ninia sebae punctulata* (Bocourt,1883)	102	-
Streptophorus atratus sebœ, fasciatus	*Ninia s. sebae* (Duméril, Bibron & Duméril, 1854)	102	-

Günther's Names (*new)	Current Names	Page	Plate
Streptophorus atratus var.	*Ninia s. sebae* (Duméril, Bibron & Duméril, 1854)	102	-
Streptophorus atratus maculatus	*(Ninia m. maculata* (Peters, 1861))	102	-
	Ninia maculata tessellata Cope, 1876		
Streptophorus diadematus	*Ninia d. diademata* Baird & Girard, 1853	102	-
Chersodromus liebmanni	*Chersodromus liebmanni* Reinhardt, 1860	103	-
Chersodromus nigricans	*Chersodromus liebmanni* Reinhardt, 1860	103	-
Contia episcopa	*Sonora semiannulata* Baird & Girard, 1853	104	-
Contia calligaster	*Rhadinaea calligaster* (Cope, 1876)	104	-
Contia pachyura	*Rhadinaea pachyura* (Cope, 1876)	104	-
Ablabes gracilis	*Trimetopon gracile* (Günther, 1872)	104	-
*Ablabes decipiens**	*Rhadinaea decipiens* (Günther, 1893)	105	37
Ablabes sumichrasti	*Sibynophis collaris* (Gray, 1853)	105	-
Ablabes serperaster	*Rhadinaea serperaster* Cope, 1871	105	-
Elapochrus deppii	*Pliocercus e. elapoides* Cope, 1860 (M)	106	-
	Pliocercus elapoides diastemus (Bocourt, 1886)(G)		
Elapochrus æqualis	*Pliocercus a. aequalis* Salvin, 1861 (G)[17]	106	36
	Pliocercus bicolor Smith, 1941 (M)[17]		
Elapochrus dimidiatus	*Pliocercus aequalis dimidiatus* Cope, 1865[17]	107	-
Elapochrus euryzona	*Pliocercus aequalis euryzonus* Cope, 1862[17]	107	-
Henicognathus annulatus	*Scaphiodontophis a. annulatus* (Duméril, Bibron & Duméril, 1854)(G)[18]	107	-
	Scaphiodontophis annulatus nothus Taylor & Smith, 1943 (M)[18]		
*Henicognathus venustissimus**	*Scaphiodontophis annulatus venustissimus* (Günther, 1894)[18]	144	51
Liophis epinephelus	*Leimadophis e. epinephelus* (Cope, 1862)(P, SA)	107	37
	Leimadophis epinephelus juvenalis Dunn, 1937 (C)		
Liophis regalis	*Diadophis punctatus regalis* Baird & Girard, 1853	108	-
	Diadophis punctatus dugesii Villada, 1875		
Liophis lætus	*(Diadophis punctatus regalis* Baird & Girard, 1853)	108	-
	Diadophis punctatus dugesii Villada, 1875		
Coronella annulata	*Lampropeltis triangulum annulata* Kennicott, 1860	109	38
	Lampropeltis triangulum arcifera (Werner, 1903)		
	Lampropeltis triangulum blanchardi Stuart, 1935		
	Lampropeltis triangulum conanti Williams, 1978		
	Lampropeltis triangulum gaigeae Dunn, 1937		
	Lampropeltis triangulum nelsoni Blanchard, 1920		
	Lampropeltis triangulum polyzona Cope, 1860		
	Lampropeltis triangulum sinaloae Williams, 1978		
Coronella mexicana	*Lampropeltis mexicana* (Garman, 1883)	110	-
*Coronella leonis**	*Lampropeltis mexicana* (Garman, 1883)	110	39
Coronella godmani	*Rhadinaea godmani* (Günther, 1865)	110	39
Coronella decorata	*Rhadinaea decorata* (Günther, 1858)	111	-
Coronella ignita	*Rhadinaea decorata* (Günther, 1858)	111	-
Coronella vermiculaticeps	*Rhadinaea vermiculaticeps* (Cope, 1860)	111	-
Coronella quinquelineata	*Rhadinaea quinquelineata* Cope, 1886	111	-
	Rhadinaea gaigeae Bailey, 1937?		
Coronella fulviceps	*Rhadinaea fulviceps* Cope, 1886	111	-
Dromicus pleii	*Dryadophis pleei* (Duméril, Bibron & Duméril, 1854)(SA)	112	-
	Dryadophis dorsalis (Bocourt, 1890)? (M)		
Dromicus fugitivis	*Dromicus cursor* (Lacépède, 1789)	112	-
Dromicus laureatus	*Rhadinaea laureata* (Günther, 1868)	112	40
Dromicus clavatus	*Coniophanes imperialis clavatus* (Peters, 1864)	113	-
*Dromicus omiltemanus**	*Rhadinaea omiltemana* (Günther, 1894)	113	40

Günther's Names (*new)	Current Names	Page	Plate
Dromicus tæniatus	*Rhadinaea t. taeniata* (Peters, 1863)	113	-
	Rhadinaea taeniata aemula Bailey, 1940		
Dromicus fulvivittis	*Rhadinaea fulvivittis* Cope, 1876	113	-
Dromicus lachrymans	*Rhadinaea lachrymans* (Cope, 1869)	114	-
Xenodon rhabdocephalus	*Xenodon r. rabdocephalus* (Wied-Neuwied, 1824)(SA)	114	-
	Xenodon rabdocephalus mexicanus Smith, 1940 (B, G, M)[19]		
Xenodon angustirostris	*Xenodon r. rabdocephalus* (Wied-Neuwied, 1824)(C, P)	114	-
	Xenodon rabdocephalus mexicanus Smith, 1940 (G)		
Coluber guttatus	*Elaphe g. guttata* (Linnaeus, 1766)(US)	114	-
	Elaphe guttata emoryi (Baird & Girard, 1853)(M)		
Coluber triaspis	*Elaphe t. triaspis* (Cope, 1866)(B, G, M)	115	-
	Elaphe triaspis mutabilis (Cope, 1885)(C)		
	Elaphe triaspis intermedia (Boettger, 1883)(M)		
Coluber flavirufus	*Elaphe f. flavirufa* (Cope, 1867)(M)	115	-
	Elaphe flavirufa pardalina (Peters, 1868)(G)		
	Elaphe flavirufa polysticha Smith & Williams, 1966 (H)		
*Coluber chlorosoma**	*Elaphe triaspis intermedia* (Boettger, 1883)	115	41
Spilotes corais	*(Drymarchon c. corais* (Boie, 1827))	116	-
	Drymarchon corais erebennus (Cope, 1860)(M, US)		
	Drymarchon corais melanurus (Duméril, Bibron & Duméril, 1854)(B, C, G, M, P)		
	Drymarchon corais rubidus Smith, 1941 (M)		
	Drymarchon corais unicolor Smith, 1941 (G)		
Spilotes salvini	*Spilotes pullatus mexicanus* (Laurent, 1768)	116	42
Spilotes pæcilonotus	*Pseustes p. poecilonotus* (Günther, 1858)(H)	117	43
	Pseustes poecilonotus argus (Bocourt, 1888)(G, M)		
Spilotes argus	*Pseustes poecilonotus argus* (Bocourt, 1888)	118	44
Spilotes chrysobronchus	*Pseustes p. poecilonotus* (Günther, 1858)	119	-
Zamenis constrictor	*Coluber c. constrictor* Linnaeus, 1758 (US)	119	-
	Coluber constrictor oaxaca Jan, 1863 (M)		
Zamenis oaxacæ	*Coluber constrictor oaxaca* Jan, 1863	120	-
Zamenis rhombifer	*Drymobius rhombifer* (Günther, 1860)	120	45
Zamenis flavigularis	*(Masticophis flagellum testaceus* (Say, 1823))	120	-
	Masticophis m. mentovarius (Duméril, Bibron & Duméril, 1854)(G, M)		
	Masticophis mentovarius striolatus (Mertens, 1934)(M)		
Zamenis flagelliformis	*Masticophis f. flagellum* (Shaw, 1802)(US)	121	-
	Masticophis flagellum testaceus (Say, 1823)(M, US)		
Zamenis semilineatus	*Masticophis bilineatus* Jan, 1863	121	46
Herpetodryas bilineatus	*Masticophis bilineatus* Jan, 1863	122	-
Zamenis ornatus	*Masticophis taeniatus girardi* (Stejneger & Barbour, 1917)	122	46
Zamenis melanolomus	*Dryadophis m. melanolomus* (Cope, 1868)	122	-
Zamenis pulcherrimus	*Leptodrymus pulcherrimus* (Cope, 1874)(N)	123	-
	Salvadora lemniscata (Cope, 1896)(M)		
Salvadora grahamiæ	*Salvadora g. grahamiae* Baird & Girard, 1853 (M, US)	123	-
	Salvadora grahamiae lineata Schmidt, 1940 (M, US)		
	Salvadora bairdi Jan, 1860 (M)		
Salvadora mexicana	*Salvadora mexicana* (Duméril, Bibron & Duméril, 1854)	123	-
Pituophis deppii	*Pituophis d. deppii* (Duméril, 1853)	124	-
Pituophis lineaticollis	*Pituophis deppii lineaticollis* (Cope, 1861)	124	47
Pituophis elegans	*Arizona e. elegans* Kennicott, 1859 (US)	125	-
	Arizona elegans expolita Klauber, 1946 (M)		

Günther's Names (*new)	Current Names	Page	Plate
Pituophis melanoleuca	*Pituophis m. melanoleucus* (Daudin,1803)(US)	125	-
	Pituophis melanoleucus sayi (Schlegel, 1837)(M)		
	Pituophis d. deppii (Duméril, 1853)(M)		
Drymobius margaritiferus	*Drymobius m. margaritiferus* (Schlegel, 1837)(C, M, P)	125	-
	Drymobius margaritiferus fistulosus Smith, 1942 (M)		
	Drymobius margaritiferus occidentalis Bocourt, 1890 (G)		
Drymobius boddærtii	*Dryadophis b. boddaerti* (Sentzen, 1796)(SA)	125	-
	Dryadophis dorsalis (Bocourt, 1890)(G)		
	Dryadophis melanolomus alternatus (Bocourt, 1884) (C, N, P)		
	Dryadophis melanolomus laevis (Fischer, 1881)(G)		
	Dryadophis melanolomus slevini (Stuart, 1933)(M)		
	Dryadophis melanolomus veraecrucis (Stuart, 1933)(M)		
Herpetodryas rappii	*Dryadophis b. boddaerti* (Sentzen, 1796)	126	-
Drymobius cæruleus	*Dryadophis melanolomus laevis* (Fischer, 1881)	126	-
Drymobius dendrophis	*Dendrophidion dendrophis* (Schlegel, 1837)(SA)	127	-
	Dendrophidion nuchalis (Peters, 1863)(C?, G, SA)		
	Dendrophidion percarinatum (Cope, 1893)(C?, N?, SA)		
	Dendrophidion vinitor Smith, 1941 (C?, M, N?)		
Herpetodryas fuscus	*Chironius fuscus* (Linnaeus, 1758)	128	-
Herpetodryas carinatus	*Chironius carinatus* (Linnaeus, 1758)	128	-
Herpetodryas grandisquamis	*Chironius grandisquamis* (Peters, 1868)	128	-
Herpetodryas melas	*Chironius melas* (Cope, 1886)	128	-
Leptophis mexicanus	*Leptophis m. mexicanus* Duméril, Bibron & Duméril, 1854 (B, C, G, M, N)	129	-
	Leptophis mexicanus septentrionalis Mertens, 1972 (M)		
	Leptophis mexicanus yucatanensis Oliver, 1942 (M)		
Leptophis occidentalis	*Leptophis ahaetulla occidentalis* (Günther, 1859)	129	-
Leptophis modestus	*Leptophis modestus* (Günther, 1872)	129	48
Leptophis depressirostris	*Leptophis depressirostris* (Cope, 1861)	130	-
Leptophis saturatus	*Leptophis depressirostris* (Cope, 1861)	130	-
Leptophis diplotropis	*Leptophis diplotropis* (Günther, 1872)	130	-
Leptophis præstans	*Leptophis ahaetulla praestans* (Cope, 1869)(B, G, M)	130	-
	Leptophis ahaetulla occidentalis (Günther, 1859)(C)		
Leptophis bilineatus	*Leptophis depressirostris* (Cope, 1861)(C, N)	130	-
	Masticophis bilineatus Jan, 1863? (M)		
Leptophis æruginosus	*Leptophis depressirostris* (Cope, 1861)	130	-
Synchalinus corallioides	*Pseustes poecilonotus chrysobronchus* (Cope, 1876)	131	-
Tropidonotus ordinatus	(*Thamnophis s. sirtalis* (Linnaeus, 1766))	131	-
	Thamnophis spp. (M)[20]		
	Thamnophis cyrtopsis collaris (Jan, 1863)(G)		
	Thamnophis fulvus (Bocourt, 1893)(G)		
	Thamnophis marcianus praeocularis (Bocourt, 1892)(G)		
Tropidonotus saurita	(*Thamnophis s. sauritus* (Linnaeus, 1766))	132	-
	Thamnophis proximus orarius Rossman, 1963		
	Thamnophis proximus rutiloris (Cope, 1885)		
Tropidonotus chrysocephalus	*Thamnophis chrysocephalus* (Cope, 1885)	132	-
Tropidonotus præocularis	*Thamnophis marcianus praeocularis* (Bocourt, 1892)	132	-
Tropidonotus multimaculatus	*Thamnophis rufipunctatus* (Cope, 1875)	132	-
Tropidonotus scaliger	*Thamnophis scalaris scaliger* (Jan, 1863)	133	-
*Tropidonotus godmani**	*Thamnophis godmani* (Günther, 1894)[21]	133	-
Tropidonotus variabilis	*Storeria storerioides* (Cope, 1865)	133	-
Tropidonotus melanogaster	*Thamnophis m. melanogaster* (Peters, 1864)	133	-
	Thamnophis melanogaster canescens Smith, 1942		

Günther's Names (*new)	Current Names	Page	Plate
Tropidonotus validus	*Nerodia v. valida* (Kennicott, 1860)(M)[22]	134	-
	Nerodia valida celaeno (Cope, 1860)(M)		
	Nerodia valida isabelleae Conant, 1953 (M)		
	Thamnophis spp. indet. (US)		
Tropidonotus grahami	*Regina grahamii* Baird & Girard, 1853 (US)	134	-
	Nerodia valida thamnophisoides Conant, 1961? (M)		
Tropidonotus sipedon	*Nerodia s. sipedon* (Linnaeus, 1758)(US)	134	-
	Nerodia rhombifera blanchardi (Clay, 1938)(M)[23]		
	Gen. sp. indet. (C)		
Tretanorhinus nigroluteus	*Tretanorhinus n. nigroluteus* Cope, 1861	134	-
Tretanorhinus dimidiatus	(*Atretium schistosum* (Daudin, 1803)?)	135	-
	Gen. sp. indet.[24]		
Tretanorhinus mocquardi	*Tretanorhinus mocquardii* Bocourt, 1891	135	-
Helicops septemvittatus	*Limnophis bicolor* Günther, 1872?[25]	135	-
Ischnognathus dekayi	*Storeria d. dekayi* (Holbrook, 1839)(US)	136	-
	Storeria dekayi anomala Dugès, 1888 (M)		
	Storeria dekayi temporalinea Trapido, 1944 (M)		
	Storeria dekayi tropica Cope, 1885 (G)		
	Storeria occipitomaculata hidalgoensis Taylor, 1942 (M)		
Ischnognathus occipito-maculatus	*Storeria o. occipitomaculata* (Storer, 1839)(US)	136	-
	Storeria occipitomaculata hidalgoensis Taylor, 1942 (M)		
Ischnognathus storerioides	*Storeria storerioides* (Cope, 1865)	136	-
Ischnognathus copii	*Adelophis copei* Dugès, 1879	136	-
Hydromorphus concolor	*Hydromorphus concolor* Peters, 1859	137	-
Hypsiglena torquata	*Hypsiglena t. torquata* (Günther, 1860)(M)[26]	137	-
	Hypsiglena torquata affinis Boulenger, 1894 (M)		
	Hypsiglena torquata janii (Dugès, 1865)(M,US)		
	Hypsiglena torquata klauberi Tanner, 1946 (US)		
	Hypsiglena torquata chlorophaea Cope, 1860 (M)		
Hypsiglena discolor	*Tantalophis discolor* (Günther, 1860)	137	49
*Hypsiglena latifasciata**	*Pseudoleptodeira latifasciata* (Günther, 1894)	138	49
Hypsiglena ornata	*Leptodeira septentrionalis ornata* (Bocourt, 1884)	139	-
Petalognathus nebulatus	*Sibon n. nebulata* (Linnaeus, 1758)(C, G, M, N, SA)	139	-
	Sibon nebulata leucomelas (Boulenger, 1896)(SA)		
Tropidodipsas fasciata	*Sibon f. fasciata* (Günther, 1858)	139	50
Tropidodipsas dumerilii	*Sibon s. sartorii* (Cope, 1863)(G, M)	140	50
	Sibon sartorii annulata (Peters, 1870)(G)		
Tropidodipsas philippii	*Sibon philippii* (Jan, 1863)	140	-
Tropidodipsas anthracops	*Sibon anthracops* (Cope, 1868)	140	-
*Neopareas** *bicolor**	*Dipsas bicolor* (Günther, 1895)	178	56
Leptognathus mikanii	*Sibynomorphus m. mikanii* (Schlegel, 1837)(SA)	141	-
	Dipsas elegans (Boulenger, 1896)(M)		
	Dipsas catesbyi (Sentzen, 1796)?(SA)		
Leptognathus annulatus	*Sibon annulata* (Günther, 1872)	141	49
Leptognathus argus	*Sibon argus* (Cope, 1876)	142	-
Leptognathus pictiventris	*Sibon pictiventris* (Cope, 1876)[27]	142	-
Leptognathus articulatus	*Dipsas articulata* (Cope, 1868)	142	-
Leptognathus viguieri	*Dipsas viguieri* (Bocourt, 1884)	142	-
Leptognathus brevis	*Sibon n. nebulata* (Linnaeus, 1758)?	142	-
Leptognathus brevifacies	*Dipsas brevifacies* (Cope, 1866)	142	-
Leptognathus semicinctus	*Sibon s. sartorii* (Cope, 1866)	142	-
Mesopeltis sanniolus	*Sibon s. sanniola* (Cope, 1866)	143	-

Günther's Names (*new)	Current Names	Page	Plate
Mesopeltis dimidiatus	*Sibon dimidiata* (Günther, 1872)	143	51
Homalocranium planiceps	*Tantilla planiceps* (Blainville, 1835)	145	-
	Tantilla nigriceps Kennicott, 1860		
Homalocranium miniatum	*Tantilla miniata* Cope, 1863	146	-
*Homalocranium atriceps**	*Tantilla atriceps* (Günther, 1895)	146	52
Homalocranium gracile	*Tantilla gracilis* Baird & Girard, 1853	146	-
Homalocranium melanocephalum	*Tantilla melanocephala* (Linnaeus, 1758)	147	-
*Homalocranium melanocephalum fraseri**	*Tantilla melanocephala* (Linnaeus, 1758)	148	-
*Homalocranium melanocephalum pernambucense**	*Tantilla melanocephala* (Linnaeus, 1758)	148	-
*Homalocranium jani**	*Tantilla jani* (Günther, 1895)	148	52
*Homalocranium boulengeri**	*Tantilla miniata* Cope, 1863	148	52
*Homalocranium bocourti**	*Tantilla b. bocourti* (Günther, 1895)	149	-
Homalocranium armillatum	*Tantilla melanocephala* (Linnaeus, 1758)	149	52
*Homalocranium breve**	*Tantilla canula* Cope 1876	150	-
Homalocranium michoacanense	*Sonora m. michoacanensis* (Dugès, 1885)	150	36
	Sonora michoacanensis mutabilis (Stickel, 1943)		
Homalocranium annulatum	*Tantilla annulata* Boettger, 1892	150	-
Homalocranium deppii	*Tantilla deppii* (Bocourt, 1883)	151	-
Homalocranium tæniatum	*Tantilla taeniata* (Bocourt, 1883)(G)	151	-
	Tantilla taeniata tritaeniata Smith & Williams, 1967 (H)		
Homalocranium mœstum	*Tantilla moesta* (Günther, 1863)	152	52
Homalocranium schistosum	*Tantilla schistosa* (Bocourt, 1883)	152	-
Homalocranium reticulatum	*Tantilla reticulata* Cope, 1860	152	-
Homalocranium canulum	*Tantilla canula* Cope, 1876	153	-
Homalocranium mexicanum	*Tantilla melanocephala* (Linnaeus, 1758)	153	-
Homalocranium virgatum	*Tantilla reticulata* Cope, 1860	154	52
Homalocranium bimaculatum	*Tantilla calamarina* Cope, 1866	154	-
Tantilla calamarina	*Tantilla calamarina* Cope, 1866	155	-
Homalocranium rubrum	*Tantilla r. rubra* Cope, 1876	155	-
Lioninia vermiformis	*Tantilla vermiformis* (Hallowell, 1860)	155	-
Symphimus leucostomus	*Symphimus leucostomus* Cope, 1869	155	-
Scolecophis atrocinctus	*Scolecophis atrocinctus* (Schlegel, 1837)(G)	156	-
	Sonora michoacanensis mutabilis Stickel, 1943		
Procinura æmula	*Procinura aemula* Cope, 1879[28]	156	-
Pogonaspis ruficeps	*Tantilla ruficeps* (Cope, 1894)	157	-
Enulius murinus	*Enulius f. flavitorques* (Cope, 1869)	157	-
Cercocalamus collaris	*Aparallactus capensis* Smith, 1849[29]	157	-
Stenorhina degenhardti[30]	*Stenorrhina degenhardtii* (Berthold, 1846)(C, G, M, N, P, SA)	158	-
	Stenorrhina freminvillii (Duméril, Bibron & Duméril, 1854) (C, E, G, H, M)		
Tachymenis piceivittis	*Coniophanes p. piceivittis* Cope, 1870	160	-
	Coniophanes piceivittis schmidti Bailey, 1937		
Tachymenis bipunctata	*Coniophanes b. bipunctatus* (Günther, 1858)(B)	161	-
	Coniophanes bipunctatus biseriatus Smith, 1940 (M)		
Tachymenis fissidens	*Coniophanes f. fissidens* (Günther, 1858)(B, G, M, N)	161	-
	Coniophanes fissidens convergens Shannon & Smith, 1950 (M)		
	Coniophanes fissidens obsoletus Minton & Smith, 1962 (C)		
	Coniophanes fissidens proterops Cope, 1860 (M)		
	Coniophanes fissidens punctigularis Cope, 1860 (G)		
Tachymenis lateritia	*Coniophanes lateritius* Cope, 1862	162	-
Tachymenis grammophrys	*Rhadinaea laureata* (Günther, 1868)	162	-

Günther's Names (*new)	Current Names	Page	Plate
*Tachymenis decipiens**	*Rhadinaea guentheri* Dunn, 1938[31]	163	53
Hydrocalamus quinquevittatus	*Coniophanes quinquevittatus* (Duméril, Bibron & Duméril, 1854)	164	-
Tomodon nasutus	*Manolepis putnami* (Jan, 1863)	164	-
Conophis lineatus	(*Conophis l. lineatus* (Duméril, Bibron & Duméril, 1854))	165	-
	Conophis lineatus concolor Cope, 1867 (M)		
	Conophis lineatus dunni Smith, 1941 (C, G, N)		
	Conophis pulcher Cope, 1869 (G)		
Conophis vittatus	*Conophis v. vittatus* Peters, 1860	165	-
	Conophis vittatus viduus Cope, 1876		
Philodryas putnami	*Manolepis putnami* (Jan, 1863)	166	-
Erythrolamprus æsculapii	*Erythrolamprus a. aesculapii* (Linnaeus, 1766)(SA)	166	-
	Erythrolamprus bizona Jan, 1863 (C)		
	Erythrolamprus mimus impar Schmidt, 1936 (N)		
	Erythrolamprus mimus micrurus Dunn & Bailey, 1939 (P)		
Oxyrrhopus plumbeus	*Clelia clelia plumbea* (Wied, 1820)(SA)	167	-
	Clelia scytalina (Cope, 1867)(C?, G?, M)		
Oxyrrhopus cloelia	*Clelia c. clelia* (Daudin, 1803)(P, SA)	167	-
	Clelia scytalina (Cope, 1867)(M)		
Oxyrrhopus occipitoluteus	*Clelia occipitolutea* (Duméril, Bibron & Duméril, 1854)	167	-
Oxyrrhopus petolarius	*Oxyrhopus p. petola* (Linnaeus, 1758)(SA)	167	-
	Oxyrhopus petola aequifasciata Werner, 1909 (G)		
	Oxyrhopus petola sebae Duméril, Bibron & Duméril, 1854 (C, N, SA)		
Scytale coronatum	*Pseudoboa coronata* Schneider, 1801	168	-
Leptodira nigrofasciata	*Leptodeira n. nigrofasciata* Günther, 1868	169	-
Leptodira pacifica	*Leptodeira punctata* (Peters, 1866)	169	-
Leptodira mystacina	*Leptodeira nigrofasciata mystacina* Cope, 1869	169	-
Leptodira annulata	(*Leptodeira a. annulata* (Linnaeus, 1758))	170	-
	Leptodeira s. septentrionalis (Kennicott, 1859)		
*Leptodira affinis**	*Leptodeira septentrionalis ornata* (Bocourt, 1884)	170	-
Leptodira yucatanensis	*Leptodeira frenata yucatanensis* Duellman & Werler, 1955	171	-
*Leptodira splendida**	*Leptodeira s. splendida* Günther, 1895	171	53
Leptodira personata	*Leptodeira annulata cussiliris* Duellman, 1958	171	54
*Leptodira polysticta**	*Leptodeira septentrionalis polysticta* Günther, 1895	172	55
*Leptodira ocellata**	*Leptodeira annulata rhombifera* Günther, 1872	172	55
Leptodira frenata	*Leptodeira f. frenata* (Cope, 1886)	173	-
Leptodira rhombifera	*Leptodeira annulata rhombifera* Günther, 1872	173	54
Trimorphodon biscutatus	*Trimorphodon b. biscutatus* (Duméril, Bibron & Duméril, 1854)(G, M)	174	-
	Trimorphodon biscutatus quadruplex Smith, 1943 (N, P)		
Trimorphodon upsilon	*Trimorphodon t. tau* Cope, 1869	175	-
Dipsas cenchoa	*Imantodes c. cenchoa* (Linnaeus, 1758)(SA)	175	-
	Imantodes cenchoa leucomelas (Cope, 1861)(G, M)		
	Imantodes cenchoa semifasciatus (Cope, 1894)(C, P)[32]		
Dipsas gemmistrata	*Imantodes g. gemmistratus* (Cope, 1861)(C, E, G)[33]	175	-
	Imantodes gemmistratus latistratus (Cope, 1887)(M)		
	Imantodes gemmistratus oliveri Smith, 1942 (M)		
*Dipsas splendida**	*Imantodes gemmistratus splendidus* (Gunther, 1895)[33]	176	56
Dipsas tenuissima	*Imantodes tenuissimus* (Cope, 1866)	176	-
*Dipsas gracillima**	*Imantodes gemmistratus gracillimus* (Günther, 1895)[33]	177	56
Dryiophis acuminata	*Oxybelis aeneus* (Wagler, 1824)	177	-

Günther's Names (*new)	Current Names	Page	Plate
Dryiophis fulgida	*Oxybelis fulgidus* (Daudin, 1803)	178	-
Dryiophis brevirostris	*Oxybelis brevirostris* (Cope, 1861)	178	-
Nothopsis rugosus	*Nothopsis rugosus* Cope, 1871	179	-
Loxocemus bicolor	*Loxocemus bicolor* Cope, 1861	179	-
Epicrates cenchris	*Epicrates c. cenchria* (Linnaeus, 1758)(SA)	180	-
	Epicrates cenchria maurus Gray, 1849 (C, P)		
Corallus cooki	*Corallus enydris cookii* Gray, 1842	180	-
Corallus annulatus	*Corallus a. annulatus* (Cope, 1876)	181	-
Ungalia moreleti	*Tropidophis semicinctus* (Gundlach & Peters, 1865)[34]	181	-
Ungaliophis continentalis	*Ungaliophis continentalis* Müller, 1882	181	-
Boa imperator	*Boa constrictor imperator* Daudin, 1803	181	-
Boa mexicana	*Boa constrictor imperator* Daudin, 1803	182	-
Elaps fulvius	(*Micrurus f. fulvius* (Linnaeus, 1766))	182	-
	Micrurus d. diastema (Duméril, Bibron & Duméril, 1854)(M)		
	Micrurus diastema affinis (Jan, 1858)(M)		
	Micrurus diastema alienus (Werner, 1903)(M)		
	Micrurus diastema apiatus (Jan, 1858)(G)		
	Micrurus diastema sapperi (Werner, 1903)(B, G, M)		
	Micrurus distans michoacanensis (Dugès, 1891)(M)		
	Micrurus fulvius fitzingeri (Jan, 1858)(M)		
	Micrurus nigrocinctus divaricatus (Hallowell, 1855)(B, H)		
	Micrurus nigrocinctus zunilensis Schmidt, 1932 (G)[35]		
Elaps bernardi	*Micrurus bernadi* (Cope, 1887)	184	-
Elaps aglæope	*Micrurus diastema aglaeope* (Cope, 1860)(H)	184	-
	Micrurus diastema apiatus (Jan, 1858)(G)		
	Micrurus diastema macdougalli Roze, 1967 (M)		
	Micrurus diastema sapperi (Werner, 1903)(G)		
Elaps ephippifer	*Micrurus ephippifer* (Cope, 1886)	184	-
Elaps multifasciatus	*Micrurus m. multifasciatus* (Jan, 1858)(P)	184	-
	Micrurus multifasciatus hertwigi (Werner, 1890)(C, N, P)		
	Micrurus m. mipartitus (Duméril, Bibron & Duméril, 1854) (SA)		
	Micrurus mipartitus semipartitus (Jan, 1858)(SA)		
*Elaps ruatanus**	*Micrurus nigrocinctus ruatanus* (Günther, 1895)	185	57
Elaps lemniscatus	*Micrurus l. lemniscatus* (Linnaeus, 1758)(SA)	185	-
	Micrurus l. laticollaris (Peters, 1869)(M)		
Elaps decoratus	*Micrurus decoratus* (Jan, 1858)	185	-
Elaps elegans	*Micrurus e. elegans* (Jan, 1858)(M)	186	-
	Micrurus elegans veraepacis Schmidt, 1933 (G)		
Pelamis bicolor	*Pelamis platurus* (Linnaeus, 1766)	186	-
Ancistrodon bilineatus	*Agkistrodon b. bilineatus* (Günther, 1863)(G, M)	186	58
	Agkistrodon bilineatus russeolus Gloyd, 1972 (B, M, N)[36]		
Bothrops undulatus	*Ophryacus undulatus* (Jan, 1859)	187	-
Bothrops atrox	*Bothrops atrox* (Linnaeus, 1758)(SA)	187	-
	Bothrops asper (Garman, 1883)(B, C, G, M, N, P)		
Lachesis mutus	*Lachesis m. muta* (Linnaeus, 1766)(SA)	188	-
	Lachesis muta stenophrys Cope, 1876 (C, P)		
Bothriechis schlegeli	*Bothriechis schlegelii* (Berthold, 1846)[37]	189	-
Bothriechis nigroviridis	*Bothriechis n. nigroviridis* Peters, 1859[37]	189	-
Bothriechis lateralis	*Bothriechis lateralis* Peters, 1863[37]	189	-
Bothriechis bicolor	*Bothriechis bicolor* (Bocourt, 1868)[37]	189	-
Bothriechis aurifera	*Bothriechis nigroviridis aurifer* (Salvin, 1860)[37]	189	-

Günther's Names (*new)	Current Names	Page	Plate
Bothriechis lansbergii	*Porthidium l. lansbergii* (Schlegel, 1841)(SA)[37]	190	-
	Porthidium lansbergii venezuelensis (Sandner, 1952)(SA)		
	Porthidium dunni (Hartweg & Oliver, 1938)(M)		
	Porthidium nasutum (Bocourt, 1868)(C, G, N, P, SA)		
	Porthidium yucatanicum (Smith, 1941)(M)		
Bothriechis godmani	*Porthidium godmani* (Günther, 1863)[37]	190	57
Bothriechis ophryomegas	*Porthidium ophryomegas* (Bocourt, 1868)[37]	191	-
Bothriechis nummifera	*Porthidium nummifer* (Rüppell, 1845)(M)[37]	191	-
	Porthidium nummifer occiduum (Hoge, 1966)(G)		
	Porthidium mexicanum (Duméril, Bibron & Duméril, 1854) (C, G, N)		
Bothriechis scutigera	*Porthidium godmani* (Günther, 1863)[37]	191	-
Crotalophorus miliarius	(*Sistrurus m. miliarius* (Linnaeus, 1766))	192	-
	Sistrurus r. ravus (Cope, 1865)		
Crotalus triseriatus	*Crotalus t. triseriatus* (Wagler, 1830)	192	-
	Crotalus triseriatus anahuacus Gloyd, 1940		
Crotalus polystictus	*Crotalus polystictus* (Cope, 1865)	192	-
Crotalus intermedius	*Crotalus i. intermedius* Fischer, 1882	192	-
*Crotalus omiltemanus**	*Crotalus intermedius omiltemanus* Günther, 1895	192	58
Crotalus palmeri	*Crotalus l. lepidus* (Kennicott, 1861)(M)	193	-
	Crotalus mitchellii pyrrhus (Cope, 1867)(US)		
	Crotalus mitchellii stephensi Klauber, 1930 (US)		
	Crotalus tigris Kennicott, 1859 (US)		
*Crotalus pallidus**	*Crotalus t. triseriatus* (Wagler, 1830)	193	59
*Crotalus salvini**	*Crotalus scutulatus salvini* Günther, 1895	193	59
Crotalus horridus	(*Crotalus h. horridus* Linnaeus, 1758)	194	-
	Crotalus d. durissus Linnaeus, 1758 (B)		
	Crotalus durissus culminatus Klauber, 1952 (M)		
Crotalus basiliscus	*Crotalus b. basiliscus* (Cope, 1864)[38]	194	-
	Crotalus molossus nigrescens Gloyd, 1936		
Crotalus adamanteus	*Crotalus adamanteus* Beauvois, 1799 (US)	195	-
	Crotalus atrox Baird & Girard, 1853 (M)		

SALIENTIA

Günther's Names (*new)	Current Names	Page	Plate
Rana montezumæ	*Rana montezumae* Baird, 1854 (M)	197	-
	Rana spp. indet. (G, M)[39]		
Rana halecina	*Rana pipiens* Schreber, 1872 (US)	198	-
	Rana berlandieri Baird, 1854 (M)		
	Rana forreri Boulenger, 1888 (C, G, M)		
	Rana macroglossa Brocchi, 1877 (G)		
	Rana magnaocularis Frost & Bagnara, 1976 (M)		
	Rana omiltemana Günther, 1900 (M)		
Rana macroglossa	*Rana macroglossa* Brocchi, 1877	199	-
Rana forreri	*Rana forreri* Boulenger, 1883	199	60
*Rana omiltemana**	*Rana omiltemana* Günther, 1900	200	61
*Rana bonaccana**	*Rana palmipes* Spix, 1824	201	60
Rana maculata	*Rana m. maculata* Brocchi, 1877	201	-
Rana pustulosa	*Rana pustulosa* Boulenger, 1883	202	61
Rana palmipes	*Rana palmipes* Spix, 1824	202	-
*Rana melanosoma**	*Rana palmipes* Spix, 1824	203	63
*Rana godmani**	*Rana vibicaria* (Cope, 1894)	204	63

Günther's Names (*new)	Current Names	Page	Plate
Rana cæruleopunctata	*Rana warschewitschii* (Schmidt, 1857)	205	-
Lævirana vibicaria	*Rana vibicaria* (Cope, 1894)	206	-
Hylaplesia tinctoria	(*Dendrobates tinctorius* (Schneider, 1799))	207	-
	Dendrobates auratus (Girard, 1855)		
Hylaplesia trivittata	(*Dendrobates trivittatus* (Spix, 1824))	207	-
	Dendrobates maculatus Peters, 1873		
Hylaplesia typographa	*Dendrobates pumilio* Schmidt, 1857	207	-
Hylaplesia talamancæ	*Colostethus talamancae* (Cope, 1875)	208	-
Hylaplesia truncata	*Dendrobates truncatus* (Cope, 1861)	208	-
Atelopus varius	*Atelopus varius* (Lichtenstein & Martens, 1856)	208	62
Engystoma ustum	*Gastrophryne u. usta* (Cope, 1866)	209	-
	Gastrophryne usta gadovii (Boulenger, 1903)		
Engystoma pictiventre	*Gastrophryne pictiventris* (Cope, 1886)	209	-
Engystoma elegans	*Gastrophryne elegans* (Boulenger, 1882)	210	62
Glossostoma aterrimum**	*Glossostoma aterrimum* Günther, 1900	210	64
Hypopachus variolosus	*Hypopachus variolosus* (Cope, 1866)	211	-
Hypopachus inguinalis	*Hypopachus variolosus* (Cope, 1866)(G)	211	-
	Glossostoma aequatoriale (Peracca, 1904)(SA)		
Leptodactylus pentadactylus	*Leptodactylus pentadactylus* (Laurent, 1768)	212	-
Leptodactylus albilabris	*Leptodactylus albilabris* (Günther, 1859)(WI)	213	-
	Leptodactylus fragilis (Brocchi, 1877)(M)		
Leptodactylus fragilis	*Leptodactylus fragilis* (Brocchi, 1877)	213	-
Leptodactylus microtis	*Leptodactylus melanonotus* (Hallowell, 1861)	213	-
Leptodactylus perlævis	*Leptodactylus melanonotus* (Hallowell, 1861)	214	-
Leptodactylus caliginosus	*Leptodactylus melanonotus* (Hallowell, 1861)(C, M, N)	214	-
	Leptodactylus spp. indet. (SA, WI)[40]		
Borborocœtes mexicanus	*Hylactophryne occidentalis* (Taylor, 1941)	215	-
Syrrhaphus verruculatus	*Syrrhophus verruculatus* (Peters, 1870)	215	-
*Syrrhaphus omiltemanus**	*Eleutherodactylus omiltemanus* (Günther, 1900)	216	-
Syrrhaphus ridens	*Eleutherodactylus ridens* (Cope, 1866)	216	-
Syrrhaphus verrucipes	*Syrrhophus verrucipes* Cope, 1885	216	-
Syrrhaphus leprus	*Syrrhophus leprus* Cope, 1879	217	-
Syrrhaphus guttulatus	*Syrrhophus guttilatus* (Cope, 1879)	217	-
Syrrhaphus cystignathoides	*Syrrhophus cystignathoides* (Cope, 1877)	218	-
Syrrhaphus hylæformis	*Eleutherodactylus hylaeformis* (Cope, 1875)	218	-
Paludicola nitida	*Tomodactylus nitidus* (Peters, 1869)	218	-
Paludicola (?) *mexicana*	*Eleutherodactylus mexicanus* (Brocchi, 1877)	219	-
Tomodactylus amulæ**	*Tomodactylus nitidus* (Peters, 1869)	219	64
Liohyla guentheri	*Eleutherodactylus fitzingeri* (Schmidt, 1858)[41]	220	66
	Eleutherodactylus andi Savage, 1974[41]		
Liohyla rugulosa	*Eleutherodactylus rugulosus* (Cope, 1870)	221	-
	Eleutherodactylus vocalis Taylor, 1940		
Liohyla bocourti	*Eleutherodactylus bocourti* (Brocchi, 1877)(G)	222	-
	Eleutherodactylus escoces Savage, 1975 (C)[42]		
Liohyla fleischmanni	*Eleutherodactylus fleischmanni* (Boettger, 1892)	222	64
*Liohyla pittieri**	*Eleutherodactylus rugulosus* (Cope, 1870)[42]	223	65
*Liohyla engytympanum**	*Eleutherodactylus fleischmanni* (Boettger, 1892)[42]	224	65
Liohyla ranoides	*Eleutherodactylus rugulosus* (Cope, 1870)[42]	225	-
Liohyla berkenbuschi	*Eleutherodactylus berkenbuschii* Peters, 1870	225	-
Hylodes sallæi	*Eleutherodactylus rhodopis* (Cope, 1867)	227	-
*Hylodes plicatus**	*Eleutherodactylus rhodopis* (Cope, 1867)	228	66
Hylodes polyptychus	*Eleutherodactylus bransfordii* (Cope, 1886)	228	66

Günther's Names (*new)	Current Names	Page	Plate
Hylodes underwoodi	*Eleutherodactylus bransfordii* (Cope, 1886)	229	66
*Hylodes calcitrans**	*Eleutherodactylus omiltemanus* (Günther, 1900)	230	67
*Hylodes platyrhynchus**	*Eleutherodactylus melanostictus* (Cope, 1876)	230	67
Hylodes cerasinus	*Eleutherodactylus cerasinus* (Cope, 1876)	231	-
Hylodes biporcatus	*Eleutherodactylus biporcatus* (Peters, 1863)(P)	231	-
	Eleutherodactylus sp. indet. (SA)[43]		
Hylodes euryglossus	*Eleutherodactylus punctariolus* (Peters, 1863)	232	-
Hylodes rhodopis	*Eleutherodactylus rhodopis* (Cope, 1867)(M)	232	67
	Eleutherodactylus bransfordii (Cope, 1886)(C)		
Hylodes rugosus	*Eleutherodactylus biporcatus* (Peters, 1863)[44]	233	-
*Hylodes venustus**	*Eleutherodactylus rhodopis* (Cope, 1867)	234	68
Hylodes florulentus	*Eleutherodactylus biporcatus* (Peters, 1863)	234	-
Hylodes augusti	*Hylactophryne a. augusti* (Dugès, 1879)	235	68
Hylodes laticeps	*Eleutherodactylus laticeps* (Duméril, 1853)	235	-
	Hylactophryne augusti cactorum (Taylor, 1939)		
Hylodes melanostictus	*Eleutherodactylus melanostictus* (Cope, 1876)(C)	236	-
	Eleutherodactylus sp. indet. (M)[43]		
Hylodes brocchii	*Eleutherodactylus brocchi* (Boulenger, 1882)(G)	236	68
	Eleutherodactylus melanostictus (Cope, 1876)(C)		
*Hylodes nubilus**	*Eleutherodactylus fitzingeri* (Schmidt, 1858)[41]	237	69
Hylodes alfredi	*Eleutherodactylus alfredi* (Boulenger, 1898)	237	-
Hylodes bransfordi	*Eleutherodactylus bransfordii* (Cope, 1886)	238	-
Hylodes diastema	*Eleutherodactylus diastema* (Cope, 1876)	238	-
Hylodes lanciformis	*Eleutherodactylus gollmeri* (Peters, 1863)	239	-
Hylodes megacephalus	*Eleutherodactylus biporcatus* (Peters, 1863)[44]	239	-
Hylodes gulosus	*Eleutherodactylus biporcatus* (Peters, 1863)[44]	240	-
Hylodes griseus	*Eleutherodactylus fitzingeri* (Schmidt, 1858)	240	-
Hylodes longipes	*Syrrhophus longipes* (Baird, 1859)	240	-
Hylodes lineatus	*Eleutherodactylus lineatus* (Brocchi, 1879)	241	-
Hylodes latrans	*Hylactophryne augusti latrans* (Cope, 1880)	241	-
Calyptocephalus testudiniceps	(*Caudiverbera caudiverbera* (Linnaeus, 1758))	241	-
	Gen. sp. indet. (P)[45]		
Engystomops pustulosus	*Physalaemus pustulosus* (Cope, 1864)	242	-
Bufo melanogaster	*Bufo haematiticus* Cope, 1862	242	-
Bufo hæmatiticus	*Bufo haematiticus* Cope, 1862	243	-
Bufo punctatus	*Bufo punctatus* Baird & Girard, 1852	244	-
Bufo mexicanus	*Bufo compactilis* Wiegmann, 1833	244	-
Bufo monksiæ	*Bufo occidentalis* Camerano, 1879	245	-
Bufo compactilis	*Bufo compactilis* Wiegmann, 1833 (M)	245	-
	Bufo speciosus Girard, 1854 (US)		
Bufo politus	*Bufo boreas (boreas* ?) Baird & Girard, 1852	246	-
Bufo marmoreus	*Bufo marmoreus* Wiegmann, 1833	246	70
Bufo sternosignatus	*Bufo sternosignatus* Günther, 1859 (SA)	247	-
	Bufo v. valliceps Wiegmann, 1833 (C, M)		
Bufo canaliferus	*Bufo canaliferus* Cope, 1877	248	-
Bufo debilis	*Bufo d. debilis* Girard, 1854 (M, US)	248	-
	Bufo debilis insidior Girard, 1854 (US)		
	Bufo retiformis Sanders & Smith, 1951 (M)		
	Bufo kelloggi Taylor, 1938 (M)		
Bufo marinus	*Bufo marinus* (Linnaeus, 1758)	249	-
Bufo occipitalis	*Bufo cristatus* Wiegmann, 1833	250	69
Bufo coccifer	*Bufo coccifer* Cope, 1866	250	-

Günther's Names (*new)	Current Names	Page	Plate
Bufo coniferus	*Bufo coniferus* Cope, 1862	251	69
Bufo valliceps	*Bufo v. valliceps* Wiegmann, 1833 (C, G, M, N, US)	252	-
	Bufo mazatlanensis Taylor, 1940 (M)		
Bufo melanochlorus	*Bufo melanochloris* Cope, 1878	252	-
Bufo luetkenii	*Bufo luetkenii* Boulenger, 1891	253	-
Bufo typhonius	*Bufo t. typhonius* (Linnaeus, 1758)	253	-
Bufo simus	*Bufo simus* Schmidt, 1857 (P)	254	-
	Bufo occidentalis Camerano, 1879 (M)		
Bufo intermedius	*Bufo intermedius* Günther, 1859 (SA)	255	-
	Bufo occidentalis Camerano, 1879 (M)		
Bufo bocourti	*Bufo bocourti* Brocchi, 1877	255	-
Bufo veraguensis	*Bufo veraguensis* Schmidt, 1857[46]	256	-
Bufo auritus	*Bufo coniferus* Cope, 1862	256	-
Nannophryne fastidiosa	*Bufo fastidiosus* (Cope, 1875)	257	-
Nannophryne epiotica	*Crepidophryne epiotica* (Cope, 1876)	257	-
Nannophryne cœrulescens	*Bufo fastidiosus* (Cope, 1875)	257	-
Rhinophrynus dorsalis	*Rhinophrynus dorsalis* Duméril & Bibron, 1851	258	-
Scaphiopus couchii	*Scaphiopus couchii* Baird, 1854	259	-
Scaphiopus rectifrenis	*Scaphiopus couchii* Baird, 1854	259	-
Scaphiopus hammondii	*Spea hammondii* (Baird, 1859)(US)	259	-
	Spea bombifrons (Cope, 1863)(US)		
	Spea multiplicata (Cope, 1863)(M, US)		
Scaphiopus multiplicatus	*Spea multiplicata* (Cope, 1863)	260	-
Hyla eximia	*Hyla eximia* Baird, 1854	261	-
	Ololygon staufferi (Cope, 1865)[47]		
Hyla plicata	*Hyla plicata* Brocchi, 1877	261	-
Hyla bocourti	*Hyla bocourti* (Mocquard, 1899)(G)	263	-
	Hyla eximia Baird, 1854 (M)		
*Hyla nana**	*Hyla smithii* Boulenger, 1901	263	73
Hyla cherrii	*Hyla microcephala underwoodi* Boulenger, 1898	264	-
Hyla microcephala	*Hyla m. microcephala* Cope, 1886	265	-
Hyla elæochroa	*Ololygon elaeochroa* (Cope, 1876)	265	-
Hyla bistincta	*Hyla bistincta* Cope, 1877	265	-
Hyla splendens	*Gastrotheca splendens* (Schmidt, 1857)	266	-
Hyla copii	*Hyla arenicolor* Cope, 1866	266	70
Hyla boulengeri	*Ololygon boulengeri* (Cope, 1887)	267	-
*Hyla mocquardi**	*Hyla* sp. dubium[48]	268	73
Hyla quinquevittata	*Ololygon elaeochroa* (Cope, 1876)	268	-
Hyla phæota	*Smilisca phaeota* (Cope, 1862)	269	-
*Hyla tæniopus**	*Hyla taeniopus* Günther, 1901	269	72
Hyla puma	*Smilisca puma* (Cope, 1885)	270	-
Hyla baudini	*Smilisca baudinii* (Duméril & Bibron, 1841)(C, G, M, N, US)	270	71
	Smilisca phaeota (Cope, 1862)(SA)		
Hyla venulosa	*Phrynohyas venulosa* (Laurent, 1768)	272	-
Hyla sordida	*Smilisca sordida* (Peters, 1863)	273	-
Hyla gabbi	*Smilisca sordida* (Peters, 1863)	274	70
*Hyla pseudopuma**	*Hyla p. pseudopuma* Günther, 1901	274	72
*Hyla godmani**	*Hyla godmani* Günther, 1901	275	72
Hyla uranochroa	*Hyla uranochroa* Cope, 1876	276	70
Hyla leucophyllata	*Hyla leucophyllata* (Beireis, 1783)(SA)	277	72
	Hyla ebraccata Cope, 1874 (C, N, P)		
Hyla underwoodi	*Hyla microcephala underwoodi* Boulenger, 1898	278	-

Günther's Names (*new)	Current Names	Page	Plate
Hyla nigripes	*Smilisca sordida* (Peters, 1863)	278	-
Hyla molitor	*Hyla molitor* Schmidt, 1857[49]	279	-
Hyla miotympanum	*Hyla miotympanum* Cope, 1863	279	-
Hyla prosoblepon	*Centrolenella prosoblepon* (Boettger, 1892)	280	-
Hyla pulverata	*Centrolenella pulverata* (Peters, 1873)	280	-
Hyla glandulosa	*Plectrohyla glandulosa* (Boulenger, 1883)	281	-
Hyla guatemalensis	*Plectrohyla guatemalensis* Brocchi, 1877	281	-
Hyla crassa	*Hyla crassa* (Brocchi, 1877)	281	-
Hyla punctariola	*Eleutherodactylus punctariolus* (Peters, 1863)	282	-
Hyla spilomma	*Phrynohyas venulosa* (Laurent, 1768)	282	-
Hyla miliaria	*Hyla miliaria* (Cope, 1886)	283	-
Hyla crepitans	*Hyla crepitans* Wied-Neuwied, 1824	283	-
Hyla albomarginata	*Hyla albomarginata* Spix, 1824 (SA)	284	-
	Hyla rufitela Fouquette, 1961 (C, N)		
Hyla rudis	*Pternohyla fodiens* Boulenger, 1882	285	-
Hylella sumichrasti	*Hyla sumichrasti* (Brocchi, 1879)	286	-
*Hylella picta**	*Hyla picta* (Günther, 1901)	286	73
Hylella fleischmanni	*Centrolenella fleischmanni* (Boettger, 1893)	287	73
Nototrema oviferum	*Gastrotheca ovifera* (Lichtenstein & Weinland, 1854)(SA)	288	74
	Anotheca spinosa (Steindachner, 1864)(C, M)		
Agalychnis moreletii	*Agalychnis moreletii* (Duméril, 1853)(G)	289	-
	Agalychnis annae (Duellman, 1963)(C)		
	Agalychnis callidryas (Cope, 1862)(N)		
Agalychnis callidryas	*Agalychnis callidryas* (Cope, 1862)	290	-
Agalychnis helenæ	*Agalychnis callidryas* (Cope, 1862)	290	-
Agalychnis dacnicolor	*Pachymedusa dacnicolor* (Cope, 1864)	291	-
Agalychnis lemur	*Phyllomedusa lemur* Boulenger, 1882	291	-
Pternohyla fodiens	*Pternohyla fodiens* Boulenger, 1882	292	74
Triprion petasatus	*Triprion petasatus* (Cope, 1865)	293	-
Triprion spatulatus	*Triprion s. spatulatus* Günther, 1882	293	74

CAUDATA

Günther's Names (*new)	Current Names	Page	Plate
Molge meridionalis	*Notophthalmus meridionalis* (Cope, 1880)	294	-
Amblystoma tigrinum	*Ambystoma t. tigrinum* (Green, 1825)(US)	295	-
	Ambystoma tigrinum velasci (Dugès, 1888)(M)		
	Ambystoma dumerilii (Dugès, 1870)(M)		
	Ambystoma flavipiperatum Dixon, 1963? (M)		
	Ambystoma mexicanum (Shaw, 1898)(M)		
	Ambystoma rosaceum Taylor, 1941 (M)		
Amblystoma altamirani	*Ambystoma altamirani* (Dugès, 1895)[50]	297	-
Spelerpes chiropterus	*Chiropterotriton chiropterus* (Cope, 1863)	298	-
Spelerpes gibbicaudus	*Pseudoeurycea leprosa* (Cope, 1869)	298	-
Spelerpes cephalicus	*Pseudoeurycea c. cephalica* (Cope, 1865)	299	-
Spelerpes bellii	*Pseudoeurycea b. bellii* (Gray, 1850)	299	-
	Pseudoeurycea bellii boneti Alvarez & Martín, 1967		
Spelerpes leprosus	*Pseudoeurycea leprosa* (Cope, 1869)	299	-
Spelerpes laticeps	*Pseudoeurycea c. cephalica* (Cope, 1865)	300	-
Spelerpes rostratum	*Bolitoglossa rostrata* (Brocchi, 1883)	300	-
Spelerpes sulcatum	*Pseudoeurycea c. cephalica* (Cope, 1865)	300	-
Spelerpes orizabensis	*Pseudoeurycea leprosa* (Cope, 1869)	300	-

Günther's Names (*new)	Current Names	Page	Plate
Spelerpes bocourti	(*Bolitoglossa morio* (Cope, 1869) or *B. rostrata* (Brocchi, 1883))[51]	300	76
	Bolitoglossa rostrata (Brocchi, 1883)(G)[51]		
	Bolitoglossa robusta (Cope, 1894)(C)[51]		
Spelerpes subpalmatus	*Bolitoglossa subpalmata* (Boulenger, 1896)	300	-
Spelerpes morio	*Bolitoglossa morio* (Cope, 1869)(G?)	301	-
	Bolitoglossa robusta (Cope, 1894)(C)[51]		
	Pseudoeurycea c. cephalica (Cope, 1865)(M)[51]		
Spelerpes robustus	*Bolitoglossa robusta* (Cope, 1894)	301	-
Spelerpes variegatus	*Bolitoglossa platydactyla* (Gray, 1831)(M)[51]	302	75
	Bolitoglossa alvaradoi Taylor, 1954 (C)[51]		
	Bolitoglossa flaviventris (Schmidt, 1936)(G)[51]		
	Bolitoglossa mexicana Duméril, Bibron & Duméril, 1854 (B)[51]		
	Bolitoglossa mulleri (Brocchi, 1883)(G)[51]		
	Bolitoglossa salvinii (Gray, 1868)(G)[51]		
Spelerpes lignicolor	*Bolitoglossa lignicolor* (Peters, 1873)	302	-
Spelerpes rufescens	*Bolitoglossa rufescens* (Cope, 1869)	303	-
Spelerpes yucatanus	*Bolitoglossa yucatana* (Peters, 1882)	303	-
Spelerpes lineolus	*Lineatriton lineolus* (Cope, 1865)	304	-
Spelerpes uniformis	*Oedipina uniformis* Keferstein, 1868 (C)	304	-
	Lineatriton lineolus (Cope, 1865)(M)		
Thorius pennatulus	*Thorius pennatulus* Cope, 1869	304	-
	APODA		
Dermophis mexicanus	*Dermophis mexicanus* (Duméril & Bibron, 1841)	305	-
*Dermophis gracilior**	*Dermophis mexicanus* (Duméril & Bibron, 1841)	306	76
Cœcilia isthmica	*Caecilia tentaculata* Linnaeus, 1758	306	-
Herpele ochrocephala	*Oscaecilia ochrocephala* (Cope, 1866)	307	-
Gymnopis multiplicata	*Gymnopis multiplicata* Peters, 1874 (C, P)	308	-
	Dermophis parviceps (Dunn, 1924)(SA)		
Gymnopis proxima	*Gymnopis multiplicata* Peters, 1874	308	-
Gymnopis oligozona	*Gymnopis multiplicata* Peters, 1874	309	-

NOTES

1 Placed in the subgenus *Clemmys.*

2 According to Legler, in Moll and Legler (1971).

3 Pensacola, Texas, is also mentioned as a locality. No such locality is listed for Texas in available gazetteers. The city in Florida by that name may have been intended. No taxonomic allocation is given here.

4 Placed in the subgenus "*Rhinoclemmys* or *Chelopus.*"

5 Echternacht's excellent review (1971) of geographic variation of Middle American species of *Ameiva* does not preclude recognition of subspecies, although they were not distinguished in that work. Vanzolini (1986b) states that currently recognized subspecies of *Ameiva ameiva* "are not biologically meaningful."

6 All species of *Ameiva* known from Nicaragua were included by Günther, hence the identity of, or locality data for, his material from Nicaragua cited for this species, not known from there, is problematical.

7 Corrected from *tessellata* to *tessellatus* in Günther's

list of errata.

8 Existence of perhaps two at present unknown species in the Platte River valley (cited by Günther) and in Arizona has been suggested (and rejected) several times; see review in Smith and Holland (1981), and a more recent account in Stebbins (1985: 243).

9 See Vanzolini (1986b: 17) and works cited therein.

10 The specimen from "Tehuantepec" is not *Barisia i. planifrons*, as might be expected on geographic grounds, but is *B. i. imbricata* (Smith, 1984: 153). The cited locality is erroneous.

11 Nomenclatural "currency" for the content of *Anolis sensu lato* is controversial, as both radical and conservative arrangements have authoritative advocates. At present it seems best to retain the inclusive interpretation until the controversy is resolved. Localities listed by Günther for several of the following species are outside the now known range of those species; some cases, in which the identity of the extralimital material is too uncertain to guess, are indicated by "sp. indet." In the case of *Anolis godmani*, the "Guatemala" specimen is thought to have come actually from Costa Rica.

12 See Smith (1984: 156) for justification.

13 The description of two species partitioned from populations formerly allocated to *Sceloporus grammicus microlepidotus*, by Lara-Góngora (1983), built upon pioneer studies by Hall, Sites and Dixon, has opened a nomenclatural "can of worms" affecting the entire group. Nomenclatural stability awaits much more study.

14 Allocation of this taxon to *Sceloporus variabilis*, as a subspecies (Sites and Dixon, 1982), does not take into consideration the absence of the subnasal, nor parietal peritoneal pigmentation, both unique to *Sceloporus teapensis*.

15 *Sphaerodactylus torquatus* is a synonym of *S. glaucus* (*fide* Harris and Kluge, 1984: 27-38, figs. 10-12, 17, 18), but I regard the evidence as inconclusive, and retain the species as valid. Localities listed by Günther for several species of this genus are outside their known range; some cases, in which the identity of the extralimital material is too uncertain to guess, are indicated by "sp. indet.".

16 See Vanzolini (19816a: 10) for range; identity assumed for South America on the basis of Günther's "tropical America."

17 Nomenclature *fide* Pérez-Higareda and Smith (1986).

18 According to Smith, Fitzgerald, Pérez-Higareda and Chiszar (1987).

19 Accepted by Dixon (1983: 259).

20 Quite understandably, Günther foreswore discrimination among the less easily distinguished species of *Thamnophis* of Guatemala and México, lumping specimens of six species and ten species-group taxa under the name of *Tropidonotus ordinatus*. He did not even list the Mexican localities, stating simply that the "species" occurs "throughout the country." A review of the Middle American *Thamnophis* in the British Museum (Nat. Hist.)(Smith, Nixon and Smith, 1950), updated by more recent taxonomic studies (Webb, 1966, 1976, 1982), shows that the specimens from México assigned by Günther to *T. ordinatus* include *T. chrysocephalus, T c. cyrtopsis, T. c. collaris, T. c. pulchrilatus, T. elegans errans, T. e. eques, T. e. megalops, T. s. scalaris, T. s. scaliger*, and *T. sumichrasti*. Two other species, as noted, are represented by material from Guatemala (Rossman, 1971; Webb, 1982).

21 Shown by Rossman et al. (1982) to be a full species, although long interpreted previously as a subspecies of *Thamnophis scalaris*.

22 The San Blas material is referable to an intergrade population, *Nerodia v. valida* X *N. v. isabelleae* X *N. v. thamnophisoides* (Conant, 1969: 93). The Utah and Arizona localities presumably pertain to two or more species of *Thamnophis*, probably including *T. rufipunctatus* (Cope, 1875).

23 The Misantla material, now lost, is presumed to represent an intergrade population, *Nerodia rhombifera blanchardi* X *N. r. werleri* (Conant, 1969: 69). The identity of the material from Cartago, Costa Rica, is unknown. The genus does not occur there, and no confusingly similar species are readily apparent. Perhaps the locality data are wrong.

24 Cope's material from "Jalapa," the basis for Günther's inclusion of this species, has been lost, and may have had erroneous locality data.

25 Tentative allocation, courtesy Van Wallach (pers.

comm.); an African species, reference to México erroneous.

26 The type, said to be from Nicaragua, must have erroneous locality data.

27 New combination, ex *Leptognathus pictiventris* Cope, 1876; replaces *Sibon annulata* (Günther, 1872), ex *Leptognathus annulatus,* a junior secondary homonym of *Sibon sartorii annulata* (Peters, 1870), ex *Geophis annulatus.*

28 Placed in *Sonora* by Bogert and Oliver (1945: 371) and Hardy and McDiarmid (1969: 199).

29 A South African species, erroneously cited from Central America.

30 The nine varieties, alpha through iota, treated by Günther are mixed combinations of both equivalent species here listed; Günther indicated names applied previously to some of those varieties, but he adopted none.

31 A replacement for Günther's name, preoccupied.

32 The thorough study (Myers, 1982) of Panamanian *Imantodes cenchoa* convincingly demonstrates that only one taxon of the species can be recognized there, but that conclusion does not preclude valid recognition of two or more subspecies elsewhere. Panamá, like some other parts of Central America, suffers from the "funnel-effect" character-summation that effectively obscures the character-dispersal evident in less confined, adjacent regions (cf. Pérez-Higareda and Smith, under review).

33 Both Wilson and Meyer (1985: 57-58) and Vanzolini (1986a: 9) question the validity of currently accepted (*e.g.,* Zweifel, 1959) subspecies of *Imantodes gemmistratus.*

34 A Cuban species, erroneously cited from Guatemala.

35 Undoubtedly other taxa are represented judging from the localities listed for *Micrurus fulvius.* Those names given were taken from the summaries of the genus by Roze (1967, 1980, 1983).

36 Allocations kindly provided by Roger Conant (pers. comm.), who noted that the species does not occur at Guadalajara (which may simply have been a shipping point or a collecting headquarters), and that the provenience of the specimen purportedly from Nicaragua is unacceptable.

37 *Porthidium* and *Bothriechis* have commonly been regarded as synonyms of *Bothrops*; see Pérez-Higareda et al. (1985: 102-104) for rationale in reviving these names, and for interpretation of material cited by Günther under the name *Bothriechis nummifera.*

38 Only for the localities Presidio and Colima.

39 The identity of the material from Tehuantepec, Tabasco and Guatemala is uncertain; none represents *Rana montezumae.*

40 The identity of materials from South America and the West Indies is uncertain; none represent *L. melanonotus.*

41 According to Savage (1974: 295, 296).

42 According to Savage (1975).

43 *Eleutherodactylus biporcatus* is not known from Ecuador, or *E. melanostictus* from Veracruz. The materials cited under those names from those localities are noted as "sp. indet."

44 Allocation courtesy John D. Lynch (pers. comm.).

45 No reasonable guess can be made of identity of this South American genus and species supposedly from Panamá.

46 Recorded from Panamá in error (Savage, 1969).

47 According to Savage and Heyer (1969: 24).

48 Duellman (1968: 204-205) regarded the two syntypes as representative of the *Hyla pulchella* group of South America, but of uncertain species, not of Central American origin.

49 Duellman (1970: 650) regarded this as a nomen dubium, not assignable to any known species, and erroneously attributed to Panamá.

50 *Rhyacosiredon* is an untenable genus (Ronald Brandon, pers. comm.).

51 Allocation courtesy David B. Wake (pers. comm.).

LITERATURE CITED

Bogert, Charles M. and James A. Oliver. 1945. A preliminary analysis of the herpetofauna of Sonora. Bull. Amer. Mus. Nat. Hist., 83: 297-426, 8 plates, 2 maps.

Brocchi, Paul. 1881-1883. Mission Scientifique au Mexique et dans l'Amérique Centrale. . . Recherches Zoologiques. Part 3, Sect. 2. Études des Batraciens de l'Amérique Centrale. Paris, Imprimerie Impériale. 122 pages, 24 plates (reprinted 1978, Arno Press, division New York Times Publ. Co., New York, now Ayer Co., Salem, Massachusetts).

Conant, Roger. 1969. A review of the water snakes of the genus *Natrix* in México. Bull. Amer. Mus. Nat. Hist., 142: 1-140, 22 plates, 6 maps.

Dixon, James R. 1983. Taxonomic status of the Brazilian colubrid snake, *Xenodon suspectus* Cope. Texas Jour. Sci., 35: 257-260.

Duellman, William E. 1968. The taxonomic status of some American hylid frogs. Herpetologica, 24: 194-209.

----------. 1970. The hylid frogs of Middle America. Lawrence, Kansas, Monogr. Mus. Nat. Hist., Univ. Kansas, 1: xi, 753 pages.

Duméril, Auguste Henri André, Marie-Firmin Bocourt and François Mocquard. 1870-1909. Mission Scientifique au Mexique et dans l'Amérique Centrale. . . Recherches Zoologiques. Part 3, Sect. 1. Études sur les Reptiles. Paris, Imprimerie Impériale. xiv, 1012 pages, 99 plates (reprinted 1978, Arno Press, division New York Times Publ. Co., New York, now Ayer Co., Salem, Massachusetts).

Echternacht, Arthur C. 1971. Middle American lizards of the genus *Ameiva* (Teiidae) with emphasis on geographic variation. Misc. Publs. Univ. Kansas Mus. Nat. Hist., 55: 1-86.

Frost, Darrel R. (ed.) 1985. Amphibian Species of the World, a Taxonomic and Geographical Reference. Lawrence, Kansas, Allen Press and Assoc. Syst. Colls. v, 732 pages.

Godman, Frederick DuCane. 1915. Introduction, p. 1-86. *In* Biologia Centrali-Americana, Introductory Volume. London, for the Editor by Porter and Dulau. viii, (2), 149, (1) pages, 2 plates, 8 maps.

Hardy, Laurence M. and Roy W. McDiarmid. 1969. The amphibians and reptiles of Sinaloa, México. Univ. Kansas Publs. Mus. Nat. Hist., 18: 39-252, 8 plates.

Harris, Dennis M. and Arnold G. Kluge. 1984. The *Sphaerodactylus* (Sauria: Gekkonidae) of Middle America. Occas. Pap. Mus. Zool. Univ. Michigan, 706: 1-59.

Lara-Góngora, Guillermo. 1983. Two new species of the lizard genus *Sceloporus* (Reptilia, Sauria, Iguanidae) from the Ajusco and Ocuilan sierras, México. Bull. Maryland Herpetol. Soc., 19: 1-14.

Moll, Edward O. and John M. Legler. 1971. The life history of a Neotropical slider turtle, *Pseudemys scripta* (Schoepff) in Panamá. Bull. Los Angeles Co. Mus. Nat. Hist., 11: 1-102.

Myers, Charles W. 1982. Blunt-headed vine snakes (*Imantodes*) in Panamá, including a new species and other revisionary notes. Amer. Mus. Novit., 2738: 1-50, 2 maps.

Pérez-Higareda, Gonzalo and Hobart M. Smith. 1986. The status of the Los Tuxtlas (México) false coral snakes (*Pliocercus*). Bull. Maryland Herpetol. Soc., 22: 125-130.

---------- and ----------. 1987. An annotated checklist and key to the snakes of Veracruz, México. Under review.

----------, ----------, and Jordi Juliá-Zertuche. 1985. A new jumping viper, *Porthidium olmec*, from southern Veracruz, México (Serpentes: Viperidae). Bull. Maryland Herpetol. Soc., 21: 97-106.

Peters, James A. and Roberto Donoso-Barros. 1970. Catalogue of the Neotropical Squamata. Part II. Lizards and Amphisbaenians. Bull. U.S. Nat. Mus., 297(B): i-viii, 1-293.

---------- and Braulio Orejas-Miranda. 1970. Catalogue of the Neotropical Squamata. Part I. Snakes. Bull. U.S. Nat. Mus., 297(A): i-viii, 1-347.

Rossman, Douglas A. 1971. Systematics of the Neotropical populations of *Thamnophis marcianus* (Serpentes: Colubridae). Occas. Pap. Mus. Zool. Louisiana St. Univ., 41: 1-13.

Rossman, Nita J., Douglas A. Rossman and Nancy K. Keith. 1982. Comparative visceral topography of the New World snake tribe Thamnophiini (Colubridae, Natricinae). Tulane Stud. Bot. Zool., 23: 123-164.

Roze, Janis. 1967. A check list of the New World venomous coral snakes (Elapidae) with descriptions of new forms. Amer. Mus. Novit., 2287: 1-60.

----------. 1970. *Micrurus*, p. 196-220. *In* J. A. Peters and B. Orejas-Miranda, Catalogue of Neotropical Squamata: Part I, Snakes. Bull. U.S. Nat. Mus., 297: i-viii, 1-347.

----------. 1983. New World coral snakes (Elapidae): a taxonomic and biological summary. Mem. Inst. Butantan, 46: 205-338.

Savage, Jay M. 1969. Clarification of the status of the toad *Bufo veraguensis* O. Schmidt, 1857. Copeia, 1969: 178-179.

----------. 1974. On the leptodactylid frog called *Eleutherodactylus palmatus* (Boulenger) and the status of *Hylodes fitzingeri* O. Schmidt. Herpetologica, 30: 289-299.

----------. 1975. Systematics and distribution of the Mexican and Central American stream frogs related to *Eleutherodactylus rugulosus*. Copeia, 1975: 254-306.

---------- and W. Ronald Heyer. 1969. The tree-frogs

(family Hylidae) of Costa Rica: diagnosis and distribution. Rev. Biol. Trop., 16: 1-127.

---------- and Jaime Villa R. 1986. Introduction to the Herpetofauna of Costa Rica. Soc. Study Amphib. Rept., Contrib. Herpetol., 3. viii, 207 pages.

Scott, Norman J., Jay M. Savage and Douglas C. Robinson. 1983. Checklist of reptiles and amphibians, p. 367-374. *In* Daniel H. Janzen (ed.), Costa Rican Natural History. Chicago, Univ. Chicago Press. xi, 816 pages.

Sites, Jack W., Jr. and James R. Dixon. 1982. Geographic variation in *Sceloporus variabilis* and its relationship to *S. teapensis* (Sauria: Iguanidae). Copeia, 1982: 14-27.

Smith, Hobart M. 1984. Notes on the enigmatic *Barisia imbricata* of the British Museum, and on its collection of reptiles from Amula, Guerrero, México. Bull. Maryland Herpetol. Soc., 20: 152-158.

----------, Kevin Fitzgerald, Gonzalo Pérez-Higareda and David Chiszar. 1986 (1987). A taxonomic rearrangement of the snakes of the genus *Scaphiodontophis*. Bull. Maryland Herpetol. Soc., 22: 159-166.

---------- and Richard L. Holland. 1981. Still more on *Bipes*. Herpetol. Rev., 12: 8-9.

----------, C. William Nixon and Philip W. Smith. 1950. Mexican and Central American garter snakes (*Thamnophis*) in the British Museum (Natural History). Jour. Linnean Soc. London, 41: 571-584.

Stebbins, Robert C. 1985. A Field Guide to Western Reptiles and Amphibians. Second edition, revised. Boston, Houghton Mifflin. xvi, 336 pages, 48 plates.

Stuart, Laurence C. 1963. A checklist of the herpetofauna of Guatemala. Misc. Publs. Mus. Zool. Univ. Michigan, 122: 1-150, map.

Vanzolini, Paulo E. 1986a. Addenda and corrigenda to the catalogue of Neotropical Squamata. Part I. Snakes. Smithsonian Herpetol. Inform. Serv., 70A: 1-26.

----------. 1986b. Addenda and corrigenda to the catalogue of Neotropical Squamata. Part II. Lizards and Amphisbaenia. Smithsonian Herpetol. Inform. Serv., 70B: 1-25.

Webb, Robert G. 1966. Resurrected names for Mexican populations of black-necked garter snakes, *Thamnophis cyrtopsis* (Kennicott). Tulane Stud. Zool., 15: 55-70.

----------. 1976. A review of the garter snake *Thamnophis elegans* in Mexico. Nat. Hist. Mus. Los Angeles Co. Contr. Sci., 284: 1-13.

----------. 1982. Taxonomic status of some neo-tropical garter snakes (genus *Thamnophis*). Bull. South. Calif. Acad. Sci., 81: 26-40.

Wilson, Larry David and John R. Meyer. 1985. The Snakes of Honduras. 2nd edition. Milwaukee, Wisconsin, Milwaukee Public Mus. x, 150 pages.

Zweifel, Richard G. 1959. Snakes of the genus *Imantodes* in western Mexico. Amer. Mus. Novit., 1961: 1-18.

A SKETCH OF THE LIFE AND WORK OF ALBERT GÜNTHER, M.D., F.R.S. (1830-1914)

by A. E. Gunther,* M.A., D.Sc.

In southwest Germany, in the state of Württemberg, there still are villages like those in the 1830s in which Albert Carl Ludwig Gotthilf Günther** and his brother Theodor were brought up. Albert was born in Esslingen on 3 October 1830 where his father practiced as a solicitor and charity commissioner, but he died when Albert was only six and his mother retreated to her father's rectory at Vaihingen, a village situated on the Filder Plateau which commands Stuttgart from the south. It was there that her children enjoyed country life, being driven round the parish by their grandfather, himself something of a naturalist. Thus, even before he went to school Albert was collecting butterflies and moths, selecting the oddest looking caterpillars, those with horns, to see what they would turn into.

GÜNTHER'S EDUCATION

School was the Gymnasium at Stuttgart, free or almost free for a widow's son. His objective was the University of Tübingen, which was also free if, coming from a rectory, he read theology. So it was to the "Evangelisches Stift" that he went in 1847 at the age of 17, but though he passed into holy orders in the Lutheran Church four years later, he could make nothing of subjects like philosophy, Kant being quite meaningless to him. Yet happily the "Stift" insisted on students taking some off-beat subject and Albert, choosing zoology, fell under the inspiration of Professor Wilhelm von Rapp. Thus by 1851 Albert had not only turned medical student but was already compiling a "Taschenbuch für Naturgeschichte" (Notes for Natural History). During vacations he was also engaged in systematic work at the Königliches Naturalien-Kabinet in Stuttgart under Dr. Kraus. The year 1853 saw him with a doctorate of Philosophy and Arts, a half dozen articles primarily on the local fauna, a draft thesis on *Die Fische des Neckars* and matriculation for entrance to the Medical Faculty at Berlin. Among his professors there was that great biologist Johannes Müller.

The year at Berlin was an unsettled one. The southerner did not take easily to the hard atmosphere of the place. There were worries of his own and of his mother's finances. A breakdown in health seems to have come from malnutrition; his eyes gave way under rheumatism and his diary, obscured by shorthand, suggests a severe neurosis. The offer of a school job at Bonn the next year led to the university there where he came into contact with Professor F. H. Troschel and later to his first visit to England where his mother had gone to earn enough money to educate his brother. It was at Brighton where his mother was a governess that he got his first taste of the sea. Over that Easter of 1855 he called on Professor Richard Owen at the College of Surgeons and on Dr. John Edward Gray at the British Museum and met a most friendly response. Back at Tübingen from 1855 to 1857 to complete his medical studies he felt unable altogether to keep off zoology and produced a *Handbuch der Medicinischen Zoologie* (1858) which was in press during the final weeks of his examinations.

---

*35 Rudall Crescent, Hampstead, London NW3 1RR, United Kingdom.

**Günther adhered to the original German form of his family name, as it is still cited in the scientific literature. On his death in 1914 his family adopted the Anglicized version without the umlaut.

BRITISH MUSEUM CATALOGUES AND RELATED WORK

No sooner qualified in medicine than he returned to his mother at Brighton and called on Gray to learn if there was anything to be done. By November 1857 Gray offered him a job at the British Museum identifying specimens in the so-called Spirit Room. This comprised a suite of three long, half-subterranean rooms in the east wing of the museum where the collection was stored. They were next door to the coal cellars and far from clean. Such light as the windows gave was dimmed by buildings outside and in winter there was little light at all by early afternoon. There was no ventilation to dry the floor's large damp patches; and the prevailing scent came from thousands of bottled spirits with their preserved contents. This was to be his work place for the next fifteen years.

There he was put on to completing a *Catalogue of the Colubrine Snakes* which Gray had left unfinished since 1849. He was to be paid £40 per thousand specimens or £75 the lot. Gray's original plan was to be followed exactly; species not held by the British Museum were not to be included. To Gray's great surprise four months later, by March 1858, the job was done: 3,100 specimens described at a rate of about 25 a day and completed with a geographical index that was useful to travellers, a total of 281 pages. The annoyance Gray registered at the inclusion of the index was mollified by his having a snake named after him! Unfortunately for Günther, Gray had overspent his budget and he had to wait payment until the next quarter.

Before the *Catalogue of Colubrine Snakes* had come off the press Gray directed Günther's attention to the Museum's frogs. This time he was allowed to try his hand at classification. After another four months, by July 1858, Günther had produced a *Catalogue of the Batrachia Salientia*, this time illustrated by G. H. Ford. Output at this rate gave Gray anxiety. It was outrunning his finances. So it was agreed that production should be limited to £50 a quarter, and it was on that basis that a start was made with the great *Catalogue of Fishes* (1859-1870), an enormous work covering 29,275 specimens and 8525 species. In the meantime, a paper on the problem of "animal zones" by P. L. Sclater, using birds, brought Günther to consider the greater use of more sedentary animals in his "Geographical Distribution of Reptiles" in the *Proceedings of Zoological Society of London* in 1858.

By that time Günther had established what he himself meant by a catalogue. For Gray a catalogue was a list of the specimens in the Museum. For Günther a catalogue was a frame into which an order of life could be fitted. It must be of use wherever it stood on a shelf--in London, Paris or Berlin--and must include every known specimen at the time of its publication.

Before starting on the *Catalogue of Fishes* Günther's concern had been largely with snakes and frogs and his great interest in reptiles next took him to those of British India. With the army's control by 1858 of the Indian Mutiny, Günther asked himself in what way he could, as a zoologist, best make some contribution to that country's development. A volume entitled the *Reptiles of British India* would, he felt, not be out of place. Because Gray had not given his formal authority to the work he was unwilling to approve its production and Günther was forced to work on it in his spare time at home. After some delay the volume was issued by the Ray Society in 1864.

Günther's single most important herpetological discovery (published 1867) was his recognition that the tuatara of New Zealand was the sole living representative of the otherwise extinct order Rhynchocephalia. The species *Sphenodon punctatus* had been described in 1831 as a lizard, by Gray. Perhaps Günther's most important contribution to herpetology was his choice of George A. Boulenger as his successor in charge of reptiles and fish (in 1879). It was Boulenger, of course, who went on to write the nine-volume catalogue of the world's amphibians and reptiles which provides the foundation for systematic herpetology.

HONORS, THE *ZOOLOGICAL RECORD* AND OTHER ACTIVITIES

If a young medical student of Günther's temperament could not resist writing zoology during four to five years of exacting medical study he was hardly capable of allowing a dozen years of *Catalogue of Fishes* to exclude other interests. And so it proved. There was nothing to prevent his contributing to other orders of animals--Amphibia, reptiles, birds, mammals--the lot, if he wished. He was secure in his philosophy that if one was a naturalist one must be able to cover the whole field. Although primarily an ichthyologist his was perhaps the last generation whose speciality did not exclude wider horizons. Gray, whatever his achievement, was no specialist; and by 1881 when the natural history departments moved from the old British Museum at Bloomsbury to the new Natural History Museum building at South Kensington specialization was accepted.

In July 1862, after five years at the Museum, Günther was appointed to the staff, taking the place of one who had retired. In 1867, after another five years, he was elected a Fellow of the Royal Society. From 1868 to 1905 he was a member of the Zoological Society's Council. In 1875 he became a Vice-President of the Royal Society, and in 1878 he received its Royal Medal.

Of all Günther's activities the most considerable in terms of zoological science was the initiation in 1865 of the *Record of Zoological Literature*, six years later re-named *Zoological Record* as it is known to this day. It was the logical extension of his concept of a catalogue but made to embrace the world's zoological literature, and patterned perhaps after the German counterpart in *Archiv für Naturgeschichte* with which Günther was already familiar. The idea was that the *Record* should not be merely a list of titles that might appear as an appendix to the *Annals and Magazine of Natural History* but that it should be compiled by authoritative recorders adding a critical review of new books and more important papers. Günther's ultimate objective was the coordination of the varied methods of international classification through precise description and terminology. As Günther's brainchild, it was his first job to select the recorders; but it was one thing to recruit an expert, another to be assured of an objective assessment. It was not easy to suggest improvements to authors's texts without seeming to criticize their content. It required a tact that Günther's habitual terseness of expression was not adapted to and it was natural that many took offense at the "messages" they found in the text aimed at them! Günther held the editorship of the *Record* for six years, from 1865 to 1870, when it passed to Alfred Newton, Professor of Anatomy at Cambridge, who had remained from the first a critic and a true friend.

Among Günther's interests in the 1870s was the work being done by Charles Darwin on the *Descent of Man* (1871) and by Alfred Wallace on the *Geographical Distribution of Animals* (1876). Darwin was drawing on Günther's work for the sexual characters of fish, Amphibia and reptiles, for which he was indebted, in his own words, for "four-fifths of the facts quoted from you and from your works." Wallace's requirements seemed almost to extend to much of the contents of his book acknowledging the help, among others, of Alfred Newton, W. H. Flower, and of Günther. While Wallace was prepared to confess his impatience with detail he largely relied for his basic information on the work of others. Where Wallace's thanks were perfunctory, Darwin's were not only courteous but sought to acknowledge the smallest service.

Albert Günther in 1867 (left), in about 1900 (center), and about 1912 (right). All photographs courtesy British Museum (Nat. Hist.) Library.

ADMINISTRATIVE ACTIVITIES

By 1870, at the age of 40, Günther's future was assured. The completion of the *Catalogue of Fishes* and retirement from editorship of *Zoological Record* left him free, or as free as he ever allowed himself to be, for wider departmental duties. In May 1869 Gray, having suffered a stroke more severe than what he'd already had, left the running of the department to his brother, G. R. Gray, the Assistant Keeper, who rated himself an ornithologist, and to Günther. In May 1872 G. R. Gray died unexpectedly and Günther was made Assistant Keeper leaving the signing of official documents to J. E. Gray in a shaky left hand.

On Gray's death in February 1875 Günther, who had been naturalized the year before, was appointed Keeper of the Zoological Department, and he also joined the *Annals and Magazine of Natural History* as co-editor (until 1905). He did not actually move into residence in the Keeper's Lodgings at the British Museum until 1878.

RELATIONSHIPS WITH GRAY AND OWEN

For 25 of the 38 years Günther was at the Museum he was subject to two unusual men who were adversaries of each other: to J. E. Gray as Keeper of his Department and to Sir Richard Owen as Superintendent of the Natural History Departments as a whole. His relations with them are therefore of some interest. But to neither, beyond the fact of his employment, did he owe anything in terms of his own distinctive attributes of character, initiative or power of work. In any case, in those days it was not the habit to ease the path of a younger man.

With Gray, Günther's relations were of the most formal kind. Once Gray had accepted Günther's methods of work there was little reason for differences. Gray did not have the scientific knowledge for argument. He was the dependable boss, caustic on occasion through ill-health, but with feet firmly on the ground.

On the other hand, Owen's approach, from the day of Günther's arrival, was one of incomprehensible kindness, to his mother as much as to himself. Both were frequently invited by Owen to Speen Lodge (a royal property on life loan from the Queen) in Richmond Park. There were frequent games of chess. Owen would stay with Albert's brother, Theodor, at Hampton Wick and became godfather to one of Theodor's sons. However, Günther's reverence for Owen's greatness never developed into the respect he felt for Gray. Owen was not one to assume any attitude without good reason. His reputation was that of a social genius who had worked his way up into the Royal family. At the Museum Günther could be useful to him: there was Günther's knowledge of the Continent and of languages. Günther could give him an insight into Gray's work over Gray's head. In later years Günther was to become Owen's chief of staff. However when in 1875 Günther succeeded Gray as Keeper and he found himself formally under Owen's authority the relations between the two men became markedly strained.

TRANSFER OF THE COLLECTIONS

Sir Richard Owen, a man of great reputation, must, however, be given full credit for persuading the Prime Minister that a new building for natural history was essential. To be situated in the planned exhibition area of South Kensington, it was started in 1873. The story of the transfer of the collections from the old museum at Bloomsbury between 1880 and 1883 to the new is told in Günther's *History of the Collections* (1912). As the Keeper of the largest of the collections, Zoology, Günther was more involved than any of the other keepers. The smaller collections--botany, mineralogy and geology--were transferred in 1880, in time for the Museum's formal opening in April 1881. The zoological collection was moved from 1882, when the galleries were ready to receive it, and, with a winter break, in summer of 1883. All this accounted for Günther's few papers appearing in 1883-84.

Whatever reputation Günther had acquired in the Old Museum the move to South Kensington, a milestone in his life, was greatly to enhance it. If the awkwardness of the new building's Byzantine interior and the administrative work it involved added labor to the Keepership, the public interest in the new institution was reflected in a flood of accessions. With the expansion of the British Empire overseas the Department of Zoology became something of a world center; indeed for many zoologists Günther *was* the British Museum. Between then and his retirement a dozen years later, in 1895, the department's staff doubled from eight to 16; the number of students who visited the museum from 4,000 to 8,000 per year and specimens from 1.3 to 2.2 million.

To the Keeper's normal work was added essential chores. The Old Museum's claim to retain all except the duplicates of books made a new zoological library necessary. This Günther built up himself spending many hours up ladders, pulling dusty volumes down from the top shelves of Bernard Quaritch's bookshop. The first edition of his Zoological Department's *Catalogue*, with its 1700 titles, appeared in 1880; the third of 1884 contained 5309 titles or 8586 volumes. Equally as important were the *Guides* to the new show-cases, that on reptiles being made together with G. A. Boulenger in 1885. A series of two dozen articles, including that on "Reptiles," appeared in the ninth edition of the *Encyclopaedia Britannica* of 1875-1889.

It may be recalled that when in 1857 Gray took Günther on it was for reptiles as well as for fish. Apart from editing reptiles and Amphibia in the *Zoological Record*, and helping O'Shaughnessy with it until 1879, he produced nearly 200 papers on reptiles. His most outstanding work on reptiles and amphibians of his later years was, of course, the contribution to Salvin and Godman's *Biologia Centrali-Americana* published over the period 1885 to 1902 and here reprinted.

RETIREMENT

Günther's retirement did little to stem his flow of papers. It was followed by the bestowal of various honors, the first of which was as President of the Linnean Society from 1896-1900, and its Linnean Medal in 1904. His Presidential Address of 1898 returned to gigantic land tortoises, the apples of Lord Rothschild's eye. In 1903 Günther was accorded an honorary doctoral degree by the University of Tübingen, and another in 1906 by the University of St. Andrews.

The last of his publications was an appendix to Ray Lankester's two volumes on the *History of the Natural History Departments of the British Museum* (1906), a history which he found unsatisfactory. Günther's own version of *The General History of the Department of Zoology from 1856 to 1895* (1912) is up to the standard of anything he wrote.

In spite of his dedication, perhaps indeed because of it, Albert never lacked friends. Some were wealthy landowners seeking information, from whom he would accept nothing unless specimens for the Museum.

Staff of the Department of Zoology, British Museum (Nat. Hist.) in 1895, the year of Günther's retirement; as Keeper he occupies the seat of honor at front center. Günther's hand-picked successor in charge of reptiles and fish, George A. Boulenger, stands at far left. Courtesy British Museum (Nat. Hist.) Library.

From the staff, with whom his relations were firm but friendly, he received nothing but respect. The essence of his character was loyalty to anyone who worked with him. The classic case was that of Edward Gerrard who as a junior assistant had been kind to him on arrival in 1857 and whose obituary he wrote for the Linnean Society on Gerrard's death at age 99. His relations with G. H. Ford, his primary artist, were at the same level.

That Günther's monosyllabic way of speaking, in a broken Teutonic English, gave rise to jokes was only natural. On one evening the attendant leaving him working late advised that, surrounded as he was by bottled spirits, he should put out the fire before going home, whereupon Günther seized a large bottle and poured its ample contents onto the gleaming coals which, objecting, hissed and went out. "Dere!" said Günther. He never lost his German accent and in papers presented at the Zoological Society he would refer to "dese leedle frosches."

It was fortunate for Günther's peace of mind that he died before the outbreak of the First World War in 1914. In the years before that his son, Robert, then Science Tutor at Magdalen College, Oxford, used to spend his vacations at Robert's wife's cottage in Norfolk. On his way there Robert used to call with his sons on his parents at Kew, where I personally remember Albert well. A feature of grandfather's study was a monster bookcase he had designed to hold a multitude of papers which supplied his needs for much that he wrote by midnight oil. That bookcase, depicted in the frontispiece to this book, is now in my possession. The other memory, even more striking, was of his love for birds. The garden aviary extended into his study to which the birds had access. The many varieties, especially the hoopoe and the droll grackle, recaptured for him memories of the countryside about Vaihingen. They came when called by name, and nestled in their master's lap. They regarded his study as their home and their cages merely as sleeping quarters. They knew just how far they could go with the dog or cat sleeping before the fire. Earlier, when Günther lived at Surbiton, there were tree frogs in his aviary, a pet alligator in the bedroom, and giant tortoises and a tuatara in his outdoor garden.

Albert Günther's death at Kew on the first day of February 1914 at the age of 84 was something of a Museum occasion. He was buried in the Old Cemetery at Richmond where the headstone reads: "I shall not die but live; and declare the works of the Lord."

Apart from his house, he left little more than £2,500.

BIBLIOGRAPHY AND OBITUARIES

Anonymous. 1914. Dr. Albert Günther, F.R.S. Nature, 92: 664-666.

----------. 1914. Albert C. L. G. Günther. Geol. Mag., ser. 6, 1: 141-142.

----------. 1914. A. C. L. G. Günther. The Times (London), February 2.

Camerano, L. 1914. Alberto Günther. Atti Reale Accad. Sci. Torino, 49: 825-833.

Gunther, A. E. 1975. A Century of Zoology at the British Museum Through the Lives of Two Keepers, 1815-1914. London, Dawsons of Pall Mall. 533 pages (part 2, comprising pp. 213-469, is an extensive biography of Albert Günther).

----------. 1980. The Founders of Science at the British Museum, 1753-1900. Halesworth, Suffolk, Halesworth Press. ix, 219 pages.

Gunther, R. T. 1930. Bibliography of the works of Dr. Albert Günther, M.A., Ph.D., M.D., LL.D., F.R.S., Etc. Ann. Mag. Nat. Hist., ser. 10, 6 (suppl.): 233- 286.

M'Intosh, W. C. 1914. Dr A. C. L. G. Günther, M.A., Ph.D., M.D., LL.D., F.R.S., etc. Proc. Royal Soc. Edinburgh, 34: 269-277 (M'Intosh, zoologist at University of St. Andrews and brother- in-law of Albert Günther, was also an eminent ichthyologist).

----------. 1915. A. C. L. G. Günther, 1830- 1914. Proc. Royal Soc. London, ser. B, 88: xi-xxvi, 1 portrait.

Smith, E. A. 1914. Dr. Albert Günther. Zoologist, ser. 4, 18: 113-116.

Stapf, O. and A. Alcock. 1914. [Obituaries of Albert Günther]. Proc. Linnean Soc. London, session 126: 48-52, 1 plate.

Thomson, J. A. 1927. Albert Charles Lewis Gotthilf Günther, p. 233-234. *In* Dictionary of National Biography, 1912-1921. London, Oxford Univ. Press.

ALBERT GÜNTHER'S HERPETOLOGICAL PUBLICATIONS

compiled by Kraig Adler*

All papers and books partly or wholly concerning amphibians and reptiles are listed here, according to the nominal date of publication. Where actual date of publication is known to differ this information is provided in parentheses. Order of titles within years does not necessarily imply order of publication. Titles prefixed by a dagger (†) have not been examined personally and are cited on authority of R. T. Gunther ("Bibliography of the Works of Dr. Albert Günther, M.A., Ph.D., M.D., LL.D., F.R.S., Etc." A.M.N.H., ser. 10, 6 (suppl.): 233-286, 1930) or A. E. Gunther (*in litt.*).

Abbreviations

A.M.N.H. = Annals and Magazine of Natural History (London).[1**]

E.B. = Encyclopaedia Britannica (British version, Edinburgh).[2]

P.Z.S. = Proceedings of the Zoological Society of London (London).[3]

R.Z.L. = Record of Zoological Literature (London).[4]

Z.R. = Zoological Record (London).[4]

1853

Ueber den Puppenzustand eines Distoma [concerning flukes of frogs]. Jahreshefte Vereins Vaterländ. Naturk. Württemberg, 9: 95-99, 1 plate.

†Thierische Gifte. *In* Kalender der Natur. Stuttgart, Becher (†later edition, 1859).

1857

[Ueber Regeneration des Schwanzes in Froschlarven]. Jahreshefte Vereins Vaterländ. Naturk. Württemberg, 13: 54.

1858

Neue Schlangenarten in der Sammlung des britischen Museums. Arch. Naturg., 24(1): 221-243.

Neue Batrachier in der Sammlung des britischen Museums. Arch. Naturg., 24(1): 319-328.

Catalogue of Colubrine Snakes in the Collection of the British Museum. London, British Museum (Nat. Hist.). xvi, 281 pages (published 9 October 1858, *fide* C. D. Sherborn, A.M.N.H., ser. 10, 13: 309, 1934; reprinted by British Museum (Nat. Hist.), London, 1971, and again by Linnaeus Press, Amsterdam, 1976).

On the systematic arrangement of the tailless batrachians and the structure of *Rhinophrynus dorsalis*. P.Z.S., 1858: 339-352.

On the geographical distribution of reptiles. P.Z.S., 1858: 373-398.

Catalogue of the Batrachia Salientia in the Collection of the British Museum. London, British Museum (Nat. Hist.). xvi, 160 pages, 12 plates (published 12 February 1859, *fide* C. D. Sherborn, A.M.N.H., ser. 10, 13: 309, 1934).

1859

On the systematic arrangement of the tailless batrachians and the structure of *Rhinophrynus dorsalis*. A.M.N.H., ser. 3, 3: 61-74.

On the geographical distribution of reptiles. A.M.N.H., ser. 3, 3: 221-237.

On the genus *Elaps* of Wagler. P.Z.S., 1859: 79-89, 3 plates.

List of the cold-blooded vertebrates collected by Mr. Fraser in the Andes of western Ecuador. P.Z.S., 1859: 89-93.

---

*Section of Neurobiology and Behavior, Cornell University, Seeley G. Mudd Hall, Ithaca, New York 14853-2702, U.S.A.

**Endnotes begin on page LXVII.

On sexual differences found in bones of some Recent and fossil species of frogs and fishes. A.M.N.H., ser. 3, 3: 377-387, 2 plates.

Description of a new genus of West African snakes, and revision of the South American *Elaps*. A.M.N.H., ser. 3, 4: 161-174, 1 plate.

On the reptiles from St. Croix, West Indies, collected by Messrs. A. and E. Newton. A.M.N.H., ser. 3, 4: 209-217, 1 plate.

Second list of cold-blooded Vertebrata collected by Mr. Fraser in the Andes of western Ecuador. P.Z.S., 1859: 402-420, 1 plate (probably published 1860[3]).

Description of a new species of *Anolis* from Central America. P.Z.S., 1859: 421-422 (probably published 1860[3]).

On the reptiles and fishes collected by the Rev. H. B. Tristram in northern Africa. P.Z.S., 1859: 469-474, 1 plate (probably published 1860[3]).

1860[5]

Description of *Uriëchis microlepidotus*, a new snake from South Africa. A.M.N.H., ser. 3, 5: 168-169, 1 plate (see correction, A.M.N.H., ser. 3, 5: 342, 1860).

Description of *Leptodeira torquata*, a new snake from Central America. A.M.N.H., ser. 3, 5: 169-171, 1 plate.

Description of a new species of *Anolis* from Central America. A.M.N.H., ser. 3, 5: 493-494.

On a new snake from the Galapagos Islands. A.M.N.H., ser. 3, 6: 78-79.

Description of *Homalocranium laticeps*, a new snake from Carthagena. A.M.N.H., ser. 3, 6: 149-150.

Note on *Psammophis Perroteti*, D. & B. A.M.N.H., ser. 3, 6: 428-429, 1 plate.

On new reptiles and fishes from Mexico. A.M.N.H., ser. 3, 6: 442-445.

Skizzen aus dem zoologischen Garten in London. Arch. Naturg., 26(1): 29-56 (part 1, cold-blooded vertebrates).

On a new snake from the Galapagos Islands. P.Z.S., 1860: 97-98.

On the reptiles of Siam. P.Z.S., 1860: 113-117, 1 plate.

Contribution to a knowledge of the reptiles of the Himalaya Mountains. P.Z.S., 1860: 148-175, 4 plates.

Third list of cold-blooded Vertebrata collected by Mr. Fraser in Ecuador. P.Z.S., 1860: 233-240, 1 plate.

Description of *Homalocranium laticeps*, a new snake from Carthagena. P.Z.S., 1860: 240-241.

On new reptiles and fishes from Mexico. P.Z.S., 1860: 316-319.

On a West-African genus of snakes (*Meizodon*). P.Z.S., 1860: 427-430 (probably published 1861[3]).

1861[5,6]

On the dentition of *Herpeton tentaculatum*. A.M.N.H., ser. 3, 7: 195.

On a West African genus of snakes (*Meizodon*). A.M.N.H., ser. 3, 7: 223-226.

Account of the reptiles sent by Dr. Wucherer from Bahia. A.M.N.H., ser. 3, 7: 412-417.[6]

On the ophidian genus *Helicops*. A.M.N.H., ser. 3, 7: 425-428.

On a new species of the family Boidae. A.M.N.H., ser. 3, 8: 189.

Second list of Siamese reptiles. A.M.N.H., ser. 3, 8: 266-268.

List of Ceylon reptiles, p. 321-322. *In* J. E. Tennent, Sketches of the Natural History of Ceylon; with Narratives and Anecdotes Illustrative of the Habits and Instincts of the Mammalia, Birds, Reptiles, Fishes, Insects, &c., Including a Monograph of the Elephant. London, Longman, Green, Longman, and Roberts. xxiii, (1), 500 pages, 10 plates (list co-authored with J. E. Gray; chapter 9, "Reptiles," comprises pp. 271-322, 3 plates).

Account of the reptiles sent by Dr. Wucherer from Bahia. P.Z.S., 1861: 12-18.[6]

On the anatomy of *Regenia ocellata*. P.Z.S., 1861: 60-62.

On the anatomy of *Monitor niloticus* from western Africa, and of *Regenia albogularis*. P.Z.S., 1861: 109-113.

On a new species of the family Boidae. P.Z.S., 1861: 142, 1 plate.

Second list of Siamese reptiles. P.Z.S., 1861: 187-189.

List of the cold-blooded Vertebrata collected by B. H. Hodgson, Esq., in Nepal. P.Z.S., 1861: 213-227.

[Notes on snakes from China sent by R. Swinhoe, Esq.]. P.Z.S., 1861: 390-391 (published 1862).

1862

[Remarks on the incubation of the female python in the Society's Gardens]. P.Z.S., 1862: 1-2.

[A list of snakes from Bahia]. P.Z.S., 1862: 23.

On new species of snakes in the collection of the British Museum [first account]. A.M.N.H., ser. 3, 9: 52-59, 124-132, 2 plates.

Descriptions of new species of reptiles and fishes in the collection of the British Museum. P.Z.S., 1862: 188-194, 3 plates.

Eine neue Art von *Coronella* mit Furchenzahn. Arch. Naturg., 28(1): 48-52.

1863

On *Atheris burtonii*, a new snake from west Africa. P.Z.S., 1863: 16-17, 1 plate.

Note on *Diemennia superciliosa*. P.Z.S., 1863: 17-18.

Addition to Dr. Wucherer's article on the ophidians of Bahia. P.Z.S., 1863: 56-58.

Contribution to the herpetology of Ceram. P.Z.S., 1863: 58-60.

On new species of snakes in the collection of the British Museum [second account]. A.M.N.H., ser. 3, 11: 20-25, 1 plate.

On new species of batrachians from Australia. A.M.N.H., ser. 3, 11: 26-28, 1 plate.

Descriptions of new species of reptiles and fishes in the collection of the British Museum. A.M.N.H., ser. 3, 11: 134-140.

On some species of tree-snakes (*Ahœtulla*). A.M.N.H., ser. 3, 11: 283-287.

Observations on Australian tree-frogs living in the Society's Menagerie. P.Z.S., 1863: 249-251, 1 plate.

On *Atheris burtonii*, a new snake from west Africa. A.M.N.H., ser. 3, 12: 239.

Note on *Diemennia superciliosa*. A.M.N.H., ser. 3, 12: 239-240.

Addition to Dr. Wucherer's article on the ophidians of Bahia. A.M.N.H., ser. 3, 12: 325-326.

Third account of new species of snakes in the collection of the British Museum. A.M.N.H., ser. 3, 12: 348-365, 2 plates.

Contribution to the herpetology of Ceram. A.M.N.H., ser. 3, 12: 397-399.

1864

Observations on Australian tree-frogs living in the Society's Menagerie. A.M.N.H., ser. 3, 13: 250-252.

List of the new species of reptiles discovered by the late M. Mouhot in Siam and Cambodia, p. 169-173. *In* H. Mouhot, Travels in the Central Parts of Indo-China (Siam), Cambodia, and Laos During the Years 1858, 1859, and 1860 [ed. by C. Mouhot]. London, J. Murray. Volume 2, viii, 301 pages, 6 plates.

Third contribution to our knowledge of batrachians from Australia. P.Z.S., 1864: 46-49, 1 plate.

Third contribution to our knowledge of batrachians from Australia. A.M.N.H., ser. 3, 14: 311-315.

On the reptiles and fishes obtained by Capt. Speke during the East-African expedition.[7] P.Z.S., 1864: 115, 1 plate.

The Reptiles of British India.[8] London, for the Ray Society by R. Hardwicke. (1), xxvii, 452 pages, 26 plates (reprinted by Oxford & IBH Publ. Co., New Delhi, c. 1982).

Report on a collection of reptiles and fishes made by Dr. Kirk in the Zambesi and Nyassa regions. P.Z.S., 1864: 303-314, 2 plates.

Description of a new species of *Eublepharis*. A.M.N.H., ser. 3, 14: 429-430.

Descriptions of new species of batrachians from West Africa. P.Z.S., 1864: 479-482, 1 plate (published 1865).

Report on a collection of reptiles and fishes from Palestine. P.Z.S., 1864: 488-493 (published 1865).

Reptilia, p. 99-132. *In* R.Z.L., 1 (published 1865).

1865

[A list of reptiles from Madagascar]. P.Z.S., 1865: 469.

Fourth account of new species of snakes in the collection of the British Museum. A.M.N.H., ser. 3, 15: 89-98, 2 plates.

Reptilia, p. 139-162. *In* R.Z.L., 2 (published 1866).

1866

Fifth account of new species of snakes in the collection of the British Museum. A.M.N.H., ser. 3, 18: 24-29, 2 plates.

Reptilia, p. 117-130. *In* R.Z.L., 3 (published 1867).

1867

Additions to the knowledge of Australian

reptiles and fishes. A.M.N.H., ser. 3, 20: 45-68.

Contribution to the anatomy of *Hatteria* (*Rhynchocephalus*, Owen). Philos. Trans. Royal Soc. London, 157: 595-630, 3 plates (titlepage of volume dated 1868 but Günther's monograph was issued separately in 1867).

Contribution to the anatomy of *Hatteria* (*Rhynchocephalus*, Owen). Proc. Royal Soc. London, 15: 460-462.

Contribution to the anatomy of *Hatteria* (*Rhynchocephalus*, Owen). A.M.N.H., ser. 3, 20: 128-129.

Reptilia, p. 126-149. *In* R.Z.L., 4 (published 1868).

1868

[Exhibition of specimens of the ova and young of the axolotl (*Siredon mexicanum*)]. P.Z.S., 1868: 319.

Sixth account of new species of snakes in the collection of the British Museum. A.M.N.H., ser. 4, 1: 413-429, 3 plates.

First account of species of tailless batrachians added to the collection of the British Museum. P.Z.S., 1868: 478-490, 4 plates (published 1869; see correction, P.Z.S., 1870: 150).

Reptilia, p. 114-131. *In* R.Z.L., 5 (published 1869).

1869

Report on two collections of Indian reptiles. P.Z.S., 1869: 500-507, 3 plates (published 1870).

Reptilia, p. 105-122. *In* R.Z.L., 6 (published 1870).

1870[9]

Note on the locality of *Megalixalus infrarufus*. P.Z.S., 1870: 150.

Second account of species of tailless batrachians added to the collection of the British Museum. P.Z.S., 1870: 401-402, 1 plate.

Description of a new Indian lizard of the genus *Calotes*. P.Z.S., 1870: 778-779, 1 plate (published 1871).

Reptilia, p. 65-80. *In* Z.R., 7 (published 1871).

1871

List of the lizards belonging to the family Sepidae, with notes on some of the species. P.Z.S., 1871: 240-244.

Description of a new species of *Tejus* (*Tejus rufescens*) from Mendoza. P.Z.S., 1871: 541-543.

Reptilia, p. 77-88. *In* Z.R., 8 (probably published 1873[4]).

1872

Seventh account of new species of snakes in the collection of the British Museum. A.M.N.H., ser. 4, 9: 13-37, 4 plates.

Descriptions of some Ceylonese reptiles and batrachians. A.M.N.H., ser. 4, 9: 85-88.

On two species of *Hydrosaurus* from the Philippine Islands. P.Z.S., 1872: 145-146, 2 plates.

On the reptiles and amphibians of Borneo. P.Z.S., 1872: 586-600, 6 plates.

Note on *Hyla punctata* and *Hyla rhodoporus*. P.Z.S., 1872: 662-663.

Description of three new species of *Eremias*. A.M.N.H., ser. 4, 9: 381-382.

Notice of a new species of lizard (*Eumeces albofasciolatus*) from North Australia. A.M.N.H., ser. 4, 10: 370-371.

On some new species of reptiles and fishes collected by J. Brenchley, Esq. A.M.N.H., ser. 4, 10: 418-426.

Note on the black snake of Robben Island, South Africa. P.Z.S., 1872: 836-837 (published 1873).

Reptilia, p. 61-81. *In* Z.R., 9 (probably published 1874[4]).

1873

Description of two new species of frogs from Australia. A.M.N.H., ser. 4, 11: 349-350.

Description of a new saurian (*Hyalosaurus*) allied to *Pseudopus*. A.M.N.H., ser. 4, 11: 351.

Notes on some reptiles and batrachians obtained by Dr. Adolf Bernhard Meyer in Celebes and the Philippine Islands. P.Z.S., 1873: 165-172, 2 plates.

Description of a new snake from Madagascar. A.M.N.H., ser. 4, 11: 374-375.

Preliminary notice of some extinct tortoises from the islands of Rodriguez and Mauritius. A.M.N.H., ser. 4, 11: 397.

Contribution to our knowledge of *Ceratophrys* and *Megalophrys*. A.M.N.H., ser. 4, 11: 417-419.

Notes on, and descriptions of, some lizards with rudimentary limbs in the British Museum. A.M.N.H., ser. 4, 12: 145-148.

Reptiles, p. 395-407, 4 plates. *In* J. L. Brenchley, Jottings During the Cruise of H.M.S. Curaçoa Among the South Sea Islands in 1865. London, Longmans, Green, and Co. xxviii, 487 pages, 59 plates, 1 map.

1874

A contribution to the fauna of Savage Island. P.Z.S., 1874: 295-297, 1 plate.

Descriptions of some new or imperfectly known species of reptiles from the Camaroon Mountains. P.Z.S., 1874: 442-445, 2 plates.

Description of the living and extinct races of gigantic land-tortoises.--Parts I. and II. Introduction, and the tortoises of the Galapagos Islands. Proc. Royal Soc. London, 22: 421-422.

Description of a new European species of *Zootoca*. A.M.N.H., ser. 4, 14: 158-159.

Description of the living and extinct races of gigantic land-tortoises.--Parts I. and II. Introduction, and the tortoises of the Galapagos Islands. A.M.N.H., ser. 4, 14: 311-313.

1875

Das Sammeln von Reptilien und Fischen, p. 389-402. *In* G. Neumayer (ed.), Anleitung zu wissenschaftlichen Beobachtungen auf Reisen [ed. 1]. Berlin, R. Oppenheim. viii, 696 pages, 3 foldout maps.

Second report on collections of Indian reptiles obtained by the British Museum. P.Z.S., 1875: 224-234, 5 plates.

A list of the saurians of Australia and New Zealand, p. 9-19, 20 plates.[10] *In* J. Richardson and J. E. Gray (eds.), The Zoology of the Voyage of H.M.S. Erebus & Terror, Under the Command of Captain Sir James Clark Ross, R.N., F.R.S., During the Years 1839 to 1843. Vol. 2. Reptiles, Fishes, Crustacea, Insects, Mollusca. London, E. W. Janson (volume published in parts over the period 1844-1875 with Günther's section entirely in 1875).

The gigantic land tortoises of the Mascarene and Galapagos islands. Nature, 12: 238-239, 259-261, 296-297.

The gigantic land tortoises of the Mascarene and Galapagos islands. Zoologist, ser. 2, 10: 4679-4685.

Description of the living and extinct races of gigantic land-tortoises. Parts I. & II. Introduction, and the tortoises of the Galapagos Islands. Philos. Trans. Royal Soc. London, 165: 251-284, 13 plates.

Note on *Herpeton tentaculatum*. A.M.N.H., ser. 3, 15: 159.

Third report on collections of Indian reptiles obtained by the British Museum. P.Z.S., 1875: 567-577, 4 plates (published 1876).

1876

Chameleon, p. 381-382. *In* E.B., ed. 9, vol. 5 (American version published 1878[2]).

Statement regarding dr. Welwitsch's Angola reptiles. Jorn. Sci. Math. Phys. Nat., Acad. Real Sci., Lisboa, 5: 275-276 (republished in same journal, ed. 2, 1885).

Notes on the mode of propagation of some Ceylonese tree-frogs, with description of two new species. A.M.N.H., ser. 4, 17: 377-380, 1 plate.

Description of a new frog from north-eastern Asia. A.M.N.H., ser. 4, 17: 387.

[Letters from Commander W. E. Cookson, R.N., concerning the Galapagos tortoises, with Günther's remarks thereon]. P.Z.S., 1876: 422, 520-526.

Notes on a small collection brought by Lieut. L. Cameron, C.B., from Angola. P.Z.S., 1876: 678-679.

Descriptions of new species of reptiles from Australia, collected by Hr. Dämel for the Godeffroy Museum. Jour. Mus. Godeffroy, 5: 45-47.

Description of a new species of lizard from Asia Minor. P.Z.S., 1876: 818 (published 1877).

1877

Descriptions of some new species of reptiles from Madagascar. A.M.N.H., ser. 4, 19: 313-317, 1 plate.

Description of three new species of lizards from islands of Torres Straits. A.M.N.H., ser. 4, 19: 413-415.

Description of the living and extinct races of gigantic land-tortoises.--Parts III. & IV. The races of the Aldabra Group and Mascarene Islands (Conclusion.). A.M.N.H., ser. 4, 20: 146-147.

Description of the living and extinct races of gigantic land-tortoises.--Parts III. and IV. The races of the Aldabra Group and Mascarene Islands. Proc. Royal Soc. London, 25: 506-507.

Reptiles, p. 66-67. *In* A. Günther (ed.), Account of the Zoological Collection Made During the Visit of H.M.S. 'Peterel' to the Galapagos Islands. P.Z.S., 1877: 64-93, 3 plates.

On a collection of reptiles and fishes from Duke-of-York Island, New Ireland, and New Britain. P.Z.S., 1877: 127-132, 2 plates.

Notice of two large extinct lizards, formerly inhabiting the Mascarene Islands. Jour. Linnean Soc. London, Zool., 13: 322-327 (†also Trans. Soc. Roy. Arts Sci. l'île Maurice [Mauritius], n.s., 11: 110-115, 1883).

The Gigantic Land-Tortoises (Living and Extinct) in the Collection of the British Museum. London, British Museum (Nat. Hist.). v, 96 pages, 55 plates (1-54, 31B, of which seven are folding; published 12 January 1878, *fide* C. D. Sherborn, A.M.N.H., ser. 10, 13: 309, 1934).

1878

On reptiles from Midian collected by Major Burton. P.Z.S., 1878: 977-978, 1 plate (published 1879).

1879

The extinct reptiles of Rodriguez, p. 452-456; and [Recent] reptiles, p. 470. *In* J. D. Hooker and A. Günther (eds.), An Account of the Petrological, Botanical, and Zoological Collections Made in Kerguelen's Land and Rodriguez During the Transit of Venus Expeditions, Carried Out By Order of Her Majesty's Government in the Years 1874-75. Philos. Trans. Royal Soc. London, 168 (extra vol.): ix, 1-580, 55 plates.

Notice of a collection of reptiles from islands of Torres Straits. A.M.N.H., ser. 5, 3: 84-87.

List of the mammals, reptiles, and batrachians sent by Mr. Everett from the Philippine Islands. P.Z.S., 1879: 74-79, 1 plate.

Description of four new species of *Chamæleon* from Madagascar. P.Z.S., 1879: 148-150, 3 plates.

On mammals and reptiles from Johanna, Comoro Islands. A.M.N.H., ser. 5, 3: 215-219.

Description of a new species of *Chamæleon* from Madagascar. A.M.N.H., ser. 5, 4: 246-247, 1 plate.

Notice of a collection of mammals and reptiles from Cyprus. P.Z.S., 1879: 741 (published 1880).

1880

Catalogue of the Books in the Department of Zoology, British Museum [ed. 1]. London, British Museum (Nat. Hist.). 124 pages.

On the occurrence of *Tachymenis vivax* in Cyprus. A.M.N.H., ser. 5, 5: 436.

Description of new species of reptiles from eastern Africa. A.M.N.H., ser. 5, 6: 234-238.

Description of *Ophites japonicus*, a new snake from Japan. A.M.N.H., ser. 5, 6: 462.

[Notice of a memoir on some rare reptiles and batrachians now or lately living in the Society's Menagerie]. P.Z.S., 1880: 648 (published 1881).

1881

Reptiles, batrachians, and fishes, p. 18-22, 2 plates. *In* A. Günther (ed.), Account of the Zoological Collections Made During the Survey of H.M.S. 'Alert' in the Straits of Magellan and on the Coast of Patagonia. P.Z.S., 1881: 2-141, 11 plates.

Descriptions of the amphisbaenians and ophidians collected by Prof. I. Bayley Balfour in the Island of Socotra. P.Z.S., 1881: 461-463, 2 plates.

Seventh contribution to the knowledge of the fauna of Madagascar. A.M.N.H., ser. 5, 7: 357-360, 1 plate (the previous six contributions have variable titles and are listed on p. 357; four are herpetological).

Herpetology, p. 329-330, 2 plates. *In* F. Oates, Matabele Land and the Victoria Falls. A Naturalist's Wanderings in the Interior of South Africa From the Letters and Journals of the Late Frank Oates, F.R.G.S. [ed. 1; ed. by C. G. Oates]. London, C. Kegan Paul & Co. xliii, 383 pages, 15 plates, 4 maps (also issued as a †separate pamphlet in 1882 and in a second revised edition, 1889).

1882

Lizard, p. 732-738. *In* E.B., ed. 9, vol. 14 (American version published 1882[2]).

[Exhibition of, and remarks upon, the shell of a new tortoise of the genus *Geoemyda* from Siam]. P.Z.S., 1882: 312.

Description of a new species of tortoise

(*Geoemyda impressa*) from Siam. P.Z.S., 1882: 343-346.

Ninth contribution to the knowledge of the fauna of Madagascar. A.M.N.H., ser. 5, 9: 262-266.

Notice of a second species of *Triprion*. A.M.N.H., ser. 5, 10: 279.

Observations on some rare reptiles and a batrachian now or lately living in the Society's Menagerie. Trans. Zool. Soc. London, 11: 215-222, 5 plates.

Catalogue of the Books in the Department of Zoology, British Museum. Second Edition. London, British Museum (Nat. Hist.). 192 pages.

1883

Description of two snakes from the 'Challenger' collections. A.M.N.H., ser. 5, 11: 136-137.

1884

Contribution to our knowledge of *Hydromedusa*, a genus of South-American freshwater turtles. A.M.N.H., ser. 5, 14: 421-425, 1 plate.

Reptilia, Batrachia, and Pisces, p. 29-33. *In* A. Günther (ed.), Report on the Zoological Collections Made in the Indo-Pacific Ocean During the Voyage of H.M.S. 'Alert' 1881-2. London, British Museum (Nat. Hist.). xxv, 684 pages, 54 plates.

1885

Guide to the Gallery of Reptilia [ed. 1]. London, British Museum (Nat. Hist.). 30 pages (with assistance of G. A. Boulenger).

Reptilia and Batrachia. *In* O. Salvin and F. D. Godman (eds.), Biologia Centrali-Americana; or, Contributions to the Knowledge of the Fauna and Flora of Mexico and Central America. London, for the editors by R. H. Porter and Dulau & Co. xx, 326 pages, 76 plates (published in parts from 1885 to 1902; for dating of each signature and plate see Editor's Note in this volume)(for critical review see R. Lydekker, A.M.N.H., ser. 7, 11: 127-130, January 1903).

1886

Python, p. 144; Rattlesnake, p. 293; Reptiles, p. 432-473. *In* E.B., ed. 9, vol. 20 (American versions published 1886[2]; the section on anatomy in "Reptiles," pp. 445-464, is by St. George Mivart.)

1887

Snakes, p. 189-199. *In* E.B., ed. 9, vol. 22 (American version published 1887[2]).

Guide to the Galleries of Reptiles and Fishes. Second Edition. London, British Museum (Nat. Hist.). v, 119 pages (with assistance of G. A. Boulenger).

Notes on batrachians from Perak. A.M.N.H., ser. 5, 20: 312-316, 1 plate.

1888

Tortoise, p. 455-460. *In* E.B., ed. 9, vol. 23 (American version published 1888[2]).

Das Sammeln von Reptilien, Batrachiern und Fischen, p. 405-427. *In* G. Neumayer (ed.), Anleitung zu wissenschaftlichen Beobachtungen auf Reisen [ed. 2]. Berlin, R. Oppenheim. Volume 2, 627 pages, 2 plates (republished by Gebrüder Jänecke, Hannover, 1891).

Report on a collection of reptiles and batrachians sent by Emin Pasha from Monbuttu, Upper Congo. P.Z.S., 1888: 50-51.

On a collection of reptiles from China. A.M.N.H., ser. 6, 1: 165-172, 1 plate.

Contribution to the knowledge of snakes of tropical Africa. A.M.N.H., ser. 6, 1: 322-335, 2 plates.

Notes on reptiles and frogs from Dominica, West Indies. A.M.N.H., ser. 6, 2: 362-366.

1889

Third contribution to our knowledge of reptiles and fishes from the upper Yangtsze-Kiang. A.M.N.H., ser. 6, 4: 218-229.

Herpetology, p. 336-337, 2 plates. *In* F. Oates, Matabele Land and the Victoria Falls . . . Second edition [revised; ed. by C. G. Oates]. London, Kegan Paul, Trench, & Co. (expanded from 1881 edition).

1890

Tenth contribution to the knowledge of the fauna of Madagascar. A.M.N.H., ser. 6, 5: 69-72, 1 plate.

1891

†Instructions for Collecting Reptiles, Batrachians, and Fishes for the British

Museum. London, British Museum (Nat. Hist.). 18 pages.

Eleventh contribution to the knowledge of the fauna of Madagascar. A.M.N.H., ser. 6, 8: 287-288, 1 plate.

1892

List of the species of reptiles and fishes collected by Mr. A. E. Pratt on the upper Yang-tze-kiang and in the province Sze-chuen, with description of the new species, p. 238-250, 4 plates. *In* A. E. Pratt, To The Snows of Tibet Through China. London, Longmans, Green, and Co. xviii, 268 pages, 32 plates, 1 map.

Report on a collection of reptiles and batrachians transmitted by Mr. H. H. Johnston, C. B., from Nyassaland. P.Z.S., 1892: 555-558, 3 plates (published 1893).

1893[11]

Guide to the Galleries of Reptiles and Fishes. Third Edition. London, British Museum (Nat. Hist.). v, 119 pages, 1 map (with assistance of G. A. Boulenger).

Second report on the reptiles, batrachians, and fishes transmitted by Mr. H. H. Johnston, C.B., from British Central Africa. P.Z.S., 1893: 616-628, 5 plates (published 1894).

Descriptions of the reptiles and fishes collected by Mr. E. Coode-Hore on Lake Tanganyika. P.Z.S., 1893: 628-632, 1 plate (published 1894).

1894

Report on the collection of reptiles and fishes made by Dr. J. W. Gregory during his expedition to Mount Kenia. P.Z.S., 1894: 84-91, 4 plates.

1895

Notice of reptiles and batrachians collected in the eastern half of tropical Africa. A.M.N.H., ser. 6, 15: 523-529, 1 plate.

Introduction of a West Indian frog into the Royal Gardens, Kew. Nature, 52: 643.

West Indian frog at Kew. Kew Bull. Misc. Infor., 107: 301-302.

The reptiles and batrachians of the Natuna Islands. Novitates Zool., 2: 499-502.

1896

Numbers of zoological species known in the years 1830 and 1881. A.M.N.H., ser. 6, 17: 180.

Descriptions of two new species of snakes from Sarawak. A.M.N.H., ser. 6, 17: 228-229.

Report on a collection of reptiles and fishes made by Miss M. H. Kingsley during her travels on the Ogowe River and in Old Calabar. A.M.N.H., ser. 6, 17: 261-285, 3 plates.

Report on the collections of reptiles, batrachians and fishes made by Messrs. Potanin and Berezowski in the Chinese Provinces Kansu and Sze-chuen. Annuaire Mus. Zool. Acad. Impér. Sci. St.-Pétersbourg [Akad. Nauk SSSR], 1: 199-219, 2 plates.

Description of a new toad (*Xenorhina*) from New Guinea. Novitates Zool., 3: 184-185, 1 plate.

†Reptiles, Fishes, Insects, etc., p. 315-322. *In* The Illustrated Bible Treasury.

Testudo ephippium. Novitates Zool., 3: 329-334, 3 plates.

1897

Description of a new species of *Delma* from western Australia. A.M.N.H., ser. 6, 19: 170-172.

Note on some reptiles and a frog from Argentina. A.M.N.H., ser. 6, 20: 365-366.

Descriptions of new species of lizards and of a tree-frog from north-eastern Queensland. Novitates Zool., 4: 403-406, 2 plates.

Report on a collection of reptiles and fishes made by Miss M. H. Kingsley during her travels on the Ogowé River and in Old Calabar, p. 692-717, 2 plates. *In* M. H. Kingsley, Travels in West Africa. Congo Français, Corisco and Cameroons. London, Macmillan and Co. xvi, 743 pages (American edition New York, Macmillan Co., 1897).

1898

Notes on Indian snakes in captivity. A.M.N.H., ser. 7, 1: 30-31.

The president's anniversary address [on gigantic land tortoises]. Proc. Linnean Soc. London, 1897-1898: 14-31.

1902

Testudo galapagoensis. Novitates Zool., 9: 184-192, 6 plates.

[Exhibition of, and remarks upon, some living tadpoles of the North-American bull-frog].

P.Z.S., 1902, 2: 227 (published 1903).

1903

Reptiles from Rio de Oro, western Sahara. Novitates Zool., 10: 298-299.

1906

†Reptilia, Amphibia, Pisces, p. 10-12. *In* The Wild Fauna and Flora of the Royal Botanic Gardens, Kew. Kew Bull. Misc. Infor., [old] addit. ser. 5.

†A snake story. Observer [London], issues of 28 January and 13 February.

Das Sammeln von Reptilien, Batrachiern und Fischen, p. 561-594. *In* G. Neumayer (ed.), Anleitung zu wissenschaftlichen Beobachtungen auf Reisen [ed. 3]. Hannover, M. Jänecke. Volume 2.

1908

Agamidae and Iguanidae [concerning Prof. Kaup's claim to priority]. A.M.N.H., ser. 8, 2: 523-524.

1911

Reptiles, p. 136-176. *In* E.B., ed. 11, vol. 23 (with Hans F. Gadow and Arthur S. Woodward[2]; American version published 1911).

1912

General history of the Department of Zoology from 1856 to 1895. *In* The History of the Collections Contained in the Natural History Departments of the British Museum. London, British Museum (Nat. Hist.). Volume 2, Appendix, 109 pages (herpetology, pp. 19, 27, 52-53, 70).

NOTES

[1]Exact dating of articles in A.M.N.H. can be determined by examining the first page of each number where the month and year of issue is printed. Cornell University Library's set has the original wrappers intact, from which these dates of issue can be confirmed. The date of receipt of each number was stamped on the wrapper by the library and in virtually every instance these dates correspond to the printed dates. In those few exceptions it is the month following, but given the time necessary for transatlantic crossings it is reasonable to suppose that the dates of issue in London were as printed on the titlepages.

During Günther's career A.M.N.H. had the practice to reprint or summarize the more important papers presented at London-based scientific meetings under the heading "Proceedings of Learned Societies." Thus, in this bibliography one will find titles in A.M.N.H. that are identical to those of some of Günther's papers published in P.Z.S. and by the Royal Society. Whereas the titles are usually the same, the contents are sometimes different and these occasionally include corrections to the original version. In at least one instance, publication of the A.M.N.H. summmary antedates the "original" paper.

[2]Credited using the initials "A.C.G.," Günther published at least four sections on herpetological subjects in the ninth British edition of *Encyclopaedia Britannica* (published in Edinburgh, A. and C. Black, 1875-1889), of which the article "Lizard" (published 1882) is the earliest where authorship is certain. The other three initialled by him were published in 1886, 1887 and 1888. R. T. Gunther (A.M.N.H., ser. 10, 6: 273, 1930) claims that "Python" and Rattlesnake," published in 1886, are Günther's and I suspect that "Chameleon" (published 1876) is also, but no author's initials are given, as was the encyclopedia's practice for shorter articles. However, other herpetological subjects in this edition are initialled by John Gibson, Thomas H. Huxley and St. George Mivart.

Günther's articles were republished in the tenth (1902-1903) and eleventh (1910-1911) British editions, but these are generally omitted in this bibliography. As R. T. Gunther points out, these republications were sometimes changed only slightly and Günther's initials were omitted or those of another author substituted. For example, in the eleventh edition all seven articles here attributed to Günther's authorship are signed by Hans F. Gadow or Mivart, except for that on "Reptiles" which is co-authored by Günther, Arthur S. Woodward and Gadow (see 1911). It is clear that Gadow and Mivart simply reworked Günther's original versions, updated them and in a few instances credited him in the text, but large sections are *verbatim* from the ninth edition.

Günther's articles in the ninth edition were republished in the American ninth edition of *Encyclopaedia Britannica* which was issued by at least seven different publishers in Boston, Chicago, New York and Philadelphia. The titles, texts, volume numbers and pagination are exactly the same, the American versions having been printed from the same stereotype plates as the British. The dates of publication, however, are sometimes later than for the British version and, therefore, the American versions are cited here only by date of publication, following the citations to the original British version. The publication dates of the American versions, since they were issued by different publishers, are variable; the dates cited here are from the ninth edition published by Charles Scribner's Sons (New York, 1878-1889).

[3]Dating of articles published in P.Z.S. can be determined from the listing by F. M. Duncan et al. (P.Z.S., ser. A, 107: 71-84, 1937) covering the period 1830-1926. The last part of a given year's volume was almost always published early in the year following. However, the Zoological Society's information for the volumes for 1859 and 1860 is inconclusive. Four of Günther's papers in those volumes were read at November meetings, the results of which were probably published early in the following year since this was the case for November papers published in the volumes immediately preceding and following these two volumes.

[4]Günther was recorder for the Reptilia section of *Zoological Record* (entitled *Record of Zoological Literature* for volumes 1-6) covering the years 1864 to 1872. In those days this publication was not merely a list of literature but also included reviews of major works and extensive commentaries by the recorders.

Günther's sections are here listed according to the year of literature coverage since that is the way in which these are typically cited in the scientific literature. In fact, volumes 1-7 were issued in the following year. Volumes 8 and 9, however, were issued later still, due to late receipt of literature from Europe resulting from the Continental war. According to the titlepages of these volumes they were issued two years later, but in each case the Reptilia section is contained in part 1 of the volume which might have been published earlier than the date given on the title.

[5]Osbert Salvin published three papers on Guatemalan amphibians and reptiles in 1860-1861 (P.Z.S., 1860: 451-461, 1861; P.Z.S., 1861: 227-229, 1861; A.M.N.H., ser. 3, 7: 325-326, 1861) which are sometimes incorrectly cited as co-authored by Günther. However, Salvin does acknowledge Günther for "determining the species."

[6]According to R. T. Gunther (A.M.N.H., ser. 10, 6: 249, 1930) Otho Wucherer's paper on Brazilian snakes, part 2 (P.Z.S., 1861: 322-325, 1862) was edited and annotated by Günther.

[7]According to R. T. Gunther (A.M.N.H., ser. 10, 6: 252, 261, 1930) additional notes on reptiles by Günther are contained in J. A. Grant's article on the Speke and Grant Expedition (Jour. Royal Geogr. Soc. London, 42: 243-342, 1872; herpetology, pp. 312-315) but no section in that publication is specifically authored by him.

[8]No precise information on the date of publication of *The Reptiles of British India* is available. Günther's preface is dated 31 May 1864 and the book is cited in a list of major natural history works published during 1864 in a magazine issued 7 January 1865 (London Review Pol., Soc., Liter., Arts Sci., 10: 52) and was reviewed, anonymously, in the same serial six months later (vol. 10, pp. 670-671, 24 June). Moreover, a stinging criticism of Günther's book was published in May 1865 (A.M.N.H., ser. 3, 15: 416-418), written by T. C. Jerdon, surgeon-major of the Madras Army, in a letter to the editor dated "Kurnal, [India], Feb. 24, 1865." It seems reasonable to conclude that the book was published during the latter half of 1864.

[9]Günther is sometimes listed, incorrectly, as author of the reptile section (pp. 444-459) in W. T. Blanford's *Observations on the Geology and Zoology of Abyssinia. . . .* London, Macmillan and Co. xii, 487 pages, 8 plates, 1 map, 1870), but on p. 444 Blanford, the actual author, acknowledges Günther for "determinations of the majority of the reptiles."

[10]All plates had been previously published in J. E. Gray's section of this voyage (entitled "The Reptiles of Australia," pp. 1-8, 10 plates, 1845, of which Günther's article is a continuation) or in Gray's *The Lizards of Australia and New Zealand in the Collection of the British Museum* (London, B. Quaritch. 8 pages, 18 plates, 1867), except for plates 5 and 6 which were published for the first time with Günther's article which cites all 20 plates.

[11]Günther is sometimes incorrectly cited as author of the herpetological portion (pp. 34-35, 415) of H. Lansdell's *Chinese Central Asia. A Ride to Little Tibet*, volume 2 (London, S. Low, Marston & Co., 1893; American edition New York, C. Scribner's Sons, 1894) but he was, in fact, apparently responsible only for the identifications of species.

BIOLOGIA
CENTRALI-AMERICANA

REPTILIA AND BATRACHIA
by Albert Günther

BIOLOGIA

CENTRALI-AMERICANA.

REPTILIA AND BATRACHIA.

BY

ALBERT C. L. G. GÜNTHER, M.A., M.D., PH.D., F.R.S., F.L.S., &c.

1885–1902.

CONTENTS.

INTRODUCTION.

More than fifteen years have elapsed since the first sheets of this Volume of the 'Biologia Centrali-Americana' were published. The slow progress of the work was due partly to the demand made upon the author's time by other, chiefly official, duties, and partly to the anticipation of receiving important additions to the material that was then available at the British Museum, and which included the rich collections brought home by Salvin, and later by Godman, from Guatemala and Mexico. A staff of collectors was at work for the 'Biologia' in various parts of Central America, and they had received instructions from the Editors to pay attention to Reptiles and Batrachians, besides the special branches to which they devoted themselves principally. To wait as long as possible for the result of these instructions seemed to be all the more desirable as the great work of the 'Mission Scientifique au Mexique et dans l'Amérique Centrale,' in which the Herpetology was entrusted to the able hands of M. Bocourt, was in course of publication, and, like the numerous important contributions by the late Mr. Cope, contained a multitude of forms not represented in the English collections.

Of the material received from the above-mentioned sources during the progress of the work I have to refer specially to the following:—from Mr. H. H. Smith, who sent several large collections from Atoyac in Vera Cruz, Teapa in Tabasco, Omilteme and Amula in Guerrero, and Cuernavaca in Morelos; from Hr. A. Forrer, who visited the Tres Marias Islands, and thoroughly searched several localities in Western Mexico, particularly Presidio, near Mazatlan; from Dr. A. C. Butler, who selected some localities on the plateau of Central Mexico as his collecting-grounds; from Hr. C. T. Höge and Señor M. Trujillo, who sent not a few interesting species from Jalapa and Misantla in Vera Cruz; from Mr. G. F. Gaumer, who collected in Yucatan and Honduras, visiting Cozumel, Bonacca, and Ruatan Islands; from Mr. W. B. Richardson and Hr. Ernst Rothschuh, both of whom made valuable collections at Matagalpa in Nicaragua; from

Mr. G. C. Champion, who collected Reptiles in Guatemala and the State of Panama, and to whom we are indebted for some important specimens from Chiriqui; and from Mr. H. Rogers and Mr. C. F. Underwood, who largely collected Reptiles and Batrachians in various localities in Costa Rica (Irazu, Rio Sucio, La Palma, the neighbourhood of San José, &c.).

All these collections yielded new or valuable material, extending our knowledge of the species and of their range; but also clearly showing that, so far as Reptiles and Batrachians are concerned, our information is not nearly so extensive as it is in the case of the Birds and certain orders of Insects. The collections contained some of the forms described by Cope, Bocourt, Brocchi, and others (as well as many unknown to those herpetologists), but a large proportion, perhaps the majority, continued to be known to me from descriptions only. In fact, the districts whose Reptilian fauna is satisfactorily known are but few in number, our acquaintance with them being chiefly due to the labours of Naturalists who have resided in the country for a lengthened period, and who have devoted themselves specially to this branch of study. From among them I must mention the late F. Sumichrast, who thoroughly explored the Reptilian Fauna of the Isthmus of Tehuantepec; Dr. A. Dugès, of Guanajuato, who not only published numerous papers on the Reptiles of his district, but also supplied Cope and the Paris Museum with a complete series of the species observed by him; the late A. Sallé and A. Boucard, who collected for many years in the country round Oaxaca, Orizaba, and Vera Cruz; O. Salvin, who paid three visits to Guatemala, collecting chiefly in the neighbourhood of Antigua and in Alta Vera Paz. Perhaps no other part of Central America is better known than the plateau and highlands round Mexico city, whence numerous collections have reached Europe; and the Republic of Costa Rica, whose Government has done so much for the physical exploration of the country. The Reptiles of the latter region have formed the subject of important contributions by Cope and Peters, and collections more recently received through the liberality of the authorities of the Museo Nacional de Costa Rica, at San José, and of M. Pittier de Fahéga, have considerably added to our knowledge of this Fauna.

According to the general plan laid down by the Editors of the 'Biologia,' descriptive diagnoses were to be given of those species only which had not been satisfactorily described elsewhere. I have followed this plan throughout the part treating of Reptilia, but I thought it better to depart from it in the Batrachian portion of the work, and to accompany all the species, as distinguished by me, with a diagnosis, not only on account of the numerous additions made since the publication of Mr. Boulenger's Catalogues of Batrachia, but also because the views of Herpetologists differ only too frequently as to the taxonomic value of characters used by them for specific distinction. In this latter part I have received valuable assistance from the authorities of the United States National Museum, Dr. Franz Werner of Vienna, Professor Böttger of Frankfurt, Professor Camerano and Count Peracca of Turin, Dr. A. Dugès, Professor Möbius, and M. Mocquard, either by the loan of or information concerning typical specimens: assistance for which I beg to return my best thanks *.

The geographical limits fixed by the Editors for the scope of the 'Biologia' are, in the North, the Rio Grande del Norte and the political boundary of Sonora and Chihuahua, and, in the South, the Isthmus of Darien. Lower California is excluded. The *general* features of the Reptilian and Amphibian Faunas of this area have been satisfactorily ascertained †. Forming the connecting link between the two Neogæan regions, Central America possesses a Reptilian and Batrachian Fauna with the various constituent elements so mixed that, if only certain families or genera were taken into consideration, almost every district of this area could be associated with either the North- or South-American region. The tropical Fauna, of course, gradually changes into, or is replaced by, that of the temperate region, as we proceed from lower to higher latitudes; but this change is not uniform throughout the breadth of the land; the two faunas overlap each other in deep and manifold indentations. Tropical types are found to preponderate in the low lands of the Atlantic side, which expand into the broad Yucatan peninsula, and on the humid slopes of moderate elevation; some extend to, and even reach northwards of, the Rio Grande. On the Western side they are found in similar localities, but in a narrower belt, along the Pacific coast. On the other hand, numerous types of the southern North-American Fauna are spread over Northern

* Unfortunately I have not had an opportunity of examining any of the specimens of Reptiles named by Cope. An application which I made to the authorities of the U. S. National Museum, at the commencement of the work, could not be acceded to at the time, as the materials were in Mr. Cope's hands and had not been returned by him.

† *Cfr.* Cope, "Zoological Position of Texas," Bull. U. S. Nat. Mus. no. 17 (1880); Boulenger, Catalogue of Batrachia Gradientia, 1882, p. 104; Ann. & Mag. Nat. Hist. 1885, xvi. p. 77 (Lacertilia); Günther, Encycl. Brit. ed. ix. art. Reptiles (1885); Cope, Amer. Nat. 1896, pp. 886, 993; von Martens, Naturwiss. Wochenschrift, vol. i. (1901).

Mexico, extending along the Central-American plateau to the extreme limits of our area, and even beyond. This southward extension of northern types is due partly to the identical physical conditions of the arid tableland of Sonora and Chihuahua, which is merely a continuation of that of Arizona and New Mexico, and partly to the great altitude and temperate climate of the Central-American plateau.

Thus, a boundary-line between the North- and South-American regions cannot be drawn; the whole of Central America is a transition-tract which, unlike any other part of the world, shows the most extraordinary diversity of climatic, physical, and meteoric conditions within comparatively small areas, favouring the evolution of a great variety of types of genera and species, and influencing the dispersal of immigrants from the North and South. To divide Central America into zoological provinces delimited by the distribution of the genera and species of Reptiles and Batrachians is a task which, in the present state of our knowledge, would appear to be rather premature. Cope attempted it for a part of Mexico as recently as 1896. In a paper on the geographical distribution of the Batrachia and Reptilia of North America he allotted Central and Northern Mexico to his Sonoran, Austroriparian, Toltecan (and Tamaulipan) subregions, including in the Sonoran Chihuahua, and "perhaps" Durango and Sinaloa, and subdividing the Toltecan into an Austroriental, Austrocentral, and Austroccidental district. But this division cannot be accepted with confidence, because, as he himself points out, large unexplored districts intervene between these various divisions, and because our knowledge of the range of the species, and of the individual species themselves, is too fragmentary.

To show the range of the 695 species enumerated in the present volume, I have prepared the subjoined Table (pp. x–xvii) on the plan adopted by Mr. Godman for the genera of Rhopalocera, with one or two modifications. This shows in two columns whether a species extends beyond the southern or northern limit of the Central-American area: Mexico is divided into Northern and Southern Mexico *, Tampico on the Atlantic and Mazatlan on the Pacific coast being grouped with Northern Mexico, as in Mr. Godman's table; but I have given to the Yucatan peninsula a separate column, on account of its large extent and its markedly different features, the country being uniformly low, very flat, and dry; British Honduras or Belize I have grouped

* The species recorded from "Mexico," without definite locality, are placed under Southern Mexico, excepting those of known northern distribution.

with Guatemala; and Salvador, of which little is known beyond M. Bocourt's collection from La Union, on the Pacific coast, with Honduras. However, the Table registers those facts only which are contained in the text. Gladly would I have added the large amount of information recorded in the numerous publications that have appeared subsequently to the dates of issue of this work, by Bocourt, Böttger, Boulenger, Cope, Dugès, Mocquard, Stejneger, and others; but as this supplementary work will require much more attention and time than is consistent with the Editor's desire to bring the 'Biologia' to a close, it has been decided not to delay the publication of the present volume. Thus, all supplementary information, together with a report on collections received too late to be utilized for the present volume, must be reserved for a paper or papers to be published, I hope, at no very distant date.

A. G.

August 1902.

Geographical Distribution of the Reptiles and Batrachians represented in Mexico and Central America.

	Page.	Northern range.	N. Mexico.	S. Mexico.	Yucatan.	Guatemala & British Honduras (Belize).	Honduras & Salvador.	Nicaragua.	Costa Rica.	Panama.	Southern range.
CHELONIA.											
Testudinidæ.											
1. Cistudo mexicana, *Gray*	1	...	*	*							
2. Emys ornata, *Gray*	2	...	*	*	...	...	*	...	*	*	
3. —— cataspila, *Gthr.*	4	...	...	*							
4. —— salvini, *Gthr.*	4	...	...	...	...	*					
5. —— mobiliensis, *Holbr.*	5	*	*								
6. —— callirostris, *Gray*	5	...	...	?							
7. —— umbra, *O'Sh.*	5	...	...	*	...	*					
8. —— pulcherrima, *Gray*	6	...	*								
9. —— incisa, *Boc.*	7	...	...	*	...	...	*				
10. —— gabbii, *Cope*	7	...	...	...	...	...	...	...	*		
11. —— funerea, *Cope*	7	...	...	...	...	...	...	...	*		
12. —— areolata, *Dum.*	8	...	...	...	...	*					
13. —— rubida, *Cope*	8	...	...	*							
14. —— punctularia, *Daud.*	9	...	...	*	*	...	...	...	...	...	*
15. Dermatemys mawii, *Gray*	10	...	...	*	*	*	*				
16. Chelydra serpentina, *L.*	10	*	*	*	*	*	*	*	*	*	*
17. —— rossignoni, *Dum.*	10	...	...	*	...	*					
18. Staurotypus triporcatus, *Wiegm.*	11	...	...	*	*	*					
19. —— salvini, *Gray*	11	...	...	*	...	*					
20. Claudius angustatus, *Cope*	12	...	...	*							
21. Cinosternum sonoriense, *Lec.*	13	*	*								
22. —— cruentatum, *Dum.*	13	...	...	*							
23. —— triliratum, *Lec.*	14	...	...	*							
24. —— albogulare, *Boc.*	14	...	...	...	...	...	...	...	*		
25. —— hirtipes, *Wagl.*	15	*	*								
26. —— integrum, *Lec.*	16	...	...	*							
27. —— effeldtii, *Ptrs.*	16	...	...	*	...	*					
28. —— leucostomum, *Dum.*	17	*	...	*	...	*	...	...	*	*	*
29. —— brevigulare, *Gthr.*	17	...	...	*							
30. —— cobanum, *Gthr.*	18	...	...	...	...	*					
CROCODILIA.											
1. Crocodilus americanus, *Schn.*	19	*	*	*	*	*	*	*	*	*	*
2. Alligator punctulatus, *Spix*	21	...	...	*	...	*	...	...	...	*	*

	Page.	Northern range.	N. Mexico.	S. Mexico.	Yucatan.	Guatemala & British Honduras (Belize).	Honduras & Salvador.	Nicaragua.	Costa Rica.	Panama.	Southern range.
LACERTILIA (*con.*).											
Anguidæ (*con.*).											
45. Gerrhonotus rudicollis, *Wiegm.*	38	...	...	*							
46. —— imbricatus, *Wiegm.*	38	...	...	*							
47. —— planifrons, *Boc.*	38	...	...	*							
48. —— cæruleus, *Wiegm.*	38	*	*	*	...	*	*	*	*		
49. —— obscurus, *Gthr.*	40	...	...	*							
50. —— moreletii, *Boc.*	41	...	...	*	...	*					
51. —— liocephalus, *Wiegm.*	41	*	...	*							
52. —— rhombifer, *Ptrs.*	42	...	...	...	...	...	...	...	...	*	
53. —— modestus, *Cope*	42	...	...	...	...	*					
54. —— webbii, *Baird*	35	*									
55. —— fulvus, *Cope*	35	...	...	...	...	...	...	...	*		
Xenosauridæ.											
56. Xenosaurus grandis, *Gray*	43	...	...	*							
Helodermatidæ.											
57. Heloderma horridum, *Wiegm.*	43	...	*	*							
58. —— suspectum, *Baird*	44	*	*								
Iguanidæ.											
59. Anolis sagræ, *D. & B.*	45	...	...	*	...	*	...	...	...	...	*
60. —— beckeri, *Boul.*	45	...	...	...	*						
61. —— rodriguezi, *Boc.*	45	...	...	...	...	*					
62. —— aureolus, *Cope*	45	...	...	...	*						
63. —— limifrons, *Cope*	45	...	...	...	...	...	...	...	...	*	
64. —— bransfordii, *Cope*	45	...	...	...	...	...	...	*			
65. —— trochilus, *Cope*	45	...	...	..	...	...	...	*	*	*	
66. —— ortonii, *Cope*	45	...	...	...	...	*	...	...	...	...	*
67. —— schiedi, *Cope*	46	...	...	...	...	*	...	...	...	...	*
68. —— polylepis, *Ptrs.*	46	...	...	...	...	...	...	...	...	*	
69. —— tropidolepis, *Boul.*	46	...	...	...	...	...	...	...	*		
70. —— pulchripes, *Ptrs.*	46	...	...	...	...	...	...	...	...	*	
71. —— baccatus, *Boc.*	46	...	...	*							

LACERTILIA.											
TEJIDÆ.											
1. Ameiva surinamensis, *Laur.*	22	...	...	...	...	...	...	*	*	*	*
2. —— gabbiana, *Cope*	23	...	...	...	...	...	...	...	*		
3. —— undulata, *Wiegm.*	23	...	...	*	...	*	*	*	*		
4. —— festiva, *Licht.*	24	...	...	...	...	*	...	...	*	*	
5. —— quadrilineata, *Hallow.*	24	...	...	...	...	...	...	*			
6. Cnemidophorus sexlineatus, *L.*	25	*	*	*	*	*					
7. —— costatus, *Cope*	26	...	...	*							
8. —— deppii, *Wiegm.*	27	...	...	*	...	*	*	?	...	...	*
9. —— lineatissimus, *Cope*	27	...	...	*							
10. —— mariarum, *Gthr.*	28	...	...	*							
11. —— guttatus, *Wiegm.*	28	...	...	*							
12. —— inornatus, *Baird*	29	...	*								
13. Gymnophthalmus sumichrasti, *Cope*	29	...	...	*	...	*	...	...	...	...	*
14. Tretioscincus lævicauda, *Cope*	29	...	...	...	...	...	...	*			
15. Argalia pœcilochilus, *Licht.*	29	...	...	...	...	...	...	...	...	*	*
16. Anadia metallica, *Cope*	30	...	...	...	...	...	...	...	*		
LEPIDOPHYMIDÆ.											
17. Lepidophyma flavomaculatum, *Dum.*	30	...	...	*	...	*	...	...	...	*	
AMPHISBÆNIDÆ.											
18. Chirotes canaliculatus, *Bonnat.*	30	*	*								
19. Amphisbæna fuliginosa, *L.*	31	...	...	...	...	...	...	...	...	*	*
SCINCIDÆ.											
20. Mocoa lateralis, *Say*	31	*	...	*							
21. —— assata, *Cope*	31	...	...	*	...	*	*				
22. Eumeces lynxe, *Wiegm.*	32	...	...	*							
23. —— brevirostris, *Gthr.*	32	...	*	*							
24. —— bocourti, *Boul.*	32	...	*								
25. —— callicephalus, *Boc.*	32	...	...	*							
26. —— sumichrasti, *Cope*	32	...	...	*							
27. —— dugesii, *Thom.*	32	...	...	*							
28. —— schwartzii, *Fisch.*	33	...	...	...	*						
29. —— furcirostris, *Cope*	33	...	...	*							
30. Mabuia agilis, *Raddi*	33	...	...	*	*	*	...	...	...	*	*
31. —— alliacea, *Cope*	33	...	...	...	...	...	...	...	*		
ANELYTROPIDÆ.											
32. Anelytropsis papillosus, *Cope*	34	...	...	*							
ANGUIDÆ.											
33. Diploglossus bilobatus, *O'Sh.*	34	...	...	...	...	...	...	...	*		
34. —— steindachneri, *Cope*	34	...	...	*	.	*					
35. —— monotropis, *Kuhl*	34	...	...	...	...	...	...	...	*	*	*
36. Ophisaurus ventralis, *L.*	35	*	...	*							
37. Gerrhonotus deppii, *Wiegm.*	35	...	...	*							
38. —— gramineus, *Cope*	36	...	...	*							
39. —— oaxacæ, *Gthr.*	36	...	...	*							
40. —— tæniatus, *Wiegm.*	36	...	...	*							
41. —— vasconcelosii, *Boc.*	37	...	...	...	...	*					
42. —— auritus, *Cope*	37	...	...	...	...	*					
43. —— fimbriatus, *Boc.*	37	...	...	...	...	*					
44. —— antauges, *Cope*	37	...	...	*							

72. —— guentheri, *Boc.*	46	...	...	*							
73. —— squamulatus, *Ptrs.*	47	...	...	...	...	...	...	...	...	*	*
74. —— latifrons, *Berth.*	47	...	...	...	...	...	...	...	...	*	*
75. —— microtus, *Cope*	47	...	...	...	...	...	...	...	*		
76. —— insignis, *Cope*	47	...	...	...	...	...	...	...	*		
77. —— pachypus, *Cope*	47	...	...	...	...	...	...	...	*		
78. —— pentaprion, *Cope*	47	...	...	...	...	...	...	...	...	*	*
79. —— copii, *Boc.*	47	...	...	...	...	*	...	...	*		
80. —— petersii, *Boc.*	47	...	...	*	...	*					
81. —— heliactin, *Cope*	48	...	...	*							
82. —— ustus, *Cope*	48	...	...	...	*	...	*				
83. —— cymbops, *Cope*	48	...	...	*							
84. —— concolor, *Cope*	48	...	...	...	...	...	...	*			
85. —— bourgeaui, *Boc.*	48	...	...	*							
86. —— salvini, *Boul.*	48	...	...	...	...	*					
87. —— læviventris, *Wiegm.*	49	...	...	*							
88. —— nebulosus, *Wiegm.*	49	...	*	*							
89. —— nebuloides, *Boc.*	49	...	...	*							
90. —— palpebrosus, *Ptrs.*	49	...	...	...	...	...	...	...	...	*	
91. —— intermedius, *Ptrs.*	49	...	...	*	...	*	...	...	*	*	
92. —— sallæi, *Gthr.*	49	...	...	*	..	*					
93. —— binotatus, *Ptrs.*	50	...	...	...	...	...	*	...	...	...	*
94. —— cumingii, *Ptrs.*	50	...	...	*							
95. —— cupreus, *Hallow.*	50	...	...	...	...	*	...	*	*		
96. —— crassulus, *Cope*	50	...	...	*	...	*					
97. —— humilis, *Ptrs.*	50	...	...	...	...	*	...	...	...	*	
98. —— quaggulus, *Cope*	50	...	...	...	...	...	...	*			
99. —— tropidonotus, *Ptrs.*	51	...	...	*							
100. —— uniformis, *Cope*	51	...	...	...	*	*					
101. —— metallicus, *Boc.*	51	...	...	*							
102. —— oxylophus, *Cope*	51	...	...	...	...	...	...	...	*		
103. —— godmani, *Boul.*	51	...	...	...	...	*	...	...	*		
104. —— radulinus, *Cope*	52	...	...	...	...	...	...	...	...	*	*
105. —— pœcilopus, *Cope*	52	...	...	...	...	...	...	...	...	*	*
106. —— lemurinus, *Cope*	52	...	...	...	...	...	...	...	...	*	
107. —— biporcatus, *Wiegm.*	52	...	...	...	...	*	...	...	...	*	*
108. —— chrysolepis, *D. & B.*	52	...	...	*	...	...	*	...	...	...	*
109. —— capito, *Ptrs.*	52	...	...	*	...	*	...	...	*	*	
110. —— obtusirostris, *Ptrs.*	52	...	...	...	...	...	...	...	...	*	
111. Polychrus gutturosus, *Berth.*	53	...	...	...	...	...	...	...	*	...	*
112. Corythophanes cristatus, *Merr.*	53	...	...	...	...	*	...	...	*	*	
113. —— percarinatus, *Dum.*	53	...	...	...	...	*	...	...	.	...	*
114. —— hernandesii, *Wiegm.*	53	...	...	*	*						
115. Læmanctus longipes, *Wiegm.*	54	...	...	*							
116. —— serratus, *Cope*	54	...	...	*	*						
117. —— deborrii, *Boul.*	54	...	...	*	...	*					
118. —— alticoronatus, *Cope*	54	...	...	...	*						
119. Basiliscus americanus, *Laur.*	54	...	...	...	...	...	...	...	*	*	
120. —— plumifrons, *Cope*	55	...	...	...	...	...	...	...	*		
121. —— vittatus, *Wiegm.*	55	...	...	*	*	*	*	*	*	...	*
122. Enyalius heterolepis, *Boc.*	56	...	...	...	...	...	...	...	...	*	
123. Iguana rhinolophus, *Wiegm.*	56	...	*	*	*	*	...	...	*	*	*
124. —— tuberculata, *Laur.*	56	...	...	...	...	*	*	*	...	*	*
125. Ctenosaura acanthura, *Shaw*	56	*	*	*	*						
126. —— completa, *Boc.*	58	...	...	...	*	*	*				
127. —— quinque-carinata, *Gray*	58	...	...	*	...	...	*				
128. —— defensor, *Cope*	58	...	...	...	*						
129. Crotaphytus wislizenii, *B. & G.*	59	*	*								

Geographical Distribution (*continued*).

	Page.	Northern range.	N. Mexico.	S. Mexico.	Yucatan.	Guatemala & British Honduras (Belize).	Honduras & Salvador.	Nicaragua.	Costa Rica.	Panama.	Southern range.
LACERTILIA (*con.*).											
IGUANIDÆ (*con.*).											
130. Crotaphytus collaris, *Say*	59	*	*								
131. Holbrookia texana, *Trosch.*	59	*	*								
132. —— maculata, *Gir.*	60	*	*								
133. —— propinqua, *B. & G.*	60	*	*								
134. Uta lateralis, *Boul.*	61	...	*								
135. —— gularis, *Cragin*	61	...	*								
136. —— auriculata, *Cope*	61	...	*								
137. —— irregularis, *Fisch.*	62	...	*								
138. —— bicarinata, *Dum.*	62	...	*	*							
139. Sceloporus spinosus, *Wiegm.*	63	*	*	*							
140. —— viviparus, *Cope*	64	...	...	*							
141. —— poinsetti, *B. & G.*	65	*	*								
142. —— acanthinus, *Boc.*	65	...	...	*	...	*					
143. —— torquatus, *Wiegm.*	66, 77	...	*	*							
144. —— omiltemanus, *Gthr.*	66	...	...	*							
145. —— undulatus, *Daud.*	67	*	*	...	...	*					
146. —— lunæi, *Boc.*	67	...	...	...	...	*					
147. —— irazuensis, *Gthr.*	67	...	...	...	...	...	...	...	*		
148. —— salvini, *Gthr.*	68	...	...	*	...	*					
149. —— smaragdinus, *Boc.*	68	...	...	...	*	*	...	...	*		
150. —— yarrovii, *Cope*	69	*	*								
151. —— consobrinus, *B. & G.*	69	*	...	*							
152. —— pyrrhocephalus, *Cope*	70	..	...	*							
153. —— chrysostictus, *Cope*	70	...	...	...	*						
154. —— grammicus, *Wiegm.*	71	...	...	*	...	*					
155. —— gratiosus, *B. & G.*	71	*	...	*							
156. —— ornatus, *Baird*	72	...	*								
157. —— rubriventris, *Gthr.*	72	...	...	*							
158. —— microlepidotus, *Wiegm.*	72	...	*	*							
159. —— couchii, *Baird*	73	*	*								
160. —— scalaris, *Wiegm.*	73	*	*	*	*						
161. —— æneus, *Wiegm.*	74	...	...	*							
162. —— jalapæ, *Gthr.*	74	...	...	*							
163. —— pleurolepis, *Gthr.*	74	...	...	*							
164. —— variabilis, *Wiegm.*	75	*	*	*	...	*	*				
165. —— teapensis, *Gthr.*	75	...	...	*							
166. —— fulvus, *Boc.*	76	...	...	...	...	...	*	*			
167. —— siniferus, *Cope*	76	...	...	*							
168. —— utiformis, *Cope*	76	...	...	*							

	Page.	Northern range.	N. Mexico.	S. Mexico.	Yucatan.	Guatemala & British Honduras (Belize).	Honduras & Salvador.	Nicaragua.	Costa Rica.	Panama.	Southern range.
OPHIDIA (*con.*).											
COLUBRIDÆ AGLYPHÆ (*con.*).											
16. Geophis dugesii, *Boc.*	91	...	...	*							
17. —— bicolor, *Gthr.*	91	...	...	*							
18. —— redimita, *Cope*	92	...	*	*							
19. —— omiltemana, *Gthr.*	92	...	...	*							
20. —— mutitorques, *Cope*	93	...	...	*							
21. —— longiceps, *Cope*	93	...	...	*							
22. —— latifrontalis, *Garm.*	93	...	*								
23. —— fasciata, *Fisch.*	93	...	...	...	...	*					
24. —— psephota, *Cope*	94	...	...	...	...	...	...	...	*		
25. —— zebrina, *Jan*	94	...	...	*							
26. Adelphicus quadrivirgatus, *Jan*	94	...	...	*	...	*	*				
27. Toluca lineata, *Kennic.*	95	...	...	*							
28. Pseudoficimia frontalis, *Cope*	96	...	*	*							
29. Sympholis lippiens, *Cope*	96	...	...	*							
30. Conopsis nasus, *Gthr.*	97	*	*	*							
31. Ficimia olivacea, *Gray*	98	...	...	*							
32. —— publia, *Cope*	98	...	...	*	*						
33. —— cana, *Cope*	99	*	*								
34. —— quadrangularis, *Gthr.*	99	...	*								
35. Rhinochilus lecontei, *B. & G.*	100	*	*								
36. —— antonii, *Dugès*	100	...	*								
37. Symphimus leucostomus, *Cope*	155	...	...	*							
38. Leptocalamus unicolor, *Fisch.*	100	...	...	*							
39. —— torquatus, *Gthr.*	157	...	...	...	...	...	...	*			
40. Streptophorus atratus, *Hallow.*	101	...	...	*	...	*	*	*	*	...	*
41. —— diadematus, *B. & G.*	102	...	...	*							
42. Chersodromus liebmanni, *Reinh.*	103	...	...	*							
43. —— nigricans, *Reinh.*	103	...	...	?							
44. Contia episcopa, *Kennic.*	104	*	*								
45. —— calligaster, *Cope*	104	...	...	...	...	...	...	...	*		
46. —— pachyura, *Cope*	104	...	...	...	...	...	...	...	*		
47. Ablabes gracilis, *Gthr.*	104	...	...	...	...	...	...	...	*		
48. —— decipiens, *Gthr.*	105	...	...	...	...	...	...	...	*		
49. —— sumichrasti, *Boc.*	105	...	...	*							
50. —— serperaster, *Cope*	105	...	...	...	...	...	...	...	*		
51. Elapochrus deppii, *Ptrs.*	106	...	...	*	...	*					
52. —— æqualis, *Salvin*	106	...	...	*	...	*					
53. —— dimidiatus, *Cope*	107	...	...	...	...	...	...	*	*		
54. —— euryzona, *Cope*	107	...	...	...	...	...	...	...	...	*	*
55. Henicognathus annulatus, *D. & B.*	107	...	...	*	...	*					

169. —— melanogaster, *Cope*	77	...	...	*							
170. —— ferrari-perezi, *Cope*	77	...	...	*							
171. —— melanorhinus, *Boc.*	77	...	...	*							
172. —— squamosus, *Boc.*	77	...	...	...	...	*					
173. —— serrifer, *Cope*	77	...	...	...	*						
174. Phrynosoma orbiculare, *Wiegm.*	78	...	*	*							
175. —— cornutum, *Harl.*	79	*	*								
176. —— asio, *Cope*	79	...	...	*	...	*					
177. —— modestum, *Gir.*	79	*	*								
178. —— braconnieri, *Boc.*	79	...	...	*							
179. —— taurus, *Boc.*	79	...	...	*							
GECKONIDÆ.											
180. Gonatodes albogularis, *D. & B.*	80	...	...	...	...	...	*	*	...	*	*
181. —— caudiscutatus, *Gthr.*	80	...	...	...	...	...	...	...	...	*	*
182. Phyllodactylus tuberculosus, *Wiegm.*	80	*	*	*	...	*	*	*			
183. —— ventralis, *O'Sh.*	80	...	...	...	...	...	...	*			
184. Thecadactylus rapicauda, *Houtt.*	81	...	...	*	*	*	...	*	*	...	*
185. Aristelliger præsignis, *Hallow.*	81	...	...	...	*	*	...	*			
186. Gehyra mutilata, *Wiegm.*	81	...	*								
187. Sphærodactylus homolepis, *Cope*	82	...	...	...	...	...	...	*			
188. —— torquatus, *Strauch*	82	...	*								
189. —— glaucus, *Cope*	82	...	...	*	*	*	...	...	*		
190. —— lineolatus, *Licht.*	82	...	...	*	...	*	...	*	...	*	*
191. —— casicolus, *Cope*	82	...	...	...	...	...	...	...	...	*	
192. —— fantasticus, *D. B.*	83	...	...	*	...	...	...	...	...	...	*
193. —— anthracinus, *Cope*	83	...	...	*							
194. Eublepharis variegatus, *Baird*	83	*	*								
195. —— dovii, *Boul.*	83	...	...	...	...	...	...	...	...	*	
196. —— fasciatus, *Boul.*	84	...	*								
197. Coleonyx elegans, *Gray*	84	...	...	*	*	*	...	...	*		
OPHIDIA.											
STENOSTOMATIDÆ.											
1. Glauconia albifrons, *Wagl.*	85	...	..	*	...	*	*	...	*	*	*
2. —— myopica, *Garm.*	85	...	*								
3. —— dulcis, *B. & G.*	85	*	*								
4. —— humilis, *B. & G.*	86	*	*	*							
TYPHLOPIDÆ.											
5. Typhlops tenuis, *Salv.*	86	...	...	*	...	*					
6. —— microstomus, *Cope*	86	...	...	...	*						
7. Helminthophis frontalis, *Ptrs.*	86	...	...	...	...	...	...	...	*		
8. Liotyphlops albirostris, *Ptrs.*	87	...	...	...	...	...	...	...	...	*	
9. Anomalolepis mexicanus, *Jan*	87	...	...	?*							
COLUBRIDÆ AGLYPHÆ.											
10. Geophis dolichocephala, *Cope*	87	...	...	...	...	...	...	...	*		
11. —— chalybæa, *Wagl.*	87	...	...	*	*	*	...	...	*	*	
12. —— rostralis, *Jan*	89	...	...	*							
13. —— semidoliata, *D. & B.*	90	...	...	*							
14. —— mœsta, *Gthr.*	90	...	...	*	...	...	...	...	*		
15. —— brachycephala, *Cope*	91	...	...	...	...	...	...	...	*		

56. —— venustissimus, *Gthr.*	144	...	...	...	...	...	...	*			
57. Liophis epinephelus, *Cope*	107	...	...	...	...	...	...	...	*	*	*
58. —— regalis, *B. & G.*	108	...	*								
59. —— lætus, *Jan*	108	...	...	*							
60. Coronella annulata, *Kennic.*	109	...	*	*	*	*	...	...	*	*	
61. —— mexicana, *Garm.*	110	...	*								
62. —— leonis, *Gthr.*	110	...	*								
63. —— godmani, *Gthr.*	110	...	...	...	...	*					
64. —— decorata, *Gthr.*	111	...	...	*	...	*	...	*	*	*	
65. —— ignita, *Cope*	111	...	...	...	...	...	...	...	...	*	
66. —— vermiculaticeps, *Cope*	111	...	...	...	...	...	...	...	*	*	
67. —— quinquelineata, *Cope*	111	...	...	*							
68. —— fulviceps, *Cope*	111	...	...	...	...	...	...	...	...	*	
69. Dromicus fugitivus, *Donnd.*	112	...	...	...	...	?					
70. —— laureatus, *Gthr.*	112	...	...	*							
71. —— clavatus, *Ptrs.*	113	...	...	*							
72. —— omiltemanus, *Gthr.*	113	...	...	*							
73. —— tæniatus, *Ptrs.*	113	...	...	*							
74. —— fulvivittis, *Cope*	113	...	...	*							
75. —— lachrymans, *Cope*	114	...	...	?							
76. Xenodon rhabdocephalus, *Wied*	114	...	...	*	...	*	...	...	...	...	*
77. —— angustirostris, *Ptrs.*	114	...	...	...	...	*	...	...	*	*	
78. Coluber guttatus, *L.*	114	*	*								
79. —— triaspis, *Cope*	115	...	...	*	*	*	...	...	*		
80. —— flavirufus, *Cope*	115	...	...	*	*	*					
81. —— chlorosoma, *Gthr.*	115	...	...	*							
82. Spilotes corais, *Cuv.*	116	*	*	*	*	*	...	...	*	*	
83. —— salvini, *Gthr.*	116	...	...	*	*	*	...	*			
84. —— pœcilonotus, *Gthr.*	117	...	...	*	...	*	*				
85. —— argus, *Boc.*	118	...	...	*							
86. —— chrysobronchus, *Cope*	119	...	...	...	...	...	...	...	*		
87. Zamenis constrictor, *L.*	119	*	*								
88. —— oaxacæ, *Jan*	120	...	...	*							
89. —— rhombifer, *Gthr.*	120	...	...	...	...	...	...	...	...	*	*
90. —— flavigularis, *Hallow.*	120	...	*	*	...	*					
91. —— flagelliformis, *Laur.*	121	*	?								
92. —— semilineatus, *Cope*	121	*	*								
93. —— ornatus, *B. & G.*	122	*	*								
94. —— melanolomus, *Cope*	122	...	...	...	*						
95. —— pulcherrimus, *Cope*	123	...	...	*	...	...	...	*			
96. Salvadora grahamiæ, *B. & G.*	123	*	*	*							
97. —— mexicana, *D. & B.*	123	...	...	*							
98. Pituophis deppii, *D. & B.*	124	...	...	*							
99. —— lineaticollis, *Cope*	124	...	...	*	...	*					
100. —— elegans, *Kennic.*	125	*	*								
101. —— melanoleuca, *Daud.*	125	*	*								
102. Drymobius margaritiferus, *Schleg.*	125	...	*	*	...	*	...	...	*	*	
103. —— boddærtii, *Seetzen*	125	...	*	*	...	*	...	*	*	*	*
104. —— cæruleus, *Fisch.*	126	...	...	...	...	*					
105. —— dendrophis, *Schleg.*	127	...	...	*	...	*	...	*	*	...	*
106. Herpetodryas fuscus, *L.*	128	...	...	...	...	...	...	...	...	*	*
107. —— carinatus, *L.*	128	...	...	...	...	...	...	...	*	*	*
108. —— grandisquamis, *Ptrs.*	128	...	...	...	...	...	...	...	*		
109. —— melas, *Cope*	128	...	...	...	...	...	...	*			
110. Leptophis mexicanus, *D. & B.*	129	...	*	*	*	*	...	*	*		
111. —— occidentalis, *Gthr.*	129	...	...	...	...	...	...	...	...	*	*
112. —— modestus, *Gthr.*	129	...	...	...	...	*					
113. —— depressirostris, *Cope*	130	...	...	...	...	...	...	...	...	*	
114. —— saturatus, *Cope*	130	...	...	...	...	...	...	*	*		

Geographical Distribution (*continued*).

	Page.	Northern range.	N. Mexico.	S. Mexico.	Yucatan.	Guatemala & British Honduras (Belize).	Honduras & Salvador.	Nicaragua.	Costa Rica.	Panama.	Southern range.
OPHIDIA (*con.*).											
COLUBRIDÆ AGLYPHÆ (*con.*).											
115. Leptophis diplotropis, *Gthr.*	130	...	*	*							
116. —— præstans, *Cope*	130	...	...	*	...	*	...	...	*		
117. —— bilineatus, *Gthr.*	130	...	...	*	...	...	...	*	*		
118. —— æruginosus, *Cope*	130	...	...	...	...	...	...	...	*		
119. Synchalinus corallioides, *Cope*	131	...	...	...	...	...	...	...	*		
120. Tropidonotus ordinatus, *L.*	131	*	*	*	*	*					
121. —— saurita, *L.*	132	*	*	*	*	*					
122. —— chrysocephalus, *Cope*	132	...	...	*							
123. —— præocularis, *Boc.*	132	...	...	...	...	*					
124. —— multimaculatus, *Cope*	132	...	*								
125. —— scaliger, *Jan*	133	...	...	*							
126. —— godmani, *Gthr.*	133	...	...	*							
127. —— variabilis, *Dugès*	133	...	...	*							
128. —— melanogaster, *Ptrs.*	133	...	...	*							
129. —— validus, *Kennic.*	134	*	*								
130. —— grahami, *B. & G.*	134	*	?								
131. —— sipedon, *L.*	134	*	*	*	...	...	...	...	*		
132. Tretanorhinus nigroluteus, *Cope*	134	...	...	...	...	...	...	*	...	*	
133. —— dimidiatus, *Boie*	135	...	...	*							
134. —— mocquardi, *Boc.*	135	...	...	?							
135. Helicops septemvittatus, *Fisch.*	135	...	...	*							
136. Ischnognathus dekayi, *Holbr.*	136	*	...	*	...	*					
137. —— occipito-maculatus, *Stor.*	136	*	...	*							
138. —— storerioides, *Cope*	136	...	*	*							
139. —— copii, *Dugès*	136	...	...	*							
140. Hydromorphus concolor, *Ptrs.*	137	...	...	...	...	...	...	...	*		
141. Hypsiglena torquata, *Gthr.*	137	*	*	*	...	...	...	*			
142. —— discolor, *Gthr.*	137	...	...	*							
143. —— latifasciata, *Gthr.*	138	...	...	*							
144. —— ornata, *Boc.*	139	...	...	...	...	...	...	...	...	*	
145. Petalognathus nebulatus, *L.*	139	...	...	*	...	*	...	*	*	...	*
146. Tropidodipsas fasciata, *Gthr.*	139	...	...	*							
147. —— dumerilii, *Jan*	140	...	..	*	*	*					
148. —— philippii, *Jan*	140	...	*								
149. —— anthracops, *Cope*	140	...	...	...	...	...	...	*			
150. Neopareas bicolor, *Gthr.*	178	...	...	...	...	...	...	*			
151. Leptognathus mikanii, *Schleg.*	141	...	...	*	...	...	...	...	...	...	*
152. —— annulatus, *Gthr.*	141	...	...	...	...	...	...	...	*		
153. —— argus, *Cope*	142	...	...	...	...	...	...	...	*		
154. —— pictiventris, *Cope*	142	...	...	...	...	...	...	...	*		
155. —— articulatus, *Cope*	142	...	...	...	...	...	...	...	*	*	
156. —— viguieri, *Boc.*	142	...	...	...	...	...	...	...	..	*	

	Page.	Northern range.	N. Mexico.	S. Mexico.	Yucatan.	Guatemala & British Honduras (Belize).	Honduras & Salvador.	Nicaragua.	Costa Rica.	Panama.	Southern range.
OPHIDIA (*con.*).											
COLUBRIDÆ GLYPHODONTES (*con.*).											
214. Leptodira yucatanensis, *Cope*	171	...	...	...	*						
215. —— splendida, *Gthr.*	171	...	...	*							
216. —— personata, *Cope*	171	...	*	*							
217. —— polysticta, *Gthr.*	172	...	...	*	*	*	*	...	...	*	
218. —— ocellata, *Gthr.*	172	...	...	...	..	...	...	*	*		
219. —— frenata, *Cope*	173	...	...	*							
220. —— rhombifera, *Gthr.*	173	...	...	...	...	*					
221. Trimorphodon biscutatus, *D. & B.*	174	...	*	*	...	*	...	*	...	*	
222. —— upsilon, *Cope*	175	...	*	*							
223. Dipsas cenchoa, *L.*	175	...	...	*	...	*	...	...	*	*	*
224. —— gemmistrata, *Cope*	175	...	...	*	...	*	*	...	*		
225. —— splendida, *Gthr.*	176	...	...	...	*						
226. —— tenuissima, *Cope*	176	...	...	*	*						
227. —— gracillima, *Gthr.*	177	...	...	*							
228. Dryiophis acuminata, *Wied*	177	...	*	*	*	*	*	*	*	*	*
229. —— fulgida, *Daud.*	178	...	...	*	*	*	*	...	...	*	*
230. —— brevirostris, *Cope*	178	...	...	...	...	...	...	*	*	*	
NOTHOPSIDÆ.											
231. Nothopsis rugosus, *Cope*	179	...	...	...	...	...	...	...	...	*	
BOIDÆ.											
232. Loxocemus bicolor, *Cope*	179	...	...	*	...	*					
233. Epicrates cenchris, *L.*	180	...	...	...	...	...	...	...	*	*	*
234. Corallus cooki, *Gray*	180	...	...	..	...	...	...	...	...	*	*
235. —— annulatus, *Cope*	181	...	...	...	...	...	...	...	*		
236. Ungalia moreletii, *Boc.*	181	...	...	..	...	*					
237. Ungaliophis continentalis, *Müll.*	181	...	...	...	...	*					
238. Boa imperator, *Daud.*	181	...	*	*	*	*	...	*	*	*	*
239. —— mexicana, *Rapp*	182	...	...	*							
ELAPIDÆ.											
240. Elaps fulvius, *L.*	182	*	*	*	*	*	*	*	*	*	*
241. —— bernardi, *Cope*	184	...	...	*							
242. —— aglæope, *Cope*	184	...	...	?	...	*	*				
243. —— ephippifer, *Cope*	184	...	...	*							
244. —— multifasciatus, *Jan*	184	...	...	...	...	...	...	*	*	*	*
245. —— ruatanus, *Gthr.*	185	...	...	...	...	...	*				
246. —— lemniscatus, *L.*	185	...	...	*	...	...	...	...	...	...	*
247. —— decoratus, *Jan*	185	...	...	?	...	...	...	...	...	...	*
248. —— elegans, *Jan*	186	...	...	*	...	*					

157. —— brevis, *D. & B.*	142	...	...	*							
158. —— brevifacies, *Cope*	142	...	...	...	*						
159. —— semicinctus, *Boc.*	142	...	...	*	...	*					
160. Mesopeltis sanniolus, *Cope*	143	...	...	...	*						
161. —— dimidiatus, *Gthr.*	143	...	...	*	...	*	...	*			
Colubridæ glyphodontes.											
162. Homalocranium miniatum, *Cope*	146	...	...	*							
163. —— atriceps, *Gthr.*	146	...	*								
164. —— planiceps, *Blainv.*	145	*	?								
165. —— gracile, *B. & G.*	146	*	?								
166. —— melanocephalum, *L.*	147	...	...	*	...	*	...	...	*	...	*
167. —— jani, *Gthr.*	148	...	...	...	...	*	...	*			
168. —— boulengeri, *Gthr.*	148	...	...	*							
169. —— bocourti, *Gthr.*	149	...	...	*							
170. —— armillatum, *Cope*	149	...	...	...	...	*	...	...	*		
171. —— breve, *Gthr.*	150	...	...	...	...	*					
172. —— michoacanense, *Cope*	150	...	...	*							
173. —— annulatum, *Böttg.*	150	...	...	...	...	...	..	*			
174. —— deppii, *Boc.*	151	...	...	*							
175. —— tæniatum, *Boc.*	151	...	...	...	...	*	*				
176. —— mœstum, *Gthr.*	152	...	...	...	*	*					
177. —— schistosum, *Boc.*	152	...	...	*	...	*					
178. —— reticulatum, *Cope*	152	...	...	...	...	...	...	...	...	*	
179. —— canulum, *Cope*	153	...	...	...	*						
180. —— mexicanum, *Gthr.*	153	...	...	*							
181. —— virgatum, *Gthr.*	154	...	...	...	...	...	...	...	*		
182. —— bimaculatum, *Cope*	154	...	*	*							
183. —— calamarinum, *Cope*	155	...	*	*							
184. —— rubrum, *Cope*	155	...	...	*							
185. Lioninia vermiformis, *Hall*	155	...	...	...	..	...	...	*			
186. Scolecophis atrocinctus, *Schleg.*	156	...	...	*	...	*	*				
187. Procinura æmula, *Cope*	156	...	*								
188. Pogonaspis ruficeps, *Cope*	157	...	...	...	...	...	...	...	*		
189. Enulius murinus, *Cope*	157	...	...	...	...	...	...	*			
190. Cercocalamus collaris, *Gthr.*	157	...	...	?							
191. Stenorhina degenhardti, *Berth.*	158	...	...	*	*	*	*	*	*	*	*
192. Tachymenis piceivittis, *Cope*	160	...	...	*	*						
193. —— bipunctata, *Gthr.*	161	...	...	*	...	*					
194. —— fissidens, *Gthr.*	161	*	*	*	...	*	...	*	*		
195. —— lateritia, *Cope*	162	...	...	*							
196. —— grammophrys, *Dugès*	162	...	...	*							
197. —— decipiens, *Gthr.*	163	...	...	...	...	...	...	...	*		
198. Hydrocalamus quinquevittatus, *D. & B.*	164	...	...	*	...	*					
199. Tomodon nasutus, *Cope*	164	...	..	*							
200. Conophis lineatus, *Gthr.*	165	...	...	...	*	*	...	*	*		
201. —— vittatus, *Ptrs.*	165	...	...	*							
202. Philodryas putnami, *Jan*	166	...	...	*							
203. Erythrolamprus æsculapii, *L.*	166	...	...	?	...	...	...	*	*	*	*
204. Oxyrrhopus plumbeus, *Wied.*	167	...	...	*	...	*	...	...	*	...	*
205. —— cloelia, *Daud.*	167	...	...	*	*	...	...	...	...	*	*
206. —— occipito-luteus, *Cope*	167	...	...	...	...	...	...	...	...	*	
207. —— petolarius, *L.*	167	...	...	...	...	*	...	*	*	...	*
208. Scytale coronatum, *Schn.*	168	...	...	...	...	...	...	...	...	*	*
209. Leptodira nigrofasciata, *Gthr.*	169	...	...	...	...	...	...	*			
210. —— pacifica, *Cope*	169	...	*								
211. —— mystacina, *Cope*	169	...	...	*							
212. —— annulata, *L.*	170	*	*								
213. —— affinis, *Gthr.*	170	...	...	?							

Hydrophiidæ.											
249. Pelamis bicolor, *Schn.*	186	...	...	*	...	...	...	...	*	*	
Crotalidæ.											
250. Ancistrodon bilineatus, *Gthr.*	186	...	*	*	*	*	...	*			
251. Bothrops undulatus, *Jan*	187	...	...	*							
252. —— atrox, *L.*	187	...	...	*	...	*	...	*	*	*	*
253. Lachesis mutus, *L.*	188	...	...	...	...	...	...	...	*	*	*
254. Bothriechis schlegelii, *Berth.*	189	...	...	...	...	*	...	*	*	*	*
255. —— nigroviridis, *Ptrs.*	189	...	...	...	...	...	...	...	*		
256. —— lateralis, *Ptrs.*	189	...	...	...	...	...	...	...	*		
257. —— bicolor, *Boc.*	189	...	...	...	...	*					
258. —— aurifera, *Salvin*	189	...	...	..	...	*					
259. —— lansbergii, *Schleg.*	190	...	...	*	*	*	...	*	*	*	*
260. —— godmani, *Gthr.*	190	...	...	...	...	*					
261. —— ophryomegas, *Boc.*	191	...	...	...	...	*					
262. —— nummifera, *Rüpp.*	191	...	...	*	...	*	..	*	*		
263. —— scutigera, *Fisch.*	191	...	...	...	...	*					
264. Crotalophorus miliarius, *L.*	192	...	...	*							
265. Crotalus triseriatus, *Wagl.*	192	...	...	*							
266. —— polystictus, *Cope*	192	...	...	*							
267. —— intermedius, *Fisch.*	192	...	...	*							
268. —— omiltemanus, *Gthr.*	192	...	...	*							
269. —— palmeri, *Garman*	193	*	*								
270. —— pallidus, *Gthr.*	193	...	...	*							
271. —— salvini, *Gthr.*	193	...	*								
272. —— horridus, *Daud.*	194	...	...	*	...	*	...	...	...	...	*
273. —— basiliscus, *Cope*	194	...	*	*							
274. —— adamanteus, *Beauv.*	195	*	*								
BATRACHIA.											
Batrachia Salientia.											
Ranidæ.											
1. Rana montezumæ, *Baird*	197	...	...	*	...	*					
2. —— halecina, *Holbr.*	198	*	*	*	...	*	...	...	*		
3. —— macroglossa, *Brocchi*	199	...	...	...	...	*					
4. —— forreri, *Boul.*	199	...	*								
5. —— omiltemana, *Gthr.*	200	...	...	*							
6. —— bonaccana, *Gthr.*	201	...	...	...	*						
7. —— maculata, *Br.*	201	...	...	...	...	*					
8. —— pustulosa, *Boul.*	202	...	*								
9. —— palmipes, *Spix*	202	...	...	*	...	*	...	...	...	...	*
10. —— melanosoma, *Gthr.*	203	...	...	...	...	*	...	*			
11. —— godmani, *Gthr.*	204	...	...	...	...	...	...	...	*		
12. —— cæruleopunctata, *Steind.*	205	...	...	...	...	...	...	*	*		
13. Lævirana vibicaria, *Cope*	206	...	...	...	...	...	...	...	*		
Hylaplesiidæ.											
14. Hylaplesia tinctoria, *Schn.*	207	...	...	...	...	...	...	*	*	*	*
15. —— trivittata, *Spix*	207	...	...	...	...	...	...	...	...	*	*
16. —— typographa, *Keferst.*	207	...	...	...	...	...	...	*	*		
17. —— talamancæ, *Cope*	208	...	...	...	...	...	...	...	*		
18. —— truncata, *Cope*	208	...	...	...	...	...	...	...	...	?	

Geographical Distribution (*continued*).

	Page.	Northern range.	N. Mexico.	S. Mexico.	Yucatan.	Guatemala & British Honduras (Belize).	Honduras & Salvador.	Nicaragua.	Costa Rica.	Panama.	Southern range.
BATRACHIA (*con.*).											
Batrachia Salientia (*con.*).											
Brachycephalidæ.											
19. Atelopus varius, *Stann.*	208	...	...	...	...	*	...	...	*	*	
Engystomatidæ.											
20. Engystoma ustum, *Cope*	209	...	*	*							
21. —— pictiventre, *Cope*	209	...	...	...	...	...	...	*			
22. —— elegans, *Boul.*	210	...	...	*							
23. Glossostoma aterrimum, *Gthr.*	210	...	...	...	...	...	...	...	*		
24. Hypopachus variolosus, *Cope*	211	*	*	*	...	...	...	...	*		
25. —— inguinalis, *Cope*	211	...	...	...	...	*	...	...	...	...	*
Cystignathidæ.											
26. Leptodactylus pentadactylus, *Laur.*	212	...	...	...	...	...	...	...	*	...	*
27. —— albilabris, *Gthr.*	213	...	...	*							
28. —— fragilis, *Br.*	213	...	...	*							
29. —— microtis, *Cope*	213	...	...	*							
30. —— perlævis, *Cope*	214	...	..	*							
31. —— caliginosus, *Gir.*	214	...	*	*	...	...	...	*	*	...	*
32. Borborocœtes mexicanus, *Boul.*	215	..	*								
33. Syrrhaphus verruculatus, *Ptrs.*	215	...	...	*							
34. —— omiltemanus, *Gthr.*	216	...	...	*							
35. —— ridens, *Cope*	216	...	...	...	...	...	...	*	...	*	
36. —— verrucipes, *Cope*	216	...	...	*							
37. —— leprus, *Cope*	217	...	...	*							
38. —— guttulatus, *Cope*	217	...	...	*							
39. —— cystignathoides, *Cope*	218	...	...	*							
40. —— hylæformis, *Cope*	218	...	...	...	...	...	...	...	*		
41. Paludicola nitida, *Ptrs.*	218	...	...	*							
42. —— mexicana, *Br.*	219	...	...	*							
43. Tomodactylus amulæ, *Gthr.*	219	...	...	*							
44. Liohyla guentheri, *Keferst.*	220	...	...	...	...	...	...	...	*		
45. —— rugulosa, *Cope*	221	...	...	*	...	*	...	*			
46. —— bocourti, *Br.*	222	...	...	...	...	*	...	...	*		
47. —— fleischmanni, *Böttg.*	222	...	...	...	...	...	...	...	*		
48. —— pittieri, *Gthr.*	223	...	...	...	...	...	...	...	*		
49. —— engytympanum, *Gthr.*	224	...	...	...	...	...	...	...	*		
50. —— ranoides, *Cope*	225	...	...	...	...	...	...	*	*		
51. —— berkenbuschi, *Ptrs.*	225	...	...	*							
52. Hylodes sallæi, *Gthr.*	227	...	...	*	...	*					
53. —— plicatus, *Gthr.*	228	...	...	*							

	Page.	Northern range.	N. Mexico.	S. Mexico.	Yucatan.	Guatemala & British Honduras (Belize).	Honduras & Salvador.	Nicaragua.	Costa Rica.	Panama.	Southern range.
BATRACHIA (*con.*).											
Batrachia Salientia (*con.*).											
Rhinophrynidæ.											
108. Rhinophrynus dorsalis, *D. & B.*	258	...	...	*	*	*					
Scaphiopodidæ.											
109. Scaphiopus couchii, *Baird*	259	*	*								
110. —— rectifrenis, *Cope*	259	...	*								
111. —— hammondii, *Baird*	259	*	*	*							
112. —— multiplicatus, *Cope*	260	...	...	*							
Hylidæ.											
113. Hyla eximia, *Baird*	261	...	*	*							
114. —— bocourti, *Mocq.*	263	...	...	*	...	*					
115. —— nana, *Gthr.*	263	...	...	*							
116. —— cherrii, *Cope*	264	...	...	.	...	...	...	...	*		
117. —— microcephala, *Cope*	265	...	...	...	...	...	...	...	...	*	
118. —— elæochroa, *Cope*	265	...	...	...	...	...	...	...	*	*	
119. —— bistincta, *Cope*	265	...	...	*							
120. —— plicata, *Br.*	261	...	...	?							
121. —— splendens, *Schmidt*	266	...	...	...	...	...	...	...	...	*	
122. —— copii, *Boul.*	266	*	*	*							
123. —— boulengeri, *Cope*	267	...	...	...	...	...	...	*			
124. —— mocquardi, *Gthr.*	268	...	...	...	...	?	...	...	...	?	
125. —— quinquevittata, *Cope*	268	...	...	...	...	...	...	*			
126. —— phæota, *Cope*	269	...	...	...	...	...	...	...	...	*	*
127. —— tæniopus, *Gthr.*	269	...	...	*							
128. —— puma, *Cope*	270	...	...	...	...	...	...	*			
129. —— baudini, *D. & B.*	270	*	*	*	*	*	...	*	*	...	*
130. —— venulosa, *Laur.*	272	...	*	*	...	*	...	*	*	*	*
131. —— sordida, *Ptrs.*	273	...	...	...	...	...	...	...	...	*	
132. —— gabbi, *Cope*	274	...	...	...	...	...	...	...	*		
133. —— pseudopuma, *Gthr.*	274	...	...	...	...	...	...	...	*		
134. —— godmani, *Gthr.*	275	...	...	*	...	...	...	...	?		
135. —— uranochroa, *Cope*	276	...	...	...	...	...	...	...	*		
136. —— leucophyllata, *Beiris*	277	...	...	...	...	...	...	*	*	*	*
137. —— underwoodi, *Boul.*	278	...	...	...	...	...	...	...	*		
138. —— nigripes, *Cope*	278	...	...	...	...	...	...	...	*		
139. —— molitor, *Schmidt*	279	...	...	...	...	...	...	...	...	*	
140. —— miotympanum, *Cope*	279	...	...	*							
141. —— prosoblepon, *Böttg.*	280	...	...	...	...	...	...	...	*	*	
142. —— pulverata, *Ptrs.*	280	...	...	...	...	...	...	...	...	*	

54. —— polyptychus, *Cope*	228							*	*		
55. —— underwoodi, *Boul.*	229								*		
56. —— calcitrans, *Gthr.*	230			*							
57. —— platyrhynchus, *Gthr.*	230								*		
58. —— cerasinus, *Cope*	231								*		
59. —— biporcatus, *Ptrs.*	231									*	*
60. —— euryglossus, *Cope*	232								*		
61. —— rhodopis, *Cope*	232			?					*		
62. —— rugosus, *Ptrs.*	233							*		*	
63. —— venustus, *Gthr.*	234			*							
64. —— florulentus, *Cope*	234								*		
65. —— augusti, *Br.*	235		*	*							
66. —— laticeps, *Dum.*	235			*	*						
67. —— melanostictus, *Cope*	236			*					*		
68. —— brocchii, *Boul.*	236					*			*		
69. —— nubilus, *Gthr.*	237								*		
70. —— alfredi, *Boul.*	237			*							
71. —— bransfordi, *Cope*	238							*			
72. —— diastema, *Cope*	238							*		*	
73. —— lanciformis, *Cope*	239							?			
74. —— megacephalus, *Cope*	239								*		
75. —— gulosus, *Cope*	240								*		
76. —— griseus, *Hallow.*	240							?			*
77. —— longipes, *Baird*	240			*							
78. —— lineatus, *Br.*	241			*							
79. —— latrans, *Cope*	241	*	?								
80. Calyptocephalus testudiniceps, *Cope*	241									*	
Bufonidæ.											
81. Engystomops pustulosus, *Cope*	242			*						*	*
82. Bufo hæmatiticus, *Cope*	243							*	*	*	*
83. —— punctatus, *B. & G.*	244	*	*	*							
84. —— mexicanus, *Br.*	244		*								
85. —— monksiæ, *Cope*	245			*							
86. —— compactilis, *Wiegm.*	245	*		*							
87. —— politus, *Cope*	246							*			
88. —— marmoreus, *Wiegm.*	246			*							
89. —— sternosignatus, *Gthr.*	247			*					*		*
90. —— canaliferus, *Cope*	248			*							
91. —— debilis, *Gir.*	248	*	*								
92. —— marinus, *L.*	249		*	*	*	*	*	*	*	*	*
93. —— occipitalis, *Camer.*	250			*							
94. —— coccifer, *Cope*	250			*		*			*		
95. —— coniferus, *Cope*	251								*	*	*
96. —— valliceps, *Wiegm.*	252	*	*	*		*		*	*		
97. —— melanochlorus, *Cope*	252								*		
98. —— luetkenii, *Boul.*	253								*		
99. —— typhonius, *L.*	253									*	*
100. —— simus, *Schmidt*	254		*	*						*	
101. —— intermedius, *Gthr.*	255			*							*
102. —— bocourti, *Br.*	255					*					
103. —— veraguensis, *Schmidt*	256									*	
104. —— auritus, *Cope*	256								*		
105. Nannophryne fastidiosa, *Cope*	257								*		
106. —— epiotica, *Cope*	257								*		
107. —— cœrulescens, *Cope*	257								*		

143. —— glandulosa, *Boul.*	281					*					
144. —— guatemalensis, *Br.*	281					*					
145. —— crassa, *Br.*	281			*							
146. —— punctariola, *Ptrs.*	282								?	*	
147. —— spilomma, *Cope*	282			*							
148. —— miliaria, *Cope*	283							*			
149. —— crepitans, *Wied*	283									*	*
150. —— albomarginata, *Spix*	284							*	*		*
151. —— rudis, *Mocq.*	285			*							
152. Hylella sumichrasti, *Br.*	286			*							
153. —— picta, *Gthr.*	286			*							
154. —— fleischmanni, *Böttg.*	287								*		
155. Nototrema oviferum, *Weinl.*	288			*					*		
156. Agalychnis moreletii, *Dum.*	289					*		*	*		
157. —— callidryas, *Cope*	290			*						*	
158. —— helenæ, *Cope*	290			*				*	*	*	
159. —— dacnicolor, *Cope*	291		*	*							
160. —— lemur, *Boul.*	291								*		
161. Pternohyla fodiens, *Boul.*	292		*								
162. Triprion petasatus, *Cope*	293				*						
163. —— spatulatus, *Gthr.*	293		*								
Batrachia Caudata.											
Salamandridæ.											
1. Molge meridionalis, *Boul.*	294	*	*								
2. Amblystoma tigrinum, *Green*	295	*	*	*							
3. —— altimirani, *Dugès*	297			*							
4. Spelerpes chiropterus, *Cope*	298			*							
5. —— gibbicaudus, *Blatch.*	298			*							
6. —— cephalicus, *Cope*	299		*								
7. —— bellii, *Gray*	299			*							
8. —— leprosus, *Cope*	299			*							
9. —— laticeps, *Br.*	300			*							
10. —— rostratus, *Br.*	300					*					
11. —— sulcatus, *Br.*	300			*							
12. —— orizabensis, *Blatch.*	300			*							
13. —— subpalmatus, *Boul.*	300								*		
14. —— morio, *Cope*	301			*		*			*		
15. —— robustus, *Cope*	301								*		
16. —— variegatus, *Gray*	302			*		*			*		
17. —— lignicolor, *Ptrs.*	302									*	
18. —— rufescens, *Cope*	303			*							
19. —— yucatanus, *Ptrs.*	303				*						
20. —— lineolus, *Cope*	304			*							
21. —— uniformis, *Keferst.*	304			*					*		
22. Thorius pennatulus, *Cope*	304			*							
Batrachia Apoda.											
1. Dermophis mexicanus, *D. & B.*	305			*		*				*	
2. —— gracilior, *Gthr.*	306									*	
3. Cœcilia isthmica, *Cope*	306									*	*
4. Herpele ochrocephala, *Cope*	307									*	*
5. Gymnopis multiplicata, *Ptrs.*	308								*	*	*
6. —— proxima, *Cope*	308								*		
7. —— oligozona, *Cope*	309					*					

LIST OF PLATES.

ERRATA ET CORRIGENDA.

Page	Line	
25	3	*for* Guyaquil *read* Guayaquil.
26	15	*for* **tessellata** *read* **tessellatus.**
39	34	for *C.* read *G.*
81	7	for *Platydactytus* read *Platydactylus.*
86	2	for *Rana* read *Rena.*
87	30	*for* Tab. XXXII. *read* Tab. XXXIII.
164	26	for *Monolepis* read *Manolepis.*
191	16	for *mummifer* read *nummifer.*
194	36	for *Caudisoma* read *Caudisona.*
227	12	for *petriculus* read *pelviculus.*
242	17	*for* VENEZUETA *read* VENEZUELA.
267	22	for *Scytopsis* read *Scytopis.*

BIOLOGIA CENTRALI-AMERICANA.

ZOOLOGIA.

Class REPTILIA.

Order CHELONIA.

Fam. TESTUDINIDÆ.

CISTUDO.

Cistudo, Gray, Cat. Shield Rept. p. 39.

1. Cistudo mexicana.

Cistudo mexicana, Gray, P. Z. S. 1849, p. 16, t. ii.; Bocourt, Miss. Sc. Mex., Rept. p. 17.

Hab. MEXICO, Tampico and Mexico city (*Bocourt*).

Gray was the first who specifically separated from the common type of North-American Box-Tortoise a form whose hind feet are armed with three large claws instead of four. However, in examining the value of this character, Agassiz (Contr. Nat. Hist. Amer. i. p. 444) came to a different conclusion, stating that the outer toe of the hind foot fades away gradually; and, whilst he distinguished not less than four different species of *Cistudo*, these species included three- as well as four-toed specimens.

In a case in which one naturalist, though he has examined many hundred specimens, does not feel justified in expressing a decided opinion, it would be most hazardous for another to do so who has much less material at hand. Yet, having seen specimens of Agassiz's four species, I may be allowed to say that I incline to the belief that there is, in fact, one species of Box-Tortoise only, and that one of its hind toes becomes aborted in the south-western and southern range of its distribution. Nevertheless, I have considered it best to retain the name given by Gray, the more so as both the typical specimens are singularly distinguished by having a small additional scute intercalated between the fourth and fifth, so that there are six vertebrals altogether. This, of course, may be an abnormal condition peculiar to the locality where the specimens were obtained, and which unfortunately is not known. They are simply marked "Mexico."

The species seems to penetrate rather far southwards into Mexico, as Bocourt mentions it from Tampico and the city of Mexico.

EMYS.

Emys, Duméril, Elém. d'Hist. Nat.

α. Clemmys.

In Mexico two freshwater Turtles occur (*Emys ornata* and *E. cataspila*) in considerable abundance, which are distinguished by a short head, a short and obtuse snout, and a more or less depressed shell, which is only slightly and indistinctly rugose, and in which the costal scutes are ornamented each by a large ocellated spot more or less distinctly surrounded by yellow and darker concentric rings. The sternum is marked by symmetrical figures formed by black lines arranged in pairs. These figures never form rings on the front part of the sternum, as is the case in an allied species from the United States, to which the names of *Emys holbrookii* and *E. elegans* have been given. On the anal scutes this sternal ornamentation terminates with a more or less straight double transverse line. The markings on the sternum are most distinct in the young, become pale with age, and may disappear altogether.

In the adult stage these two Turtles differ merely by the position of the ocellus on the two middle costal scutes; if each of the scutes be imagined to be divided into four quadrangles, in one species the spot occupies the postero-inferior quadrangle, in the other the postero-superior. Gray (Cat. Shield Rept. p. 25) has already noticed this distinction; but unfortunately he referred the young of one of the species to the adult of the other, so that his *Emys venusta* is synonymous with his *E. ornata.*

Some herpetologists may be inclined to consider both forms as variations; but, until the true relations between the two forms are clearly ascertained, it will be better to distinguish them by separate names. Most probably *Emys ornata* will prove to be the more southern form.

1. **Emys ornata.** (Tab. I.)

Emys ornata, Gray, Syn. Rept. p. 30 (1831); ? Dum. & Bibr. Erp. Gen. ii. p. 286 (1835); Gray, Zool. Beech. Voy. p. 93, t. 29. fig. 2 (1839); Dum. & Boc. Miss. Sc. Mex. p. 13, t. 3. fig. 1 (1870); Sowerby and Lear, Tort. t. 44.

Emys ornata (part.), Gray, Cat. Tort. B. M. 1844, p. 22; Gray, Cat. Shield Rept. 1855, p. 24 (not fig.).

Emys venusta, Gray, Cat. Shield Rept. p. 24, t. 12 A.

? *Emys valida,* Leconte, Proc. Ac. N. Sc. Philad. 1859, p. 7.

Callichelys ornata (part.), Gray, Ann. & Mag. Nat. Hist. 1863, xii. p. 176; Suppl. Cat. Shield Rept. 1870, p. 48.

Callichelys venusta (part.), Gray, Ann. & Mag. Nat. Hist. 1863, xii. p. 181; Suppl. Cat. Shield Rept. p. 49.

Pseudemys ornata, Cope, Journ. Ac. N. Sc. Philad. viii. p. 153.

Hab. MEXICO, Mazatlan (*coll. Brit. Mus., Colley, Forrer*), Presidio (*Forrer*); HONDURAS (*Dyson*); COSTA RICA (*Cope*); PANAMA, Chiriqui (*Druce*).

Of this species we have a tolerably complete series, and, fortunately, from the same locality, namely Mazatlan, so that there is not the least doubt about the pertinence of old and young specimens to the same species. Two of the Mazatlan specimens are the original types of Gray's first description.

In the young (2–3 inches long) each costal scute is ornamented by yellow, green, and blackish concentric rings, which form a large ocellus occupying the greater part of the surface of the scute; a minute yellow spot occupies the centre of the ocellus. The sternum is ornamented by more or less symmetrical black lines placed in pairs, and forming along the middle of the sternum a rather irregular pattern without complete rings in front, and with a more or less straight double transverse line behind.

The longitudinal stripes and bands of the head form an extremely pretty pattern, and are much multiplied by each yellow or green streak having blackish borders; the band along the temple is broad in the middle, becoming narrower in front and behind, and is of a bright orange colour. The ornamental band on the chin is generally interrupted on the throat by a transverse blackish border.

As the specimens grow older, the principal yellow ring of the costal plates appears smaller in proportion to the size of the scute, and the black spot is moved towards the hinder and upper part of the scute.

Beside a number of specimens which have the locality indicated in a more or less vague manner, I have examined the two young specimens collected by Colley at Mazatlan (one of which has been coarsely figured in Capt. Beechey's 'Voyage'), two adult specimens with the shell 10 and 12 inches long (from the same locality, collected by Mr. Forrer), an adult (stuffed) and two young specimens in spirits respectively 2 and 5 inches in length, collected by Forrer at Presidio: the young are figured on Tab. I. A young specimen obtained by Mr. Herbert Druce from Chiriqui, and two specimens from Honduras collected by Mr. Dyson, which are the types of Gray's *Emys venusta*. Cope records the species from Costa Rica.

It would appear that the specific name of *ornata* was suggested to Gray by Bell, who, however, has forfeited the authorship by never having published either a figure or a description. Gray was the first who described the species under that name in the 'Synopsis Reptilium,' 1831, p. 30. Duméril and Bibron described it under the same name from a specimen given to them by Bell; they erroneously refer to it *Emys annulifera* of Gray, which is quite a distinct species. Bell's specimen in the Paris Museum is that also figured by Aug. Duméril and Bocourt in 'Mission Scientifique au Mexique.'

2. **Emys cataspila,** sp. n. (Tab. VI. fig. B, young.)

Emys ornata (part.), Gray, Cat. Tort. B. M. 1844, p. 22; Gray, Cat. Shield Rept. 1855, p. 24, t. 12 (not good).

Callichelys ornata (part.), Gray, Ann. & Mag. Nat. Hist. 1863, xii. p. 176; Suppl. Cat. Shield Rept. p. 48 (1870).

Hab. MEXICO (*coll. Brit. Mus.*).

Although our series of specimens of this form is pretty complete, it is far from being satisfactory, inasmuch as they are derived from different sources, with no more precise information as to locality than the term "Mexico" in about half of their number.

The young (2 inches long) shows very much the same pattern and style of coloration of the head, back, and sternum as *Emys ornata*; but the principal ocellus, with its yellow border, is small, occupying only a small part in the postero-inferior portion of each of the costal scutes. This position and size of the ocellus is not greatly changed in three other specimens from 5 to 7 inches in length; in larger specimens the ocellus is advanced more towards the dorsal line, but never occupies the upper half of the costal scutes as in *Emys ornata.* The ocellus consists now of a brownish oval spot with a deep black pupil and a broad yellowish border (see figure given by Gray).

The largest of our specimens has a shell 13 inches in length.

3. **Emys salvini,** sp. n. (Tabb. II. & III.)

Callichelys venusta (part.), Gray, Suppl. Cat. Shield Rept. p. 49.

Hab. GUATEMALA (*Salvin*).

This species is closely allied to *Emys ornata* and *E. cataspila,* but differs in the shape of the costal ocellus and of the figure on the sternum. The shape and coloration of the head seem to be the same as in those two species.

The principal yellow ring on the costal scutes is of a suboval shape, and occupies nearly the whole of the depth of the scute. There are several similar, but less distinct concentric rings within and without the principal ring. The ornamentation on the sternum is very symmetrical, ending behind in a pointed angle, and not in a straight and transverse line as in the other two species.

Of this species I have seen only one specimen, the shell of which is 9 inches long. It was obtained by Mr. Salvin in Guatemala, and was referred by Gray to his *Emys venusta.*

The specimen is figured on Tabb. II. and III. of the natural size; and separate figures are given of the vertebral and second costal scutes, which in the main figure are very much foreshortened.

4. Emys mobiliensis.

Emys mobiliensis, Holbr. N.-Am. Herp. i. p. 71, t. 9.
Ptychemys mobiliensis, Agass. Nat. Hist. U. S. Amer. i. p. 433.
? *Emys ventricosa,* Gray, Cat. Shield Rept. p. 28, t. 14.

Hab. NORTH AMERICA, Pensacola, Texas.—MEXICO, Nuevo Leon.

Agassiz (*l. c.*) mentions that he had received specimens of this species from Pensacola, Texas, and from Nuevo Leon, near Cadereita, Mexico. I have not seen an example of the species, unless the type of *Emys ventricosa*, the origin of which is unknown, belongs to it.

5. Emys callirostris.

Emys callirostris, Gray, Cat. Shield Rept. p. 25, t. 12 B.
Callichelys callirostris, Gray, Ann. & Mag. Nat. Hist. 1863, xii. p. 181.

Hab. West coast of CENTRAL AMERICA.

This, perhaps the most beautiful, species of this section of *Emys*, has been well figured by Gray, so that a further description seems to be unnecessary. Only the typical specimen is known, and there is every reason to believe that it came into the collection of Haslar Hospital with a number of other objects from the west coast of Central America; at any rate, it will be well to include it provisionally in the present list, as it has been confounded with the following species.

6. Emys umbra. (Tabb. IV., V., & VI. fig. A).

Emys grayi, Aug. Dum. & Boc. Miss. Sc. Mex., Rept. p. 13, t. 3. fig. 2 (not Günth.).
Emys callirostris, Bocourt, Journ. Zool. 1876, p. 387 (not Gray).
Emys concinna, Gray, Ann. & Mag. N. H. 1873, xii. p. 110.
Emys umbra, (Bocourt,) O'Shaughnessy, Zool. Record, xiii., Rept. p. 6 *.
Callichelys callirostris, Sumichrast, Bull. Soc. Zool. France, 1880, p. 166.

Hab. MEXICO (*coll. Brit. Mus.*), Tapana and San Mateo near Tehuantepec (*Sumichrast, coll. B. M.*); GUATEMALA, Rio Nagualate (*Bocourt*).

This species differs from all the preceding species by the acutely-pointed snout, which is conspicuously longer than the eye. The hind feet, as well as the fore feet, are broadly webbed; the claws sharp and long. The shell is much depressed, slightly rugose or smooth, and generally very similar to that of *Emys ornata.* The type of the coloration of the soft parts also very much resembles that of the latter species, but the

* The author of the Record on Reptiles for 1876 refers to p. 26 of the 'Journal de Zoologie.' This is incorrect; he merely used a separate copy of M. Bocourt's paper, which cannot legitimately be considered a publication, and at the end of which a note is added in which the change of name from *grayi* to *umbra* is proposed. As this note does not occur in the really published paper in the 'Journal de Zoologie,' 1876, p. 411, the publication of the name *umbra* rests solely with the notice in the 'Zoological Record' as quoted above.

longitudinal bands on the upper part of the head are less numerous and less distinct, the lateral band over the temple is not dilated in the middle, and the band on the chin is bifurcate behind and continued as one of the lateral bands of the throat. The shell of the adult is brownish horn-coloured, each costal scute being marked with an oblong blackish spot situated high above, and very close to the vertebral scutes; sometimes a yellowish ring passes at some distance round the spot. The last vertebral scute bears also a small blackish spot in the middle of its suture with the costal scutes. The sternum is marbled with black, and only now and then faint traces of symmetrical markings may be discerned.

I have not seen the young of this species; it is probably much more brightly ornamented than the adult, and there are indications of the sternum having similar symmetrical markings at this age as are observed in the other allied species.

We have received six specimens, from 10 to 14 inches in length, through Mr. Sumichrast, who obtained them at Tapana and San Mateo, near Tehuantepec. A seventh specimen from "Mexico" has been in the British Museum for many years, and was referred by Dr. Gray to his *Emys venusta.*

I have not the least doubt that the specimen from Rio Nagualate on the west coast of Guatemala, described by Duméril and Bocourt as *Emys grayi*, belongs to the same species, the costal spots having become obsolete. Specimens sent by Sumichrast to M. Bocourt were confounded by the latter with *Emys callirostris*, a short-snouted totally distinct species.

This species is aquatic, but herbivorous.

β. Rhinoclemmys or Chelopus.

7. **Emys pulcherrima.** (Tabb. VII. & VIII. fig. A.)

Emys pulcherrima, Gray, Cat. Shield Rept. p. 25, t. 25. figg. 1, 2 (this figure is faulty in representing six longitudinal bands on the neck instead of five).
Callichelys pulcherrima, Gray, Ann. & Mag. N. H. 1863, xii. p. 181 (and Suppl. Cat. Shield Rept. p. 49, where the species is confounded with a specimen said to come from South Africa).
Rhinoclemmys pulcherrima, Gray, Ann. & Mag. N. H. 1873, xi. p. 145.

Hab. Mexico, Presidio (*Forrer*).

The neck is ornamented with five parallel rather narrow red bands edged with blackish; occipital region with some detached spots of similar coloration. A narrow stripe, similarly coloured, runs along the edge of the upper part of the snout and stops behind the supraciliary region. Two similar bands from the eye, round the snout, *above* the nostrils, and finally another round the upper jaw. A short stripe along the median line of the upper part of the snout. Sides of the neck and throat finely punctulated with black. Each scute with some short, curved, yellow, dark-edged bands. Central parts of the sternum blackish or black. Shell depressed, oblong, with the lateral edges sharp, strongly curved upwards; subcircular and dentated behind in the young. Nuchal narrow; vertebrals distinctly keeled. Toes very slightly webbed.

This species was hitherto known from a very young stuffed example only, 2 inches long, and occupied a very precarious position in systematic works. However, Herr Forrer rediscovered it at Presidio, in Western Mexico, the youngest of the three specimens collected by him being absolutely identical with the type. The largest specimen, an adult female, is 7 inches long.

8. **Emys incisa.**

Emys incisa, Bocourt, Ann. Sc. Nat. 1868, x. p. 121; Miss. Sc. Mex., Rept. p. 11, tt. 1. & 2.
Rhinoclemmys frontalis, Gray, Ann. & Mag. N. H. 1873, xi. p. 144.
Chelopus incisus, Bocourt, Journ. Zool. 1876, v. p. 387 (not synon.); Sumichrast, Bull. Soc. Zool. France, 1880, p. 165.

Hab. MEXICO, Tehuantepec and Chiapas (*Sumichrast*); SALVADOR, La Union (*Bocourt*).

A narrow yellow (?) line runs along the upper edge of the snout over the eye to the tympanum; two other shorter concentric lines on the side of the snout, above the nostrils; a fourth on the horny sheath of the jaw. A median single line on the upper side of the snout is sometimes absent. Side of the head with some irregular small ocellated spots; a few similar spots sometimes on the occipital portion of the head. Upper part of the neck without bands or spots, olive-coloured; throat white, immaculate; sides of the neck, and the hinder part of the underside, punctulated with black. Upper part of the shell sometimes uniformly coloured, sometimes with a small, yellow, black-edged ocellus on the areola of each costal scute. Sternum black in the middle, yellowish on the sides. Shell moderately depressed, rather broad, with the lateral margins sharp and strongly reverted; hind margin distinctly serrated, the serrature becoming obsolete in very old specimens. Vertebral keel very distinct. Nuchal small, occasionally absent. Toes very slightly webbed.

The figure given by M. Bocourt is sufficiently accurate to ensure the recognition of this species. The largest specimen I have seen had a shell 8 inches long. It is a terrestrial herbivorous species.

9. **Emys gabbi.**

Chelopus gabbii, Cope, Journ. Ac. N. Sc. Philad. viii. 1875, p. 153.

Hab. COSTA RICA, Limon (*Cope*).

I have not seen this species, which seems to take its place near to *E. incisa.*

10. **Emys funerea.**

Chelopus funereus, Cope, Journ. Ac. N. Sc. Philad. viii. 1875, p. 154.

Hab. COSTA RICA, Limon (*Cope*).

Distinguished by the absence of markings on the upper side of the head and neck; this species will be easily recognized. I have not seen specimens of it.

11. **Emys areolata.** (Tab. VIII. fig. B.)

Emys areolata, Aug. Dum. Cat. Rept. p. 10; Arch. Mus. vi. 1852, p. 223, t. 14 (the artist introduced into this figure six vertebral and five costal scutes); A. Dum. & Bocourt, Miss. Sc. Mex., Rept. p. 13.

? *Chelopus areolatus,* Cope, Proc. Ac. Sc. Phil. 1866, p. 123.

Hab. GUATEMALA, Lake of Peten (*Salvin, coll. Brit. Mus.*).

Neck with four parallel, white, black-edged bands, the outer of which are far down on the side, and continued forwards over the temple into the supraciliary region; the inner commence from the occipital region. Crown of the head indistinctly marbled; a whitish line along the median line of the upper part of the snout. A minute white speck in front of the upper angle of the orbit (much smaller than in *E. scabra*). Cheek and tympanic region with irregular blackish markings; throat sparsely and minutely spotted with black. Central parts of the sternum blackish or black. Shell depressed, oblong, with the lateral edges sharp, curved upwards. Nuchal equilateral, triangular; vertebrals slightly keeled. Toes scarcely webbed.

The typical specimen came from Lake Peten; the specimen in the British Museum was collected by Mr. Salvin in the same district, and referred by Gray first to *E. scabra* and afterwards to *E. annulata.*

Emys pulcherrima of A. Duméril and Bocourt (Miss. Sc. Mex., Rept. p. 15, t. 7) is not the species described under that name by Gray; it represents a menagery-specimen without precise information as to locality, which has been named *Rhinoclemmys bocourti* by Gray (Ann. & Mag. N. H. 1873, xii. p. 111), and which possibly may prove to be identical with *E. areolata.*

12. **Emys rubida.** (Tab. VIII. figg. C, C′, C″.)

Chelopus rubidus, Cope, Proc. Am. Philad. Soc. (1869) xi. 1871, p. 148.

Rhinoclemmys mexicana, Gray, Proc. Zool. Soc. 1870, pp. 659, 723; Suppl. Cat. Shield Rept. p. 30; Proc. Zool. Soc. 1871, p. 296, t. 28.

Chelopus mexicanus, Bocourt, Journ. Zool. 1876, v. p. 387; Sumichrast, Bull. Soc. Zool. France, 1880, p. 166.

Hab. MEXICO, San Juan del Rio (*Rébouch, coll. B. M.*), Tehuantepec (*Sumichrast, coll. B. M.*).

The coloration of the soft parts, although varying somewhat in details, is characteristic enough. A narrow yellow (?) band runs round the edge of the snout, from one orbit to the other; a second much broader one, horseshoe-shaped, with the convexity towards the front, occupies the crown of the head, and is mixed up posteriorly more or less with the markings of the occiput, which, like those of the neck, consist of blackish vermiculations which are darkest on the edges and light in the middle. A short, oblique, yellow band behind the eye is very constant. Throat punctulated. The scutes are sometimes without markings, sometimes with indistinct yellowish figures. The areolar portion of each costal scute is frequently ornamented with a small, oval, yellow spot and a narrow annulus. The middle of the sternum blackish. Shell moderately depressed, oblong, with the lateral edges not or slightly recurved. Nuchal small, occasionally absent. Only a trace of a vertebral keel. Toes not webbed.

Of this species we have a series of examples from both localities named. The shell of the largest example does not exceed 8 inches in length. It is a terrestrial frugivorous species. Cope's description was evidently prior to that of Gray; and Bocourt refers, by some error, Cope's name of "*rubidus*" to his *E. incisus.*

13. **Emys punctularia.**

Testudo punctularia, Daud. Hist. Nat., Rept. ii. p. 249.
Emys scabra, Bell, Testud. cum tab.
Geoclemmys callocephalus, Gray, Proc. Zool. Soc. 1863, p. 254.
Chelopus punctularius, Cope, Proc. Ac. Nat. Sc. Phil. 1865, p. 185.
Rhinoclemmys scabra, Gray, Proc. Zool. Soc. p. 722, cum fig. capitis.

Hab. MEXICO, Yucatan, Tabasco.—SOUTH AMERICA, Guiana, &c.

This is a South-American species, common in Dutch and British Guiana and other tropical parts of South America. However, Cope (*l. c.*) refers to it two Central-American specimens, one from Yucatan, and the other from Tabasco. I have not seen these specimens, which are preserved in the Smithsonian Institution, and am therefore unable to confirm Mr. Cope's identification.

Geoclemys callocephalus, or *Rhinoclemmys callocephala*, Gray, Proc. Zool. Soc. 1863, p. 254, is based on a dry specimen, which, after having lived for some time in captivity, came into the collection of the British Museum in the year 1853. Its shell is deteriorated; and of the coloration of the soft parts only so much is preserved as clearly shows that the specimen is identical with *Emys scabra* or *punctularia.* The figure given by Gray of the head of this species is very inaccurate and misleading.

NOTE.—*Geoclemmys melanosterna*, Gray, Proc. Zool. Soc. 1861, p. 205, is an extra-limital species. Of the two specimens known, the type bears its original label "Chirambira, Darien," altered in Gray's paper to "Cherunha, Gulf of Darien." Chirambira is a place a short distance north of Buenaventura, where the second specimen was obtained. Thus there is no doubt that this species belongs to the Colombian fauna.

Geoclemmys annulata, Gray, Proc. Zool. Soc. 1860, p. 231, t. 29, is likewise extra-limital, from the coast of Ecuador. It is true that Gray in the Suppl. Cat. Shield Rept. p. 30, and in the Hand-list of Shield Rept. p. 27, refers to Mr. Salvin as the discoverer of the species in Ecuador and on the Gulf of Darien. But this mistake is due to his having confounded (in the last years of his life) Mr. Salvin with Mr. Fraser. In fact, the spirit-specimen said to be from the "Gulf of the Darien" is one of the types, and the very specimen figured. The locality "Darien" became attached to it through the insufficient care of a person who assisted Gray.

DERMATEMYS.

Dermatemys, Gray, Proc. Zool. Soc. 1847, p. 55.

1. Dermatemys mawii.

Dermatemys mawii, Gray, Proc. Zool. Soc. 1847, p. 56; Cat. Shield Rept. p. 49, t. 21*; Proc. Zool. Soc. 1864, p. 125; Dum. & Boc. Miss. Sc. Mex., Rept. p. 17, t. 7. fig. 2; Cope, Proc. Ac. Nat. Sc. Philad. 1865, p. 187, and 1868, p. 119; Preudhomme de Borre, Bull. Ac. Belg. 1869, p. 116 (young).
Emys berardii, A. Dum. Arch. Mus. d'Hist. Nat. vi. p. 231, t. 15.
Dermatemys berardii, Cope, l. c. p. 120.
? *Dermatemys abnormis,* Cope, l. c. (young).
Dermatemys salvinii, Gray, Suppl. Cat. Shield Rept. p. 50.
? *Chloremys abnormis,* Gray, Proc. Zool. Soc. 1870, p. 711, t. 42 (young).

Hab. Mexico, Tabasco (*Cope*), Laguna de Terminos, Yucatan (*Zool. Soc.*), Vera Cruz (*Bérard*); British Honduras, Belize (*Cope, Bocourt, P. de Borre*); Guatemala (*Salvin*).

A vegetable-feeder, called "Tortuga blanca" in Mexico.

CHELYDRA.

Chelydra, Schweigger, Prodr. Monogr. Chelon. p. 23.

1. Chelydra serpentina.

Chelydra serpentina (Linn.), Peters, MB. Berl. Ak. 1862, p. 627; Cope, l. c. 1872, p. 23.
? *Chelydra,* sp., Cope, Proc. Ac. N. Sc. Philad. 1865, p. 187.

Hab. North America.—Mexico and southwards to Ecuador.

Called "*Chiquihuan*" in Mexico.

Ranges from Canada to Ecuador. Mexican specimens have been examined by Cope.

2. Chelydra rossignoni.

Emysaurus rossignonii, Dum. & Bocourt, Miss. Sc. Mex., Rept. p. 18, t. 5. fig. 2.

Hab. Mexico; Guatemala (*Bocourt*).

* The suture dividing the gular plate is much too distinctly represented in this figure; in the original specimen (which is dry) there is no more of a suture than in another from Salvin's collection—merely a groove in the underlying bone is indistinctly visible in both specimens through the transparent substance of the epidermal scute.

I have not seen any specimens from Central America. This species is said to be distinguished from *C. serpentina* chiefly by having four chin-barbels instead of two. This seems to me quite as variable a character as the others assigned to the Central-American form. In the head of an adult specimen from Meadville, Pa., in spirits, I find some of the lateral skin-papillæ on the chin so much enlarged as to deserve almost the appellation of barbels. However, Bocourt states that he has had three specimens, showing alike the development of four barbels.

STAUROTYPUS.

Staurotypus, Wagler, Syst. Amphib. p. 137 (1830).

1. Staurotypus triporcatus.

Terrapene triporcata, Wiegm. Isis, 1828, p. 364.

Staurotypus triporcatus, Wagl. Ic. Amphib. t. 33; Gray, Cat. Shield Rept. p. 47, t. 20 B; Dum. & Bocourt, Miss. Sc. Mex., Rept. p. 21; Gray, Suppl. Cat. Shield Rept. p. 65; Cope, Proc. Ac. Nat. Sc. Philad. 1865, p. 188.

Hab. MEXICO, Tabasco (*Cope*), Rio Alvarado, Bay of Campeachy (*Deppe*), Vera Cruz (*Sallé*); GUATEMALA, Rio Usumacinta (*Morelet*).

Vernacular name at Tabasco: *Huau.*

2. Staurotypus salvini.

Staurotypus salvini, Gray, Proc. Zool. Soc. 1864, p. 127; Dum. & Boc. Miss. Sc. Mex., Rept. p. 22, t. 5. fig. 3.

Stauremys salvinii, Gray, Proc. Zool. Soc. 1869, p. 179, fig. 5; or Suppl. Cat. Shield Rept. p. 65; Cope, Proc. Ac. N. Sc. Phil. 1872, p. 28.

Staurotypus marmoratus, Fischer, Wiegm. Arch. 1872, p. 265, t. 10 (young).

Claudius severus, Cope, l. c. p. 24; Sumichrast, Bull. Soc. Zool. France, 1880, p. 167.

Claudius pictus, Cope, l. c. p. 26.

Staurotypus (*Claudius*) *severus*, Bocourt, Journ. Zool. 1876, p. 387.

Hab. MEXICO, Santa Efigenia, Tapana, Tehuantepec (*Sumichrast*); GUATEMALA, Huamuchal (*Salvin*), Vera Paz (*Bocourt*).

Having before me the type of *Stauremys salvini*, and a large series of specimens which were collected by Sumichrast, and therefore are typical of *Claudius severus*, I can positively state that there is no ground whatever for specifically separating them from each other or from the specimen figured by Bocourt and named *Claudius pictus* by Cope. Gray's assertion that the tubercles on the tail are placed in three series in *S. salvini* is erroneous; they are four-rowed as in the other specimens. The rows of tubercles are less separate and regular in the short-tailed females than in males, in which the tail is thicker and longer. The tubercles themselves are in some specimens more developed than in others, and, according to Sumichrast, may disappear altogether.

The mobility of the fore part of the sternum is extremely variable and depends on the amount of ossification which has taken place near the transverse suture, but no joint is developed. Not less variable is the width of the inguinal bridge, which is sometimes not even the same on either side of the same individual. With regard to coloration, every degree of transition occurs from a finely mottled head to one of uniform coloration without any spots.

This species is so closely allied to *Staurotypus triporcatus* that some difficulty exists in referring young examples to either of these two species. Young examples of *S. triporcatus* do not seem to possess the very broad front part of the sternum by which the old ones are distinguished. Therefore in maintaining for the present the genus *Claudius*, I cannot follow Cope in including in it his *C. severus*, but restrict it to *C. angustatus*, for which it was originally formed, although the characters assigned to it seem to me scarcely sufficient to justify a distinct generic appellation.

CLAUDIUS.

Claudius, Cope, Proc. Ac. N. Sc. Phil. 1865, p. 185.

1. **Claudius angustatus.**

Claudius angustatus, Cope, Proc. Ac. N. Sc. Phil. 1865, p. 187; 1872, p. 28; Proc. Am. Phil. Soc. 1869, t. 9; Dum. et Bocourt, Miss. Sc. Mex., Rept. p. 20, t. 4.

Claudius megalocephalus, Bocourt, Ann. Sc. Nat. Zool. 1868, p. 122; Cope, Proc. Ac. N. Sc. Phil. 1872, p. 27.

Hab. Mexico, Tabasco (*Cope*).

Called "Talmame" by the natives. Carnivorous.

CINOSTERNUM.

Cinosternum, Spix, Test. et Ran. Bras. p. 17 (1824).

Of the twelve or thirteen species known of this genus, not less than nine are peculiar to the Central-American fauna, one only occupying territories in the South, and two or three * in the North. The distinction of the species is a difficult task; and it is still more difficult to select distinctive characters from the lengthy descriptions by authors who notice the most variable details. The shape of the shell, of the individual scutes, of the beak, tail, the width of the hinder sternal lobe are all variable individually, as well as according to sex and age. In the following synopsis I have admitted such characters only as may be more or less fully relied upon.

* It is not within the scope of this work to investigate the question whether *Cinosternon hippocrepis* of Gray and *Platythyra flavescens* of Agassiz are identical or not, as there is no evidence of either occurring within the limits of Mexico.

I. Shell with three keels.
  A. Sternum with a notch behind . . . . *C. sonoriense,* p. 13.
  B. Sternum not notched behind.
    1. Jaws spotted.
      a. Axillary and inguinal scutes separate. . . . . . *C. cruentatum,* p. 13.
      b. Axillary and inguinal scutes joining each other . . *C. triliratum,* p. 14.
    2. Jaws uniform yellowish . . . . *C. albogulare,* p. 14.

II. Shell with one keel or none.
  A. Sternum with a notch behind.
    1. Jaws spotted . . . . *C. hirtipes,* p. 15.
    2. Jaws yellow . . . . *C. integrum,* p. 16.
  B. Sternum without a notch.
    1. Jaws spotted . . . . *C. effeldtii,* p. 16.
    2. Jaws uniformly coloured.
      a. A joint between posterior lobe and middle of sternum . . . . *C. leucostomum,* p. 17.
      b. A straight suture between posterior lobe and middle of sternum; axillary and inguinal scutes separate . . *C. brevigulare,* p. 17.
      c. A curved suture between posterior lobe and middle of sternum; axillary and inguinal scutes in contact . *C. cobanum,* p. 18.

1. **Cinosternum sonoriense.**

Kinosternum sonoriense, Leconte, Proc. Ac. N. Sc. Philad. 1854, p. 184.
Thyrosternum sonoriense, Agass. Amer. Testud. p. 428, t. 5. figs. 8–11 (young).

Hab. MEXICO, Tucson, Sonora (*Leconte*), Guadalupe, Cañon Sonora (*Agassiz*).

Shell with three keels, but the lateral are obsolete and scarcely to be traced, except on the two last costal scutes. Sternum emarginate behind. Head and neck black, above mottled and spotted with whitish, beneath whitish, varied in the same manner with dusky; jaws horn-colour, varied with black. The anterior joint of the sternum ligamentous, the posterior sutural, the hind lobe not entirely closing the box of the shell. Gular plate large, equilaterally triangular.

Of this species I have not seen any examples, and the characters are taken from Leconte's description. His specimen was 4·4 inches long. Gray considered his *K. punctatum,* based upon a very young example from East Florida, to be identical with the Sonora species. But this is very doubtful, and unfortunately cannot be determined from the type specimen, which is in a very bad state of preservation.

2. **Cinosternum cruentatum.** (Tabb. IX., X., XI.)

Cinosternum cruentatum, A. Dum. Cat. Méth. Rept. p. 16; and Arch. Mus. vi. p. 238, t. 16. figg. 1, 2; Dum. & Boc. Miss. Sc. Mex., Rept. p. 23; Bocourt, Journ. Zool. 1876, p. 398.
Kinosternum mexicanum, Leconte, Proc. Ac. N. Sc. Phil. 1854, p. 182.
Swanka scorpoides, Gray, Suppl. Cat. Shield Rept. p. 67 (part.).

Cinosternum shawianum, Bocourt, Journ. Zool. 1876, p. 397 (not of Bell, which is the true *C. scorpioides* from Guiana); Sumichrast, Bull. Soc. Zool. France, 1880, p. 169.

Hab. MEXICO (*coll. Brit. Mus.*), Tonala in Chiapas (*Sumichrast*).

Shell with three keels, placed close together and disappearing in very old examples; first vertebral scute as broad as, or even broader than, long. Sternum not emarginate behind; the central portion shorter than either of the two lobes; joint of the hind lobe well developed and straight; gular plate as long as the median suture of the front lobe; axillary and inguinal scutes not in contact with each other.

The tail is very thick in the male and armed with a curved claw; short, tubercular, and provided with a much smaller claw in the female. The head is greyish, mottled with yellow, especially on the cheek; jaws yellow, with black markings, of which the most constant occupies the median line of the lower jaw; middle of the throat mottled with grey.

The principal characters of this species are already indicated in a very young specimen only a few days old; its sternum is yellow with black subsymmetrical markings arranged chiefly along the median line.

Besides some specimens of the origin of which nothing is known except that they came from Mexico, I have examined four examples collected by Sumichrast at Tonala (Chiapas); the largest of them are 5 inches long, a size which this species does not seem to exceed. As far as is known at present, it does not seem to occur beyond Mexico.

The old female figured has only four claws on the fore foot.

3. Cinosternum triliratum.

Kinosternum triliratum, Leconte, Proc. Ac. N. Sc. Phil. 1859, p. 6.

Hab. MEXICO (*Pease*).

Shell with three strong keels; sternum bivalved, entire and rather pointed behind; jaws yellow, varied with black (the upper entire, not hooked*). Gular large, triangular; inguinal scutes long, rather wide, joining the axillaries.

I have not seen specimens of this species.

4. Cinosternum albogulare.

Cinosternon albogulare, Bocourt, Miss. Sc. Mex., Rept. p. 24; Journ. Zool. 1876, p. 399.

Hab. COSTA RICA (*Bocourt*).

Shell with three keels, sternum not emarginate behind. The first vertebral scute longer than broad; gular shorter than the median suture of the front lobe; fixed por-

* The hook on the upper jaw is not a specific character; it is developed with age, at least in those species of which I have been able to examine individuals of different ages.

tion of the sternum shorter than either of the two lobes; axillary and inguinal scutes touching each other by a point; jaws and throat uniform yellowish.

I have not seen a specimen of this apparently very distinct species.

5. **Cinosternum hirtipes.** (Tabb. XII., XIII., XIV., XV.)

Cinosternum hirtipes, Wagl. Descr. & Ic. Amphib. t. 30; Strauch, Vertheil. Schildkr. p. 101.

Kinosternum henrici, Leconte, Proc. Ac. N. Sc. Phil. 1859, p. 4; Yarrow in Wheeler's Report upon Expl. and Surv., Rept. p. 583, t. 16.

Hab. North America, New Mexico (*Leconte*), Arizona (*Yarrow*).—Mexico, Mazatlan and Tres Marias Islands (*Forrer*).

I consider this species to be the southern and more developed form of *Cinosternon pennsylvanicum*, but the shell is broader in young specimens, and generally more convex and raised along the vertebral line in old ones. The sternum is emarginate behind, the joint of the hind lobe forming a somewhat curved line. The fixed part of the sternum is always somewhat shorter than either the front or the hind lobes; gular plate not much shorter than the median suture of the front lobe. The development of the axillary and inguinal plates is subject to variation in this species; in a half-grown specimen from Mazatlan these two scutes are broadly in contact with each other, whilst they barely touch each other in full-grown examples; so it is also in a young specimen from Tres Marias Islands, an adult individual from this last locality having a merely rudimental axillary plate, which is separate from the inguinal.

The vertebral keel is very indistinct, even in young examples, the vertebral region of the shell being flattened, but never concave as in *Cinosternum pennsylvanicum*. The first vertebral is as broad as long, in old examples bell-shaped, the lateral margins being concave. Upper parts of the head brownish or blackish, with irregular yellowish spots; sides of the head, jaws, and throat yellow, marbled and streaked with black. The tail is very strong and long in the male, and armed with a curved claw; in the female it is much shorter and clawless.

Of this species I have examined five specimens—three (male, female, and half-grown) from Mazatlan, and two (female and young) from Tres Marias Islands. The variation of form of the shell is well represented in this series. The shell of the largest male is $6\frac{1}{2}$ inches long, the females being $\frac{1}{2}$ inch shorter. The specimens named *C. henrici* were collected in New Mexico and Arizona, and prove the close affinity of the Mexican form to *C. pennsylvanicum*.

To show the great variation of form to which some species of this genus may be subject, three views of each are given of an adult male and female from Mazatlan, and of an adult female from Tres Marias Islands, together with the arrangement of the axillary and inguinal plates in a half-grown specimen from Mazatlan.

6. Cinosternum integrum.

Kinosternum integrum, Leconte, Proc. Ac. N. Sc. Philad. 1854, p. 183.

Hab. MEXICO (*Pease*).

Shell without keel. Sternum a little emarginate behind, bivalved, entirely closing the box. Gular plate small, nearly equilateral. Axillary long and narrow, touching the inguinal by a point. Head and neck brown, spotted with yellow, a large orange spot behind the eye somewhat varied with black; jaws yellow.

The specimen described is $4\frac{1}{2}$ inches long.

Bocourt (Journ. Zool. 1876, p. 393), in his synopsis of the species of this genus, includes the description of what he considers the species named by Leconte; however, I suspect that he had a distinct species characterized by spotted jaws, and by a sternum not emarginate behind. Leconte's description also is somewhat inaccurate, inasmuch as he ascribes a keel to this species in his diagnosis, denying its presence in the more detailed description.

On the other hand, I am inclined to think that the *Cinosternum rostellum*, Bocourt, l. c. p. 391, from Guanajuato, cannot be specifically separated from *C. integrum.*

7. Cinosternum effeldtii.

Cinosternum effeldtii, Peters, MB. Ak. Wiss. Berl. 1873, p. 603, cum tab.

? *Cinosternum berendtianum,* Cope, Proc. Ac. N. Sc. Philad. 1865, p. 189; Bocourt, Journ. Zool. 1876, p. 395.

Hab. MEXICO, Vera Cruz (*Peters*); GUATEMALA (*Salvin*).

Shell with a single keel. Sternum not emarginate behind, with the two joints, by which the lobes are attached to the middle portion, well developed; middle portion shorter than the hind lobe, but almost as long as the front lobe. Gular longer than broad, and longer than the median suture of the front lobe. Inguinal and axillary scutes separate from each other. Head and jaws marked with brown and yellow.

I have examined two specimens found by Mr. Salvin in Guatemala, an adult male and female, the latter being 4 inches long. From Peters's description it would appear that sometimes the axillary and inguinal scutes are in contact with each other by a point. I am by no means assured of the identity of this species with *Cinosternum berendtianum* of Cope, although there can, I think, be no doubt that the species described by Bocourt under the latter name is, in fact, the *Cinosternon effeldtii* of Peters. Although Cope says nothing of the coloration of the jaws, he states that *Cinosternum berendtianum* is most nearly related to *C. leucostomum,* thus apparently letting us infer that both are characterized by yellow unspotted jaws.

8. **Cinosternum leucostomum.** (Tabb. XVI., XVII.)

Cinosternon leucostomum, A. Dum. Cat. Méth. Rept. 1851, p. 17; and Arch. Mus. Hist. Nat. 1855, p. 239, t. 17; Strauch, Vertheil. Schildkr. p. 99; Dum. & Boc. Miss. Sc. Mex., Rept. p. 25; Bocourt, Journ. Zool. 1876, p. 394; Cope, Proc. Ac. N. Sc. Philad. 1865, p. 189; and Journ. Ac. N. Sc. Philad. viii. 1876, p. 153.
Kinosternum scorpoides, Gray, Cat. Shield Rept. p. 44 (part.).
Swanka maculata, Gray, Suppl. Cat. Shield Rept. p. 68.

Hab. ? North America, New Orleans (*Duméril, Strauch*).—Mexico, Tabasco (*Cope*), Cosamaloapam * (*Sallé*); Guatemala, Rio Usumacinta (*Duméril*), Vera Paz (*Bocourt, Salvin*), Huamuchal (*Salvin*); Costa Rica, Sipurio, Old Harbour (*Cope*); Panama, Turbo (*Smiths. Inst.*).—Colombia, Bogota (*Duméril*).

Vernacular name: "Pochitoque camatotl" (Tabasco).

Shell with a single keel, most prominent on the two last vertebral scutes, and generally inconspicuous on the middle of the back, which is flat. Convexity of the shell variable. First vertebral scute as broad as long, sometimes bell-shaped, with the lateral margins concave. Sternum not emarginate behind, the two lobes being attached to the fixed portion by well-developed straight joints. Central part of the sternum rather shorter than either of the two lobes. Gular plate large, generally about as long as the median suture of the front lobe. Axillary and inguinal scutes barely touching each other or entirely separate.

The tail is very long and thick in the male, and armed with a large curved claw; it is much smaller, tubercular, and provided with a small claw in the female. Head greyish brown above, mottled with yellowish; jaws without spots, either uniform yellowish or horn-coloured or blackish; the throat is sparsely mottled with brownish.

Besides some specimens the origin of which is either unknown or uncertain, I have before me a series of seven examples collected by M. Sallé in Mexico and by Mr. Salvin in Guatemala; they are the types of Gray's *Swanka maculata*. The shell of the longest is $5\frac{1}{2}$ inches long.

9. **Cinosternum brevigulare,** sp. n. (Tab. XVIII. fig. A.)

Hab. Mexico, Playa Vicente (*Sallé*).

This species is distinguished, first by having a scarcely movable hind lobe, which is attached to the sternum by a suture rather than by a joint, and secondly by the shortness of the gular plate, the length of which is only two fifths of the median suture of the front lobe. Shell with a distinct broad keel; first vertebral as long as broad; axillary and inguinal scutes separate. Head brown above, marbled with whitish on the sides; jaws whitish; throat slightly mottled with brownish.

* This is a town on the banks of the Papaloapan river. The latter name has become "Papalco Apoia" in Gray's Suppl. Cat. Shield Rept. p. 68!

The type of this well-characterized species is a female $3\frac{1}{4}$ inches long, and had been placed by Dr. Gray as his *Swanka scorpoides*.

10. **Cinosternum cobanum,** sp. n. (Tab. XVIII. fig. B.)

Hab. GUATEMALA, Coban (*Salvin*), Cahabon (*Sarg*).

Shell moderately convex, with an obtuse vertebral keel. First vertebral scute a little broader than long. Hind lobe of the sternum attached to the fixed portion by a suture, the two rami of which distinctly converge towards the middle. Gular plate short, as long as the interpectoral suture, much broader than long. Axillary and inguinal in contact. Sternum without notch behind. Jaws uniform whitish.

This species seems to represent *C. leucostomum* in Alta Vera Paz; although founded on only two specimens, of which the largest is not quite 2 inches long, the distinguishing characters are sufficiently well developed. The smaller specimen from Coban is figured.

Order CROCODILIA.

Sumichrast (Bull. Soc. Zool. France, v. 1880, p. 170) gives the following account of the occurrence, &c. of the two Crocodilians found in Central America:—

"Whilst the indigenous Crocodile (*C. pacificus*) inhabits both salt and fresh waters, estuaries, rivers, and lakes, the Alligator or Huizizil * never inhabits running or brackish water, but only the bottom of shallow or muddy streams in plains or stagnant pools in woods. It lies in deep burrows along the banks under the submerged roots of the great trees. Its character is much more savage than that of the Crocodile, and when disturbed it leaves its lair and fearlessly attacks an enemy. The Crocodile only rarely does this. The female defends her young bravely, whether against the attacks of the male, who never loses an opportunity of devouring them, or that of any other enemy. Crocodiles frequently live in numerous herds in the estuaries along the edge of the sea, while the Alligator's habits are solitary and nocturnal. On finding itself captured it becomes furious, lashing out violently with its tail and giving vent to a sort of hollow threatening grunt, and though it never attains so large a size as the Crocodile † is more difficult to take. The food consists of fish, crustacea, and aquatic insects, and I have frequently found its stomach full of sand, gravel, and water-plants.

"The iris is of a greenish grey as in the Crocodile. Age does not make any great change in its form or the shape of its head."

* The word "Huizizil" is of Aztec origin; among the ancient Mexicans it was applied to the Humming-bird.

† The largest individuals I have observed do not exceed 1·72 metre.

CROCODILUS.

Crocodilus, auct.

1. **Crocodilus americanus,** Schneid.

The American Crocodile has a most extensive geographical distribution. It is found in the large West-Indian Islands, and is very common on the Atlantic side of Mexico and Central America *; on the Pacific side it extends southwards to Ecuador, whence specimens were brought by Mr. Fraser. Naturally, a species of so wide a range exhibits no inconsiderable amount of variation, especially in the comparative length of the snout, in the more or less regular arrangement of the lateral dorsal scutes, in the prominence of the keels of the scutes, and in the development (as regards number and size) of the scutes protecting the neck. The last of these characters is the least trustworthy of all, and therefore I am not prepared to admit the species named *Crocodilus mexicanus* by Bocourt. The variations in the width of the skull are all the more perplexing, as evidently a great change takes place in this respect during the growth of the individual. From specimens less than one foot in length little information is to be gained, but as they grow the snout is more produced and becomes comparatively narrow. This goes on until they have attained a length of five or six feet. About that period the growth of bone is slower in the longitudinal axis of the jaw than in the transverse, and the jaws increase in width as the animal becomes older. It is therefore necessary in the utilization of this character to compare specimens of the same size and presumably of the same age.

Induced by the labours of the French naturalists who distinguished several species of the Central-American Crocodile, I re-examined a very large series (34 specimens), and came to the conclusion that at least two forms, a long-snouted and a broad-snouted, could be distinguished, although even these two forms cannot be sharply defined, all having, besides, one important character in common by which the American Crocodile can always be recognized, namely, the longitudinal elevation of bone along the fronto-nasal region.

For the long-snouted form the subspecific term of *acutus*, and for the broad-snouted that of *moreletii* † may be chosen, and it may be useful to indicate in the accompanying table the amount of variation which attains in each of these two forms.

* Wyman (Sill. Journ. vol. 49, 1870, p. 105) records the occurrence of *Crocodilus acutus* on the coast of Florida, and describes a skull $18\frac{1}{4}$ inches long, comparing it with one of similar size from South America. As the Floridan skull is much narrower than any I have examined of *C. acutus*, and resembles in this respect the Crocodile of the Orinoco, I have some hesitation in referring the Floridan Crocodile to *C. acutus*.

† Wiegmann, Herpetol. Mex. p. 22, and Strauch, Syn. Crocodil. p. 41, maintain that *Crocodilus rhombifer* is common in Mexico and other parts of Central America. This statement rests evidently on a confusion of the Cuban species with the broad-snouted form of *C. americanus*. As far as I have been able to ascertain, the *C. rhombifer* of Cuba is peculiar to that island. It lacks the longitudinal rise of the fronto-nasal region, and the dorsal scutes form four regular longitudinal series besides one or two irregular lateral rows. Duméril's statement as to the shortness of the web between the outer hind toes in Cuban specimens, I am unable to confirm ; this membrane shrinks in dry specimens, but in our Cuban Crocodiles it is as much developed as in *C. acutus*, or nearly so. Finally, I may remark that the skull on which Cuvier founded *C. rhombifer* may prove to be nothing but a short-snouted *C. americanus*.

Variations in the form of the head in *Crocodilus americanus.*

acutus, Cuv., =*pacificus*, Boc.				*moreletii*, Dum., =*mexicanus*, Boc.			
Hab.	Size.	Length of head.	Proportion of width* of snout to length of head.	Hab.	Size.	Length of head.	Proportion of width of snout to length of head.
	in.	in.			in.	in.	
a. Jamaica	10	$1\frac{2}{3}$	$1 : 3\frac{1}{3}$				
b. Nicaragua	11	$1\frac{3}{4}$	$1 : 3\frac{1}{2}$				
c. Mazatlan	14	$2\frac{1}{2}$	$1 : 3\frac{3}{4}$				
d. West Ecuador ..	20	$3\frac{1}{4}$	$1 : 4\frac{1}{4}$				
e. Jamaica	28	$4\frac{1}{2}$	$1 : 4$				
f. Chepo	28	5	$1 : 4\frac{1}{2}$	*a.* Mexico	$29\frac{1}{2}$	4	$1 : 3\frac{2}{3}$
g. Tehuantepec	29	4	$1 : 4$	*b.* Mexico	$32\frac{1}{2}$	$4\frac{1}{3}$	$1 : 3\frac{2}{3}$
h. Presidio	37	6	$1 : 4\frac{1}{8}$	*c.* Mexico	38	5	$1 : 3\frac{1}{2}$
i. Presidio	43	7	$1 : 4\frac{1}{10}$	*d.* Mexico	$42\frac{1}{2}$	$5\frac{4}{5}$	$1 : 3\frac{1}{2}$
k. Guatemala	57	9	$1 : 4\frac{1}{2}$				
l. Jamaica	93	14	$1 : 4$				
m. ?	103	15	$1 : 3\frac{3}{4}$				
n. ?	105	16	$1 : 3\frac{2}{3}$				
o. Pacific Coast....	?	17	$1 : 3\frac{9}{10}$				
p. Greytown	$133\frac{1}{2}$	$19\frac{1}{2}$	$1 : 3\frac{1}{2}$	*e.* Mexico	?	20	$1 : 2\frac{1}{2}$

a. **Crocodilus americanus,** var. **acutus.**

Crocodilus americanus plumieri, Schneid. Hist. Amph. fasc. 2, p. 167.

Le Crocodile de Saint-Domingue, Geoffr. Ann. Mus. Hist. Nat. ii. 1803, p. 53, t. 37. fig. 1; Descourtilz, Voy. d'un Naturaliste, vol. iii. 1809, p. 11 et seqq. pls. 2–5.

Crocodilus acutus, Cuv. Ann. Mus. Hist. Nat. x. 1807, p. 55, t. 1. figs. 3 & 14, t. 2. fig. 5; Geoffr. ibid. p. 79; Dum. & Bibr. Erpét. génér. iii. p. 119; Ramon de la Sagra, Cuba, Rept. p. 62, t. 5.

Crocodilus biscutatus, Cuv. l. c. p. 53, t. 2. fig. 6.

Crocodilus americanus, Gray, Cat. Tort. 1844, p. 60; Dum. & Bocourt, Miss. Sc. Mex., Rept. p. 30, t. 8. fig. 1 † and t. 9. fig. 1.

Crocodilus pacificus, Dum. & Bocourt, Miss. Sc. Mex., Rept. p. 31, t. 9. fig. 5; Sumichrast, Bull. Soc. Zool. v. 1880, p. 171.

Hab. Mexico, Mazatlan, Presidio (*Forrer*), Tehuantepec (*Sumichrast*); Guatemala

* The width of the snout is taken on the level of the tenth upper tooth.

† This figure is again referred to as representing *C. moreletii* on p. 37, and it will be difficult to decide for which it was intended.

(*Salvin*); NICARAGUA (*Richardson*); PANAMA, Chepo river (*Arcé*).—COLOMBIA Cartagena (*Duméril & Bibron*); WESTERN ECUADOR (*Fraser*); ANTILLES, Cuba (*Ramon de la Sagra*), Jamaica (*Brit. Mus.*), San Domingo and Martinique (*Duméril & Bibron*); MARGARITA ISLAND and TRINIDAD (*Ramon de la Sagra*).

Vernacular name: "Lagarto" (Guatemala).

β. **Crocodilus americanus,** var. **moreletii.**

Crocodilus rhombifer (nec Cuv.), Wiegm. Herpetol. Mex. p. 22; Strauch, Synops. Crocodil. p. 41 (part.).

Crocodilus moreletii, A. Dum. Arch. Mus. vi. p. 255, t. 20; Dum. & Bocourt, Miss. Sc. Mex., Rept. p. 37, t. 9. figg. 2, 3.

Alligator lacordairei, Preudhomme dè Borre, Bull. Ac. Sc. Belg. xxviii. 1869, p. 109, c. t. (pull.).

Crocodilus mexicanus, Dum. & Bocourt, l. c. p. 34, t. 8. fig. 3.

Hab. MEXICO (*Brit. Mus.*), Alvarado river (*Deppe*), Tampico (*Bocourt*); BRITISH HONDURAS, Belize (*Preudh. de Borre*); GUATEMALA, Lake of Peten (*Morelet*), Rio Polochic, East Guatemala (*Bocourt*).

ALLIGATOR.

Alligator, Cuv.

1. **Alligator punctulatus.**

Jacaretinga punctulatus, Spix, Anim. Nov. p. 2, tab. 2.

Alligator punctulatus, Dum. & Bibr. p. 91; Strauch, Synops. Crocod. p. 24.

Champsa vallifrons, Natterer, Ann. Wien. Mus. ii. p. 322, t. 24.

Champsa punctulata, Natterer, l. c. p. 323, t. 25.

Jacare longiscutata, Gray, Trans. Zool. Soc. vi. p. 164, t. 34.

Jacare ocellata, Gray, l. c. pl. 33.

Jacare punctulata, Gray, l. c. p. 165.

Alligator chiapasius, Bocourt, Journ. Zool. 1876, v. p. 400; Sumichrast, Bull. Soc. Zool. 1880, v. p. 170.

Hab. MEXICO, Tehuantepec and Chiapas (*Sumichrast*); GUATEMALA, Huamuchal (*Salvin*).—VENEZUELA, Lake Valencia; BRITISH GUIANA (*Schomburgk, Brit. Mus.*); BOLIVIA, Santa Cruz de la Sierra (*Gray*); BRAZIL, Solimoens River (*Spix*), Rio Branco and Rio Negro (*Natterer*); ARGENTINE REPUBLIC (*Christie*); MARTINIQUE (*Plée*); TRINIDAD (*Mus. Paris.*).

Vernacular name: "Lagarto" (Guatemala, Panama).

This species seems to range from the head waters of the Parana and Paraguay through Brazil and Northern South America to Southern Mexico.

Order LACERTILIA.

Fam. TEJIDÆ.

AMEIVA.

Ameiva, Duméril & Bibron, Erp. Gén. v. p. 92.

1. **Ameiva surinamensis.** (Tab. XIX.)

Seps surinamensis, Laur. Syn. Rept. p. 59.
Tejus ameiva, Wied, Reise Bras. i. p. 88, ii. p. 337; Spix, Lacert. Bras. p. 21, t. 23.
Tejus lateristriga, Spix, l. c. p. 22, t. 24. fig. 1.
Tejus tritæniatus, Spix, l. c. p. 22, t. 24. fig. 2.
Ameiva vulgaris, Dum. & Bibr. v. p. 100.
Ameiva surinamensis, Gray, Cat. Liz. p. 18; Cope, Proc. Ac. N. S. Phil. 1862, p. 67.
Ameiva guttata, Gray, Cat. Liz. p. 18.
*Cnemidophorus præsignis**, Baird & Gir. Proc. Ac. N. Sc. Philad. 1852, p. 129; Girard, U. S. Naval & Astron. Exped. ii. p. 227, t. 38; Hallowell, Proc. Ac. N. Sc. Philad. 1860, p. 483.
Ameiva præsignis, Cope, Proc. Ac. N. Sc. Philad. 1862, p. 67; Journ. Ac. N. Sc. Philad. viii. 1876, p. 157.
Ameiva præsigna, Bocourt, Miss. Sc. Mex. Rept. p. 265, t. 20 B. fig. 9.
Cnemidophorus maculatus †, Fischer, Verh. Naturw. Ver. Hamb. iii. 1879, p. 95, t. 4.

Hab. NICARAGUA (*Cope*); COSTA RICA (*Brit. Mus.*); PANAMA, Chagres (*B. & G.*).

Abdominal scutes in ten longitudinal series; none of the scales on the throat are conspicuous for their size, but they increase somewhat in size in the posterior rows; a broad band of granular scales separates them from the scales of the collar-fold, which are rather small, in three or more series. Nostril between two scutes; supraorbitals four. Humeral scutes in about three longitudinal series.

The coloration of this species varies greatly, not only according to the locality, but also with age and individually. Thus the British Museum possesses specimens from Costa Rica 6 inches long (without tail), which agree entirely in coloration with the figure given by Girard of a specimen from Chagres (*C. præsignis*). Another specimen, which is figured on Pl. XIX., from the same locality, and was obtained with the others, but

* This synonymy is not intended to include the binomial names of all the varieties (real or supposed) of this widely-spread species, which have been introduced chiefly by American authors, but only such as, like '*præsignis*," have been intended for Central-American specimens.

† I am indebted to the kindness of Prof. Pagenstecher and Dr. Fischer for the opportunity of comparing one of the types with our specimens.

7 inches long (without tail), is, without any question, the same species; yet its coloration is very different. Instead of a light vertebral line, a series of bluish-white spots occupies the middle of the back; similar spots, edged with black, and arranged in irregular transverse series, ornament the flanks, very faint traces of lateral light longitudinal bands being still visible. The head and anterior part of the trunk and fore limbs are immaculate. Lower parts yellowish white, the scutes of the two outer rows black, each with a large bluish-white spot in the centre. Upper side of the hind limbs and tail with bluish-white specks.

This species ranges from Montevideo (*Darwin*), through Paraguay, Brazil, the Guianas into the southern part of Central America, and is one of the most common Lizards of the Tropical-American Region.

2. **Ameiva gabbiana.**

Ameiva gabbiana, Cope, Journ. Ac. N. Sc. Philad. viii. 1876, p. 117, t. 28. fig. 3 (head).

Hab. COSTA RICA, Old Harbour (*Cope*).

Abdominal scutes in eight longitudinal series; præanal region protected by a large round scute, with two or three smaller ones in front. Scales in the middle of the throat but little larger than the others; anterior margin of the collar with a transverse row of large scales. Nostril between two scutes; supraorbitals three. Humeral scutes in one series. "Two lateral light bands separated by a darker shade than that of the dorsal interval, and crossed, like the side below the lower line, by black bars. Below immaculate, or with black spots on the gular region."

3. **Ameiva undulata.**

Cnemidophorus undulatus, Wiegm. Herpet. Mex. p. 27; Salvin, Proc. Zool. Soc. 1860, p. 454.
Ameiva vulgaris, Dum. & Bibr. iv. p. 100.
? *Ameiva pulchra*, Hallow. Proc. Ac. N. Sc. Philad. 1860, p. 483; Cope, ibid. 1862, p. 62.
Ameiva undulata, Cope, Proc. Ac. N. Sc. Philad. 1862, p. 62; Bocourt, Miss. Sc. Mex., Rept. p. 254, t. 20 A. fig. 7, and 20 B. fig. 1; Sumichrast, Bull. Soc. Zool. 1880, p. 177.

Hab. MEXICO, Oaxaca (*Bocourt*), Vera Cruz (*Cope*), Isthmus of Tehuantepec (*Sumichrast*); GUATEMALA, Peten and Isabal (*Bocourt*), Dueñas, Coban, and Lanquin (*Salvin*); HONDURAS (*Brit. Mus.*); SALVADOR (*Bocourt*); NICARAGUA (*Hallowell*); COSTA RICA, Irazu (*Brit. Mus.*).

Abdominal scutes in eight longitudinal series; the præanals of small size, in several transverse rows. A patch of enlarged scales occupies the middle of the throat, and scales of similar size occupy the anterior margin of the collar. Nostril between two scutes; supraorbitals three. Humeral scutes in one main series, accompanied by some smaller scutes, but separated by a granular interspace from the pectoral scutes.

One of the commonest and most generally distributed species of Central America, extending southwards into Venezuela and Ecuador.

Very common in woods about the Isthmus of Tehuantepec (*Sumichrast*); common at Dueñas, Coban, and Lanquin, in Guatemala (*Salvin*).

4. Ameiva festiva.

Ameiva festiva, Lichtenst. Nomencl. Rept. Mus. Berol. p. 13; Cope, Journ. Ac. N. Sc. Philad. viii. 1876, p. 117.

Ameiva eutropia, Cope, Proc. Ac. N. Sc. Philad. 1862, p. 62.

Ameiva edwardsii, Bocourt, Ann. Sc. Nat. xvii. 1873, art. 17.

Ameiva festivus, Bocourt, Miss. Sc. Mex., Rept. p. 260, t. 20 A. fig. 10, and t. 20 D. fig. 3.

Hab. GUATEMALA, lower forests of Vera Paz (*Salvin*), Rio de la Pasion, Rio Polochic; COSTA RICA (*Cope*); PANAMA, Veragua (*Berl. Mus.*).

Abdominal scutes in eight longitudinal series; three large præanals, one in front of the two others. A patch of enlarged scales occupies the middle of the throat, and is surrounded by very small scales. Supraorbitals three. Anterior margin of the collar covered with large scales. Nostril between two scutes. Humeral scutes in one series, separated by a granular interspace from the pectoral scutes.

This species is scarcer and less generally distributed than *A. undulata*, and extends southwards into Colombia (*Cope*).

5. Ameiva quadrilineata.

Cnemidophorus quadrilineatus, Hallowell, Proc. Ac. N. Sc. Philad. 1860, p. 483.

Ameiva quadrilineata, Cope, Proc. Ac. N. Sc. Philad. 1862, p. 61.

Hab. NICARAGUA (*Hallowell*).

Abdominal scutes in eight longitudinal series; the præanals disposed in a single series, the posterior being the largest. A patch of enlarged scales occupies the middle of the throat; they are numerous, and graduate into the smaller ones. Supraorbitals three, sometimes a minute posterior fourth. Collar bordered by two series of enlarged scales. Humeral scutes in one main series. "Two narrow yellow lines on each side; the superior from the superciliary margin, convergent on the nuchal region, the inferior from the orbit, interrupted by the femur, continued on the base of the tail, bounded above and beneath by black. Inferior lateral region black, greenish vermiculated. Back posteriorly varied with black."

A. quadrilineata was founded upon young specimens, the bodies of which are only 16 lines long; and Bocourt is quite right in regarding the species as a very doubtful one (Miss. Sc. Mex., Rept. p. 295); he himself notices, under the name proposed by Hallowell, a young specimen from Veragua, but thinks that this as well as the types may prove to be young *A. undulata.*

[**Ameiva edracantha,** Bocourt (Ann. Sc. Nat. 1873, xix. art. 4; Miss. Sc. Mex., Rept. p. 263, t. 20 A. fig. 6), founded on a specimen in the Paris Museum which is said to have been brought from Mexico, occurs, in fact, near Guyaquil, where two specimens were collected by Fraser. This species, therefore, is extralimital, and I may mention on this occasion that the spinous lateral præanal scales are a sexual character, the female having flat præanals of the usual form.]

CNEMIDOPHORUS.

Cnemidophorus, Duméril & Bibron, Erp. Gén. v. p. 123.

While the Ameivas rarely leave the interior of woods, and are never seen in barren places, the Mexican species of *Cnemidophorus* are generally found in open and sunny spots, on the sandy banks of rivers, or among clear thickets (*Sumichrast*).

1. Cnemidophorus sex-lineatus*.

Lacerta sex-lineata, Linn. Syst. Nat. ed. 12, i. p. 364; Harlan, Journ. Ac. Nat. Sc. Philad. vi. 1827, p. 18.
Ameiva sex-lineata, Holbrook, N.-Amer. Herpet. 1836, p. 63, t. 6, and 2nd edit. 1842, p. 109, t. 15.
Cnemidophorus sex-lineatus, Dum. & Bibr. v. p. 131; Gray, Lizards, p. 21; Cope, Proc. Ac. N. Sc. Philad. 1866, p. 303; Bocourt, Miss. Sc. Mex., Rept. p. 273, t. 20 c. fig. 11.
Cnemidophorus gularis, Baird & Gir. Proc. Ac. N. Sc. Philad. 1852, p. 128; and U. S. Mex. Bound. Surv., Rept. t. 34. figg. 1–6.
Cnemidophorus perplexus, Baird & Gir. Proc. Ac. N. Sc. Philad. 1852, p. 128.
Cnemidophorus mexicanus, Peters, Berl. MB. 1869, p. 62 †.
Cnemidophorus communis, Cope, Proc. Am. Phil. Soc. xvii. 1877, p. 95.
Cnemidophorus angusticeps, Cope, l. c.

Hab. NORTH AMERICA, Southern United States.—MEXICO, Presidio (*Forrer*), Colima (*Bocourt*), Matamoras (*U. S. Nat. Mus.*), Guadalaxara, Cordova, and Yucatan (*Cope*); GUATEMALA, Coban (*Cope*).

Abdominal scutes in eight longitudinal series; scales in the middle of the throat somewhat enlarged, and gradually passing into the smaller lateral scales. A marginal

* Bocourt includes in the synonymy of this species a number of other names which have been created by American authors on very slight grounds. I fully agree with him as to their value, but it will be difficult to determine to which species they should be referred. From the character of the descriptions and figures I regard the task as impossible, the more so as it appears from Yarrow's 'Check-list of North-American Reptilia in the U. S. National Museum,' that the majority of the types of Baird's and Girard's species have not been preserved.

† Having some doubts as to the propriety of including this among the synonyms of *C. sex-lineatus,* I asked Prof. von Martens for information; and from his report it appears not only that the specimens of *C. mexicanus* are identical with those in the British Museum which were collected by Herr Forrer at Presidio, but also that Peters himself left a note on the bottle containing them that they were not distinct from *C. gularis* (B. & G.)

row of rather small scutes along the collar-fold; other smaller scutes or scales in front of it. Nostril in one scute; supraorbitals four. Humeral scutes in four longitudinal series, passing by somewhat smaller scales into the pectoral scutes. Three large præanals surrounded by smaller ones. The scales of the tail strongly keeled, the keels being distinctly subdiagonal, and terminating in a posterior projection.

The typical coloration consists in three straight bluish-white lines on each side of the body, of which the middle starts from the superciliary margin. In some localities this coloration is common to the adult male as well as to the female, the former having, however, often a dark blue or blackish chest or abdomen. In other localities the male assumes a gayer style of coloration; the longitudinal lines are more or less incomplete or disappear, and the body is prettily ornamented with bluish or greenish spots on a black ground-colour on the sides, and on a brown ground on the back. Throat and abdominal scutes variegated with bluish-black.

Common and widely spread in the Southern United States.

Var. **tessellata.**

Cnemidophorus tessellatus, Say in Pittsb. Exped. Rock. Mount. 1823, ii. p. 351.
Cnemidophorus sackii, Wiegm. Herpet. Mex. p. 29; Peters, Berl. MB. 1869, p. 63.
Cnemidophorus tigris, Baird & Gir. Proc. Ac. N. Sc. Phil. 1852, p. 69; U. S. Mex. Bound. Surv., Rept. t. 33.
Cnemidophorus gracilis, Baird & Gir. Proc. Ac. N. Sc. Phil. 1852, p. 128; U. S. Mex. Bound. Surv., Rept. t. 34. figg. 7–14.
Cnemidophorus melanostethus, Cope, Proc. Ac. N. Sc. Phil. 1863, p. 104.

Hab. North America, California, Texas, Colorado.—Mexico (*Wiegmann, U. S. Nat. Mus.*).

This variety grows to a larger size than the typical *C. sex-lineatus,* and in large specimens (5 inches long, without tail) the longitudinal lines may disappear altogether, leaving merely a pattern consisting of black irregular transverse spots. The snout also is longer, which is especially visible in the greater length of the postnasal scute; but this is by no means a constant character.

2. **Cnemidophorus costatus.**

Cnemidophorus costatus, Cope, Journ. Am. Phil. Soc. 1877, p. 95.

Hab. Mexico.

"A few rows of large scales on the collar, of which the marginal is the largest; two preoculars and a freno-ocular, four supraorbitals; similar to *C. communis* (*C. sex-lineatus*), but head shorter and scuta wider; brown, with black cross bands on sides, which join across the middle on the lumbar region; sacral region and femora white spotted."

"The short head of this species is accompanied by an abbreviation of the sutures of mutual contact of the fronto-nasal and internasal pairs of scuta, which is not seen in the other allied species."

3. Cnemidophorus deppii.

Cnemidophorus deppii, Wiegm. Herpet. Mex. p. 28; Peters, Berl. MB. 1869, p. 62; Bocourt, Miss. Sc. Mex., Rept. p. 281, t. 20. figg. 3, t. 20 c. figg. 5; Sumichrast, Bull. Soc. Zool. Fr. 1880, p. 177.

Cnemidophorus octolineatus, Baird, Proc. Ac. N. Sc. Philad. 1858, p. 255 *.

? *Cnemidophorus decem-lineatus*, Hallowell, Proc. Ac. N. Sc. Philad. 1860, p. 482.

Cnemidophorus immutabilis, Cope, Journ. Am. Phil. Soc. 1877, p. 93.

Cnemidophorus lativittis, Cope, l. c. p. 94.

Hab. Mexico, Colima (*Bocourt*), Tehuantepec (*Sumichrast*); Eastern Guatemala (*Bocourt*); Salvador (*Bocourt*); ? Nicaragua (*Hallowell*).—Venezuela, Caracas (*Brit. Mus.*).

Abdominal scutes in six series anteriorly and in eight posteriorly; scales in the middle of the throat somewhat enlarged, and gradually passing into the smaller lateral scales. Collar covered by three or four series of enlarged scales. Nostril in one scute; supraorbitals three, rarely four †. Humeral scutes in four longitudinal series, separated by a narrow granular fold from the pectoral scutes; three large præanals. The scales of the tail are strongly keeled, the keels being very slightly oblique. Body with eight or nine parallel bluish-white lines, nearly as broad as the interspaces between them.

4. Cnemidophorus lineatissimus.

Cnemidophorus lineatissimus, Cope, Journ. Am. Phil. Soc. 1877, p. 94.

Hab. Mexico, Colima, Guadalaxara (*Cope*).

"Several rows of large scales on the collar, the border row not larger nor much smaller; two preoculars; no freno-ocular; three supraorbitals; larger gulars extending across throat; superior preocular not descending to labials; postbrachials large, continuous with brachials; femorals with eight rows; black, with ten or eleven pale bands; sides and femora pale spotted; throat black. The muzzle is rather acute, but not elongate, and the fronto-nasals have considerable mutual contact. The larger and smaller gulars are abruptly distinguished from each other, and the former are smaller than the scales of the collar. The frenal plates form a circle surrounding a large median scute, of which the two posterior plates are the largest. There is a space between the two submedian lines, which is often divided by a median line. Below the lowest line the sides are black, with large light spots, open below."

* Said to have four supraorbitals (Cope, Journ. Am. Phil. Soc. 1877, p. 95).

† A specimen in the Brussels Museum from Mexico shows distinctly a small fourth supraorbital.

5. **Cnemidophorus mariarum.** (Tab. XX.)

Hab. Mexico, Tres Marias Islands (*Forrer*).

Abdominal scutes in eight longitudinal and in thirty-eight transverse series. Scales of the middle of the throat but little larger than the others. Collar covered by three or four series of enlarged scales. Nostril in one scute. Snout narrow and long in adult specimens, and comparatively shorter in young ones. Loreal with the postero-inferior angle detached. Supraorbitals four; temporal scales granular; scales of the body unusually small; the scales of the tail with a rather weak keel, which runs nearly parallel to the lateral margins of the scale. Humeral scutes in five or six longitudinal series, separated from the pectoral scutes by a granular fold; femoral scutes in ten or twelve series. Femoral pores twenty-one. Three præanals.

The young have six narrow whitish lines, of which the dorsal pair are very faint; the two outer lines take their origin from behind the eye, and are separated by a blackish band; the lower parts are uniform white. In very large specimens ($4\frac{1}{2}$ inches long, without tail) the bands have nearly disappeared, the sides being coarsely marbled with greenish-white and black; upper side of the hind legs with round greyish-white spots; lower parts nearly uniform black, or black marbled with greenish-white.

Herr Forrer brought five specimens of different sizes and ages.

6. **Cnemidophorus guttatus.** (Tab. XXI.)

Cnemidophorus guttatus, Wiegm. Herpet. Mex. p. 29; Peters, Berl. MB. 1869, p. 62.

Cnemidophorus microlepidopus, Cope, Proc. Amer. Phil. Soc. xvii. 1877, p. 93; Sumichrast, Bull. Soc. Zool. Fr. 1880, p. 177.

Cnemidophorus unicolor, Cope, l. c.; Sumichrast, l. c.

Hab. Mexico, Tehuantepec and Huamelula (*Sumichrast*), Vera Cruz (*Brit. Mus.*).

Abdominal scutes in eight series; scales of the middle of the throat somewhat enlarged, and gradually passing into the small lateral scales. Collar covered by three or four series of enlarged scales. Nostril in one scute; supraorbitals three. Humeral scutes in three or four longitudinal series, separated by a narrow granular fold from the pectoral scutes; three large præanals. The scales of the tail are strongly keeled, the keels not projecting behind, and having a subdiagonal direction. Young specimens have the lines of the body very indistinct, but there is generally a white line running along the middle of the posterior side of the thigh, and continued along the side of the tail. They have also frequently four spots on the edge of the ear-opening, and one or two others between the ear and the eye. In old specimens the longitudinal lines may disappear altogether, or be indicated by series of white spots only.

7. Cnemidophorus inornatus.

Cnemidophorus inornatus, Baird, Proc. Ac. N. Sc. Philad. 1858, p. 255; Cope, Proc. Am. Phil. Soc. xvii. 1877, p. 93.

Hab. NORTH MEXICO, Nuevo Leon (*Baird*).

Four supraorbitals. Scales on the gular fold smaller than those on the breast anteriorly, and scarcely larger than those on the middle of the chin. Scales of the back tuberculated and elevated. Light greenish-olive, paler beneath. No lines on the body.

GYMNOPHTHALMUS.

Gymnophthalmus, Merrem, Tent. Syst. Amphib. p. 74.

Of the six species known, three are West-Indian, two South-, and one Central-American.

1. Gymnophthalmus sumichrasti.

Epaphelus sumichrasti, Cope, Journ. Ac. N. Sc. Philad. 1876, p. 115.
Gymnophthalmus sumichrasti, Bocourt, Miss. Sc. Mex., Rept. p. 471, t. 22 H. fig. 2.

Hab. MEXICO, Tehuantepec; WEST GUATEMALA.—VENEZUELA (*Brit. Mus.*).

Sumichrast says that he has but rarely met with this lizard near Cacoprieto under rotting trunks in the forest.

TRETIOSCINCUS.

Tretioscincus, Cope, Proc. Ac. N. Sc. Philad. 1862, p. 184.

Beside the Nicaraguan species, only one other, from Colombia (*T. bifasciatus*, A. Dum.), is known.

1. Tretioscincus lævicauda.

Tretioscincus lævicauda, Cope, Proc. Am. Phil. Soc. xi. 1870, p. 557.

Hab. NICARAGUA.

ARGALIA.

Argalia, Gray, Ann. & Mag. Nat. Hist. 1846, xviii. p. 67.

1. Argalia pœcilochilus.

Gerrhonotus pœcilochilus, Lichtenst. Nomencl. Rept. Mus. Berol. p. 16.
Ecpleopus (*Argalia*) *pœcilochilus*, Peters, Abhandl. Berl. Ak. 1863, p. 213.

Hab. PANAMA, Veragua.—VENEZUELA.

ANADIA.

Anadia, Gray, Cat. Liz. p. 58.
Chalcidolepis, Cope, Journ. Ac. N. Sc. Philad. 1876, viii. p. 116.

1. Anadia metallica.

Chalcidolepis metallicus, Cope, l. c. t. 24. fig. 5, t. 28. fig. 2.
Ecpleopus (*Chalcidolepis*) *metallicus*, Bocourt, Miss. Sc. Mex., Rept. p. 371.

Hab. Costa Rica, Aguacate Mountains.

Fam. LEPIDOPHYMIDÆ.

LEPIDOPHYMA.

Lepidophyma, A. Duméril, Cat. Méth. Rept. Mus. Paris, 1851, p. 137.
Xanthusia, Baird, Proc. Ac. N. Sc. Philad. 1858, p. 255.
Poriodogaster, Gray, Proc. Zool. Soc. 1863, p. 154.

1. Lepidophyma flavomaculatum.

Lepidophyma flavomaculatum, A. Dum. l. c. p. 138; Rev. & Mag. Zool. 1852, p. 409, t. 17; Bocourt, Miss. Sc. Mex., Rept. p. 306, t. 20 f. figg. 2; Sumichrast, Bibl. Univ. et Suisse, 1873, p. 251.
Poriodogaster grayi, Gray, l. c. t. 21; Peters, Proc. Zool. Soc. 1874, p. 307.
Lepidophyma smithii, Bocourt, Journ. Zool. 1876, p. 403; Miss. Sc. Mex., Rept. p. 309, t. 20 f. figg. 3, and t. 20 g. figg. 2; Sumichrast, Bull. Soc. Zool. Fr. 1880, p. 178.
Lepidophyma, sp., F. Müller, Verh. naturf. Ges. Basel, 1878, p. 390, t. 1 & 2.

Hab. Mexico, Tehuantepec (*Sumichrast*); Guatemala, Peten (*Morelet*), Mazatenango (*Müller*); Panama (*Peters*).

Sumichrast (*l. c.*) has given an account of the habits of this very singular type of Lizards.

Fam. AMPHISBÆNIDÆ.

CHIROTES.

Chirotes, Duméril & Bibron, Erp. Gén. v. p. 472.

1. Chirotes canaliculatus.

Bipes canaliculatus, Bonnat. Encycl. Erpét. p. 68, t. 12.
Lacerta lumbricoides, Shaw, Nat. Misc. vi. t. 212.
Lacerta mexicana, Donndorf, Zool. Beitr. iii. p. 135.
Chirotes canaliculatus, Merr. Tent. p. 181; Dum. & Bibr. Erp. Gén. p. 474; Bocourt, Miss. Sc. Mex., Rept. p. 487, t. 21 c. fig. 8.
Chirotes lumbricoides, Fleming, Phil. Zool. ii. p. 278; Long, Exped. Rock. Mount. i. p. 484; Gray, Cat. Tort. &c. 1884, p. 74.

Hab. North America, Platte River (*Long*).—Mexico.

Very rare.

AMPHISBÆNA.

Amphisbæna, Linnæus.

Of the numerous South-American species only one is known to penetrate northwards into the southernmost part of Central America. It is perhaps the most common of South-American Amphisbænas.

1. **Amphisbæna fuliginosa.**

Amphisbæna fuliginosa, Linn. Syst. Nat.; Bocourt, Miss. Sc. Mex., Rept. p. 489, t. 21. figg. 5 & 6.

Hab. PANAMA (*Bocourt*).

Fam. SCINCIDÆ.

MOCOA.

Mocoa, Gray, Cat. Liz. p. 80.

1. **Mocoa lateralis.**

Scincus lateralis, Say in Long's Exped. Rocky Mount. 2nd ed. ii. 1823, p. 349.

Lygosoma lateralis, Dum. & Bibr. v. p. 719; Holbr. N. Amer. Rept. 2nd ed. ii. p. 133, t. 19; Bocourt, Miss. Sc. Mex., Rept. p. 446, t. 22 F. figg. 3.

Mocoa lateralis, Gray, Cat. Liz. p. 83.

Oligosoma gemmingerii, Cope, Proc. Ac. N. Sc. Philad. 1864, p. 180.

Lygosoma (Mocoa) gemmingerii, Bocourt, Miss. Sc. Mex., Rept. p. 449.

Hab. NORTH AMERICA, Southern States.—MEXICO, Orizaba and Tehuantepec (*Sumichrast*), Jalapa (*Höge*).

Found in temperate forests (*Sumichrast*).

2. **Mocoa assata.**

Lampropholis assatus, Cope, Proc. Ac. N. Sc. Philad. 1864, p. 179.

Mocoa lateralis (sp. *affinis*), Müller, Kat. Mus. Basel, 1878, p. 628.

Lygosoma (Mocoa) assata, Bocourt, l. c. p. 450, t. 22 F. fig. 7.

Mocoa assata, Cope, Journ. Ac. N. Sc. Philad. viii. 1876, p. 115; Sumichrast, Bull. Soc. Zool. Fr. v. 1880, p. 179.

Hab. MEXICO, Tehuantepec (*Sumichrast*); GUATEMALA, Rio Motagua (*Salvin & Godman*), Atitlan, Solola, Alta Vera Paz (*Bocourt*); SALVADOR, Volcan de Isalco (*Dow*).

Very common in the neighbourhood of Santa Efigenia and Tapana (*Sumichrast*).

EUMECES.

Eumeces (Wiegmann), Peters, MB. Berl. Akad. 1864, p. 48.

1. **Eumeces lynxe.**

Euprepes lynxe, Wiegm. Herpet. Mex. p. 36.
Plestiodon quinquelineatum, Dum. & Bibr. Erp. Gén. v. p. 707; Dugès, La Naturaleza, 1870, p. 144.
Eumeces lynxe, Peters, l. c. p. 49; Bocourt, Miss. Sc. Mex., Rept. p. 437, t. 22 E.

Hab. MEXICO, mountain districts to 12,000 feet (*Sumichrast*), Guanajuato (*Dugès*).

2. **Eumeces brevirostris.** (Tab. XXII. fig. B.)

Mabouia brevirostris, Günth. Proc. Zool. Soc. 1860, p. 316.
Eumeces brevirostris, Bocourt, l. c. p. 439, t. 22 A. figg. 7, and t. 22 E. fig. 1 *a.*

Hab. MEXICO, Ciudad in Durango (*Forrer*), Oaxaca (*Sallé*).

Scales round the middle of the trunk in twenty-four series. Postnasal absent, postmental not divided. Generally two pairs of enlarged nuchal scales; subcaudals large. Back brown; along each side a black band, commencing from the nostril and disappearing at the root of the tail; it is bordered by a yellowish-white line above and below, the upper border running along the canthus rostralis, the lower along the upper labials. Tail with a bluish tinge; lower parts whitish.

3. **Eumeces bocourti.** (Tab. XXII. fig. C.)

Eumeces bocourtii, Bouleng. Ann. & Mag. N. H. 1883, xi. p. 342.

Hab. MEXICO, Presidio (*Forrer*).

Scales round the middle of the trunk in twenty-six series. Postnasal present, but small; postmental undivided. Two pairs of enlarged nuchal scales; subcaudals scarcely enlarged. Bronze-coloured above; sides with a brownish band, which disappears on the hinder part of the body; anteriorly it is bordered above by a faint yellowish line. Lower parts whitish.

4. **Eumeces callicephalus.**

Eumeces callicephalus, Bocourt, Miss. Sc. Mex., Rept. p. 431, t. 22 D. fig. 2, and t. 22 E. fig. 2.

Hab. MEXICO, Guanajuato (*Dugès*).

5. **Eumeces sumichrasti.**

Plistodon sumichrasti, Cope, Proc. Ac. N. Sc. Philad. 1866, p. 321.

Hab. MEXICO, Orizaba (*Sumichrast*).

Sumichrast found only two specimens of this species in the oak-woods at a height of 1800 feet.

6. **Eumeces dugesii.**

Eumeces dugèsii, Thominot, Bull. Soc. Philomath. vii. 1883, p. 138.

Hab. MEXICO, Guanajuato (*Dugès*).

7. Eumeces schwartzii.

Eumeces schwartzei, Fischer, Abhandl. naturwiss. Ver. Hamburg-Altona, viii. 1884, Heft 2, p. 3, t. 7. fig. 1.

Hab. MEXICO, Island in the Laguna de los Terminos, Bay of Campeche.

8. Eumeces furcirostris.

Eumeces furcirostris, Cope, Proc. Am. Phil. Soc. xxii. 1885, p. 169.

Hab. MEXICO, Vera Cruz, Jalapa (*Cope*).

MABUIA.

Mabuya (Fitzinger), Gray, Cat. Liz. p. 93.

1. Mabuia agilis *.

Scincus agilis, Raddi, Mem. Soc. Ital. xix. 1823, p. 62.
Mabuya agilis, Fitz. Neu. Classif. Rept. 1826, p. 23; Bocourt, Miss. Sc. Mex., Rept. p. 395, t. 22 B. figg. 2.
Eumeces spixii, Dum. & Bibr. Erp. Gén. v. p. 642.
Mabouya æenea, Gray, Cat. Liz. p. 94.
Mabuia unimarginata, Cope, Proc. Ac. N. Sc. Phil. 1862, p. 187.

Hab. MEXICO (*Sumichrast*, *Sallé*, *Peters*), Yucatan (*Boucard*); GUATEMALA (*Bocourt*); PANAMA (*Cope*), Veragua (*Peters*).—TROPICAL SOUTH AMERICA, common and generally distributed.

Sumichrast (*in litt.*) writes from Mexico:—"Common in the tierra caliente of both coasts; it lives under the bark of dead trees, under stones, and among the impenetrable thickets formed by spinous piñuelas (*Fourcroya*, sp.)."

2. Mabuia alliacea.

Mabuia alliacea, Cope, Journ. Ac. N. Sc. Phil. viii. 1876, p. 115.

Hab. COSTA RICA, low country (*Cope*).

Fam. ANELYTROPIDÆ.

The few species of this family known hitherto are African. The following species has recently been discovered in Mexico.

ANELYTROPSIS.

Anelytropsis, Cope, Proc. Am. Phil. Soc. xxii. † 1885.

* The following synonymy includes references to Central-American specimens only.

† Thanks to the kindness of Mr. Boulenger I am enabled to take this reference from an advance sheet of a part of the volume quoted which has not yet been published.

1. Anelytropsis papillosus.

Anelytropsis papillosus, Cope, l. c.

Hab. Mexico, Jalapa.

Fam. ANGUIDÆ.

DIPLOGLOSSUS.

Diploglossus, Wiegm. Herpet. Mex. p. 36.

1. Diploglossus bilobatus.

Celestus bilobatus, O'Shaughn. Ann. & Mag. N. H. 1874, xiii. p. 257.
Diploglossus bilobatus, Bocourt, Miss. Sc. Mex., Rept. p. 382, t. 22 a. figg. 1.

Hab. Costa Rica (*Brit. Mus.*).

2. Diploglossus steindachneri. (Tab. XXII. fig. A.)

Diploglossus steindachneri, Cope, Proc. Ac. N. Sc. Phil. 1864, p. 179; Bocourt, l. c. p. 383, t. 22. figg. 3.
Diploglossus chalybæus, Cope, l. c. 1866, p. 321; and Proc. Am. Phil. Soc. xxii. 1885, p. 170.

Hab. Mexico, Orizaba (*Sumichrast*), Jalapa (*Cope*); Guatemala, Vera Paz, low forest (*Godman & Salvin*), Atitlan (*Bocourt*).

The specimen from Vera Paz, which we figure here, is coloured differently from the type. The back is nearly uniform greenish olive, with an indistinct blackish longitudinal line following the middle of the series of scales, and composed of short streaks, one to each scale. The sides are so densely mottled with whitish and blackish specks as to almost obliterate the ground-colour. Lower parts bluish white.

Siderolampus enneagrammus, Cope, l. c. 1860, p. 368, with a body 15 lines long, is probably the young of this species.

Sumichrast found this Lizard near Orizaba, at an elevation of from 4000 to 6000 feet.

3. Diploglossus monotropis. (Tab. XXIII.)

Scincus monotropis, Kuhl, Beiträg. p. 18.
Diploglossus monotropis, Wiegm. Herpet. Mex. p. 36; Peters, Berl. MB. 1864, p. 48; Cope, Proc. Ac. N. Sc. Phil. 1868, p. 98; Journ. Ac. N. Sc. Phil. viii. 1876, p. 118; and Proc. Amer. Phil. Soc. xvii. 1878, p. 92; Bocourt, l. c. p. 391, t. 22. figg. 6.
Camilia jamaicensis, Gray, Cat. Liz. p. 119.

Hab. Costa Rica and Isthmus of Darien (*Cope*).—Ecuador (*Mus. Brit.*, *Peters*, *Cope*).

This species, to which Jamaica has been erroneously assigned as its native country, does not seem to be uncommon in Ecuador, and has been recorded from the east side of the Isthmus of Darien and Costa Rica by Cope. It is well and fully described by Bocourt.

OPHISAURUS.

Ophisaurus, Daudin, Hist. Rept. vii. p. 346.

1. **Ophisaurus ventralis.**

Glass-Snake, Catesby, Carol. ii. t. 59.

Anguis ventralis, Linn. Syst. Nat. 12 ed. i. p. 391.

Ophisaurus ventralis, Daud. l. c. p. 352, t. 61. figg. 40, 41, & t. 88; Dum. & Bibr. Erp. Gén. v. p. 425; Holbrook, N. Amer. Herpet. ii. p. 139; Yarrow, Bull. U. S. Nat. Mus. no. 24, p. 46.

Hab. North America, Southern States.—Mexico, Jalapa (*U. S. Nat. Mus.*).

GERRHONOTUS.

Gerrhonotus, Wiegmann, Herpet. Mex. p. 29.

Abronia, Gerrhonotus, Elgaria, Barisia, Gray, Ann. Nat. Hist. 1838, i. pp. 389, 390.

Pterogasterus (Peale & Green), *Gerrhonotus, Mesaspis, Barissia,* Cope, Proc. Am. Phil. Soc. 1878, xvii. pp. 96, 97.

This genus is a truly Central-American, or rather Mexican, type, two only out of seventeen species occurring beyond the northern boundary of Mexico, and only one of them being more widely and generally spread north than south of that boundary. I have therefore appended short diagnoses to all the species, except the two following, about the relations of which to the other species I have not been able to form an opinion:—

Gerrhonotus webbii, Baird, Proc. Ac. N. Sc. Phil. 1858, p. 255; and U. S. and Mex. Bound. Exped., Rept. p. 11, t. 24. figg. 1–10.—*Hab.* San Diego.

Gerrhonotus fulvus, Cope, Journ. Ac. N. Sc. Phil. 1865, p. 118 (not Bocourt), = *Gerrhonotus monticolus,* Cope, Proc. Am. Phil. Soc. 1877, p. 97.—*Hab.* Eastern Cordillera of Costa Rica (11,500 feet).

The majority of the species are inhabitants of the mountainous or alpine districts, where they live in oak- and pine-forests, at an altitude varying from 6000 to 12,000 feet, under decaying trunks, or among the large floating tufts of *Tillandsia usneoides.* When caught they bite most furiously (*Sumichrast*).

1. **Gerrhonotus deppii.**

Gerrhonotus deppii, Wiegm. Herpet. Mex. p. 31, t. 9. fig. 2; Dum. & Bibr. Exp. Gén. v. 398; Bocourt, Miss. Sc. Mex., Rept. p. 325, t. 21 a. figg. 3.

Abronia deppii, Gray, Cat. Liz. p. 53.

Hab. Mexico (*Wiegmann*).

Dorsal scales smooth, in ten or eleven longitudinal series. Nasal shield not in contact with the rostral. Seven shields on the upper surface of the snout, between the vertical and rostral, disposed thus $\left\{\begin{matrix}a-a'\\b-b'\\c\\d-d'\end{matrix}\right\}$; a supranasal is present besides.

2. **Gerrhonotus gramineus.** (Tab. XXIV. fig. B.)

Gerrhonotus gramineus, Cope, Proc. Ac. N. Sc. Phil. 1864, p. 179; Bocourt, l. c. p. 330, t. 21 A. figg. 6 (part. ?).

Hab. MEXICO, Orizaba (*Sumichrast*).

Dorsal scales with a very obtuse keel, in fourteen, nuchal scales in four, ventral in twelve longitudinal series. Nasal shield not in contact with the rostral. Seven shields on the upper surface of the snout, disposed as in *G. tæniatus*; supranasals present. The groove between the ear and shoulder is covered with very small scales. Green, dorsal scales blackish at the base and yellowish at the tip; greenish white below. (Iris of the eye pale-blue: *Sumichrast.*)

3. **Gerrhonotus oaxacæ.** (Tab. XXIV. figg. A, A′, A″.)

? *Gerrhonotus gramineus*, Bocourt, l. c. p. 330 (part.).

Hab. MEXICO, Oaxaca (*Boucard*).

Closely allied to *G. gramineus.*

Dorsal scales with a very obtuse keel, in fifteen, nuchal scales in four, ventral in fourteen longitudinal series. Nasal shield not in contact with the rostral. Seven shields on the upper surface of the snout, between the vertical and rostral, disposed thus $\left\{\begin{matrix}a-a'\\b-b'\\c\\d-d'\end{matrix}\right\}$; a supranasal is present besides. Rather large and prominent scales on the side of the neck, behind the ear. Uniform olive-bronze-coloured above (ad. ♀); young with irregular brown transverse bands across the neck, body, and tail. Uniform greenish white below.

4. **Gerrhonotus tæniatus.**

Gerrhonotus tæniatus, Wiegm. Isis, 1828, p. 379; and Herpetol. Mex. p. 32, t. 9. fig. 1; Dum. & Bibr. Erp. Gén. v. p. 399; Bocourt, l. c. p. 327, t. 21 A. figg. 4, 5.

Hab. MEXICO (*Wiegmann, Bocourt*).

Dorsal scales with a very obscure keel, in fourteen longitudinal series; ventral scales in twelve, nuchal scales in six series. Nasal shield not in contact with the rostral. Seven shields on the upper surface of the snout, between the vertical and rostral, disposed thus $\left\{\begin{matrix}a-a'\\b-b'\\c\\d-d'\end{matrix}\right\}$; a supranasal is present besides. Small granular scales on the sides of the neck, behind the ear.

5. Gerrhonotus vasconcelosii.

Gerrhonotus (Abronia) vasconcelosi, Bocourt, Nouv. Arch. Mus. 1872, p. 107; Miss. Sc. Mex., Rept. p. 334, t. 21. fig. 3, and t. 21 A. figg. 8.

Hab. GUATEMALA, Argueta in Los Altos (*Bocourt*).

Dorsal scales with a very obtuse keel, in twelve or thirteen, nuchal scales in six, ventral in fourteen longitudinal series. Nasal shield not in contact with the rostral. Scales above the tympanum subconical and short. Seven shields on the upper surface of the snout, disposed as in *G. tæniatus*; supranasals present. Greenish, with dark cross bands.

6. Gerrhonotus auritus.

Gerrhonotus auritus, Cope, Proc. Ac. N. Sc. Phil. 1868, p. 306.

Hab. GUATEMALA, in the neighbourhood of Peten and Coban.

Dorsal scales strongly but finely keeled, in sixteen, nuchal scales in six, ventral in fourteen longitudinal series. Scales above the tympanum long and pointed. Shields on the upper surface of the snout as in *G. gramineus.* Pea-green, with dark angular cross bands.

7. Gerrhonotus fimbriatus.

Gerrhonotus auritus, Bocourt, Miss. Sc. Mex., Rept. p. 337, t. 21. fig. 2, and t. 21 A. fig. 7.
Barissia fimbriata, Cope, Proc. Am. Phil. Soc. xxii. 1885, p. 171.

Hab. GUATEMALA, pine-forests of Alta Vera Paz (*Bocourt*).

Dorsal scales obtusely keeled, in fourteen, nuchal scales in six, ventral in fourteen longitudinal series. Nasal shield not in contact with the rostral. Scales above the tympanum long and pointed. Four pairs of shields on the upper surface of the snout, between the vertical and rostral; no supranasals. Olive-coloured, with dark angular cross bands.

8. Gerrhonotus antauges.

Barissia antauges, Cope, Proc. Ac. N. Sc. Phil. 1866, p. 132.
Gerrhonotus viridiflavus, Bocourt, Ann. Sc. Nat. 1873, xvii. art. 2.
Gerrhonotus bocourti, Peters, MB. Berl. Ak. 1876, p. 297.
Gerrhonotus antauges, Bocourt, Miss. Sc. Mex., Rept. p. 346, t. 21. fig. 4, and t. 21 B. figg. 7.

Hab. MEXICO, Tehuantepec (*Sumichrast*).

Dorsal scales smooth, or very obtusely keeled, in fourteen, ventral in twelve longitudinal series. Nasal shield not in contact with the rostral. Eight shields on the upper surface of the snout, between vertical and rostral, disposed in pairs, of which the second are the supranasals, sometimes meeting in the middle, at other times separated from each other by the third pair. Lower parts blackish, anteriorly mottled with yellow.

9. Gerrhonotus rudicollis.

Gerrhonotus rudicollis, Wiegm. Isis, 1828, p. 379; and Herpet. Mex. p. 31, t. 10. fig. 1; Bocourt, Miss. Sc. Mex., Rept. p. 367, t. 21 B. figg. 3.
Barissia rudicollis, Gray, Cat. Liz. p. 55.

Hab. Mexico (*Deppe*).

Dorsal scales strongly keeled and, like the ventral, in twelve or fourteen, nuchal in four or six longitudinal series. Nasal in contact with the rostral; supranasals none. Three pairs of shields on the upper surface of the snout, between vertical and rostral. Sides with brown irregular cross bands.

One specimen only is known.

10. Gerrhonotus imbricatus.

Gerrhonotus imbricatus, Wiegm. Isis, 1828, p. 379; and Herpet. Mex. p. 34, t. 10. figg. 2 & 5; Dum. & Bibr. Erp. Gén. v. p. 407; Bocourt, Miss. Sc. Mex., Rept. p. 363, t. 21 B. figg. 2.
Gerrhonotus lichenigerus, Wagl. Ic. Amph. t. 34. fig. 2; Wiegm. Herpet. Mex. p. 35, t. 10. fig. 6 (*G. adspersus*); Bocourt, l. c. p. 365, t. 21 B. figg. 1 (*G. imbricata,* var.).
Barissia imbricata, Gray, Cat. Liz. p. 55.

Hab. Mexico, Cuernavaca and Guanajuato (*Bocourt*), City of Mexico (*Brit. Mus.*), Vera Cruz (*Boucard*), Tehuantepec (*Sumichrast*).

Dorsal scales keeled, in twelve or fourteen longitudinal series. The nasal is not in contact with the rostral; supranasals none. Three pairs of shields on the upper surface of the snout, all very convex.

This well-known species is subject to variation of colour, white dots being irregularly scattered about the scutes of the head and body. The mental scute is generally followed by a pair of scutes, more rarely by a single one.

11. Gerrhonotus planifrons.

Gerrhonotus planifrons, Bocourt, l. c. p. 361, t. 21 c. figg. 1.

Hab. Mexico, Oaxaca (*Boucard*).

This form, if it should prove to be a constant modification, differs from the typical *G. imbricatus* in having the head-shields flat, and sixteen longitudinal series of dorsal scales.

12. Gerrhonotus cæruleus. (Tab. XXV. figg. A–D.)

Gerrhonotus cæruleus, Wiegm. Isis, 1828, p. 379; Dum. & Bibr. Erp. Gén. v. p. 397; Bocourt, Miss. Sc. Mex., Rept. p. 353, t. 21 c. figg. 3.
Gerrhonotus burnettii, Gray in Griff. An. King., Rept. p. 64; and in Beechey's Voy. Rept. p. 96, t. 31. fig. 2 (bad); and in Cat. Liz. p. 54; Bocourt, l. c. p. 356, t. 21 c. figg. 4.

Gerrhonotus multicarinatus, Blainv. Nouv. Ann. Mus. d'Hist. Nat. 1835, p. 289, t. 25. fig. 2 (bad); Dum. & Bibr. Erp. Gén. v. p. 404; Bocourt, l. c. p. 357, t. 21 c. figg. 5.
Elgaria kingii, Gray, Ann. Nat. Hist. i. 1838, p. 390, and Cat. Liz. p. 54.
Gerrhonotus multifasciatus, Dum. & Bibr. Erp. Gén. v. p. 401.
Gerrhonotus wiegmanni, Gray, Cat. Liz. p. 54.
Tropidolepis scincicauda, Skilton, Amer. Journ. Sc. 1849, vii. p. 202, figg. 1–3 (bad).
Elgaria formosa, Baird & Gir. Proc. Ac. N. Sc. Phil. 1852, p. 175; and in Wilkes's U. S. Explor. Exped., Herp. p. 206, t. 23. figg. 10–17.
Elgaria nobilis, Baird & Gir. Proc. Ac. N. Sc. Phil. 1852, p. 129.
Gerrhonotus nobilis, Baird & Gir. in U. S. & Mex. Bound. Exped., Rept. p. 11, t. 25. figg. 1–8.
Elgaria scincicauda, Baird & Gir. in Stansbury, Exped. Utah, p. 348, t. 4. figg. 1–3; and in Wilkes's U. S. Expl. Exped., Herp. p. 210, t. 23. figg. 1–9.
Elgaria grandis, Baird & Gir. Proc. Ac. N. Sc. Phil. 1852, p. 176; and in Wilkes's Explor. Exped., Herp. p. 212, t. 22. figg. 1–9.
Elgaria marginata, Hallow. Proc. Ac. N. Sc. Phil. 1852, p. 179; and in Sitgreave, Exped. Zuni & Colorado, p. 114, t. 3 (bad).
Gerrhonotus kingii, Bocourt, l. c. p. 339, t. 21 c. figg. 2.

Hab. NORTH AMERICA, British Columbia, California, and New Mexico.—MEXICO and along the Pacific side of CENTRAL AMERICA to COSTA RICA, Cartago (*Mus. Brit.*).

Dorsal scales keeled, in from fourteen to sixteen (seventeen), nuchal and ventral in twelve longitudinal series. Nasal shield in contact with the rostral. Normally seven shields on the upper surface of the snout, between the vertical and rostral, which are disposed thus $\left\{\begin{matrix} a\text{—}a' \\ (a'') \\ b\text{—}b' \\ c \\ d\text{—}d' \end{matrix}\right\}$; but this number may be increased by the intercalation of a central shield a'' between the a and b series, or diminished by the confluence of two shields. Supranasals absent. Also the coloration varies considerably, but generally the sides, and sometimes also the back, are ornamented with dark or black transverse stripes, many of the scales composing the bands having yellow or white tips.

The variations in the scutellation of the snout do not indicate local races, but occur in specimens from the same locality, or an identical scutellation may be found in very distant localities; being unaccompanied by other characters, these variations are useless for specific distinction.

The size of the unpaired shield c varies very much, and is sometimes much increased at the expense of the paired shields in front and behind. It is smallest in the type of *C. burnettii* (exact locality unknown) and in a specimen from Vancouver's Island *; six other specimens from the same locality and from the mainland opposite have this

* This would be probably the *Elgaria principis*, Baird & Gir. Proc. Ac. N. Sc. Phil. 1852, p. 175, and in Wilkes's U. S. Explor. Exped., Herp. p. 214, t. 22. figg. 9–16.

shield much larger and so much produced backward as to touch the vertical; and the same peculiarity occurs in a specimen from some part of the Pacific coast of Central America, whilst in two other specimens from the same locality the shield is of the usual size.

Figures of these three last specimens are given on Tab. XXV. figs. A–C, to show the variation of the shields:—

In one (figg. A, A′) shield *c* is of moderate size, a small unpaired shield is intercalated between *a* and *a*′, *b* and *b*′ are in contact with each other.

In the second (fig. C) *c* is so much enlarged as to come in contact with the front central, and to separate *b* and *b*′.

In the third (fig. B) *c* is produced backwards, separating *d* and *d*′, and *b* and *a*″ are confluent into one shield.

The last-named anomaly, viz. the confluence of *b* with *a*″, is of very common occurrence in Californian specimens.

In specimens called *E. grandis*, from Oregon, the central shield is much enlarged at the expense of *b*—*b*′, which are small and widely apart from each other.

A specimen from Cartago (Costa Rica) with normal scutellation is figured on Tab. XXV. fig. D.

All the figures are of the natural size.

13. **Gerrhonotus obscurus.** (Tab. XXV. figg. E, E′.)

Hab. Mexico (*Brit. Mus.*).

Dorsal scales keeled, in fourteen longitudinal and fifty (between occiput and root of tail) transverse series; nuchal scales in eight, ventral in twelve longitudinal; the latter in forty-two (between front of humerus and vent) transverse series. Nasal shield not in contact with the rostral; supranasals present. Seven shields on the upper surface of the snout, between vertical and rostral, disposed thus $\left\{\begin{matrix} a\text{—}a' \\ b\text{—}b' \\ c \\ d \quad d' \end{matrix}\right\}$; *c* is produced backwards, meeting the vertical and separating shields *d* from each other; *b* and *d*, and *b*′ and *d*′ are in contact. Brown above, with a black median line; sides and abdomen black, with greenish-white specks and vermiculations. Each of the large gulars with a greenish-white spot.

Length to the tympanum	14 millim.
„ humerus	22 „
„ vent	65 „

(Tail partly reproduced.)

The specimen is figured of the natural size, but the views of the head are magnified.

14. **Gerrhonotus moreletii.** (Tab. XXIV. fig. C.)

Gerrhonotus wiegmanni, Salvin, Proc. Zool. Soc. 1860, p. 455.

Gerrhonotus moreletii, Bocourt, Nouv. Arch. Mus. Bull. 1872, p. 102; and Miss. Sc. Mex., Rept. p. 349, t. 21. fig. 1 and t. 21 B. figg. 5.

Gerrhonotus fulvus, Bocourt, Miss. Sc. Mex., Rept. p. 352, t. 21 B. figg. 6.

Hab. MEXICO, Vera Cruz (*Bocourt*); GUATEMALA, Lanquin, Dueñas, and Rio Chisoy, below the town of Cubulco (*Salvin*), Peten (*Morelet*), pine-forests of Totonicapam (*Bocourt*).

Dorsal scales keeled, in twenty, nuchal scales in fourteen or sixteen, ventral in twelve longitudinal series. Nasal shield not in contact with the rostral. Normally seven shields on the upper surface of the snout, between the vertical and rostral, which are disposed thus $\left\{\begin{matrix} a—a' \\ b—b' \\ c \\ d—d' \end{matrix}\right\}$, shield c being always large, in contact with the vertical, and sometimes confluent with d and d'. The a and b series also show sometimes irregularities of arrangement. Supranasals present. Brownish olive, irregularly marbled and spotted with black and yellowish.

15. **Gerrhonotus liocephalus.**

Gerrhonotus liocephalus, Wiegm. Isis, 1828, p. 379; Bocourt, Miss. Sc. Mex., Rept. p. 342, t. 21 A. figg. 1, 2, 2*a*.

Scincus (*Pterogastenes*) *ventralis*, Peale & Green, Journ. Ac. N. Sc. Phil. 1838, vi. p. 233.

Gerrhonotus tessellatus, Wiegm. Herpetol. Mex. p. 32, t. 10. fig. 3; Dum. & Bibr. Erp. Gén. v. 1839, p. 405; Gray, Cat. Liz. p. 53; Cope, Proc. Ac. N. Sc. Phil. 1866, p. 321.

Gerrhonotus infernalis, Baird, Proc. Ac. N. Sc. Phil. 1858, p. 255; Cope, l. c. p. 322.

Gerrhonotus ventralis, Cope, l. c. p. 321.

Gerrhonotus ophiurus, Cope, l. c. p. 321.

Gerrhonotus lemniscatus, Bocourt, Nouv. Arch. Mus. 1872, p. 105.

Hab. NORTH AMERICA, Texas, Devil's River (*Baird*).—MEXICO, Oaxaca, Guanajuato (*Bocourt*), Cordova (*Sallé*).

Dorsal scales keeled, in sixteen, ventral in twelve (ten) longitudinal series. Nasal shield not in contact with the rostral. Ten shields on the upper surface of the snout, between the vertical and rostral, of which one pair are the supranasals (s), and which are disposed thus $\left\{\begin{matrix} a \quad a' \\ a'' \\ s \quad s' \\ b—b' \\ c \\ d—d' \end{matrix}\right\}$. Brownish-olive, with a series of blackish spots along the lateral fold, and with scattered triangular or subquadrangular darker spots of the size of a scale; these spots are sometimes contiguous and form irregular transverse bands, the bands being in other specimens composed of white spots and edged with black (*lemniscatus*).

This species varies somewhat in the length of the tail, width of the snout, and length of the limbs.

Sumichrast also found this species near Cordova (Vera Cruz) at an altitude of 2100 feet.

16. **Gerrhonotus rhombifer.**

Gerrhonotus rhombifer, Peters, MB. Berl. Akad. 1876, p. 298; Bocourt, Miss. Sc. Mex., Rept. p. 332, t. 21 B. figg. 4.

Hab. PANAMA, Veragua, Chiriqui (*Peters*).

Dorsal scales smooth, in sixteen or eighteen, ventral in ten longitudinal series. Nasal shield not in contact with the rostral. Eleven shields * on the upper surface of the snout, between the vertical and rostral, of which one pair are the supranasals (s) and which are disposed thus $\left\{\begin{matrix} a—a''—a' \\ s \quad s' \\ b'' \\ b—b' \\ c \\ d—d' \end{matrix}\right\}$. Upper parts with brown cross bands.

Only one specimen, which is young, the head and body being only 49 millim. long, is known.

17. **Gerrhonotus modestus.**

Pterogasterus modestus, Cope, Proc. Am. Phil. Soc. xvii. 1877, p. 97.

Hab. ? GUATEMALA (*U.S. Nat. Mus.*).

Dorsal scales not keeled, slightly convex above, in ten, ventral in twelve longitudinal series. Nasal shield not in contact with the rostral. Ten shields on the upper surface of the snout, between vertical and rostral, apparently disposed thus $\left\{\begin{matrix} a—a' \\ a'' \\ s—s' \\ b—b' \\ c \\ d—d' \end{matrix}\right\}$. Colour brown above; sides reddish brown, bordered above by a blackish line. Abdomen olive.

A single specimen, the body of which is 72 millim. long, is known.

* Thus it is shown in the figure given by Bocourt, which is taken from the typical specimen, but Peters omits to mention the shields of the a series.

Fam. XENOSAURIDÆ.

XENOSAURUS.

Cubina sp., Gray, Ann. & Mag. Nat. Hist. 1856, xviii. p. 270
Xenosaurus, Peters, MB. Ak. Berl. 1861, p. 453.

1. Xenosaurus grandis.

Cubina grandis, Gray, l. c.
Xenosaurus fasciatus, Pet. l. c.
Xenosaurus grandis, Cope, Proc. Ac. N. Sc. Phil. 1866, p. 322; Bocourt, Miss. Sc. Mex., Rept. p. 303, t. 20 F. figg. 1.

Hab. MEXICO, Cordova and Oaxaca (*Sallé*), Huanusco (*Peters*), Orizaba, rare in the alpine region (*Sumichrast*).

Fam. HELODERMATIDÆ.

HELODERMA.

Heloderma, Wiegmann, Isis, 1829, p. 627.

1. Heloderma horridum. (Tab. XXVI.)

Acaltetepon, Hernandez, Hist. Nov. Hispan. p. 315.
Heloderma horridum, Wiegm. l. c.; and Herp. Mex. p. 24, fig. 1; Wagl. Ic. Amphib. fasc. 2; Troschel, Arch. Naturgesch. 1853, p. 294, tt. 13 & 14; Sumichrast, Bibl. Univ. xix. 1864, p. 45 (Ann. & Mag. N. H. 1864, xiii. p. 497); Kaup, Wiegm. Arch. 1864, p. 33, t. 3. figg. 1–2; Gervais, Journ. Zool. 1873, p. 453, t. 12. figg. 4–8; Sumichrast, Compt. Rend. 1875, lxxx. p. 676, and Bull. Soc. Zool. Fr. 1880, p. 178; Bocourt, Miss. Sc. Mex., Rept. p. 297, tt. 20 E & 20 G; Fischer, Verh. Ver. Hamb. v. pp. 2, 16.

Hab. MEXICO, Presidio, near Mazatlan (*Forrer*), Sinaloa (*Bocourt*), Guanajuato (*Dugès*), Tapana, Tehuantepec (*Sumichrast*).

"Escorpion" of the Mexicans; "Talachini" of the Zapotec Indians.

The external and anatomical characters of this most interesting Lizard have been fully described and the poisonous nature of its bite proved beyond doubt. Adult specimens vary more or less in coloration; and it is therefore of interest to ascertain the normal type, as it appears in the young; the more so, as this species seems to constantly differ in this respect from its short-tailed northern congener. The size and division of the plates or scales also vary in the adult.

The young specimen figured here of the natural size is from Tapana. It is black, with yellow spots and bands. The head is entirely black; a series of five or six yellow angular spots along the median dorsal line, each with a brown margin. Sides of the body with three cross bands; single yellow scales are scattered over the black interspaces. Legs with a yellow ring round the root, and tail with five yellow rings.

2. **Heloderma suspectum.**

Heloderma horridum, Baird, U. S. & Mex. Bound. Exped., Rept. p. 11, t. 26.
Heloderma suspectum, Cope, Proc. Ac. N. Sc. Phil. 1869, p. 5.

Hab. North America, Arizona, Texas.—Mexico, Sonora.

On its poisonous qualities see Shufeldt, Am. Nat. xvi. p. 907; Boulenger, Proc. Zool. Soc. 1882, p. 631.

Fam. IGUANIDÆ.

ANOLIS.

Anolis, Daudin, Hist. Rept. iv. p. 50.

The species of *Anolis* have been quite recently subjected to a thoroughly critical examination by Boulenger, who is about to publish a complete account of them in his forthcoming second volume of the 'Catalogue of Lizards.' In that account the specific characters of the 112 species admitted or distinguished by the author are worked out with such precision that it will remain the basis for the future study of this group for many years to come; and there is no need for repeating here his detailed diagnoses of the species which I have included in the following list as occurring in Central America.

With the aid of Boulenger's work it is now also possible to give a more definite idea of the distribution of this genus which is so characteristic of the fauna of the northern portion of Tropical America. The majority of the species are evidently confined to more or less narrow limits, only a few having a wider distribution, and none extending over the entire range of the genus. The greatest amount of variety is met with in the West Indies; for although the number of island species is slightly exceeded by those occurring in Central America, several are sufficiently differentiated to be referred to distinct genera, whilst all the Central-American species belong to the genus *Anolis* proper.

The numerical representation of these Lizards in the various districts of the region is as follows:—

Species peculiar to the West Indies	34
„ common to the West Indies and North America	1
„ common to the West Indies and Central America	1
„ common to the West Indies and South America	1
„ peculiar to Central America	44
„ found south of the Isthmus of Darien only	24
„ found north as well as south of the Isthmus of Darien	5
„ of uncertain locality	2
	112

1. **Anolis sagræ.**

Anolis sagræi, Dum. & Bibr. Erp. Gén. iv. p. 149; Cocteau in Sagra, Hist. Cub., Rept. p. 131, t. 13; Bocourt, Miss. Sc. Mex., Rept. p. 80, t. 15. fig. 14.

Anolius nebulosus, Gray, Cat. Liz. p. 203.

Draconura catenata, Gosse, Ann. & Mag. N. H. 1850, vi. p. 346.

Anolis ordinatus, Cope, Proc. Ac. N. Sc. Phil. 1864, p. 175.

Hab. Mexico (*Bocourt*); British Honduras (*Brit. Mus.*), Belize, Half Moon Cay (*Brit. Mus.*), St. George's Cay; Guatemala, Yzabal (*Salvin*).—Cuba; Jamaica; Venezuela, Caracas.

2. **Anolis beckeri.**

Anolis beckeri, Bouleng. Proc. Zool. Soc. 1881, p. 921.

Hab. Mexico, Yucatan.

3. **Anolis rodriguezi.**

Anolis rodriguezii, Bocourt, Miss. Sc. Mex., Rept. p. 62, t. 13. fig. 1.

Hab. Guatemala, Panzos (*Bocourt*).

4. **Anolis aureolus.**

Anolis aureolus, Cope, Proc. Am. Phil. Soc. xxii. 1885, p. —*.

Hab. Mexico, Yucatan.

5. **Anolis limifrons.**

Anolis (*Draconura*) *limifrons*, Cope, Proc. Ac. N. Sc. Phil. 1862, p. 178; Bocourt, Miss. Sc. Mex., Rept. p. 65, t. 14. fig. 20.

? *Anolis rivieri*, Thominot, Bull. Soc. Philomat. 1882, vi. p. 251.

Hab. Panama, Veragua (*Cope*, *Bocourt*).

6. **Anolis bransfordii.**

Anolis bransfordii, Cope, Proc. Ac. N. Sc. Phil. 1874, p. 67.

Hab. Nicaragua (*Bransford*).

7. **Anolis trochilus.**

Anolis trochilus, Cope, Proc. Ac. N. Sc. Phil. 1871, p. 215; and Journ. Ac. N. Sc. Phil. viii. 1876, pp. 121, 157.

Hab. Nicaragua; Costa Rica, Talamanca (*Cope*), San Jose (*Van Patten*); Panama (*Cope*).

8. **Anolis ortonii.**

Anolis ortonii, Cope, Proc. Ac. N. Sc. Phil. 1868, p. 97.

Anolis cynocephalus, Bocourt, Nouv. Arch. Mus. vi. 1870, Bull. p. 13, and Miss. Sc. Mex., Rept. t. 14. fig. 7.

Anolis bouvieri, Bocourt, l. c. p. 58, t. 14. fig. 8.

Hab. Guatemala, Escuintla (*Bouvier*).—Upper Amazons; Ecuador; Northern Peru.

* Not yet published.

9. **Anolis schiedi.**

Dactyloa schiedii, Wiegm. Herpet. Mex. p. 48.
Anolis gibbiceps, Cope, Proc. Ac. N. Sc. Phil. 1864, p. 174.
Anolis schiedii, Bocourt, Miss. Sc. Mex., Rept. p. 64, t. 14. fig. 19.

Hab. GUATEMALA, Coban (*Bocourt*).—VENEZUELA, Caracas (*Brit. Mus.*).

10. **Anolis polylepis.**

Anolis polylepis, Peters, MB. Berl. Acad. 1873, p. 738.

Hab. PANAMA, Chiriqui.

11. **Anolis tropidolepis.** (Tab. XXVII. figg. C.)

Anolis tropidolepis, Boulenger, Cat. Lizards, ii. p. —*.

Hab. COSTA RICA, Irazu (*Rogers*).

Head a little longer than the tibia; forehead concave, no frontal ridges. Gular appendage very small in the male. Body and tail of the male slightly compressed; no fold on the neck or back, but the male possesses a double series of somewhat raised and enlarged scales along the median line of the neck. Digital dilatations but little developed; thirteen or fourteen lamellæ under the second and third phalanges of the fourth toe. The scales of the upper surface of the head are strongly keeled, those of the supraorbital row but little enlarged; occipital scute smaller than the ear-opening. Scales of the back and sides very small, granular, those in the vertebral region rather larger and strongly keeled; also the abdominal scales are keeled, imbricate, larger than the dorsal. Brownish or greenish, with an angular dark cross band between the eyes and an oblique band from the orbit to the nape. Males with a broad dark band from the eye to the side of the body; females frequently with a light dark-edged, or sometimes dark-brown, vertebral band.

The views of the head and foot are twice the natural size.

12. **Anolis pulchripes.**

Anolis pulchripes, Peters, MB. Berl. Akad. 1873, p. 739.

Hab. PANAMA, Chiriqui.

13. **Anolis baccatus.**

Anolis baccatus, Bocourt, Miss. Sc. Mex., Rept. p. 59, t. 14. fig. 14.

Hab. MEXICO (*Sallé*).

14. **Anolis guentheri.**

Anolis güntheri, Bocourt, Miss. Sc. Mex., Rept. p. 61, t. 14. fig. 15.

Hab. MEXICO (*Bocourt*).

* Not yet published.

15. **Anolis squamulatus.**

Anolis sqamulatus, Peters, MB. Berl. Ak. 1863, p. 145; Bocourt, Miss. Sc. Mex., Rept. t. 14. fig. 21; Cope, Proc. Ac. N. Sc. Phil. 1871, p. 200.

Hab. PANAMA, Puerto Cabello (*Peters*).

16. **Anolis latifrons.**

Anolis latifrons, Berthold, Abh. Ak. Götting. iii. 1847, p. 6, t. 1. fig. 2.
Anolis laticeps, Cope, Proc. Am. Phil. Soc. xvii. 1878, p. 92.

Hab. PANAMA, Darien.—COLOMBIA.

17. **Anolis microtus.**

Anolis microtus, Cope, Proc. Ac. N. Sc. Phil. 1871, p. 214; and Journ. Ac. N. Sc. Phil. viii. 1876, p. 119, t. 24. fig. 2.

Hab. COSTA RICA, San Jose (*Van Patten*), Irazu (*Rogers*).

18. **Anolis insignis.**

Anolis insignis, Cope, Proc. Ac. N. Sc. Phil. 1871, p. 213; and Journ. Ac. N. Sc. Phil. 1876, viii. p. 120, t. 24. fig. 1.

Hab. COSTA RICA, San Jose (*Van Patten, Salvin*).

19. **Anolis pachypus.**

Anolis pachypus, Cope, Journ. Ac. N. Sc. Phil. 1876, viii. p. 122, t. 24. fig. 3.

Hab. COSTA RICA, Pico Blanco (*Cope*).

20. **Anolis pentaprion.**

Anolis pentaprion, Cope, Proc. Ac. N. Sc. Phil. 1862, p. 178.

Hab. PANAMA, Darien, Truando (*Cope*).—ECUADOR, Guayaquil (*Fraser*).

21. **Anolis copii.**

Anolis copei, Bocourt, Miss. Sc. Mex., Rept. p. 77, t. 15. figg. 10; Cope, Journ. Ac. N. Sc. Phil. viii. 1876, p. 121.

Hab. GUATEMALA, Panzos (*Bocourt*), Vera Paz, low forest and high plateau (*Salvin, Sarg*); COSTA RICA (*Salvin*), Old Harbour (*Cope*).

22. **Anolis petersi.**

Anolis biporcatus, Cope, Proc. Ac. N. Sc. Phil. 1871, p. 215 (nec Wiegm.).
Anolis petersii, Bocourt, Miss. Sc. Mex., Rept. p. 79, t. 13. fig. 2, and t. 15. fig. 11.

Hab. MEXICO (*Sallé*); GUATEMALA, Alta Vera Paz (*Bocourt*).

23. **Anolis heliactin.**

Anolis heliactin, Cope, Proc. Ac. N. Sc. Phil. 1864, p. 172; Bocourt, Misc. Sc. Mex., Rept. p. 106, t. 13. fig. 4, and t. 16. fig. 32.

Hab. MEXICO, Oaxaca (*Sallé*).

24. **Anolis ustus.** (Tab. XXVII. fig. D.)

Anolis ustus, Cope, Proc. Ac. N. Sc. Phil. 1864, p. 172.

Hab. MEXICO, Yucatan (*Brit. Mus.*).—BRITISH HONDURAS, Belize (*Brit. Mus.*).

The figures of the head and foot are twice the natural size.

25. **Anolis cymbops.**

Anolis cymbops, Cope, Proc. Ac. N. Sc. Phil. 1864, p. 173.

Hab. MEXICO, Vera Cruz (*Brit. Mus.*).

26. **Anolis concolor.**

Anolis refulgens, Hallow. Proc. Ac. N. Sc. Phil. 1860, p. 480 (nec Schleg.).
Anolis (Gastrotropis) concolor, Cope, ibid. 1862, p. 180.

Hab. NICARAGUA (*Wright*).

27. **Anolis bourgeaui.**

Anolis bourgeæi, Bocourt, Miss. Sc. Mex., Rept. p. 76, t. 15. fig. 9.

Hab. MEXICO, Huatusco (*Peters*), Orizaba (*Bourgeau*).

28. **Anolis salvini.** (Tab. XXVII. fig. A.)

Anolis salvini, Boulenger, Cat. Lizards, ii. p. —*.

Hab. GUATEMALA (*Salvin & Godman*).

Head much longer than the tibia; forehead concave, no frontal ridges. Gular appendage small. Body and tail not compressed; no fold on the neck or back. Digital dilatations moderately developed; sixteen lamellæ under the second and third phalanges of the fourth toe. The scales on the upper surface of the head are smooth; only one series of scales intervenes between the supraorbital arches; supraoculars enlarged, numerous. No distinct occipital scute. Dorsal and ventral scales feebly keeled, a little larger than the granular lateral scales. Median line of the tail with a series of enlarged, strongly keeled scales. Light greyish with indistinct darker rings and vermiculated lines; indistinct bands across the tail and limbs; some **V**-shaped markings on the back.

The figure is of the natural size; fig. A′ represents a portion of the tail magnified.

* Not yet published.

29. **Anolis læviventris.**

Dactyloa læviventris, Wiegm. Herpet. Mex. p. 47.
Anolis læviventris, Peters, MB. Berl. Ak. 1863, p. 141, and 1869, p. 874; Bocourt, Miss. Sc. Mex., Rept. p. 87, t. 16. fig. 18.

Hab. MEXICO.

30. **Anolis nebulosus.**

Dactyloa nebulosa, Wiegm. Herpet. Mex. p. 47.
Anolis nebulosus, Bocourt, Miss. Sc. Mex., Rept. p. 68, t. 15. fig. 3; Sumichrast, Bull. Soc. Zool. Fr. 1880, v. p. 174.

Hab. MEXICO, Ventanas, Presidio, and Tres Marias Islands (*Forrer*), Colima (*Dugès*), Azucar and Oaxaca (*Sallé & Boucard*), Cuernavaca (*Méhédin*), Jalapa (*Höge*).

Inhabits mountain-sides and ravines, and is not found in the plains (*Sumichrast*).

31. **Anolis nebuloides.**

Anolis nebuloides (!), Bocourt, Miss. Sc. Mex., Rept. p. 74, t. 13. fig. 10.

Hab. MEXICO, Putla (*Paris Mus.*), Huamuchla (*Brit. Mus.*).

32. **Anolis palpebrosus.**

Anolis palpebrosus, Peters, MB. Berl. Ak. 1873, p. 740.

Hab. PANAMA, Chiriqui.

33. **Anolis intermedius.**

Anolis intermedius, Peters, MB. Berl. Ak. 1863, p. 143; Bocourt, Miss. Sc. Mex., Rept. p. 69, t. 15. fig. 4; Cope, Journ. Ac. N. Sc. Phil. viii. 1876, p. 124.
Anolis nannodes, Cope, Proc. Ac. N. Sc. Phil. 1864, p. 173; Bocourt, l. c. p. 71, t. 15. fig. 5.
Anolis tessellatus, O'Shaughn. Ann. & Mag. N. H. 1875, xv. p. 279.

Hab. MEXICO, Ariba (*Cope*), Jalapa (*Höge*); GUATEMALA, Coban (*Salvin*); COSTA RICA (*Cope*), Irazu (*Rogers*); PANAMA, Veragua (*Peters*).

Sumichrast (*in litt.*) states that this species is extensively distributed in the Mexican State of Vera Cruz in warm and temperate districts upwards to an altitude of at least 6000 feet. It lives on insects, but is also fond of the pulps of certain fruits, as coffee-berries, &c.

34. **Anolis sallæi.** (Tab. XXVII. fig. B.)

Anolis sallæi, Günth. Proc. Zool. Soc. 1859, p. 421; Salvin, ibid. 1860, p. 453; Bocourt, Miss. Sc. Mex., Rept. p. 90, t. 13. fig. 3, and t. 16. fig. 21; Sumichrast, Bull. Soc. Zool. Fr. v. 1880, p. 173.
Anolis williamsi, Bocourt, l. c. t. 13. fig. 9.
Anolis jacobi, Bocourt, l. c. p. 73, t. 13. fig. 8.

Hab. Mexico (*Sallé*), Oaxaca (*Brit. Mus.*), Tehuantepec (*Sumichrast*), Vera Cruz (*Jacob*), Hacienda del Holo (*Geddes*); Guatemala, Dueñas (*Salvin*), Lower Vera Paz (*Salvin & Godman*), San Augustin, Volcan de Atitlan (*Bocourt*).

The figure represents the female (type), of the natural size; B′ is twice the natural size.

35. **Anolis binotatus.**

Anolis binotatus, Peters, MB. Berl. Ak. 1863, p. 140; Bocourt, Miss. Sc. Mex., Rept. p. 92, t. 16. figg. 22, 23.

Hab. Salvador (*Bocourt*).—Ecuador, Guayaquil (*Peters*).

36. **Anolis cumingii.**

Anolis cumingii, Peters, MB. Berl. Ak. 1863, p. 140; Bocourt, Miss. Sc. Mex., Rept. p. 89, t. 16. fig. 20.

Hab. Mexico.

37. **Anolis cupreus.**

Anolis cupreus, Hallow. Proc. Ac. N. Sc. Phil. 1860, p. 481.

Anolis hoffmanni, Peters, MB. Berl. Ak. 1863, p. 142; Bocourt, Miss. Sc. Mex., Rept. p. 86, t. 15. figg. 15, 16.

? *Anolis dollfusianus*, Bocourt, l. c. p. 84, t. 16. fig. 19.

Hab. Guatemala (*Salvin & Godman*), Volcan de Atitlan (*Bocourt*); Nicaragua (*Hallowell*); Costa Rica (*Peters, Bocourt*), Irazu (*Rogers*).

38. **Anolis crassulus.** (Tab. XXVII. fig. F.)

Anolis crassulus, Cope, Proc. Ac. N. Sc. Phil. 1864, p. 173; Bocourt, Miss. Sc. Mex., Rept. p. 82, t. 16. fig. 17.

Hab. Mexico, Vera Cruz (*Sumichrast*, in litt.); Guatemala, Coban, Vera Paz (*Salv*

The views of the head are twice the natural size.

39. **Anolis humilis.**

Anolis humilis, Peters, MB. Berl. Ak. 1863, p. 138; Bocourt, Miss. Sc. Mex., Rept. p. 105, t. 13. figg. 7, and t. 16. fig. 31.

Hab. Guatemala, Vera Paz (*Salvin*); Panama, Veragua (*Peters*).

40. **Anolis quaggulus.**

Anolis quaggulus, Cope, Proc. Am. Phil. Soc. xxii. 1885, p. —*.

Hab. Nicaragua.

* Not yet published.

41. **Anolis tropidonotus.**

Anolis tropidonotus, Peters, MB. Berl. Ak. 1863, p. 135, and Ann. & Mag. N. H. 1869, iv. p. 273; Bocourt, Miss. Sc. Mex., Rept. p. 103, t. 13. fig. 6, and t. 16. fig. 30.
Norops auratus, O'Shaughn. Ann. & Mag. N. H. 1869, iii. p. 188, and iv. p. 274.
Norops tropidonotus, O'Shaughn. l. c. 1875, xv. p. 277.

Hab. Mexico, Huatusco (*Peters*), Oaxaca (*Sallé*), Orizaba (*Bourgeau*).

42. **Anolis uniformis.**

Anolis uniformis, Cope, Proc. Am. Phil. Soc. xxii. 1885, p. —*.

Hab. Mexico, Yucatan; Guatemala.

43. **Anolis metallicus.**

Anolis metallicus, Bocourt, Ann. Sc. Nat. xvii. 1872, art. 2; and Miss. Sc. Mex., Rept. t. 17 bis. fig. 1.

Hab. Mexico.

44. **Anolis oxylophus.**

Anolis oxylophus, Cope, Journ. Ac. N. Sc. Phil. viii. 1876, p. 123, t. 24. fig. 4, and t. 28. fig. 5.

Hab. Costa Rica.

45. **Anolis godmani.** (Tab. XXVII. fig. E.)

Anolis godmani, Boulenger, Cat. Lizards, ii. p. —*.

Hab. Guatemala (*Salvin & Godman*); Costa Rica, Irazu (*Rogers*).

Head a little longer than the tibia; forehead concave, with short and divergent frontal ridges. Gular appendage moderately developed. Body not compressed; no fold on the neck or back. Tail cylindrical. Digital dilatations narrow; thirteen or fourteen lamellæ under the second and third phalanges of the fourth toe. The scales of the upper surface of the head are keeled; those of the supraorbital row enlarged, and separated from the opposite row by from two to four intervening series. A few enlarged, keeled, supraocular scales. Occipital scute larger than the ear-opening, separated from the supraorbitals by from three to five series of scales. Scales of the back very small, keeled, gradually passing into the minute granules of the sides; abdominal scales larger, imbricate, keeled. Upper parts brownish; upper lip yellowish, bordered above by a sharply defined brown band passing through the eye and above the ear-opening; a dark cross band between the eyes; sometimes a few distant brown spots along the vertebral line.

The view of the head is twice the natural size.

* Not yet published.

46. **Anolis radulinus.**

Anolis (*Gastrotropis*) *radulinus*, Cope, Proc. Ac. N. Sc. Phil. 1862, p. 180.

Hab. PANAMA, Darien, R. Truando (*Schott*).

47. **Anolis pœcilopus.**

Anolis (*Dracontura*) *pœcilopus*, Cope, Proc. Ac. N. Sc. Phil. 1862, p. 179.

Hab. PANAMA, Darien, R. Truando (*Schott*).—COLOMBIA, Cartagena (*Schott*).

48. **Anolis lemurinus.**

Anolis (*Gastrotropis*) *lemurinus*, Cope, Proc. Ac. N. Sc. Phil. 1861, p. 213.

Hab. PANAMA, Veragua (*Mitchell*).

49. **Anolis biporcatus.**

Dactyloa biporcata, Wiegm. Herpet. Mex. p. 47.
Anolis vittigerus, Cope, Proc. Ac. N. Sc. Phil. 1862, p. 179.
Anolis biporcatus, Bocourt, Miss. Sc. Mex., Rept. p. 98, t. 15. fig. 8.

Hab. GUATEMALA, Dueñas, Lower Vera Paz, Rio Polochic (*Salvin*), San Miguel Tucuru, Panzos (*Bocourt*); PANAMA, Darien, R. Truando (*Schott*).—VENEZUELA; TRINIDAD; WESTERN ECUADOR.

50. **Anolis chrysolepis.**

Anolis chrysolepis, Dum. & Bibr. Erp. Gén. iv. p. 94; Gray, Cat. Lizards, p. 207; Bocourt, Miss. Sc. Mex., Rept. p. 99, t. 16. fig. 26; Peters, MB. Berl. Ak. 1863, p. 142.
Anolis planiceps (part.), Trosch. in Schomburgk, Reis. Brit. Guian. iii. p. 649.
Anolis nummifer, O'Shaughn. Ann. & Mag. N. H. 1875, xv. p. 278.
Anolis turmalis, O'Shaughn. l. c.
Anolis (*Draconura*) *chrysolepis* (part.), O'Shaughn. Proc. Zool. Soc. 1881, p. 241.

Hab. MEXICO (*Peters*); HONDURAS (*Brit. Mus.*).—W. ECUADOR, Esmeraldas; VENEZUELA; GUIANA; AMAZONS, Para; WEST INDIES, Grenada.

51. **Anolis capito.**

Anolis (*Draconura*) *capito*, Peters, MB. Berl. Ak. 1863, p. 142; Bocourt, Miss. Sc. Mex., Rept. p. 101, t. 16. fig. 27; Cope, Journ. Ac. N. Sc. Phil. viii. 1876, pp. 124, 157.
Anolis carneus, Cope, Proc. Ac. N. Sc. Phil. 1864, p. 169.

Hab. MEXICO, Tabasco (*Mus. Brux.*); GUATEMALA, lower forests of Vera Paz (*Salvin & Godman*), Alta Vera Paz (*Bocourt*); COSTA RICA (*Peters*), Old Harbour (*Cope*); PANAMA (*Cope*).

52. **Anolis obtusirostris.**

Anolis obtusirostris, Peters, MB. Berl. Ak. 1873, p. 741.

Hab. PANAMA, Chiriqui.

POLYCHRUS.

Polychrus, Cuvier, Règne Anim.

1. **Polychrus gutturosus.**

Polychrus gutturosus, Berthold, Abh. Ges. Wiss. Götting. iii. 1847, p. 5, t. 1. fig. 1.

Polychrus multicarinatus, Peters, MB. Berl. Ak. 1869, p. 786; Bocourt, Miss. Sc. Mex., Rept. p. 112, t. 17. figg. 8.

Hab. Costa Rica (*Peters*).—Colombia (*Berthold*).

CORYTHOPHANES.

Corythophanes, Boie.

An exclusively Central-American genus.

1. **Corythophanes cristatus.**

Agama cristata, Merr. Syst. Amphib. p. 50.

Corythophanes cristatus, Boie, Isis, 1827, p. —; Gravenhorst, Act. Ac. Leop. Carol. xvi. 1833, p. 938, t. 65. figg. 6–10; Dum. & Bibr. Erp. Gén. iv. p. 174; Dum. Arch. Mus. Hist. Nat. viii. 1856, p. 517, t. 20. fig. 1; Bocourt, Miss. Sc. Mex., Rept. p. 118; Cope, Journ. Ac. N. Sc. Phil. viii. pp. 124, 156.

Hab. Guatemala, Vera Paz, low forest (*Salvin & Godman*), Peten (*Morelet*); Costa Rica (*Cope, Salvin*); Panama (*Cope*).

Not common in Guatemala (*Salvin*).

2. **Corythophanes percarinatus.**

Corythophanes percarinatus, Dum. Arch. Mus. viii. p. 518, t. 20. fig. 3; Bocourt, Miss. Sc. Mex., Rept. p. 120, t. 17. fig. 2.

Corythophanes cristatus, Salvin, Proc. Zool. Soc. 1860, p. 453.

Hab. Guatemala, Escuintla (*Morelet*), Lanquin (*Salvin*), Coban (*Bocourt*).—Ecuador (*Fraser*).

3. **Corythophanes hernandesii.**

Quatapalcatl, Hernandez, Hist. Mex. ii. cap. xiii. p. 61.

Chamæleopsis hernandezii, Wiegm. Isis, 1831, p. 298; Gravenhorst, Act. Ac. Leop. Carol. xvi. 1833, p. 944, t. 65. figg. 1–5; Wiegm. Herp. Mex. pp. 15, 37, t. 6.

Corythophanes chamæleopsis, Dum. & Bibr. Erp. Gén. iv. p. 176; Sumichrast, Bibl. Univ. et Rev. Suiss. 1864, p. 58; Ann. & Mag. N. H. 1864, xiii. p. 505.

Corythophanes mexicanus, Bocourt, Miss. Sc. Mex., Rept. p. 122, t. 17. fig. 1.

Hab. Mexico, Vera Cruz, Tehuantepec, Chiapas (*Sumichrast*), Orizaba (*Brit. Mus.*), Yucatan (*Brit. Mus.*).

In woods, especially oak-forests, among rocks. Feeding on insects.

LÆMANCTUS.

Læmanctus, part., Wiegmann, Herp. Mex. p. 45.

1. Læmanctus longipes.

Læmanctus longipes, Wiegm. l. c. p. 46, t. 4; Dum. & Bibr. Erp. Gén. iv. p. 72; Boulenger, Bull. Soc. Zool. Fr. 1877, p. 464, t. 7. fig. 2.

Hab. MEXICO, Jalapa (*Wiegmann*).

Confined to the forest districts of the eastern coast, up to an altitude of 3000 feet (*Sumichrast*, in litt.).

2. Læmanctus serratus.

Læmanctus serratus, Cope, Proc. Ac. N. Sc. Phil. 1864, p. 176; Bocourt, Miss. Sc. Mex., Rept. p. 116; Boulenger, l. c. p. 463, t. 7. fig. 3.

Læmanctus longipes, Dum. Arch. Mus. viii. 1856, t. 21. fig. 4; Bocourt, l. c. p. 114, t. 17. fig. 4.

Hab. MEXICO, Orizaba (*Cope*), Oaxaca (*Ghisbreght*), Vera Cruz (*Méhédin & Boucard*), Campeche (*Boulenger*).

3. Læmanctus deborrii.

Læmanctus de Borrei, Boulenger, l. c. p. 465, t. 7. fig. 1.

Hab. MEXICO, Tabasco (*Boulenger*); GUATEMALA, Vera Paz, Coban (*Salvin*).

4. Læmanctus alticoronatus.

Læmanctus alticoronatus, Cope, Proc. Ac. N. Sc. Phil. 1865, p. 192; Proc. Am. Phil. Soc. 1869, t. 11.

Hab. MEXICO, Yucatan (*Cope*, *Brit. Mus.*).

BASILISCUS.

Basiliscus, Laurenti, Syn. Rept. p. 50.

1. Basiliscus americanus.

Lacerta basiliscus, Linn. Syst. Nat. i. 366.

Basiliscus americanus, Laur. Syn. Rept. p. 50; Gray, Cat. Liz. p. 192; Bocourt, Miss. Sc. Mex., Rept. p. 127.

Basiliscus mitratus, Merr. Syst. Amph. p. 46; Daud. Rept. p. 311, t. 43; Dum. & Bibr. Erp. Gén. iv. p. 181; Cope, Journ. Ac. N. Sc. Phil. viii. 1876, p. 125, t. 25. fig. 2.

Ophyessa bilineata, Gray in Beechey's Voy. Zool. p. 94 (young).

Thysanodactylus bilineatus, Gray, Cat. Liz. p. 194.

Lophosaura goodridgii, Gray, Ann. & Mag. N. H. 1852, x. p. 438; and in Zoology of the 'Herald,' p. 143, t. 26.

Basiliscus guttulatus, Cope, l. c. pp. 125, 156.

Basiliscus goodridgii, Cope, l. c. p. 125.

Hab. Costa Rica (*Bocourt, Rogers*); Panama (*Cope*), Isthmus of Darien (*Cope*), Island of Quibo ('*Herald*').

Doubtful localities are Guiana and Martinique (*Paris Mus.*), and Fernando do Norunha, given by Gray, Lizards, p. 194.

2. **Basiliscus plumifrons.**

Basiliscus plumifrons, Cope, Journ. Ac. N. Sc. Phil. viii. 1876, p. 125, t. 25. fig. 1.

Hab. Costa Rica.

3. **Basiliscus vittatus.** (Tab. XXVIII.)

Basiliscus vittatus, Wiegm. Isis, 1828, p. 373; Dum. & Bibr. Erp. Gén. iv. p. 187; Sumichrast, Arch. Sc. Phys. et Nat. 1864, xix. p. 56; Ann. & Mag. N. H. xiii. 1864, p. 504; Bocourt, Miss. Sc. Mex., Rept. p. 129, t. 17. fig. 3; Cope, Journ. Ac. N. Sc. Phil. viii. 1876, p. 125.
Corythæolus vittatus, Kaup, Isis, 1829, p. 1147; Wiegm. Herpet. Mex. p. 39, t. 5.
Cristasaura mitrella, Gray, Ann. & Mag. N. H. 1852, x. p. 439.
Basiliscus americanus, Salvin, Proc. Zool. Soc. 1860, p. 453.
Daconura bivittata, Hallow. Proc. Ac. N. Sc. Phil. 1860, p. 482.
Paroloma bivittata, Cope, Proc. Ac. N. Sc. Phil. 1862, p. 181.
Basiliscus nuchalis, Cope, Proc. Ac. N. Sc. Phil. 1862, p. 181.

Hab. Mexico (generally distributed in the warmer and temperate parts, near rivers), Putla (*Rébouch*), Orizaba (*Brit. Mus.*), Hacienda del Hobo (*Geddes*); Guatemala, common at Lanquin and Dueñas, and in the lower forests of Vera Paz (*Salvin*); Honduras (*Dyson*); Nicaragua, Greytown (*Caldwell*); Costa Rica, Sipurio (*Cope*).—Ecuador (*Fraser*).

"Pasarios" of the Mexicans; "Zumbichi" of the Zapotec Indians.

Found to a height of 1000 metres (*Sumichrast*).

Mr. Salvin (*l. c.*) observes:—"This Basilisk is very common at Lanquin, where I had no difficulty in obtaining plenty of specimens of both sexes and all ages, by offering the Indians a small reward for all they brought. They may frequently be seen on the low branches of a bush, and are particularly fond of basking on the boughs of a felled tree in a clearing near a stream. When running fast they carry the tail slightly elevated. At Dueñas this species is more rare; I saw but three or four specimens, and only succeeded in securing one. In some specimens of the males, the tail is much more compressed than in others. In the series of young the crest is shown in all stages of development."

Extralimital is Basiliscus galeritus, A. Dum., or *Ptenosaura seemanni*, Gray, from Ecuador and Colombia.

ENYALIUS.

Enyalius, Wagler, Syst. Amphib. p. 150.

Of the ten or eleven species of this South-American genus, only the following occur north of the Isthmus of Darien. We have not seen it.

1. **Enyalius heterolepis.**

Enyalius heterolepis, Bocourt, Ann. Sc. Nat. xix. 1874, art. 4, and Miss. Sc. Mex., Rept. t. 20 c. fig. 14.

Hab. PANAMA, Veragua.

IGUANA.

Iguana, Laurenti.

1. **Iguana rhinolophus.**

Iguana rhinolophus, Wiegm. Herp. Mex. p. 44; Dum. & Bibr. Erp. Gén. iv. p. 207; Tyler, Proc. Zool. Soc. 1850, t. 3 (*Metopoceros cornutus*); Cope, Proc. Ac. N. Sc. Phil. 1862, p. 356; Sumichrast, Ann. & Mag. N. H. 1864, xiii. p. 500; Bocourt, Miss. Sc. Mex., Rept. p. 134; Cope, Journ. Ac. N. Sc. Phil. viii. 1876, p. 124.

Hab. MEXICO, Mazatlan and Presidio (*Forrer*), Cordova (*Sallé*), Tehuantepec and Rio Coazacoalcos (*Sumichrast*), Cozumel Island on the coast of Yucatan (*G. F. Gaumer*); GUATEMALA, Chiapam (*Salvin*); COSTA RICA (*Cope*); PANAMA (*Brit. Mus.*), Darien, R. Truando (*Schott*).

" Iguana verde " of the Creoles; " Guchachi-guëla " of the Zapotec Indians.

This is the northern or Central-American race of the South-American *I. tuberculata*; it does not seem to extend southwards beyond the Isthmus of Darien, but occurs also in the island of Santa Lucia and probably other parts of the West Indies.

2. **Iguana tuberculata.**

Iguana tuberculata, Laur. Syn. Rept. p. 49.

Hab. GUATEMALA (*Salvin*); HONDURAS (*Brit. Mus.*); NICARAGUA (*Brit. Mus.*); PANAMA (*Dow*), Darien (*Schott*).—Generally distributed over the northern parts of SOUTH AMERICA and BRAZIL; occurring also in many West-Indian Islands.

CTENOSAURA.

Ctenosaura, Wiegmann, Isis, 1828, p. 371.

1. **Ctenosaura acanthura.**

Lacerta acanthura, Shaw, Zool. ii. 1802, p. 216.

Cyclura teres, Harlan, Journ. Ac. Sc. Phil. 1825, p. 246, t. 16.

Ctenosaura cycluroides, Wiegm. Isis, 1828, p. 371; Bocourt, Miss. Sc. Mex., Rept. p. 143 (part.); Sumichrast, Bull. Soc. Zool. Fr. v. 1880, p. 175.

Ctenosaura shawii, Gray, Wiegm.

Cyclura pectinata, Wiegm. Herpet. Mex. p. 42, t. 2; Dum. & Bibr. Erp. Gén. iv. p. 221.
Cyclura denticulata, Wiegm. l. c. p. 43, t. 3.
Cyclura articulata, Wiegm. l. c.
Cyclura acanthura, Sumichrast, Arch. Sc. Phys. et Nat. xix. 1864, p. 49; Ann. & Mag. N. H. 1864, xiii. p. 500.
Ctenosaura acanthura, Gray, Cat. Liz. p. 191; Cope, Proc. Ac. Nat. Sc. Phil. xviii. p. 124; Bocourt, Le Naturaliste, 1882, p. 47; Boul. Cat. Liz. ii. p. 195, varr. A, B, D.
Ctenosaura pectinata, Gray, Cat. Liz. p. 191; ? Cope, Proc. Ac. Nat. Sc. Phil. xviii. p. 124; and Proc. Am. Phil. Soc. xxii. 1885, p. 388; Bocourt, Miss. Sc. Mex., Rept. p. 140, and Le Naturaliste, 1882, p. 47; Sumichrast, Bull. Soc. Zool. v. 1880, p. 174.
Ctenosaura teres, Bocourt, Miss. Sc. Mex., Rept. p. 142, and Le Naturaliste, 1882, p. 47.
Ctenosaura multispinis, brevirostris, and *teres*, Cope, Proc. Am. Phil. Soc. 1886, p. 266.

Hab. North America, California. — Mexico, Tres Marias Is., Presidio, Mazatlan, Ciudad, and Ventanas (*Forrer*), Tierra Colorada in Guerrero (*H. H. Smith*), Colima (*Dugès, in mus. Paris*; *U.S. Nat. Mus.*), Putla (*Boucard*), Tehuantepec (*Sumichrast*), Tampico (*W. B. Richardson*), Vera Cruz (*Sallé*; *Méhédin, in mus. Paris*), Yucatan, Cozumel I. (*Cope*).

"Iguana negra" of the Creoles; "Guchachi-chévé" of the Zapotec Indians.

Tail subcylindrical; the spines of the broad rings moderately developed, the broad rings of the anterior verticelli separated by three or two narrow rings, rarely by one. Dorsal crest generally interrupted in the sacral region; if continuous, the continuity is effected merely by the low and somewhat enlarged scales of the median series. Scales of the calf of the leg very small. The coloration varies and changes with age. The ground-colour of the young is generally green, marbled with darker on the back, the dark markings forming more or less distinct, irregular cross bands, which are sometimes confluent, sometimes spotted with black, and about seven or eight in number on the back. With age the dark colour becomes more diffused and irregularly distributed over the body, at places entirely suppressing the ground-colour, which itself assumes a more olive tinge or changes into yellowish. Specimens from Tampico are uniform black when adult, and of a greenish-olive when young.

I have examined a great number of specimens, which I am unable to divide into species, although they by no means agree with one another in every point. The development of the crest, which by some authors has been used as a specific character, depends partly on age, partly on sex. The largest specimens I have seen, males as well as females, are thirty inches long, of which the tail takes two thirds. Shaw's type is still preserved in the British Museum, and said to be from California; also Bocourt has examined specimens from that country.

2. **Ctenosaura completa.** (Tabb. XXIX., XXX.)

Ctenosaura cycluroides, part., Bocourt, Miss. Sc. Mex., Rept. p. 143.

Ctenosaura completa, Bocourt, l. c. p. 145, and Le Naturaliste, 1882, p. 47; Cope, Proc. Am. Phil. Soc. 1886, p. 269.

Ctenosaura acanthura, var. C, Boul. Cat. Liz. ii. p. 197.

Hab. Mexico, Yucatan and Cozumel I. (*Cope*), Mugeres I. (*G. F. Gaumer*); British Honduras, Belize (*Dyson*); Guatemala, Peten (*Morelet*), Chiapam (*Salvin*); Honduras (*Mus. Brit.*, *Bocourt*), Bonacca I. (*G. F. Gaumer*); Salvador, La Union (*Bocourt*).

Tail subcylindrical; the spines of the tail very prominent and acute in the adult male, the scutes of the narrow rings being also strongly armed. Dorsal crest not interrupted in the sacral region, the continuity being effected by short lobes in the adult male, and by depressed scales in females and young. Scales of the calf of the leg larger than in *C. acanthura.* Greenish or greenish-olive, with six black cross-bands on the back, of which the anterior is the least developed, and, like the following, more or less distinctly paired; in old specimens these bands are narrower and more or less broken up, and the anterior may disappear altogether.

Of this species I have seen nine specimens—two adult males and several young from Honduras, an adult female and young from Belize, and an adult female from Guatemala. It grows to the same size as *C. acanthura*, from which it is not easily distinguished.

Tab. XXIX. represents a male from Bonacca Island, of half the natural size, with separate view of a part of the tail; Tab. XXX. a female from Chiapam, of the natural size, also the inner side of the calf of the leg, to show the size of the scales.

3. **Ctenosaura quinquecarinata.**

Cyclura quinquecarinata, Gray, Zool. Misc. p. 59; Cope, Proc. Am. Phil. Soc. xi. 1871, p. 161.

Enyaliosaurus quinquecarinatus, Gray, Cat. Liz. p. 192.

Ctenosaura quinquecarinata, Sumichrast, Bibl. Univ. et Rev. Suisse, 1873, p. 259; and Bull. Soc. Zool. v. 1880, p. 175.

Hab. Mexico, Oaxaca (*Boucard*), Tehuantepec (*Sumichrast*); Honduras (*Mus. Brit.*).

Tail much thickened and depressed near its base, verticillated, each verticellus formed by a ring of large strongly armed scutes, and by a ring of much narrower and unarmed scutes. The armature is confined to the upper and lateral surfaces of the tail, the lower side being comparatively smooth. The strongest spines are arranged in one median, and two or three lateral series. Dorsal crest very low, obsolete in the sacral region. Upper parts and throat yellowish, marbled with black or brown.

4. **Ctenosaura defensor.**

Cachryx defensor, Cope, Proc. Ac. N. Sc. Phil. xviii. p. 124; Proc. Am. Phil. Soc. 1869, p. 169, t. 10; Bocourt, Miss. Sc. Mex., Rept. p. 143, t. 17 bis. figg. 12, 12 *a*.

Hab. Mexico, Yucatan (*A. Schott, U.S. Nat. Mus.*).

Extralimital are the following species:—

1. *Iguana acanthura*, Blainville,=*Cyclura acanthura*, Dum. & Bibr.,=*Cyclura hemilopha*, Cope, Boulenger,=*Ctenosaura acanthura*, Bocourt (Miss. Sc. Mex., Rept. p. 138), =*Ctenosaura interrupta*, Bocourt (Le Naturaliste, 1882, p. 47).—Lower California.

2. *Ctenosaura erythromelas*, Boul. Proc. Zool. Soc. 1886, p. 241, t. 23; *Cachryx erythromelas*, Cope, Proc. U.S. Nat. Mus. 1886, p. 437.—Probably Lower California.

CROTAPHYTUS.

Crotaphytus, Holbrook, N. Amer. Herpet. ii. p. 79.

1. **Crotaphytus wislizenii.**

Crotaphytus wislizenii, Baird & Girard, Proc. Ac. N. Sc. Phil. vi. p. 69; and in Stansbury's Exped. Great Salt Lake, p. 340, t. 3; and U.S. & Mex. Bound. Survey, Rept. p. 7, t. 31; Bocourt, Miss. Sc. Mex., Rept. p. 155, t. 17 bis. figg. 4, 4 *a*; Boul. Cat. Liz. ii. p. 204.

Crotaphytus gambelii, Baird & Girard, Proc. Ac. N. Sc. Phil. vi. p. 126.

Crotaphytus fasciatus, Hallow. Proc. Ac. N. Sc. Phil. vi. p. 207; and in Sitgreave's Expedit. Zuni and Colorado R. p. 115, t. 5.

Hab. NORTH AMERICA, Oregon, California, New Mexico, Arizona and Texas.—MEXICO, Sonora.

2. **Crotaphytus collaris.**

Agama collaris, Say in Long's Exped. Rocky Mount. ii. p. 252; Harlan, Med. Phys. Res. p. 142, t. 1. fig. 4.

Crotaphytus collaris, Holbr. N. Amer. Herpet. ii. p. 79, t. 10; Cope, Proc. Ac. N. Sc. Phil. xviii. p. 302; Bocourt, Miss. Sc. Mex., Rept. p. 154, t. 17 bis. figg. 5, 5 *a*, 6, 6 *a*; Yarrow, Check-list N. Amer. Rept. p. 52; Boul. Cat. Liz. ii. p. 203.

Hab. NORTH AMERICA, Southern and South-western States.—MEXICO, Chihuahua, Nuevo Leon, Western Sonora, Sta. Catarina (*U.S. Nat. Mus.*).

HOLBROOKIA.

Holbrookia, Girard, Proc. Am. Assoc. Adv. Sci. iv. p. 201 (1851).

Cophosaurus, Troschel in Wiegm. Arch. f. Naturg. 1850, p. 389 (publ. 1852).

1. **Holbrookia texana.**

Cophosaurus texanus, Troschel, Wiegm. Arch. f. Naturg. 1850 (publ. 1852), p. 389, t. 6.

Holbrookia texana, Baird & Gir. Proc. Ac. N. Sc. Phil. vi. p. 125; and in U.S. & Mex. Bound. Surv., Rept. p. 8, t. 30; Cope, Proc. Am. Phil. Soc. xxii. 1885, p. 169; Yarrow, Check-list N. Amer. Rept. p. 50; Boul. Cat. Liz. ii. p. 208.

Holbrookia affinis, Baird & Gir. Proc. Ac. N. Sc. Phil. vi. p. 125; and in U.S. & Mex. Bound. Surv., Rept. p. 8.

Hab. NORTH AMERICA, Texas to California.—MEXICO, N. Sonora (*Clark*), Alamos, Castaneulas (*Couch, U.S. Nat. Mus.*), Monterey, Nuevo Leon (*Couch, Cope*).

Hind limbs long, if laid forwards reaching to, or nearly to, the end of the snout; and the length of the tibia equalling the distance of the postgular fold from the end of the snout. The supraoculars form an isolated patch of larger scutes surrounded by granular scales. Two subcrescentic black bands on each side of the abdomen, behind the middle of the trunk, and close in front of the hind leg. Tail strongly depressed, with about six broad, deep black cross-bands underneath.

A very young specimen does not differ in any respect from the adult.

This species is common in Texas. Mr. Forrer has sent us two specimens from Nuevo Leon, adult and young, and others from El Paso. As to *Holbrookia affinis*, which is clearly the same species as *H. texana*, Baird and Girard state at first (*l. c.* 1°) that the specimens were found with *H. texana* on the Rio San Pedro, a tributary of the Rio Grande del Norte, and afterwards (*l. c.* 2°) that they came from Sonora.

2. **Holbrookia maculata.**

Holbrookia maculata, Girard, Proc. Amer. Assoc. Adv. Sc. iv. p. 201 (1851); and in Stansbury, Exped. Gt. Salt Lake, p. 342, t. 4. figg. 1–3; Bocourt, Miss. Sc. Mex., Rept. p. 161, t. 17 bis. figg. 7, 7 *a*; Yarrow, Check-list N. Amer. Rept. p. 49; Boul. Cat. Liz. ii. p. 209.

Holbrookia approximans, Baird, Proc. Ac. N. Sc. Phil. x. p. 253.

Hab. North America, Southern States.—Northern Mexico, Sonora (*U.S. Nat. Mus.*).

Hind limbs of moderate length, if laid forwards not reaching the eye. Tail subcylindrical, shorter than the head and body, without spots underneath. One or two black spots on the anterior half of the side of the trunk.

The British Museum possesses three specimens from Mexico, besides a number from various parts of the Southern United States. These specimens differ considerably in the comparative length and slenderness of the hind foot. In some from Kansas and Northern Texas the length of the hind foot is scarcely equal to the distance of the postgular fold from the end of the snout, whilst it more or less exceeds that distance in others from Tennessee, Duval County (Texas), and Mexico. In our largest specimen (a female from Duval Co.) the lateral spots are nearly obsolete.

3. **Holbrookia propinqua.** (Tab. XXXI. fig. C.)

Holbrookia propinqua, Baird & Gir. Proc. Ac. N. Sc. Phil. vi. p. 126; Cope, Bull. U.S. Nat. Mus. 1880, no. 17, p. 15; Boul. Cat. Liz. ii. p. 208.

? *Holbrookia elegans,* Bocourt, Miss. Sc. Mex., Rept. p. 164, t. 17 bis. figg. 8, 8 *a*, head.

Hab. North America, Texas.—Mexico, Presidio (*Forrer*).

Hind limbs long, if laid forwards reaching nearly to the end of the snout; and the length of the tibia equals the distance of the postgular fold from the end of the snout. The middle supraoculars larger than those of the snout. Two oblique black bands or band-like spots descending backwards on each side of the abdomen, in the middle of the length of the trunk. Tail depressed, but less so than in *H. texana*, uniform white underneath.

In the adult female the black lateral bands are shorter, less distinct, and may disappear altogether. It seems to me very probable that *Holbrookia elegans*, from Mazatlan, is founded on such a specimen without spots.

Extralimital: *Holbrookia lacerata*, Cope, Bull. U.S. Nat. Mus. 1880, no. 17, p. 15, from Texas.

UTA.

Uta, Baird & Girard, in Stansbury's Exped. Gr. Salt Lake, p. 344 (1852).

1. **Uta lateralis.** (Tab. XXXI. fig. D.)

Uta (Phymatolepis) lateralis, Boul. Ann. & Mag. N. H. 5th ser. xi. p. 342.
? *Uta gularis*, Cragin, Bull. Washburn Labor. 1884, p. 7.

Hab. MEXICO, Presidio and Tres Marias Islands (*Forrer*).

Hind leg, if laid forwards, extending to the tympanum. The nasals are in immediate contact with the internasals. On each side of the median line, which is formed by a series of very small keeled scales, there are two rows of large scales, the scales of the inner row being twice or thrice as large as those of the outer. The remainder of the back and sides is covered with minute tubercular scales, but on either side of the back a row of larger keeled scales runs nearly along the entire length of the back. The scales of this row are not by any means regular in size and arrangement; no other row of larger scales on the sides; three or four pointed scales in front of the tympanum, but none along its upper circumference. Upper parts brownish-grey, with a double row of small indistinct black spots along the middle of the back; another row of larger subrhombic black spots along each side of the back. A whitish band runs from the end of the snout along the upper lip, and passes through the tympanum to the shoulder-joint; a black vertical bar in front of the shoulder, with a horizontal extension above and below forming a figure like the letter **Z**. Male with the throat blue and without blackish cross-bands. Abdomen blue with a whitish median band.

Of the six specimens I have examined, one of the largest and most adult is 134 millim. long, the tail measuring 80 millim.

It is probable that *Uta gularis*, Cragin, from Guaymas, Mexico, is identical with this species; but it would seem to differ by a slightly shorter hind leg, which extends forward only to the front of the shoulder-joint, and by possessing a short additional row of larger scales on each side of the abdomen.

Uta auriculata, Cope, Proc. Bost. Soc. N. H. xiv. p. 303, from Socorro, one of the Revillagigedo Islands, is also closely allied to this species, but distinguished by a crest of scales above the tympanum, and blackish cross-bands in the gular region.

2. **Uta irregularis.**

Phymatolepis (*Uta*) *irregularis*, Fischer, Abhandl. nat. Ver. Brem. vii. 3, 1882, p. 232, t. 17. figg. 1–4; Boul. Cat. Liz. ii. p. 216.

Hab. MEXICO, Highlands.

I have not examined a specimen of this species.

3. **Uta bicarinata.**

Phymatolepis bicarinatus, A. Dum. Arch. Mus. viii. p. 549, t. 23. fig. 2; Bocourt, Miss. Sc. Mex., Rept. p. 165, t. 17 bis. figg. 9, 9 *a*, *b*; Sumichrast, Bull. Soc. Zool. v. 1880, p. 176.

Uta bicarinata, Cope, Proc. Ac. N. Sc. Phil. xvi. p. 177; Boul. Cat. Liz. ii. p. 215; Cope, Proc. Am. Phil. Soc. 1885, p. 379.

Hab. MEXICO, Presidio, Ventanas, Ciudad in Durango (*Forrer*), Cuernavaca, Matamoros Izucar (*Ferrari-Perez*), Puebla (*Boucard*), Jalisco, North of Rio Santiago (*Richardson*), Tehuantepec (*Sumichrast*).

An arboreal species according to Sumichrast.

SCELOPORUS.

Sceloporus, Wiegmann, Isis, 1828, p. 369.

Tropidolepis, Cuvier, Règne Anim. 2nd edit. ii. p. 38.

The Reptiles of this genus are, after the *Anoles*, the most numerously represented type of Lizards in Central America. Northwards they extend far into the United States, but southwards they have not been found beyond the limits of Central America. Sumichrast writes about the Mexican species observed by him as follows:—"The *Scelopori* are natives of the Central Tablelands and Alpine districts, comparatively few being found on the warm plains of the coast. On the eastern or Atlantic side *Sceloporus variabilis* is the only one which is to be found in the tierra caliente of Vera Cruz. On the Pacific side of the isthmus of Tehuantepec I collected but three species, viz., *Sceloporus variabilis*, *S. melanorhinus*, and *S. siniferus*. In the vicinity of Orizaba (Vera Cruz), but to a height varying between 1200 and 2500 metres, *Sceloporus malachiticus*, *formosus*, *microlepidotus*, and *scalaris* are to be found. Others, as *Sceloporus horridus* and *S. æneus*, are at home on the dry and sterile plains of the plateau, and people, with Rattlesnakes and Tapayaxins (*Phrynosoma*), the rugged lava deposits known as *malpaës* (bad lands) or the monotonous fields of Agave. Although most of the species are strictly humivagous lizards, others dwell in trees, and spend most of their life in them: such is the case with *Sceloporus melanorhinus* and some others."

Many of the species, perhaps the majority, are ovoviviparous.

No other genus of Lizards offers greater difficulties in distinguishing the numerous species, races, or forms, and these difficulties are almost insurmountable when it is attempted to take into consideration species vaguely defined and without distinct indication of the locality where the type was found. But little difficulty is met with when the species of a certain comparatively limited district are under examination; then I have always found that no, or but little, variation obtains among the specimens. Thus I have seen from fifty to 100 examples of a species from the same locality, all agreeing in the form of the supraoculars, length of limbs and toes, size of scales, prominent markings, and showing variations only in the irregular mottling, or in the shade of the ground-colour, and in the number of femoral pores.

The difficulties are greatest among the large-scaled forms, that is among those which have some twenty-two to thirty-five scales in a line between the collar and beginning of the tail; and these difficulties have been increased by the attempts to unite as many modifications as possible under the same specific term. Consistently, either they all must be thrown together, or many more forms should be distinguished than have been admitted hitherto. Species described as distinct, and from distant localities and different altitudes, should not be thrown together without direct evidence of their absolute identity.

Our collection, like that of every other museum, and the information as to altitude of the specimens, is much too incomplete to give anything like a satisfactory account of the Central-American species of this genus. But, by testing their value by an examination of several series, each from the same locality, I come to the conclusion that the following characters can be relied on :—

1. The size of the scales.
2. The arrangement of the large supraoculars.
3. The length of the hind limbs and hind toes.
4. The presence or absence of the nuchal collar, and its extent.
5. The markings, if any, in front and behind the shoulder-joint.
6. The coloration of the abdomen of the adult male.

A. *The scales on the sides gradually passing into, and not conspicuously larger than, the dorsal scales.*

α. *No præanal pores.*

1. Sceloporus spinosus.

Sceloporus spinosus, Wiegm. Isis, 1828, p. 370; and Herp. Mex. p. 50, t. 7. fig. 3; Baird, U.S. & Mex. Bound. Surv., Rept. p. 5, t. 29. figg. 4–6; Bocourt, Miss. Sc. Mex., Rept. p. 174, t. 18. figg. 2, 2 *a*, *b*; Cope, Proc. Am. Phil. Soc. 1885, p. 399; Boul. Cat. Liz. ii. p. 226.

Sceloporus horridus, Wiegm. Herp. Mex. p. 50; Bocourt, Miss. Sc. Mex., Rept. p. 178, t. 18. figg. 8, 8 *a*, *b*; Cope, Proc. Am. Phil. Soc. 1885, p. 396.

Sceloporus clarkii, Baird & Gir. Proc. Ac. N. Sc. Phil. vi. p. 127; Yarrow, in Wheeler's Report Explor. W. 100th Merid. v. p. 575, t. 23. figg. 1, 1 *a*.

Sceloporus magister, Hallow. Proc. Ac. N. Sc. Phil. vii. p. 93.

Sceloporus oligoporus, Cope, Proc. Ac. N. Sc. Phil. xvi. p. 177.

Hab. North America, Southern States.—Mexico, Sonora (*Hallowell*).

From twenty-one to twenty-four scales in a series along the back*; scales on the back strongly keeled, the keel terminating in a long spine; lateral denticulations may be present or absent. The large supraoculars not divided. Length of the hind limb equal to, or rather less than, the distance of the vent from the collar; fourth toe longer than the head †. Femoral pores varying in number from two to fifteen. The most characteristic points in the coloration of this race are the light ground-colour of the upper parts, be it a greyish or purplish or olive tinge; secondly, a light unspotted band running along each side of the back to the root of the tail; thirdly, blackish or brownish marbled spots which often are so disposed as to give the appearance of irregular transverse bands. The throat shows often indistinct blackish longitudinal lines. The collar-fold on the side of the neck is generally uncoloured, and in females always so; but in very old males from Presidio a blackish patch appears, which extends across the throat in the shape of a blackish band, as in *Sceloporus viviparus*. This latter race, however, is never provided with dorsal bands.

Size of adult male four and a quarter inches without tail.

The great diversity in the number of pores has given rise to the application of several specific terms; but as the several numbers are not accompanied by any other peculiar feature, and as they cannot be considered to be limited to certain localities, I feel compelled to unite all these specimens under the same heading.

The following list of localities will show the want of agreement between the number of pores and the distribution of the species:—

Specimens with 2–5 pores: Ventanas (*Forrer*), Colima (*Xantus, in U.S. Nat. Mus.*; *Mus. Paris*), Jalisco, North of Rio Santiago (*Richardson*), Guadalajara (*Cope*), Cuernavaca (*Mus. Brit.*), Vera Cruz (*Mus. Paris*).

Specimens with 7–10 pores: Texas (*U.S. Nat. Mus.*), Sonora (*U.S. Nat. Mus.*), Presidio (*Forrer*), Guanajuato (*Mus. Brit. & Mus. Paris*), Puebla (*Mus. Brit.*), Atlixco (*Mus. Brit.*).

Specimens with 12–15 pores: Duval Co., Texas (*Taylor*), Colorado Bottom (*U.S. Nat. Mus.*), Arizona (*W. L. Wright*), Mexico (*Mus. Brit.*), Sonora (*U.S. Nat. Mus.*), Amula and Iguala (*H. H. Smith*), Tampico (*Richardson*).

2. **Sceloporus viviparus.**

Sceloporus viviparus, Cope, Proc. Am. Phil. Soc. 1885, p. 398.

Hab. Mexico, Omilteme 8000 feet (*H. H. Smith*), Orizaba (*Sumichrast*), Mirador (*Sartorius*).

Twenty-two or twenty-three scales in a series along the back; scales on the back keeled, with a long terminal spine and a few obsolete lateral denticulations; scales on

* The scales are counted from a line connecting the two lateral collar-folds to one connecting the hinder side of the thighs. I believe thus to obtain more constant numbers than if the scales on the neck behind the occiput were included, as these scales are subject to variation in specimens from the same locality.

† Head = distance of upper end of tympanum from end of snout.

the neck smaller than those on the back. The large supraoculars not divided. Hind limb equal to the distance of the vent from the collar; fourth toe rather longer than the head. Femoral pores fifteen. Green or greenish-bronze coloured, without transverse collar on the neck; a large black patch on the collar-fold of the side of the neck, the two patches being connected by a black band across the throat. Young males have the abdomen blue, yellowish grey in the middle, the colours separated from each other by black. In very old males the light colour in the middle disappears, and the whole abdomen is uniform blackish blue. In an adult male each dorsal scale is black, with a broad green band, also the upperside of the head is symmetrically spotted with green.

Size of adult male 3 inches without tail.

Cope (*l. c.*) says that he has seen a specimen from Tehuantepec; but as this individual has a double row of supraoculars, the locality is not inserted here.

3. **Sceloporus poinsetti.**

Sceloporus poinsettii, Baird & Gir. Proc. Ac. N. Sc. Phil. vi. p. 126; and in U. S. & Mex. Bound. Surv., Rept. p. 5, t. 29. figg. 1–3.

Hab. North America, Texas.—Mexico, Rio San Pedro and Sonora (*Clark*), Ciudad in Durango (*Forrer*).

Twenty-one or twenty-two scales in a series along the back. The large supraoculars divided into a double series. Hind limb equal to the distance of the vent from the humeral joint; fourth toe as long as the head. Femoral pores eleven or twelve. Scales on the neck not conspicuously smaller than those of the back. A broad black collar edged with light green across the neck; body and tail with alternate dark and greenish cross-bands, the darker bands being the broader. Lower parts blue and black in the adult male, a whitish band along the middle of the abdomen.

Size of adult male 4 inches without tail.

The coloration of the upper parts is probably not constant.

4. **Sceloporus acanthinus.**

Sceloporus acanthinus, Bocourt, Miss. Sc. Mex., Rept. p. 180, t. 18. figg. 10, 10 *a*, *b*, & t. 19. figg. 4, 4 *a*; Boul. Cat. Liz. ii. p. 221.

Hab. Mexico, Jalapa (*Höge*); Guatemala, San Agustin on the Volcan de Atitlan (*Mus. Paris*).

Twenty-four scales along the back. The large supraoculars not divided. Hind limb equal to the distance of the vent from the humeral joint; fourth toe longer than the head. From twelve to fourteen femoral pores. Anterior nuchal scales much smaller than those on the back. A narrow black collar only one or two scales broad across the neck.

Size of an apparently adult female $2\frac{1}{2}$ inches without tail.

Originally described from two specimens taken on the western slope of the Volcan de Atitlan, at an elevation of 610 metres.

5. **Sceloporus torquatus.**

Sceloporus torquatus, Wiegm.* Isis, 1828, p. 369; and Herpet. Mex. p. 49, t. 7. fig. 1; Bocourt, Miss. Sc. Mex., Rept. p. 171, t. 18. figg. 9, 9 *a–c*; Cope, Proc. Am. Phil. Soc. 1885, p. 401; Boul. Cat. Liz. ii. p. 219; Dugès, La Naturaleza, i. p. 113 (1878).

Agama torquata, Green & Peale, Journ. Ac. N. Sc. Phil. vi. 1830, p. 231.

Tropidolepis torquatus, Gray, in Beechey's Voy., Zool. p. 95, t. 30. fig. 2.

Hab. Mexico, Highlands, Misteca, and Hacienda del Bobo (*Geddes*), Temascaltepec (*Keating*), city of Mexico, Jalisco, north of the Rio Santiago (*Richardson*).

From twenty-two to twenty-four scales along the back. The large supraoculars not divided. Hind limb equal to the distance of the vent from the humeral joint; fourth toe as long as the head. Femoral pores 15–20. Scales on the neck conspicuously smaller than those on the back. A broad black collar, edged with yellow (white, greenish), across the neck. Lower parts of the adult male blue, blackish in the middle in old individuals, with more or less yellowish in younger ones. Back sometimes with brownish markings or irregular yellow spots.

Size of an adult male $4\frac{3}{4}$ inches without tail.

6. **Sceloporus omiltemanus**, sp. n. (Tab. XXXII. fig. A.)

Hab. Mexico, Omilteme in the State of Guerrero 8000 feet (*H. H. Smith*).

Twenty-three or twenty-four scales in a series along the back; scales on the back smooth or faintly keeled, with a short terminal spine and a few fine denticulations; scales on the neck smaller than those on the back. The large supraoculars divided into a double series. Hind limb equal to the distance of the vent from the collar; fourth toe a little longer than the head. Femoral pores twelve or thirteen. Green or dark-bronze-coloured, with a broad black collar edged with light green across the neck; some scales of the upper arm red or green. Adult male blue underneath, the middle of the abdomen being blackish anteriorly and whitish posteriorly. Tail not banded. Adult female with the throat blue, and with the abdomen uniform greenish.

Size of adult male $3\frac{3}{4}$ inches without tail.

In very young specimens the black white-edged collar is already quite distinct, but the scales on the back are strongly keeled and without lateral denticulations.

To this fine race of the *torquatus*-group none of the various specific or varietal names are applicable, and as its locality can be stated with sufficient exactness, and does not seem to have been visited by other collectors, I think it best to describe this lizard under a distinct name.

* Wiegmann has priority in giving this name to the species, Green and Peale having selected the same specific term two years later, independently and in ignorance of Wiegmann's paper in the 'Isis.'

7. Sceloporus undulatus.

Agama undulata, Daud. Rept. iii. p. 384.

Sceloporus undulatus, Fitz. Syst. Rept. p. 75; Baird & Gir. U. S. Expl. Exped., Herp. p. 379, t. 19. figg. 15–21; Bocourt, Miss. Sc. Mex., Rept. p. 195, t. 18 bis. figg. 11, 11 *a, b*; Cope, Proc. Am. Phil. Soc. 1885, p. 398; Boul. Cat. Liz. ii. p. 227.

Sceloporus thayerii, Baird & Gir. Proc. Ac. N. Sc. Phil. vi. p. 127.

Hab. North America, United States.—Mexico, Sonora (*Baird & Girard*); Guatemala, Yzabal (*Salvin*).

From twenty-six to thirty-one scales in a series along the back. The large supraoculars not divided. Hind limb as long as, or shorter than, the distance of the vent from the collar; fourth toe as long or nearly as long as the distance of the snout from the lateral collar-fold. Thirteen to seventeen femoral pores. Greyish, brownish, or bronze-coloured, with irregular dark brown bands or transverse spots across the back; no nuchal collar.

An adult male is 3 inches long without tail.

This is a northern species which scarcely enters into the limits of the Central-American fauna. The reasons which induce me to admit it, are:—(1) because Baird and Girard say expressly in the description of their *Sceloporus thayerii* that it extends into Sonora; and (2) because Mr. Salvin's collection, made at Yzabal, contains a specimen which, although it has only twenty-six scales in the dorsal series, cannot be kept distinct from *S. undulatus*.

8. Sceloporus lunæi.

Sceloporus lunæi, Bocourt, Miss. Sc. Mex., Rept. p. 184, t. 18 bis. figg. 5, 5 *a, b*.

Sceloporus formosus, part., Boul. Cat. Liz. ii. p. 222.

Hab. Guatemala, Coban (*Salvin*), plateau 1500 metres (*Bocourt*).

Twenty-five scales in a series along the back, the dorsal series distinctly convergent on the posterior part of the back; scales keeled, with a long terminal spine. A single series of large supraoculars. One canthal scale. The anterior lobes of the ear-opening not larger than the preceding scales. Hind limb nearly as long as the distance of the vent from the collar; fourth toe longer than the head. Femoral pores thirteen. Green; each dorsal scale with a yellowish spot at the base; blue underneath, the middle of the abdomen being yellowish with a black streak on each side; collar-fold deep black, this colour not extending across the neck or throat.

Size of adult male 3 inches without tail.

9. Sceloporus irazuensis, sp. n.

Sceloporus formosus, part., Boul. Cat. Liz. ii. p. 222.

Hab. Mexico, Southern States?—Costa Rica, Irazu (*Rogers*).

Twenty-six or twenty-seven scales in a series along the back, the dorsal series nearly parallel; scales keeled, with a terminal spine. The large supraoculars are broken up and form two series. Two canthal scales, but owing to the shortness of the snout the anterior is not always fully developed. Anterior lobes of the ear-opening as large as the preceding scales. Hind limb as long as the distance of the vent from the collar; fourth toe longer than the head. Femoral pores fourteen or fifteen. Green; scales of the males sometimes with a lighter basal spot, females with indistinct subangular dark markings; males blue underneath, the middle of the abdomen being yellowish; collar-fold black, this colour not extending across the neck or throat.

Size of adult male 2⅔ inches without tail.

10. **Sceloporus salvini,** sp. n.

Sceloporus formosus, part., Boul. Cat. Liz. ii. p. 222.

Hab. Mexico, Jalapa (*Höge*); Guatemala (*Salvin*).

From thirty to thirty-two scales in a series along the back; scales on the back in parallel series, keeled, with a long terminal spine. Two series of large supraoculars, the scutes of the outer series being the smaller and separated from the superciliary margin by two series of granular scales. Two canthal scales. Ear-opening with one larger and a few smaller lobes in front. Hind limb as long as the distance of the vent from the collar; fourth toe longer than the head. Femoral pores twelve. Green; with or without longitudinal rows of darker spots; a rather narrow deep black collar, which, in the middle of the neck, is interrupted for the width of one or two scales, or at least contracted, and extends across the throat; males with the throat and the sides of the abdomen blue, the middle being yellowish with a black streak on each side.

Size of adult male 3¾ inches without tail.

11. **Sceloporus smaragdinus.**

?? *Sceloporus formosus,* Wiegm. Herpet. Mex. p. 50, t. 7. fig. 2.

Sceloporus formosus, part., Bocourt, Miss. Sc. Mex., Rept. p. 182, t. 18. figg. 3, 3 *a–c*; Boul. Cat. Liz. ii. p. 222.

? *Sceloporus malachiticus,* Cope, Proc. Ac. N. Sc. Phil. xvi. p. 178; Journ. Ac. N. Sc. Phil. viii. p. 127; and Proc. Am. Phil. Soc. 1885, p. 397.

Sceloporus smaragdinus, Bocourt, Miss. Sc. Mex., Rept. p. 186, t. 18. figg. 6, 6 *a, b,* and t. 19. figg. 1, 1 *a, b.*

Sceloporus tæniocnemis, Cope, Proc. Am. Phil. Soc. 1885, p. 399.

Hab. Mexico, Yucatan (*Cope*); Guatemala (*Cope*), Dueñas and Yzabal (*Salvin*), Solola, Totonicapam, Quezaltenango (*Bocourt*); Costa Rica, Arriba, San José (*Cope*), Irazu (*Rogers*).

From thirty-two to thirty-four scales in a series along the back; scales on the back in parallel series, keeled, with a long terminal spine and more or less distinct lateral denticulations. Two series of large supraoculars, but the scutes of the outer series are sometimes unusually small. One canthal scale. Hind limb shorter than, sometimes nearly as long as, the distance of the vent from the collar; fourth toe longer than the head. Femoral pores from fifteen to seventeen. Green or greenish-bronze coloured, with a black patch on the lateral collar-fold, not extending across the neck. Males with the throat, the sides of the abdomen, and sometimes the pectoral region, blue; a median white abdominal band is separated from the blue by a black streak.

Size of adult male $3\frac{1}{2}$ inches without tail.

12. **Sceloporus yarrovii.**

Sceloporus jarrovii, Cope in Wheeler's Rep. Explor. W. 100th Mer. v. p. 569, t. 23. figg. 2, 2 *a–d*; and Proc. Am. Phil. Soc. xxii. 1885, p. 396.

Sceloporus yarrovii, Boul. Cat. Liz. ii. p. 223.

Hab. North America, Arizona (*Cope*).—Mexico, Ciudad in Durango (*Forrer*), Jalisco, north of Rio Santiago (*Richardson*).

Thirty-one scales in a series along the back; scales on the back keeled, with a terminal spine. The large supraoculars not divided. Hind limb as long as the distance of the vent from the collar; fourth toe as long as, or longer than, the head. Seventeen femoral pores. Greenish, indistinctly spotted with black; a dark black-and-white-edged collar across the neck; two light bands on each side of the neck, one above and the other through the tympanum.

A female, adult or nearly so, measures $2\frac{1}{2}$ inches without tail; an adult male 3 inches.

13. **Sceloporus consobrinus.**

Sceloporus consobrinus, Baird & Gir. in Marcy's Expl. Red River, p. 237, t. 10. figg. 5–12; Cope, Proc. Am. Phil. Soc. 1885, p. 397; Boul. Cat. Liz. ii. p. 229.

Hab. North America.—Mexico, Putla (*Boucard*).

Thirty-one or thirty-two scales in a series along the middle of the back; scales on the back strongly keeled and with a terminal spine. Supraoculars broken up into small scutes, or, if two complete series can be distinguished, the scutes of the outer series are very small. Length of the hind limb less than the distance between the vent and collar; fourth toe scarcely as long as the head. From twelve to seventeen femoral pores. Upper parts light greyish or brownish, with a white band on each side of the back, bordered laterally by a series of black spots; generally a second white band from the axil of the fore limb to that of the hind limb. No collar. A black spot in front of the humerus, which in the male extends into the collar-fold. Sides of the head with two light bands, one above and one through the tympanum, both separated by a black streak. Male with the side of the throat and the sides of the abdomen blue, the remainder of the lower surface white.

Adult specimens are 2½ inches long without tail.

The evidence of this form extending into Mexico rests, at present, upon a badly-preserved specimen in the British Museum, purchased of M. Boucard, and stated by him to have been obtained at Putla in Mexico.

14. Sceloporus pyrrhocephalus.

Sceloporus pyrocephalus, Cope, Proc. Ac. N. Sc. Phil. xvi. p. 177.
Sceloporus pyrrhocephalus, Cope, Proc. Am. Phil. Soc. 1885, p. 394.

Hab. MEXICO, Colima (*Xantus, W. Lloyd*).

The typical specimens were from the same locality.

Thirty-three or thirty-four scales in a series along the back; dorsal scales in oblique convergent series, distinctly keeled, and with a terminal spine. Supraoculars broad, in a single series. Canthal scutes two; a pair of large mentals behind the front lower labial. Hind limb equal to the distance of the vent from the fore limb; fourth toe longer than the head. Femoral pores thirteen. Tail strongly compressed. Bronze-coloured above, with a broad, black, light-edged band along the side; abdomen blue on the sides, with a row of dark blue or black cross-bars on each of the blue lateral halves; throat with oblique bluish lines; a black cross-band in front of the shoulder; an elongate black spot with yellow centre in the middle of the occiput.

Size of adult male 2¾ inches without tail.

15. Sceloporus chrysostictus.

Sceloporus chrysostictus, Cope, Proc. Ac. N. Sc. Phil. xviii. p. 125; and Proc. Am. Phil. Soc. 1885, p. 396; Boul. Cat. Liz. ii. p. 235.
? *Sceloporus cupreus,* Bocourt, Miss. Sc. Mex., Rept. p. 210, t. 18 bis. fig. 2.

Hab. MEXICO, Yucatan (*A. Schott in U.S. Nat. Mus.*).

The localities "Guatemala (*Cope*)" and "Oaxaca (*Bocourt*)" require further confirmation.

Thirty-two scales in a series along the middle of the back; scales on the back keeled, pointed, and without prominent spine. Supraoculars in a single series, large. Limbs very slender; the hind limb is longer than the distance of the vent from the collar, and the length of the fourth toe equals the distance of the snout from the shoulder. From thirteen to fifteen femoral pores. Brownish or greenish olive above, with indistinct brown markings; a broad light iridescent band along each side of the back; axil of the fore limb black; sides with one longitudinal and three or four transverse blackish bands behind; no collar on the neck, also the collar-fold is colourless. Lower parts whitish; male without distinctive coloration of the abdomen.

The adult male is 2½ inches long without tail.

16. Sceloporus grammicus.

Sceloporus grammicus, Wiegm. Isis, 1828, p. 370; and Herpet. Mex. p. 51; Bocourt, Miss. Sc. Mex., Rept. p. 192, t. 18 bis. figg. 12, 12 *a, b*; Cope, Proc. Am. Phil. Soc. 1885, p. 397; Boul. Cat. Liz. ii. p. 231.

Sceloporus heterurus, Cope, Proc. Ac. N. Sc. Phil. xviii. p. 322.

Hab. Mexico, Oaxaca (*Bocourt*), Mirador (*Sartorius*), Tehuantepec (*Sumichrast*); Guatemala (*Godman*).

Thirty-seven or thirty-eight scales in a series along the back; scales on the back in oblique series, which converge towards the median line, keeled. All the supraoculars of rather small size. Hind limb nearly as long as the distance of the vent from the collar; fourth toe rather longer than the head. Femoral pores fourteen. Brownish, with transverse black spots; a narrow black band across the nape, interrupted in the middle. Male blue on the sides of the abdomen, the blue patch being bordered with black interiorly.

Of this race I have seen two young specimens only from Guatemala.

This race is very imperfectly known, and there is great risk of confounding it with others. From Bocourt's statement as to differences in the size of the scales in the two typical specimens, it would appear to be not improbable that the specific term *grammicus* should be applied to one of them only. *Sceloporus dugesii* (Bocourt, *l. c.* p. 188, t. 18. fig. 7; Cope, *l. c.* p. 396; Boul. Cat. Liz. ii. p. 224) = *Sceloporus intermedius* (Dugès, La Naturaleza, iv. 1876, p. 29, t. 1. figs. 21–32), from Colima (*Bocourt*) and Guanajuato (*Dugès*), comes evidently very close to *S. grammicus*, but is described as having larger dorsal scales, viz. only nine in the length of the shielded part of the head, whilst in *S. grammicus* twelve or thirteen would be included in that measurement.

17. Sceloporus gratiosus.

Sceloporus graciosus, Baird & Gir. Proc. Ac. N. Sc. Phil. vi. p. 69; and in Stansb. Exped. Gr. Salt Lake, p. 346, t. 5. figg. 1–3; Cope, Proc. Am. Phil. Soc. 1885, pp. 379, 397.

Sceloporus gratiosus, Boul. Cat. Liz. ii. p 230.

Hab. North America, Utah, South-western States.—Mexico, Matamoros Izucar (*Ferrari-Perez*), Putla (*Boucard*), Jalapa (*Ferrari-Perez*).

Forty-five scales in a series along the middle of the back. Dorsal scales keeled, in parallel series. Supraocular series divided, the inner being composed of the larger scutes. Length of the hind limb less than the distance between vent and collar; fourth toe longer than the head. From sixteen to eighteen femoral pores. Light brownish above, with two light longitudinal bands on each side, a row of darker spots on each side of the median dorsal line. Male with a blue black-edged patch on each side of the abdomen.

An adult female is $1\frac{3}{4}$ inch long without tail.

18. **Sceloporus ornatus.**

Sceloporus ornatus, Baird, Proc. Ac. N. Sc. Phil. x. p. 254; Cope, Proc. Am. Phil. Soc. 1885, p. 396.

Hab. MEXICO, Patos in Coahuila (*Couch, U. S. Nat. Mus.*), Nuevo Leon (*W. Taylor*).

From forty-eight to fifty scales in a series along the middle of the back. Scales with feeble keels. The large supraoculars divided. Hind limb as long as the distance of the vent from the collar; fourth toe rather shorter than the head. Sixteen or seventeen femoral pores. Greenish; blackish along the middle of the back, with light small spots; a rather broad and complete black collar edged with light green across the neck. Male with the throat and sides of the abdomen blue.

An (adult ?) male specimen is 26 lines long without tail.

19. **Sceloporus rubriventris,** sp. n. (Tab. XXXII. fig. C.)

Hab. MEXICO, Omilteme in Guerrero (*H. H. Smith*).

Forty-eight or fifty scales in a series along the middle of the back. Dorsal scales with a small terminal spine, their keels forming lines which converge towards the median line. The larger supraoculars in a double row; two canthal scutes. Hind limb rather shorter than the distance between vent and collar; fourth toe longer than the head. From sixteen to eighteen femoral pores. Greenish-bronze coloured, marbled above; a black band from the tympanum above the collar-fold and shoulder, a black line from the eye to the tympanum. Both sexes with the throat and a broad band along each side of the abdomen orange-coloured, the middle of the abdomen being greenish.

An adult male is $2\frac{1}{4}$ inches long without tail.

This very distinct form is allied to *S. microlepidotus*, but has larger scales and a different coloration of the abdomen.

20. **Sceloporus microlepidotus.**

Sceloporus microlepidotus, Wiegm. Herpet. Mex. p. 51; Bocourt, Miss. Sc. Mex., Rept. p. 194, t. 18 bis. figg. 13, 13 *a–d*; Cope, Proc. Am. Phil. Soc. 1885, pp. 170, 379, 384, 397; Boul. Cat. Liz. ii. p. 232.

Sceloporus dispar, Baird. & Gir. Proc. Ac. N. Sc. Phil. vi. p. 127.

Hab. MEXICO, Ventanas and Ciudad in Durango (*Forrer*), Colima (*U. S. Nat. Mus.*), Guanajuato, Puebla (*Dugès, Mus. Brit.*), Oaxaca (*Boucard*), Orizaba (*Mus. Brit.*), Jalapa (*Höge*), Tezuitlan (*Ferrari-Perez*), Zacualtipan (*Cope*), Popocatepetl in pine forests 9000 feet (*F. D. G.*), Mirador, Vera Cruz (*Sartorius, U. S. Nat. Mus.*).

From sixty-two to sixty-eight scales in a series along the middle of the back; dorsal scales not mucronate, in series converging towards the median line. Supraoculars small, in several series. Hind limb rather shorter than the distance of the vent from the collar; fourth toe rather longer than the head. From fourteen to eighteen femoral pores. Brownish or greenish, marbled with lighter and darker; collar-folds black, sometimes a black collar on the throat, but never on the neck. Male with the throat and the sides of the abdomen blue, the middle of the abdomen being black with a greenish median band, or in very old specimens entirely black.

An adult male specimen is $2\frac{1}{2}$ inches long without tail.

21. Sceloporus couchii.

Sceloporus couchii, Baird, Proc. Ac. N. Sc. Phil. x. p. 254.

Hab. North America, Texas, Duval Co. (*W. Taylor*).—Mexico, Sta. Catarina in Nuevo Leon (*Couch, U.S. Nat. Mus.*).

Seventy scales in a series along the middle of the back; dorsal scales not mucronate, in series converging towards the median line. A row of large supraoculars. Hind limb longer than the distance of the vent from the collar; length of the fourth toe equal to the distance of the shoulder from the end of the snout. Eighteen femoral pores. A light greenish, sometimes interrupted, longitudinal band on each side of the back; back more or less distinctly spotted; a black spot in front of the shoulder; sides with light greenish spots. Male with the abdomen whitish, blue on the sides, and with the jaws and throat ornamented with oblique convergent blue bands.

An adult male is $2\frac{1}{4}$ inches long without tail.

β. The two series of femoral pores meeting in front of the vent.

22. Sceloporus scalaris.

Sceloporus scalaris, Wiegm. Isis, 1828, p. 370; and Herp. Mex. p. 50, t. 8. fig. 2; Bocourt, Miss. Sc. Mex., Rept. p. 202, t. 18 bis. figg. 9, 9 *a*, *b*; Cope, Proc. Am. Phil. Soc. 1885, pp. 388, 394; Boul. Cat. Liz. ii. p. 234.

Tropidolepis scalaris, Gray, in Beechey's Voy., Zool. p. 95, t. 30. fig. 3.

Hab. North America, S.W. Texas.—Mexico, Sonora (*U.S. Nat. Mus.*), Monterey, Zacualtipan in Hidalgo (*Cope*), Jalisco, north of the Rio Santiago (*Richardson*), Colima (*Dugès*), Cuernavaca (*Méhédin, Mus. Paris*), city of Mexico, Atlixco (*Mus. Brit.*), Hacienda del Bobo (*Geddes*), Orizaba (*Sumichrast*), Cozumel I. (*U.S. Fish Commission*).

Thirty-two scales in a series along the middle of the back; dorsal scales keeled, pointed. Supraoculars small, without a distinct series of large ones. Length of the hind limb much less than the distance between the vent and collar; length of the fourth toe equal to, or even a little more than, that of the head. From sixteen to nineteen femoral pores on each side, the two rows meeting in the median line. Olive or light brownish, with a yellow line on each side of the back; a series of large blackish halfmoon-shaped spots (with the convex side directed backwards) on each side of the yellow lateral line. No collar. A black spot in front of the shoulder, entering the collar-fold. Throat generally with irregular oblique blackish lines. Males with the sides of the abdomen blue, the middle being whitish.

An adult male is $2\frac{1}{2}$ inches long without tail.

23. Sceloporus æneus.

Sceloporus æneus, Wiegm. Isis, 1828, p. 370; and Herpet. Mex. p. 50; Bocourt, Miss. Sc. Mex., Rept. p. 205, t. 18 bis. figg. 4, 4 *a*, *b*; Cope, Proc. Am. Phil. Soc. 1885, pp. 170, 379, 397; Boul. Cat. Liz. ii. p. 233.

Hab. Mexico, Puebla (*Boucard*), Matamoros Izucar (*Ferrari-Perez*), Orizaba (*Mus. Brit.*), Jalapa (*U.S. Nat. Mus.*), Jalisco, north of the Rio Santiago (*Richardson*).

Thirty-one scales in a series along the middle of the back; dorsal scales keeled, pointed. Supraoculars in more than one series, rather small. Length of the hind limb less than the distance between the vent and collar, and length of the fourth toe scarcely equal to that of the head. From fourteen to nineteen femoral pores on each side, the two rows meeting in the median line. Olive, greenish, or copper coloured above, with one or two light lines along each side; the back between the lines unspotted, or with black marbled spots longitudinally arranged. A black spot between the shoulder and collar-fold. Male with the lower parts more or less extensively black.

An adult male is $2\frac{1}{2}$ inches long without tail.

24. Sceloporus jalapæ, sp. n.

Hab. Mexico, Jalapa (*Höge*).

Thirty-eight scales in a series along the middle of the back. A row of large supra-oculars, separated from the superciliary edge by two rows of very small ones. Two canthal scales; ear-opening with two lobules in front. Length of the hind limb equal to the distance between the vent and collar; length of the fourth toe considerably exceeding that of the head. From seventeen to eighteen femoral pores on each side, the two rows meeting in the median line. In a much discoloured specimen the sides are darker than the back, a differently coloured longitudinal band separating the back from the sides. An indistinct darker band runs along the median line of the back. Lateral collar-fold deep black. No nuchal collar. Belly in the male blackish on the sides, white in the middle line; throat finely mottled.

The single male specimen known is $1\frac{3}{4}$ inch long without tail.

B. *The scales on the sides much larger than the dorsal.*

25. Sceloporus pleurolepis, sp. n. (Tab. XXXII. fig. B.)

Hab. Mexico, Jalisco, north of the Rio Santiago (*Richardson*).

Thirty-three scales in a series along the middle of the back; dorsal scales rounded and, mostly, smooth. The scales on the sides are twice or thrice as large as those on the back, and keeled, each keel terminating in a strong spine which projects outwards; the scales on the limbs and tail larger than those on the back and provided with a strong spine. The series of large supraoculars is partly entire, partly divided into two. Length of the hind limb equal to the distance between the vent and collar; length of the fourth toe equal to that of the head. Eleven or twelve femoral pores on each side. Nearly uniform greenish-olive above, whitish below; throat bluish, with a white median longitudinal line.

The only specimen received, a female, is $2\frac{1}{2}$ inches long without tail.

C. *The scales on the sides imbricate, but defined from the dorsal by a sharp line.*

26. **Sceloporus variabilis.**

Sceloporus variabilis, Wiegm. Herpet. Mex. p. 51; Bocourt, Miss. Sc. Mex., Rept. p. 200, t. 18 bis. figg. 1, 1 *a*, *b*, and t. 19. fig. 2; Cope, Proc. Am. Phil. Soc. 1885, pp. 169, 379, 397; Boul. Cat. Liz. p. 236; Sumichrast, Bull. Soc. Zool. v. 1880, p. 176.

Hab. North America, Texas, Duval Co. (*W. Taylor*).—Mexico, Monterey (*Cope*), Tampico, Jalisco, north of the Rio Santiago (*Richardson*), Amula (*H. H. Smith*), Matamoros Izucar (*Ferrari-Perez*), Atoyac (*H. H. Smith*), Jalapa (*Mus. Brit.*), Hacienda del Bobo (*Höge*), Orizaba, Vera Cruz, Huatusco (*Godman*), Mirador (*Sartorius*), mountains of Huamelula *, Tehuantepec (*Sumichrast*); Guatemala, San Gerónimo (*Salvin*); Salvador, La Union (*Mus. Paris*).

From forty to forty-five scales in a series along the middle of the back. The large supraoculars in a single, undivided row. Hind limb longer than the distance between the vent and collar; fourth toe as long as the distance between the end of the snout and the shoulder. From eleven to thirteen femoral pores. Brownish or greenish, with a more or less distinct whitish band along each side of the back; the back between the bands generally with two rows of dark spots. A large black spot in the axil of the fore limb. Male with a pair of C-shaped blue bands on the abdomen, meeting in the middle of their convexity in old examples, and enclosing a reddish patch in their concavity.

Sumichrast says that this species prefers rocky ground in ravines or along the banks of rivers.

Length of an adult male 3 inches without tail.

27. **Sceloporus teapensis,** sp. n.

Hab. Mexico, Teapa in Tabasco (*H. H. Smith*).

This form agrees with *S. variabilis* in every respect, but has much larger scales on every part of the body: there are only from thirty-one to thirty-three in a series along

* In these specimens the dorsal bands are bright yellow in life, and sharply defined.

the middle of the back. The temples are covered with moderately-sized imbricate scales, and not partly granular as in *S. variabilis*.

Size of adult male 3 inches without tail.

28. **Sceloporus fulvus.**

Sceloporus fulvus, Bocourt, Miss. Sc. Mex., Rept. p. 214, t. 18 bis. figg. 8, 8 *a–c*; Boul. Cat. Liz. ii. p. 238.

Hab. SALVADOR, La Union (*Mus. Paris*); NICARAGUA (*Boulenger*).

Thirty-two scales in a series *from the occipital* to the base of the tail; a row of large supraoculars; a single canthal scale. Length of the fourth toe equal to the distance between the end of the snout and the collar-fold. Three or four femoral pores. Sides darker than the back, a yellow longitudinal band between the back and sides; the back between the bands with a double series of oblique dark brown spots. Lower parts uniform yellowish in both sexes.

Length $1\frac{3}{4}$ inch without tail.

29. **Sceloporus siniferus.**

Sceloporus siniferus, Cope, Proc. Am. Phil. Soc. xi. p. 159 (1871); Bocourt, Journ. Zool. v. p. 342 (1876); Boul. Cat. Liz. ii. p. 237; Sumichrast, Bull. Soc. Zool. v. 1880, p. 177.

Sceloporus humeralis, Bocourt, Miss. Sc. Mex., Rept. p. 206, t. 18 bis. figg. 3, 3 *a*, *b*.

Hab. MEXICO, Omilteme in Guerrero (*H. H. Smith*), Acapulco, Oaxaca (*Sallé*), Tehuantepec (*Sumichrast*).

Twenty-seven scales in a series along the middle of the back. The large supraoculars in a single, undivided row. Hind limb longer than the distance between the vent and collar; fourth toe nearly as long as the distance between the shoulder and end of the snout. From four to eight femoral pores. Brownish, sides darker, upper and lateral parts more or less distinctly spotted with darker; lower parts yellowish, throat sometimes with oblique blackish lines. Axil of the fore limb not black.

Adult specimens are 2 inches long without tail.

D. *The scales on the sides granular.*

30. **Sceloporus utiformis.**

Sceloporus utiformis, Cope, Proc. Ac. N. Sc. Phil. xvi. p. 177; Bocourt, Miss. Sc. Mex., Rept. p. 208, t. 18 bis. figg. 6, 6 *a*, *b*.

Hab. MEXICO, Colima (*Xantus*).

Forty scales between the occipital and the base of the tail; ten dorsal scales are contained in the length of the shielded part of the head. A row of large supraoculars. Femoral pores thirteen or fourteen. Blackish above, with numerous indistinct lighter cross-bars, or light brown, with a double longitudinal series of transverse black spots and a yellowish band along each side of the back.

Length $2\frac{3}{4}$ inches without tail.

The following species have not been seen by me:—

1. From Mexico and La Noria (Michoacan) and Tupataro, near Guanajuato.

Sceloporus melanogaster, Cope, Proc. Am. Phil. Soc. 1885, p. 400: according to Dugès, La Naturaleza, i. p. 114, it is not from Guanajuato. Boulenger's *Sceloporus melanogaster* is certainly a different form.

2. From Guanajuato (?).

Sceloporus ferrari-perezi, Cope, Proc. Am. Phil. Soc. 1885, p. 400.

3. From Tehuantepec and Colima.

Sceloporus melanorhinus, Bocourt, Journ. Zool. v. 1876, p. 341, and Ann. Sc. Nat. 1876, iii. art. 12; Cope, Proc. Am. Phil. Soc. 1885, p. 399; Sumichrast, Bull. Soc. Zool. v. 1880, p. 176.

4. From Vera Cruz.

Sceloporus torquatus mucronatus, Cope, Proc. Am. Phil. Soc. 1885, p. 402.

5. From Monterey, Nuevo Leon.

Sceloporus torquatus cyanogenys, Cope, Proc. Am. Phil. Soc. 1885, p. 402.

6. From Guatemala and Antigua (alt. 1500 metres), and from the mouth of the Nagualate.

Sceloporus squamosus, Bocourt, Miss. Sc. Mex., Rept. p. 212, t. 18 bis. figg. 7, 7 *a–b*, and t. 19. fig. 3.

7. From Yucatan.

Sceloporus serrifer, Cope, Proc. Ac. N. Sc. Phil. xviii. p. 124. I suspect that the specimen referred by Boulenger to this species is a distinct form; but as the exact locality of that specimen is unknown, I omit a description of it.

PHRYNOSOMA.

Phrynosoma, Wiegmann, Isis, 1828, p. 367.

These lizards are called "Caméleón" by the Mexicans *.

* Whilst in the English-speaking countries of the Southern United States the term "Horned Toad" is applied to them. Several observers have recorded a singular habit of these lizards (see, for instance, J. Wallace, Proc. Zool. Soc. 1871, p. 1), viz., under certain circumstances, apparently as a means of defence, they squirt from one of the eyes a jet of bright red liquid, very much like blood, to a distance of about six inches. Only a few individuals will do this occasionally, while many others never do it. The specimens in which it has been observed seemed to belong to *P. cornutum*. I have not been able to find the source of this secretion by dissection of fresh examples; and accurate observation of individuals which possess this power is very desirable.

1. Phrynosoma orbiculare.

α. Var. orbicularis.

Lacerta orbicularis (part.), Linn. Syst. Nat. i. p. 265.

Phrynosoma orbiculare, Wiegm. Isis, 1828, p. 367; and Herp. Mex. p. 53; Wagl. Icon. Amphib. t. 23; Sumichrast, Arch. Sc. Phys. Nat. xix. p. 60 (1864); and Ann. & Mag. N. H. 3rd ser. xiii. p. 507; Boul. Cat. Liz. ii. p. 241.

Tapaya orbicularis (Girard), Bocourt, Miss. Sc. Mex., Rept. p. 221, t. 11. figg. 1, 1 *a–g*; Dugès, La Naturaleza, i. p. 116 (1868); Sanchez, La Naturaleza, vii. p. 323 (1887).

Phrynosoma wiegmanni, Gray, in Beechey's Voy., Zool. p. 96.

Hab. Mexico, city of Mexico (*Doorman*), Ciudad in Durango (*Forrer*), Jalisco, north of the Rio Santiago (*Richardson*).

β. Var. cortezii.

Phrynosoma orbiculare, Wiegm. Herp. Mex. t. 8. fig. 1.

Tapaya orbicularis, var., Bocourt, Miss. Sc. Mex., Rept. p. 223.

Phrynosoma orbicularis, var., Boul. Cat. Liz. ii. p. 242.

Tapaya cortezii, Bocourt, Miss. Sc. Mex., Rept. t. 11. figg. 2, 2 *a–g*.

Hab. Mexico, Jalapa (*Höge*), Puebla and Orizaba (*Mus. Brit.*).

γ. Var. dugesii.

Tapaya orbicularis, var., Bocourt, Miss. Sc. Mex., Rept. p. 224.

Tapaya dugesii, Bocourt, l. c. t. 11. figg. 3, 3 *a–f*.

Phrynosoma orbicularis, var., Boul. Cat. Liz. ii. p. 243.

Hab. Mexico, Colima.

δ. Var. boucardi.

Tapaya boucardii, Bocourt, Miss. Sc. Mex., Rept. p. 225, t. 11. figg. 4, 4 *a–f*.

Hab. Mexico, plateau of Mexico.

Sumichrast says of this species:—"The *Phrynosoma*, which is peculiar to the cold and dry regions of the Mexican plateau, inhabits sandy spots exposed to the sun, such as the margins of roads and the arid ridges, where the earthy colour of its body easily conceals it from observation. Ill formed for running, it has none of that lizard-like vivacity which has become proverbial; its gait is slow and awkward. . . . Destitute of means of defence, it allows itself to be taken without even attempting to bite the hand that has seized it. . . . On several occasions when I have put females of *Phrynosoma orbiculare* into alcohol I have seen the young immediately issue from the cloaca, to the number of ten or twelve." A specimen which I kept for some time would simulate death, and although torpid on cool days, would in the heat of the day run rapidly to some place of concealment as soon as it thought itself unobserved. Dugès states that the number of young varies from fifteen to twenty-four.

2. **Phrynosoma cornutum.**

Agama cornuta, Harlan, Journ. Ac. N. Sc. Phil. iv. p. 229, t. 20 (1825).

Phrynosoma cornutum, Gray, in Griff. Anim. Kingd. ix. Syn. p. 45; Holbr. N.-Amer. Herp. ii. p. 87, t. 11; Girard, in Stansbury's Exped. Gt. Salt Lake, p. 360, t. 8. figg. 1–6, and in U.S. Expl. Exped., Herp. p. 403, t. 21. figg. 6–9; Bocourt, Miss. Sc. Mex., Rept. p. 236, t. 12. figg. 9, 9 *a–f*; Boul. Cat. Liz. ii. p. 245; Cope, Proc. Am. Phil. Soc. 1885, p. 169; Gentry, Proc. Ac. N. Sc. Phil. xxxvii. p. 145.

Phrynosoma planiceps, Hallow. Proc. Ac. N. Sc. Phil. vi. p. 178; and in Sitgreave's Exped. Zuni & Color. Riv. p. 124, t. 7; Bocourt, Miss. Sc. Mex., Rept. p. 238, t. 12. figg. 11, 11 *a–f*; Yarrow, in Wheeler's Rep. Explor. W. 100th Merid. p. 579, t. 24. fig. 1.

Hab. North America, United States.—Mexico, Monterey (*Cope*), Chihuahua (*Montague Kerr*), Tampico (*Richardson*).

Tampico is the southernmost locality known of this species at present.

3. **Phrynosoma asio.**

Phrynosoma asio, Cope, Proc. Ac. N. Sc. Phil. xvi. p. 178; Sumichrast, Arch. Sc. Phys. Nat. xlvi. p. 258 (1873); Boul. Cat. Liz. ii. p. 244.

Batrachosoma asio, Bocourt, Miss. Sc. Mex., Rept. p. 241, t. 17. figg. 9, 9 *a–c*.

Phrynosoma spinimentum, Peters, MB. Berl. Ak. 1873, p. 742.

Hab. Mexico, Colima (*Xantus, U. S. Nat. Mus.*), Tehuantepec (*Sumichrast*); Guatemala, Savana Grande (*F. D. G. & O. S.*).

Its habits are similar to those of *P. orbiculare*, but it frequents the hot plains; its food consists of beetles, ants, &c.

4. **Phrynosoma modestum.**

Phrynosoma modesta, Girard, in Stansbury's Exped. Gt. Salt Lake, pp. 361, 365, t. 6. figg. 4–8; Bocourt, Miss. Sc. Mex., Rept. p. 232.

Phrynosoma modestum, Boul. Cat. Liz. ii. p. 248.

Hab. North America, Southern United States.—Mexico, Nuevo Leon (*W. Taylor*).

5. **Phrynosoma braconnieri.**

Phrynosoma braconnieri, Bocourt, Miss. Sc. Mex., Rept. p. 233, t. 12. figg. 7, 7 *a–g*; Boul. Cat. Liz. ii. p. 248.

Hab. Mexico, Oaxaca (*Boucard*).

6. **Phrynosoma taurus.**

Phrynosoma taurus, Bocourt, Miss. Sc. Mex., Rept. p. 234, t. 12. figg. 8, 8 *a–g*; Boul. Cat. Liz. ii. p. 249.

Hab. Mexico, plateau of Puebla (*Boucard, Mus. Paris*), Matamoros (*Mus. Brit.*), Amula in Guerrero 8000 feet (*H. H. Smith*), Coscatlan (*Dugès*).

Fam. GECKONIDÆ.

GONATODES.

Gonatodes, Fitzinger, Syst. Rept. p. 91; Boulenger, Cat. Liz. i. p. 56.

1. Gonatodes albogularis.

Gymnodactylus albogularis, Dum. & Bibr. Erp. Gén. iii. p. 415.
Gonatodes albogularis, Fitz. Syst. Rept. p. 91; Boul. Cat. Liz. i. p. 59.
Stenodactylus fuscus, Hallow. Journ. Ac. N. Sc. Phil. iii. p. 33 (1855).
Gymnodactylus fuscus, Bocourt, Miss. Sc. Mex., Rept. p. 48, t. 10. figg. 5, 5 *a–c*.
Goniodactylus braconnieri, O'Shaughn. Ann. & Mag. N. H. 4th ser. xvi. p. 265.

Hab. SALVADOR, La Union (*Mus. Paris*); NICARAGUA (*Hallowell*); PANAMA (*Ida Pfeiffer*).—COLOMBIA; WEST INDIES, Cuba, Martinique (*Dum. & Bibr.*).

2. Gonatodes caudiscutatus.

Gymnodactylus caudiscutatus, Günth. Proc. Zool. Soc. 1859, p. 410.
Goniodactylus caudiscutatus, O'Shaughn. Ann. & Mag. N. H. 4th ser. xvi. p. 265.
Gonatodes caudiscutatus, Boul. Cat. Liz. i. p. 61, t. 5. fig. 2.

Hab. PANAMA (*Ida Pfeiffer*).—WESTERN ECUADOR (*Fraser*).

PHYLLODACTYLUS.

Phyllodactylus, Gray, Spicil. Zool. 1830, p. 3.

1. Phyllodactylus tuberculosus.

Phyllodactylus tuberculosus, Wiegm. N. Act. Ac. L.-C. xvii. p. 241, t. 18. fig. 2; Bocourt, Miss. Sc. Mex., Rept. p. 43, t. 10. figg. 3, 3 *a*; O'Shaughn. Ann. & Mag. N. H. 4th ser. xvi. p. 262; Sumichrast, Bull. Soc. Zool. 1880, p. 172; Boul. Cat. Liz. i. p. 79.

Hab. NORTH AMERICA, California.—MEXICO, Tres Marias Is., Ventanas, Presidio (*Forrer*), Durango, Guadalajara (*Cope*), Acapulco, Tierra Colorada (*H. H. Smith*), Tehuantepec (*Sumichrast*); GUATEMALA, Rio Motagua, San Gerónimo (*Salvin*), Coban (*Cope*); HONDURAS, Ruatan I. (*G. F. Gaumer*); SALVADOR, La Union (*Mus. Paris*); NICARAGUA (*Cope*).

Known to the inhabitants of Mexico and of the Isthmus of Tehuantepec as "Salamanquesa"; frequents houses and renders itself most useful by destroying vermin.

2. Phyllodactylus ventralis.

Phyllodactylus ventralis, O'Shaughn. Ann. & Mag. N. H. 4th ser. xvi. p. 263; Boul. Cat. Liz. i. p. 80.

Hab. NICARAGUA, Granada (*Mus. Brit.*).

It should be remembered that the type purchased for the British Museum of a dealer in Paris was stated to be from Jamaica.

THECADACTYLUS.

Thecadactylus, Cuvier, Règne Anim. ii. p. 48.

1. Thecadactylus rapicauda.

Gecko rapicauda, Houttuyn, Verh. Gen. Vlissing. ix. p. 322, t. 3. fig. 1.

Thecadactylus rapicaudus, Gray, Cat. Liz. p. 146; Cope, Journ. Ac. N. Sc. Phil. 1875, p. 119; Boul. Cat. Liz. i. p. 111.

Platydactytus theconyx, Dum. & Bibr. Erp. Gén. iii. p. 306, t. 33. fig. 2.

Hab. MEXICO, Guadalajara (*Cope*), Yucatan (*Mus. Brit.*); BRITISH HONDURAS (*Mus. Brit.*); GUATEMALA, Coban (*Cope*); NICARAGUA, Granada (*Mus. Brit.*); COSTA RICA (*Cope*).—NORTH-EASTERN TROPICAL AMERICA; WEST INDIES.

ARISTELLIGER.

Aristelliger, Cope, Proc. Ac. N. Sc. Phil. 1861, p. 496.

Idiodactylus, Bocourt, Miss. Sc. Mex., Rept. p. 41.

1. Aristelliger præsignis.

Hemidactylus præsignis, Hallow. Proc. Ac. N. Sc. Phil. 1856, p. 222.

Aristelliger præsignis, Cope, Proc. Ac. N. Sc. Phil. 1861, p. 496; Boul. Cat. Liz. i. p. 146; Garman, Proc. Am. Phil. Soc. 1887, p. 275; and Bull. Essex Inst. xx. p. 103 (1888).

Idiodactylus georgeensis, Bocourt, Miss. Sc. Mex., Rept. p. 41, t. 10. figg. 1, 1 *a–d*.

Aristelliger irregularis, Cope, Proc. Am. Phil. Soc. 1885, p. 387.

Hab. MEXICO, Cozumel I. (*U.S. Fish Commission*, *Gaumer*), Yucatan; BRITISH HONDURAS, St. George's Cay off Belize (*Scient. Commission Mex.*); NICARAGUA.—WEST INDIES, Jamaica, San Domingo, Grand Cayman and Cayman Brac (*Garman*).

Specimens from Cozumel Island and St. George's Cay differ from those from Jamaica merely in coloration, being of a uniform blackish tint, the sides of the head finely speckled with white. During life the Cozumel specimens are said to be green.

GEHYRA.

Gehyra (Gray), Boulenger, Cat. Liz. i. p. 147.

1. Gehyra mutilata.

Hemidactylus (*Peropus*) *mutilatus*, Wiegm. N. Act. Ac. L.-C. xvii. p. 238.

Gehyra mutilata, Boul. Cat. Liz. i. p. 148.

Hemidactylus peronii, Dum. & Bibr. Erp. Gén. iii. p. 352, t. 30. fig. 1.

Peripia peronii, Gray, Cat. Liz. p. 159.

Hemidactylus navarri, Dugès, La Naturaleza, vi. p. 309, t. 7 *a* (1883).

Hab. MEXICO, Presidio (*Forrer*), San Blas (*Forrer*, *Dugès*).—ISLANDS OF THE TROPICAL INDO-PACIFIC OCEAN; MALAYAN PENINSULA; BURMA.

The circumstance that this species, which is so widely distributed along the shores of the Indo-Pacific Ocean, has been discovered only within a comparatively recent period in Mexico, and still appears to be confined to a limited area, seems to show that this is a late importation into the New World.

SPHÆRODACTYLUS.

Sphærodactylus (Wagl.), Gray, in Griff. Anim. Kingd. Syn. p. 52.

1. **Sphærodactylus homolepis.**

Sphærodactylus homolepis, Cope, Proc. Am. Phil. Soc. 1886, p. 277.

Hab. NICARAGUA.

Of this species I have not seen specimens.

2. **Sphærodactylus torquatus.**

Sphærodactylus torquatus, Strauch, Mém. Ac. Sc. Pétersb. xxxv., Geckon. p. 35 (1887).

Hab. MEXICO, Mazatlan.

Not seen by me.

3. **Sphærodactylus glaucus.**

Sphærodactylus glaucus, Cope, Proc. Ac. N. Sc. Phil. 1865, p. 192; and Journ. Ac. N. Sc. Phil. 1875, p. 118; Sumichrast, Bull. Soc. Zool. Fr. 1880, p. 173; Boul. Cat. Liz. i. p. 221, t. 18. fig. 3.

Sphærodactylus inornatus, Peters, MB. Ak. Wiss. Berlin, 1873, p. 738.

Hab. MEXICO, Vera Cruz (*Mus. Brit.*), Teapa in Tabasco (*H. H. Smith*), Tehuantepec (*Sumichrast*), Merida in Yucatan (*A. Schott*); BRITISH HONDURAS, Belize (*Cope*), Stann Creek (*Robertson*); GUATEMALA, Vera Paz, low forest (*F. D. G. & O. S.*), Peten (*Cope*); COSTA RICA, Sipurio (*Cope*).

A forest species.

4. **Sphærodactylus lineolatus.**

Sphærodactylus lineolatus, Lichtenst. Nomencl. Rept. Mus. Berol. p. 6; Bocourt, Miss. Sc. Mex., Rept. p. 46; Boul. Cat. Liz. i. p. 221.

Sphærodactylus millepunctatus, Hallow. Proc. Ac. N. Sc. Phil. 1860, p. 480; Cope, Proc. Ac. N. Sc. Philad. 1861, p. 499.

Hab. MEXICO, Vera Cruz (*Mus. Brit.*); NICARAGUA, Chontales (*Mus. Brit.*); BRITISH HONDURAS, Stann Creek (*Robertson*); PANAMA, Veragua (*Mus. Berol.*).—GUIANA, Demerara (*Quelch*).

5. **Sphærodactylus casicolus.**

Sphærodactylus casicolus, Cope, Proc. Ac. N. Sc. Phil. 1861, p. 499.

Hab. PANAMA, Rio Truando (*U.S. Nat. Mus.*).

Not seen by me.

6. Sphærodactylus fantasticus.

Sphærodactylus fantasticus, Dum. & Bibr. Erpét. iii. p. 406, t. 32. fig. 2; Bocourt, Miss. Sc. Mex., Rept. p. 44, t. 10. figg. 4, 4 *a–c*; Boul. Cat. Liz. i. p. 223.

Hab. MEXICO (*Mus. Philad.*).—VENEZUELA, Caracas (*Mus. Brit.*); WEST INDIES, Martinique.

7. Sphærodactylus anthracinus.

Sphærodactylus anthracinus, Cope, Proc. Ac. N. Sc. Phil. 1861, p. 500; Peters, MB. Ak. Wiss. Berl. 1869, p. 874; ? Boul. Cat. Liz. i. p. 225; ? Strauch, Mém. Ac. Sc. Pétersb. xxxv., Geckon. p. 37 (1887).

Hab. MEXICO (*Mus. Philad.*), Jalapa (*Cope*).—WEST INDIES.

Not seen by me.

EUBLEPHARIS.

Eublepharis, Gray, Philos. Mag. iii. 2, p. 56 (1827).

This genus, together with *Psilodactylus* and *Coleonyx*, forms a very natural group in the Geckonidæ, and has a wide distribution in the Old World, like many other genera of this family. But while the latter are either represented by numerous closely allied species or are abundant in individuals, which spontaneously or accidentally have been spread and are in progress of being spread over a large area of the surface of the globe, the Eublepharidæ (as the group has been termed) consist of a very small number of species only, and are so scarce that their distribution cannot be explained by the same agencies. Those longest known are two species of *Eublepharis* from the Indian peninsula, ranging westwards into Mesopotamia. *Psilodactylus*, with one species, is West African, and finally in Central America and the Southern United States we find three species of *Eublepharis*, most closely allied to the Indian, and one species of *Coleonyx*. The not least singular point in the similarity of organization of these Lizards is that they have retained a surprising resemblance in the pattern of their coloration.

1. Eublepharis variegatus.

Stenodactylus variegatus, Baird, Proc. Ac. N. Sc. Phil. 1858, p. 254; and U. S. & Mex. Bound. Surv., Rept. p. 12, t. 23. figg. 9–27, and t. 24. figg. 11–19.
Coleonyx variegatus, Cope, Proc. Ac. N. Sc. Phil. 1866, p. 125.
Eublepharis variegatus, Boul. Cat. Liz. i. p. 233.

Hab. NORTH AMERICA, California, Rio Grande and Gila valleys (*A. Schott*), Texas, El Paso.

2. Eublepharis dovii. (Tab. XXXI. fig. B.)

Eublepharis dovii, Boul. Cat. Liz. i. p. 233.

Hab. PANAMA (*Capt. Dow*).

Skin finely granular, with numerous tubercules which are smaller than the interspaces between them, and provided with a short keel in the middle; the tubercles are very small on the head and disposed in rings on the tail. The anterior side of the legs is covered with small imbricate scales like those of the abdomen. The nasal shields form a broad suture together behind the rostral; seven upper labial shields; the middle lower labial large, rather broader than long, without mental shields behind; femoral pores none; preanal pores seven, arranged in a triangular patch. The length of the hind leg equals the distance of the snout from the shoulder, that of the fore limb equal to the distance of the shoulder from the anterior margin of the eye.

Pale coloured, with brown markings above, forming a coarse irregular network on the back; the markings are broken up into spots on the head, and ring-shaped on the tail.

The type-specimen is a male, and 5½ inches long, of which the tail takes 2½ inches.

3. **Eublepharis fasciatus.** (Tab. XXXI. fig. A.)

Eublepharis fasciatus, Boul. Cat. Liz. i. p. 234.

Hab. MEXICO, Ventanas (*Forrer*).

Skin finely granular, without larger tubercles, the lower parts and legs covered with very small imbricate scales. The nasal shields do not meet in the middle behind the rostral; seven upper labial shields; the middle lower labial large, rather broader than long, without mental shields behind; femoral and preanal pores absent (in the female). Snout rather long, longer than the distance of the ear-opening from the eye; the length of the hind leg equals the distance of the snout from the shoulder, that of the fore limb being equal to the distance of the shoulder from the anterior margin of the eye.

Pale coloured, with broad yellowish cross-bands, which are broadly margined with chocolate-brown: the first is horseshoe-shaped, and runs from the angle of the mouth round the occiput to the other side; the next is placed between the shoulders, the two following across the back of the trunk, and the fifth across the loin; on the tail they form broad rings.

The type-specimen is a female, and nearly four inches long, of which the tail takes less than one half.

COLEONYX.

Coleonyx, Gray, Ann. & Mag. N. H. xvi. p. 162 (1845).
Brachydactylus, Peters, MB. Ak. Wiss. Berl. 1863, p. 41.

1. **Coleonyx elegans.**

Coleonyx elegans, Gray, Ann. & Mag. N. H. xvi. p. 163 (1845); Cope, Proc. Ac. N. Sc. Phil. 1866, p. 125; Bocourt, Miss. Sc. Mex., Rept. p. 49, t. 10. figg. 7, 7 *a–d*; Sumichrast, Bull. Soc. Zool. Fr. 1880, p. 172; Boul. Cat. Liz. i. p. 235.
Gymnodactylus scapularis, A. Dum. Cat. Méth., Rept. p. 45.
Gymnodactylus coleonyx, A. Dum. Arch. Mus. viii. p. 483, t. 16. fig. 6.
Brachydactylus mitratus, Peters, MB. Ak. Wiss. Berl. 1863, p. 42.

Hab. MEXICO, Orizaba (*Sumichrast*), Teapa in Tabasco (*H. H. Smith*), Yucatan; BRITISH HONDURAS, Belize (*Dyson*); GUATEMALA, Peten (*Morelet*); COSTA RICA (*Peters*).

A forest species.

Order OPHIDIA.

Fam. STENOSTOMATIDÆ.

GLAUCONIA.

Stenostoma, Wagl. Serp. Bras. 1825, p. 69 (nec Latr.).
Glauconia, Gray, Cat. Liz. p. 139.

1. Glauconia albifrons.

Stenostoma albifrons, Wagl. Serp. Bras. t. 25. fig. 3; D'Orbigny, Voy. Amér. Mér., Rept. t. 6. figg. 1–6; Dum. & Bibr. Erp. Gén. vi. p. 327; Jan, Iconogr. ii. tt. 3, 6, 7. figg. 10; Bocourt, Miss. Sc. Mex., Rept. p. 505, t. 29. figg. 10, 10 *a, b.*
Glauconia albifrons, Boulenger, Cat. Sn. i. p. 63.
Typhlops undecimstriatus, Cuv. Règne Anim. ii. p. 74 (1829).
Epictia undecimstriata, Gray, Cat. Liz. p. 140.
Stenostoma fallax, Peters, MB. Ak. Wiss. Berl. 1857, p. 402.
Sabrina tessellata, Girard, Proc. Ac. N. Sc. Phil. 1857, p. 181.
Stenostoma melanostoma, Cope, ibid. 1862, p. 350.
Stenostoma phænops, Cope, Journ. Ac. N. Sc. Phil. 1875, p. 128; Bull. U. S. Nat. Mus. no. 32, p. 63 (1887); Sumichrast, Bull. Soc. Zool. Fr. 1880, p. 179.

Hab. MEXICO, Jalapa (*Höge*), Tehuantepec (*Sumichrast*), Santo Domingo de Guzman (*Buller*); GUATEMALA, Coban (*Cope*); HONDURAS, Ruatan Isl. (*G. F. Gaumer*); COSTA RICA (*Cope*); PANAMA (*Cope*).—TROPICAL AMERICA.

Common in the Isthmus of Tehuantepec, living under decaying trunks, under stones on dry soil, and even in houses.

2. Glauconia myopica.

Stenostoma myopicum, Garman, Mem. Mus. Comp. Zool. Cambr. viii. no. 3, p. 6 (1883).
Glauconia myopica, Boulenger, Cat. Sn. i. p. 69.

Hab. MEXICO, Tampico (*Garman, Richardson*).

3. Glauconia dulcis.

Rena dulcis, Baird & Girard, N. Amer. Serp. p. 142.
Stenostoma dulce, Jan, Iconogr. ii. tt. 5 & 6. figg. 5; Garman, Mem. Mus. Comp. Zool. Cambr. viii. no. 3, p. 5 (1883); Cope, Bull. U. S. Nat. Mus. no. 32, 1887, p. 63.
Leptotyphlops dulcis, Stejneger, Proc. U. S. Nat. Mus. xiv. p. 501 (1892).
Glauconia dulcis, Cope, Proc. Ac. N. Sc. Phil. 1892, p. 590.
Stenostoma rubellum, Garman, Mem. Mus. Comp. Zool. Cambr. viii. no. 3, p. 130.

Hab. NORTH AMERICA, Texas.—NORTHERN MEXICO, Batopila, Chihuahua (*Cope*).

4. Glauconia humilis.

Rana humile, Baird & Gir. N. Am. Serp. p. 143; Stejneger, Proc. U. S. Nat. Mus. xiv. p. 501 (1892).

Siagonodon dugesii, Bocourt, Miss. Sc. Mex., Rept. p. 507, t. 29. figg. 9, 9 *a–c,* and t. 30. fig. 4.

Stenostoma tenuiculum, Garman, Mem. Mus. Comp. Zool. Cambr. viii. no. 3, p. 5 (1883).

Hab. North America, California.—Mexico, San Luis Potosi (*Garman*), Presidio (*Forrer*), Colima (*Bocourt*).

Fam. TYPHLOPIDÆ.

TYPHLOPS.

Typhlops, Schneider, Hist. Amph. p. 34 (1801).

1. Typhlops tenuis.

Typhlops tenuis, Salvin, Proc. Zool. Soc. 1860, p. 454; Boulenger, Cat. Sn. i. p. 28.

Typhlops basimaculatus, Cope, Proc. Ac. N. Sc. Phil. 1866, p. 320.

Typhlops perditus, Peters, MB. Ak. Wiss. Berl. 1869, p. 435; Bocourt, Miss. Sc. Mex., Rept. p. 499, t. 29. figg. 3, 3 *a–c,* t. 30. fig. 3.

Typhlops prælongus, F. Müll. Verh. nat. Ges. Basel, vii. p. 674 (1885).

Hab. Mexico, Orizaba, Cordova, Teapa in Tabasco (*H. H. Smith*); Guatemala, Coban (*Salvin*).

2. Typhlops microstomus.

Typhlops microstomus, Cope, Proc. Ac. N. Sc. Phil. 1866, p. 125.

Hab. Mexico, Yucatan.

I have not seen a specimen of this species.

HELMINTHOPHIS.

Helminthophis, Peters, MB. Ak. Wiss. Berl. 1860, p. 517.

Idiotyphlops, Jan, Arch. per la Zool. i. p. 186 (1861).

1. Helminthophis frontalis.

Helminthophis frontalis, Peters, MB. Ak. Wiss. Berl. 1860, p. 517, fig.; Bocourt, Miss. Sc. Mex., Rept. p. 502. t. 30. fig. 2.

Hab. Costa Rica.

I have not seen a specimen of this species.

LIOTYPHLOPS.

Liotyphlops, Peters, SB. Ges. naturf. Freund. 1881, p. 69.

1. **Liotyphlops albirostris.**

Rhinotyphlops albirostris, Peters, MB. Ak. Wiss. Berl. 1857, p. 402.

Liotyphlops albirostris, Peters, SB. Ges. naturf. Freund. 1881, p. 69; Bocourt, Miss. Sc. Mex., Rept. p. 501, t. 30. figg. 1, 1 *a–c*.

Typhlops (*Idiotyphlops*) *emunctus*, Garman, Mem. Mus. Comp. Zool. Cambr. viii. no. 3, p. 3 (1883).

Hab. Panama, Veragua.

ANOMALOLEPIS.

Anomalepis, Jan, Arch. per la Zool. i. p. 185 (1861).

1. **Anomalolepis mexicanus.**

Anomalepis mexicanus, Jan, Iconogr. i. t. 5 & t. 6. fig. 1; Bocourt, Miss. Sc. Mex., Rept. p. 503, t. 29. figg. 4, 5.

Hab. Mexico.

The statement as to the origin of this species requires confirmation.

Fam. COLUBRIDÆ AGLYPHÆ.

GEOPHIS.

Geophis, Wagl. Syst. Amph. p. 342.

α. Species with the occipital in contact with a labial.

1. **Geophis dolichocephala.**

Colobognathus dolichocephalus, Cope, Proc. Ac. N. Sc. Phil. xxiii. p. 211 (1872).

Catostoma dolichocephalum, Cope, Journ. Ac. N. Sc. Phil. viii. p. 147 (1876).

Hab. Costa Rica, San José (*Van Patten*).

Head elongate, conical, scarcely distinct from neck. Anterior frontals very small; supraocular and one postocular present. Six upper labials, the third and fourth entering the orbit, the fifth forming a suture with the occipital, the sixth higher than long. Scales in thirteen rows, keeled. Ventrals 131; anal entire. Upper parts of the body and the entire tail black; a series of large distant red spots on each side, which often meet above, forming half-rings: these disappear on the tail and neck. Below red; lower lip and chin black.

I have not seen a specimen of this species.

2. **Geophis chalybæa.** (Tab. XXXII. fig. B.)

Catostoma (*Geophis*) *chalybæum*, Wagl. Syst. Amph. p. 194; Peters, MB. Ak. Wiss. Berl. 1859, p. 275; Günth. Ann. & Mag. N. H. 1872, ix. p. 16; Bocourt, Miss. Sc. Mex., Rept. p. 530, t. 31. figg. 11, 11 *a–c*.

Colobognathus hoffmanni, Peters, MB. Ak. Wiss. Berl. 1859, p. 276, fig. 2; Cope, Bull. U. S. Nat. Mus. no. 32, pp. 53, 86 *.

Geophis hoffmanni, Bocourt, Miss. Sc. Mex., Rept. p. 529, t. 31. figg. 8, 8 *a–c*.[1]

Geophidium dubium, Peters, MB. Ak. Wiss. Berl. 1861, p. 923; Cope, Bull. U. S. Nat. Mus. no. 32, p. 86.

? *Geophis dubius*, Bocourt, Miss. Sc. Mex., Rept. p. 532, t. 31. figg. 9, 9 *a–c* †.

Elapoides sieboldi, Jan, Arch. per la Zool. ii. p. 21 (1862); and Iconogr. xii. t. 1. figg. 4.

Ninia sieboldii, Garman, Mem. Mus. Comp. Zool. Cambr. viii. no. 3, p. 96 (1883).

Colophrys rhodogaster, Cope, Proc. Ac. N. Sc. Phil. 1868, p. 130 ‡; and Bull. U. S. Nat. Mus. no. 32, p. 86.

Geophis rhodogaster, Bocourt, Miss. Sc. Mex., Rept. p. 531, t. 31. figg. 12, 12 *a–d*.

Geophis fuscus, Fischer, Abhandl. naturw. Ver. Hamburg, ix. p. 11, t. 2. fig. 5 (1886).

A supraocular and one postocular, sometimes confluent into one scute, sometimes one or the other absent. Two pairs of frontals, sometimes confluent into a single pair; the anterior pair, if present, very small; two pairs of chin-shields. Generally six, rarely five, upper labials, the third and fourth entering the orbit, and the fifth in contact with the occipital. The first pair of lower labials form a suture together. Fifteen or seventeen series of scales, either all being smooth, or only those on the back or loin provided with feeble keels; anal entire. Coloration generally uniform black above; blackish, yellowish, or whitish below.

The modifications of this very variable form may be arranged thus:—

I. Scales in seventeen rows; two pairs of frontals.

A. Scales keeled. A supraocular and postocular. Ventral scutes 137 or 139. Uniform blackish above, yellowish below §.

Hab. Mexico (*Sallé*); Guatemala, Dueñas and Pacific coast (*Salvin*).

B. Scales smooth. A supraocular and postocular.—*Catostoma chalybæum* (Wagl.).

Hab. Mexico (*Deppe*).

C. Scales smooth. No supraocular; one postocular. Ventral scutes 137 or 142. Uniform blackish above, yellowish below.—*Colophrys rhodogaster* (Cope).

Hab. Mexico, Yucatan (*Smiths. Inst.*); Guatemala, Rio Chisoy below the town of Cubulco (*Salvin*); Costa Rica (*Smiths. Inst.*).

---

* Cope represents this snake as "without teeth on the palate;" they are distinctly enough shown in the figure given by Peters.

† Bocourt describes and figures this species with the first pair of infralabials not meeting in the median line, whilst Peters expressly states that they meet in front of the chin-shields. The specimens examined by myself confirm Peters' statement, and I am therefore led to suppose that there is some error in Bocourt's figure or in his determination of the specimens.

‡ According to Cope five lower labials are in contact with the chin-shields; I am unable to find more than four in any of our specimens.

§ These specimens would probably have been named *Catostoma nasale* by Mr. Cope (Proc. Ac. N. Sc. Philad. xx. p. 131); but it is rather uncertain what he understood by that name. He describes his specimens as possessing *eight* upper labials, while the figure of the head shows *seven* only. In the original description quoted the specimens are said to be "from near the city of Guatemala," and in Bull. U. S. Nat. Mus. no. 32, p. 85, from "Central Costa Rica."

II. Scales in seventeen rows; one pair of frontals only.—*Geophidium dubium.*

Hab. ——?

III. Scales in fifteen rows. Two pairs of frontals.

D. Scales on the back slightly keeled; a small supra- and a postocular; six upper labials, the third and fourth entering the orbit.

a. Uniform brownish-black above, yellowish below. Ventral scutes 129 or 134 or 135.

Hab. MEXICO (*Sallé*).

b. Back brownish-black, crossed by large quadrangular white spots, equidistant and half as broad as the distance between them, not extending downwards to the whitish coloration of the abdomen. Ventral scutes 144.—Var. *quadrangularis*, m. (Tab. XXXIII. fig. B.)

Hab. COSTA RICA, Cartago (*Brit. Mus.*).

E. Scales on the back nearly entirely smooth; a small supra- and a postocular; six, sometimes five, upper labials, the third and fourth entering the orbit.—*Colobognathus hoffmanni.*

c. Uniform brownish-black above, whitish below. Ventral scutes 143 (Mexico) or 127 (Costa Rica and Chiriqui).

Hab. MEXICO, city of Mexico (*Doorman*); COSTA RICA (*Mus. Berol.*); PANAMA, Chiriqui (*Champion*).

d. Brown, each ventral scute more or less widely brown transversely; an indistinct lighter band across the nape; sides of the back sometimes with an indistinct incomplete longitudinal band. Ventral scutes 139–147.

Hab. COSTA RICA, Irazu (*Mus. Salvin & Godman*).

F. Scales entirely smooth; no supraocular, the occipital and postfrontal meeting above the eye; a small postocular; six upper labials, the third and fourth entering the orbit. Above uniform blackish-brown. Ventral scutes 144–146.

Hab. COSTA RICA, Irazu (*Mus. Salvin & Godman*).

IV. Scales in fifteen rows. One pair of frontals only.

Scales smooth; supra- and postocular confluent. Ventral scutes 139. Blackish above, white below.—*Geophis fuscus.*

Hab. MEXICO, Jalapa (*Fischer*).

3. **Geophis rostralis.**

Elapoides rostralis, Jan, Iconogr. Ophid. xii. t. 2. fig. 2.
Geophis rostralis, Bocourt, Miss. Sc. Mex., Rept. p. 533, t. 31. figg. 10, 10 *a–c*.
? *Geophis dubius*, Bocourt, l. c. p. 532, t. 31. figg. 9, 9 *a–c*.

Hab. MEXICO (*Sallé*).

A supraocular and one postocular. Two pairs of frontal shields, of which the anterior pair are very small (and, apparently, sometimes coalesce with the posterior). Six upper labials, the third and fourth entering the orbit, and the fifth in contact with the occipital. The first pair of lower labials do not meet in the median line of the chin. Seventeen series of scales, those in the lumbar region keeled (apparently, sometimes all smooth). Ventral scutes 146. Above uniform dark coloured, below whitish.

If I am correct in supposing that Bocourt's *Geophis dubius* is distinct from the species described by Peters under that name, then it is very probable that the former is specifically identical with Jan's *Geophis rostralis*, this species being subject to variations similar to those of *Geophis chalybœa.*

4. **Geophis semidoliata.**

Rhabdosoma semidoliatum, Dum. & Bibr. Erpét. vii. p. 93; Günth. Col Sn. p. 10; Cope, Bull. U. S. Nat. Mus. no. 32, p. 85.

Catostoma semidoliatum, Cope, Proc. Ac. N. Sc. Phil. 1861 (1860), p. 339.

Elapoides semidoliatus, Jan, Iconogr. Ophid. xii. t. 2. fig. 1.

Geophis semidoliatus, Peters, MB. Ak. Wiss. Berl. 1859, p. 276; Bocourt, Miss. Sc. Mex., Rept. p. 534, t. 31. figg. 7, 7 *a–c*; Garman, Mem. Mus. Comp. Zool. viii. no. 3, 1883, p. 102.

Hab. MEXICO, Jalapa and Orizaba (*Brit. Mus.*), Misantla and Zacualtipan (*Cope*), Huatusco (*Godman*).

A supraocular and one postocular, sometimes confluent into one scute. Two pairs of frontals and two pairs of chin-shields. Five upper labials, the third entering the orbit. Fifteen series of smooth scales; anal entire. Yellowish, orange-coloured on the back, with broad black cross-bands, rather irregular in shape.

This species seems to be extremely common at Jalapa.

5. **Geophis mœsta.** (Tab. XXXIII. fig. C.)

Geophis mœstus, Günth. Ann. & Mag. N. H. 1872, ix. p. 15.

Hab. MEXICO, Amula (*H. H. Smith*); COSTA RICA, Cartago (*Mus. Brit.*).

Head rather broad, short and depressed; body and tail of moderate length. Eye small, with the pupil elliptical. Anterior frontals well developed, about one-eighth the size of posterior. Vertical as broad as long, six-sided, with the anterior angle rather obtuse, and with the posterior somewhat pointed; its lateral edges are very short, convergent. Supraocular well developed. Occipitals rounded behind, shorter than the vertical and postfrontals together. Six upper labials, the third and fourth entering the orbit; the fifth is the largest, and forms a suture with the occipital; an elongate temporal behind this suture. One postocular. The first pair of lower labials form a suture together; anterior chin-shields of moderate extent. Scales in fifteen rows, mostly smooth, but those in the dorsal region above the vent keeled. Ventrals 140–148; anal entire; subcaudals 41. Coloration entirely black, with a broad white collar, occupying a portion of the occipitals and the temple. Lower parts blackish or whitish.

Colobognathus brachycephalus, Cope, Proc. Ac. N. Sc. Phil. xxiii. p. 211 (1871), or *Catostoma brachycephalum*, Cope, Journ. Ac. N. Sc. Phil. viii. p. 147 (1876), also from Costa Rica, is probably identical with this species; but Cope describes it as having only 124 ventrals, and with a whitish lateral band on the body, which extends from the side of the neck to the tail, and is composed of two rows of alternating narrow spots, which are not always perfectly united.

6. Geophis dugesii.

Geophis dugesii, Bocourt, Miss. Sc. Mex., Rept. p. 573, t. 37. figg. 1, 1 *a–d*.

Hab. MEXICO, Tangancicuaro (*Dugès*).

Head rather narrow. Anterior frontals small; a small supraocular and a single postocular present; six upper labials, the fifth forming a suture with the occipital. The first pair of lower labials form a suture together; anterior chin-shields twice as long as posterior. Scales in fifteen rows, smooth. Ventrals 176; anal entire. Upper parts black, and the anterior part of the body with six or seven yellowish half-rings which on the posterior parts become more and more indistinct; lower parts yellowish.

I have not had the opportunity of examining specimens of this species.

7. Geophis bicolor.

Geophis bicolor, Günth. Ann. & Mag. N. H. 1868, i. p. 416 *.

Rhabdosoma guttulatum, Cope, Proc. Am. Phil. Soc. xxii. p. 385 (1885).

Hab. MEXICO, city of Mexico (*Doorman*), Jalisco, La Cumbre de los Arrastrados (*A. C. Buller*), Vera Cruz (*Sumichrast*).

Head rather broad, short, and depressed; body and tail of moderate length. Eye small, with elliptical pupil. Anterior frontals about one-fourth the size of the posterior; vertical rather longer than broad, with the anterior angle very open; supraocular well developed; occipitals as long as postfrontals and vertical together, rounded behind; six upper labials, the third and fourth entering the orbit; the fifth is the longest, and forms a long suture with the occipital †. The remainder of the temple is covered by scale-like temporals, 1+2. Two postoculars. Anterior chin-shields twice as long as posterior, relatively shorter in specimens from Jalisco, in contact with four labials. Scales in seventeen rows, smooth, without apical groove. Ventrals 151–168; anal entire; subcaudals 39–48. Upper parts uniform black, lower white; on the two or three outer series of scales the white colour appears in more or less distinct small spots, whilst the black of the upper parts extends to the angles of the ventral shields.

* Cope (Bull. U. S. Nat. Mus. no. 32, p. 85) refers to this species "*Colobognathus hoffmanni* of Jan," Iconogr. Ophid. xii. t. 2. fig. 3; but as this author figures his specimen with fifteen rows of keeled scales, and five upper labials, Mr. Cope's identification is at least doubtful.

† In the largest specimen from Jalisco a small temporal intervenes; it is a detached portion of the fifth labial.

Each subcaudal black in front. Specimens from Jalisco have the abdomen largely marbled with black.

8. Geophis redimita.

Geagras redimitus, Cope, Journ. Ac. N. Sc. Phil. 1876, p. 141.
Sphenocalamus lineolatus, Fisch. Oster Programm Akad. Gymnas. Hamburg, 1883, figg. 3–5.

Hab. MEXICO, Mazatlan (*Mus. Hamburg*), Tehuantepec (*Sumichrast*).

Head rather narrow, not distinct from neck, tail short; eye of moderate size, with round pupil. Snout depressed, projecting beyond the lower jaw, the rostral shield being broad in front, with a sharp edge. Anterior frontals narrow; vertical six-sided; supraocular well developed; one ante- and one postocular; five upper labials, of which the third only enters the orbit; the fourth is in contact with the occipital, separating the elongate temporal from the postocular. The first pair of lower labials do not meet in the median line; anterior chin-shields at least twice as large as posterior. Scales smooth, in fifteen rows. Ventrals 118 or 119; anal divided; subcaudals 25 or 30. Upper parts of a pale ground-colour, with narrow brown lines along the series of scales; head brown above, with a whitish longitudinal band on each side running from the rostral to the end of the occipital.

β. Species with the occipital separated from the upper labials by a temporal.

9. Geophis omiltemana, sp. n. (Tab. XXXIII. fig. A.)

Hab. MEXICO, Omilteme in Guerrero 8000 feet (*H. H. Smith*).

Head broad, short, and depressed, body and tail of moderate length. Eye rather small, with the pupil subelliptical. Anterior frontals well developed, one-fifth or one-sixth the size of posterior. Vertical broader than long, six-sided, with the anterior angle very obtuse, the posterior being a right one; its lateral edges are very short, convergent; supraocular rather small. Six upper labials, the third and fourth entering the orbit, but it is worthy of notice that a small portion of the third upper labial is separate, and forms a small præocular below the præocular proper; the sixth is the largest, and separated from the occipital by a large temporal, which is in contact with two postoculars. The first pair of lower labials form a suture together; anterior chin-shields not much larger than posterior. Scales in seventeen rows, of which only the two middle ones show a faint carination. Ventrals 152, anal entire, subcaudals 52. Upper parts black with narrow whitish cross-bars, of which the three or four anterior are more distinct than the following, and are dilated on the sides as they merge into the white coloration of the abdomen. The cross-bands of the middle and posterior parts of the body are very indistinct and placed in pairs.

One specimen, apparently a male, is $11\frac{1}{2}$ inches long, of which the tail takes $2\frac{1}{2}$ inches; two other specimens are of smaller size.

10. **Geophis mutitorques.**

Rhabdosoma mutitorques, Cope, Proc. Am. Phil. Soc. xxii. p. 384 (1885).

Hab. MEXICO, North-eastern Hidalgo, Zacualtipan (*Cope*).

Anterior frontals one-fourth the size of posterior. Vertical broader than long; supraocular present, but small; one postocular. Eye small. Six upper labials, the third and fourth entering the orbit, the fifth not in contact with the occipital. Temporals 1+2. The first pair of lower labials in contact with each other. Scales smooth, in seventeen rows. Ventrals 172; anal entire. Colour plum-brown, the scales of three or four lateral rows slightly pale on the edges. In old specimens the ventrals are similarly coloured with pale edges, but in younger specimens there are at intervals pale spots, and the throat and chin are pale, probably pale yellow in life. In young specimens a yellow band crosses from one angle of the mouth to the other, involving the anterior three-quarters of the parietal plates. In larger specimens this is indistinct, and in a large specimen (455 millim.) the band has disappeared.

Cope describes also a *Rhabdosoma longiceps*, from San José Acateno in the State of Vera Cruz [Proc. U. S. Nat. Mus. ix. p. 189 (1887)], which seems to be very closely allied to the above species, as stated by the author himself.

Geophis latifrontalis, Garman, Mem. Mus. Comp. Zool. viii. no. 3, p. 103 (1883), from San Luis Potosi, cannot be distinguished from the above species by means of the descriptions given.

11. **Geophis fasciata.** (Tab. XXXIV. fig. A.)

Virginia fasciata, Fisch. Jahrb. Hamb. Wiss. Anst. ii. p. 95 (1885).

Hab. GUATEMALA (*Sarg*).

Head broad, short, and depressed, distinct from neck; body and tail of moderate length. Eye rather small, with elliptical pupil. Anterior frontals well developed, one-fifth the size of posterior. Vertical longer than broad, five-sided, with the lateral edges long and nearly parallel. Supraocular large; a small separate præocular above the præocular proper; two postoculars. Six upper labials, the third and fourth entering the orbit, the fifth being the largest, and separated from the occipital by a large temporal which is in contact with both postoculars. Temporals 1+2. The first pair of lower labials form a suture together; anterior chin-shields nearly twice the size of posterior. Scales in seventeen rows, of which the dorsal are keeled. Ventrals 184–196, anal entire, subcaudals 51–56. Upper parts brownish grey, with numerous black cross-bands, which are broadest on the back and narrow on the sides, and sometimes encroach upon the ventral shields; head black above, lower parts white with some isolated black spots on the posterior parts of the abdomen, which become more numerous on the subcaudals.

The description is taken from one of the two types now in the British Museum.

12. Geophis psephota.

Catostoma psephotum, Cope, Journ. Ac. N. Sc. Phil. viii. p. 146 (1876).

Hab. Costa Rica (*Gabb*).

Anterior frontals present, four-sided; supraocular present; two postoculars; upper labials six, the third and fourth entering the eye, the sixth the longest. Temporals 1+2+2. The first pair of lower labials form a suture together. Nasal undivided. Scales keeled, in seventeen rows. Ventrals 162; anal entire. Upper parts uniform black; below black, with the half or less of an occasional ventral shield red, forming a tessellated pattern.

I have not seen a specimen of this species.

13. Geophis zebrina.

Rabdosoma zebrinum, Jan, Arch. per la Zool. ii. p. 15 (1862); Bocourt, Miss. Sc. Mex., Rept. p. 539, t. 34. figg. 1, 1 *a–e*.

Rhegnops zebrinus, Cope, Proc. Am. Phil. Soc. xxii. p. 178 (1885).

Hab. Mexico, Tehuantepec (*Bocourt*).

A supraocular and two postoculars. Two pairs of frontal shields, of which the anterior pair are very small. Seven upper labials, the third and fourth entering the orbit. Temporals 1+2, the anterior in contact with the postoculars. The first pair of lower labials in contact with each other; one pair of chin-shields. Scales smooth, in seventeen series. Ventral scutes 147–160. Upper parts light coloured, with about forty irregular brown transverse spots. Lower parts punctulated with reddish.

ADELPHICUS.

Adelphicos, Jan, Arch. per la Zool. ii. p. 18 (1862).

1. Adelphicus quadrivirgatus.

Adelphicos quadrivirgatum, Jan, Arch. per la Zool. 1862, p. 19; and Iconogr. Ophid. xi. t. 3. fig. 5; Bocourt, Miss. Sc. Mex., Rept. p. 554, t. 32. figg. 11, 11 *a–e*, 12, 12 *a–e*.

Adelphicus quadrivirgatus, F. Müller, Katal. Herpet. Samml. Basl. Mus. 1880, p. 26.

Rhegnops visoninus, Cope, Proc. Ac. N. Sc. Phil. 1866, p. 128.

Rhegnops sargii, Fisch. Jahrb. Hamb. Wiss. Anst. ii. p. 92 (1885).

Hab. Mexico (*Sallé*), Jicaltepec in Vera Cruz (*Miss. Geogr. Mex.*); British Honduras, Belize (*Cope*); Guatemala (*Sarg*), Alta Vera Paz (*Bocourt*); Honduras (*Dyson*).

Head not distinct from neck, more or less pointed; body of moderate length; tail short. Eye small, with round pupil. Anterior frontals very small; supraocular well developed; two postoculars; seven upper labials, the third and fourth entering the orbit; temporals 1+1, the anterior in contact with the oculars. The first pair of lower labials meet in the median line, but are linear, nearly suppressed by the enormously enlarged chin-shields, as are also the second and third pairs. Scales smooth, in fifteen rows. Ventrals 135–142; anal divided. Upper parts light brownish, with a dark band on each side of the back along the fifth outer series of scales, generally another line along the second outer series, and sometimes one along the median series. Lower parts yellowish, with a brown line along the middle of the tail, this line being sometimes continued along the middle of the abdomen.

This species is closely allied to *Rhabdosoma lineatum* from British Guiana, from which it differs, however, in the shape of the chin-shields, and by its divided anal; the southern species has also, as a rule, eight upper labial shields. It is singular that Fischer denies the presence of longitudinal bands in the specimens examined by him. I have received from the Stuttgart Museum one of the types of *Rhegnops sargii*, and in it five longitudinal bands are distinctly visible.

TOLUCA.

Toluca, Kennicott, U. S. & Mex. Bound. Surveys, Rept. p. 23 (1859).
Achirhina, Jan, Arch. per la Zool. ii. p. 61 (1862).

Body stout; head short and broad, wedge-shaped, almost continuous with the body; tail short; eye of moderate size. Snout pointed, the rostral turned back upon the upper side. Two pairs of frontals, of which the posterior are separated by the produced anterior angle of the vertical. Nasal single, elongate, touching the ante-orbital; no loreal. Scales smooth, in seventeen rows. None of the teeth are grooved.

1. **Toluca lineata.**

Toluca lineata, Kennicott, U. S. & Mex. Bound. Surveys, Rept. p. 23, t. 21. fig. 2; and Pac. R. R. Exp., Rept. t. 35. fig. 8.
Achirhina defilippii, Jan, Arch. per la Zool. ii. p. 61.

Hab. MEXICO, Valley of Mexico, Puebla, Toluca (*Smiths. Inst.*).

One ante- and two postoculars; seven upper labials. Upper parts light brownish grey, with three imperfect blackish longitudinal bands, each occupying a single series of scales.

I have not seen specimens of this snake.

PSEUDOFICIMIA.

Pseudoficimia, Bocourt, Miss. Sc. Mex., Rept. p. 572 (1883).

Habit of the body coronelloid; head rather short and thick, not distinct from neck; eye of moderate size. Snout pointed, with the rostral shield enlarged, but not dividing the anterior frontals; its anterior edge is rather sharp and turned upwards, the upper surface of the shield being concave. Two pairs of frontals. Nostril between two nasals; loreal replaced by posterior frontal, which is in contact with the second upper labial. Scales smooth, with one apical pore; in seventeen rows. Teeth not grooved, the hindmost rather the longest.

1. Pseudoficimia frontalis.

Toluca frontalis, Cope, Proc. Ac. Nat. Sc. Phil. 1864, p. 167.
Geagras frontalis, Cope, Journ. Ac. Phil. viii. p. 142 (1876).
Ficimia frontalis, Garman, Mem. Mus. Comp. Zool. viii. no. 3, p. 82 (1883).
Pseudoficimia pulchra, Bocourt, Miss. Sc. Mex., Rept. p. 572, t. 35. figg. 12, 12 *a–e*.

Hab. MEXICO, Ventanas and Presidio (*Forrer*), Colima and Guadalajara (*Smiths. Inst.*).

Vertical shield five-sided, broad, as large as an occipital; anterior frontal one-third the size of posterior; one anterior and two posterior oculars; seven upper labials, the third and fourth entering the orbit; temporals 1+2+3; anterior pair of chin-shields rather large, posterior small and widely separate. Ventrals 152–160; anal divided. Upper parts brownish olive, with a dorsal series of large rhombic, brown, black-edged spots, and with irregular smaller spots on the sides. A black-edged cross-band between the eyes. Lower parts uniform white.

SYMPHOLIS.

Sympholis, Cope, Proc. Ac. Nat. Sc. Phil. 1862, p. 524.
Cheilorhina, Jan, Arch. per la Zool. ii. p. 57 (1862).

Body short, stout, with the tail short and obtuse; head short, not distinct from neck; snout rounded, obtuse; eye small. Only one pair of frontals; anterior nasal confluent with the first labial; posterior nasal in contact with the ante-ocular. Scales smooth, without pores, in nineteen rows. Teeth smooth, equal in length, or but slightly increasing in strength behind.

1. Sympholis lippiens.

Sympholis lippiens, Cope, Proc. Ac. Nat. Sc. Phil. 1862, p. 524.
Geophis lippiens, Garman, Mem. Mus. Comp. Zool. viii. no. 3, p. 103 (1883).
Cheilorhina villarsii, Jan, Arch. per la Zool. ii. p. 57; and Iconogr. Ophid. xlviii. t. 1. fig. 5; cop. by Bocourt, Miss. Sc. Mex., Rept. p. 555, t. 34. figg. 5, 5 *a–e*; Garman, Mem. Mus. Comp. Zool. viii. no. 3, p. 85 (1883).

Hab. MEXICO, Guadalajara (*Smiths. Inst.*).

Vertical shield six-sided, with the anterior angle produced forward. Five upper labials, the third only entering the orbit; one ante- and one postocular; anal plate entire. Yellowish, with about eighteen black rings, which become more or less complete on the abdomen, two on the tail, and one covering the snout to behind the eyes.

CONOPSIS.

Conopsis, Günther, Col. Sn. p. 6 (1858).
Oxyrhina, sp., Jan, Arch. per la Zool. ii. p. 59 (1862).

The snakes of this genus are not strictly isodont. The posterior teeth are always somewhat stouter, if not longer than the anterior, which is particularly conspicuous in specimens exceeding ten inches in length. In specimens of that size sometimes a distinct elongate pit at the base of the tooth may be seen, apparently the commencement of a groove. Therefore *Conopsis* might be regarded as a less developed form of *Stenorhina.*

The nasal shield is invariably single; but the frontal shields may be a single or double pair in specimens from the same locality and apparently of the same brood.

1. **Conopsis nasus.** (Tab. XXXIV. figg. B–B‴.)

α. Synonymy of specimens with a single pair of frontal shields :—

Conopsis nasus, Günth. Col. Sn. p. 6; Bocourt, Miss. Sc. Mex., Rept. p. 563, t. 35. figg. 2, 2 *a–d.*
Ficimia nasus, Garman, Mem. Mus. Comp. Zool. viii. no. 3, p. 83 (1883).
Oxyrhina (*Exorhina*) *maculata,* Jan, Arch. per la Zool. ii. p. 61 (1862).
Ficimia maculata, Garman, Mem. Mus. Comp. Zool. viii. no. 3, p. 84.
Conopsis maculatus, Bocourt, Miss. Sc. Mex., Rept. p. 564, t. 35. fig. 3.

β. Synonymy of specimens with two pairs of frontal shields :—

Oxyrhina varians, Jan, Arch. per la Zool. ii. p. 60 (1862).
Conopsis varians, Bocourt, Miss. Sc. Mex., Rept. p. 566, t. 35. figg. 5, 5 *a–d.*
Ogmius varians, Cope, Proc. Am. Phil. Soc. 1869, p. 162; and Proc. U. S. Nat. Mus. ix. p. 189 (1887).
Conopsis nasus, Peters, MB. Ak. Wiss. Berl. 1869, p. 875.
Conopsis lineatus, Bocourt, Miss. Sc. Mex., Rept. p. 565, t. 35. figg. 4, 4 *a–d* (nec *Toluca lineata,* Kenn.).
Chionactis diasii, Cope, Proc. U. S. Nat. Mus. ix. p. 188 (1887).
Ogmius acutus, Cope, Proc. U. S. Nat. Mus. ix. p. 189 (1887).

Hab. North America, California (*Mus. Brit.*).—Mexico (*Sallé*), Milpas in Durango (*Forrer*), Omilteme in Guerrero (*H. H. Smith*), La Cumbre de los Arrastrados (*Buller*), Puebla (*Peters*), Tehuantepec (*Sumichrast*).

This species is subject to a considerable amount of variation; the small scutes of the head, such as the loreal, postoculars, and labials, frequently disappear by coalescence. The ground-colour may be light buff or dark brownish grey, uniform or with dark spots of variable size and shape. The lower parts may be uniform whitish, or varied with more or less numerous subquadrangular black spots; sometimes there is a black median subcaudal line. Even the physiognomy varies, the snout being in some specimens more obtusely rounded than in the type.

Several illustrations of the head, of twice the natural size, are given to show the extraordinary variations of its shape and of the scutes—fig. B representing the typical specimen from California; fig. B′ that from Milpas, B″ one from Sallé's Mexican collection, and B‴ one from La Cumbre de los Arrastrados.

FICIMIA.

Ficimia, Gray, Viper. Sn. p. 80.
Amblymetopon, Günth. Col. Sn. p. 7.
Gyalopion, Cope, Proc. Ac. N. Sc. Phil. 1861, p. 243.

Body cylindrical, rather stout; tail short; head very short, not distinct from neck. Rostral shield much enlarged, in contact with or approaching the vertical shield; its anterior edge is sharp and turned upwards, the upper surface of the shield being concave. Either there is only one pair of large frontals or their anterior portion is cut off, forming a pair of very small anterior frontals. Nostril between two shields, of which the anterior is confluent with the first labial. Scales smooth; in seventeen rows. Teeth subequal in size, not grooved.

1. **Ficimia olivacea.** (Tab. XXXV. figg. B, C.)

Ficimia olivacea, Gray, Viper. Sn. p. 80; Bocourt, Miss. Sc. Mex., Rept. p. 570, t. 35. figg. 11, 11*a*; Garman, Mem. Mus. Comp. Zool. viii. no. 3, p. 84 (1883); Cope, Proc. Ac. N. Sc. Phil. 1866, p. 132; Peters, MB. Ak. Wiss. Berl. 1869, p. 875; Steindachner, SB. Wien. Akad. lxii. p. 344, t. 6 (1870).
Amblymetopon variegatum, Günth. Col. Sn. p. 7 (young).
Ficimia variegata, Garman, Mem. Mus. Comp. Zool. viii. no. 3, p. 84.
Ficimia elaiocroma, Jan, Arch. per la Zool. ii. p. 58.

Hab. MEXICO, Puebla (*Peters*), Tuxtla (*Steindachner*), Orizaba and Tehuantepec (*Sumichrast*).

The rostral shield forms a long suture with the vertical; there is only one pair of frontals; exceptionally, a small anterior frontal may be detached on one side of the head. One anterior and two posterior oculars; loreal confluent with the posterior frontal. Seven upper labials, the third and fourth entering the orbit*. Ventrals 149–151. Adult plain slate-coloured above, whitish below. The young is gaily ornamented with black or dark brown transverse bands on the back, which, on the sides, break up into spots. There are from fifty-one to fifty-six of these cross-bars, two on the neck being broader than the remainder. The upper side of the head is ornamented with spots of different shapes, but symmetrically arranged.

Fig. B represents the adult, fig. C the young of this species. The view of the head is enlarged.

2. **Ficimia publia.**

Ficimia publia, Cope, Proc. Ac. N. Sc. Phil. 1867, p. 126.
Gyalopium publium, Cope, Am. Nat. 1884, p. 163.
Ficimia ornata, Bocourt, Miss. Sc. Mex., Rept. p. 571, t. 35. figg. 10, 10 *a–e*.

Hab. MEXICO, Cuernavaca (*Mus. Brit.*), Yucatan (*Mus. Brit.*).

* Dr. Gray states that in one of the typical specimens the nasal is separate from the first upper labial; but this separation is only artificial.

The rostral shield forms a suture with the vertical; a small pair of anterior frontals in front of the large posterior. One anterior and two posterior oculars; loreal confluent with the posterior frontal. Seven upper labials, the third and fourth entering the orbit. Ventrals 140–146. In a young specimen, from 6 to 8 inches long, the ground-colour is whitish, the back being ornamented with from thirty-six to thirty-eight dark brown cross-bars, which on the sides break up into spots. The upper part of the head is symmetrically marked with brown spots, the hindmost of which occupies the end of the occipital shields, but remains widely separate from the foremost nuchal band.

3. **Ficimia cana.**

Gyalopion canum, Cope, Proc. Ac. N. Sc. Phil. 1861, p. 243.
Gyalopium canum, Yarrow, in Wheeler's Report Geogr. & Geol. Explor. v. p. 624, t. 18. figg. 2, 2 *a* (bad) (1875).
Ficimia cana, Garman, Mem. Mus. Comp. Zool. viii. no. 3, p. 83 (1883).

Hab. North America, Arizona (*Cope*), El Paso in Texas (*Forrer*).

The rostral shield does not reach the vertical; two pairs of frontals, the anterior of which are small. One anterior and two posterior oculars; loreal confluent with the posterior frontal. Seven upper labials, the third and fourth entering the orbit; ventrals 133. In a young specimen, about 8 inches long, the ground-colour is very light, and the back crossed by forty-five brown cross-bands, which on the sides break up into spots; nine of these bands belong to the tail. The foremost band occupies the neck, and slightly encroaches upon the occipitals; upper side of the head with indistinct symmetrical transverse markings.

Found on the Mexican frontier.

4. **Ficimia quadrangularis**, sp. n. (Tab. XXXV. fig. A.)

Hab. Mexico, Presidio de Mazatlan in Sinaloa (*Forrer*).

The rostral shield does not reach the vertical; two pairs of frontals, the anterior of which are very small; vertical nearly as broad as long, occipitals not much longer. Seven upper labials, of which the hindmost is very small, scale-like, the third and fourth entering the orbit; one anterior and two posterior oculars; temporals $1+2$. One pair of chin-shields. Scales in seventeen rows; ventrals 129; subcaudals 25. Ground-colour nearly white, each scale with a narrow blackish margin. Along the back runs a row of large subquadrangular black spots; there are twenty-six on the trunk and five on the tail. Only the two anterior extend downwards on the side of the neck; they are also broader than the others, and the foremost is continued forward over the occipitals and the interorbital space, leaving the temple and the snout almost white.

One specimen 12 inches long, of which the tail takes $1\frac{1}{3}$ inch.

RHINOCHILUS.

Rhinocheilus, Baird & Girard, N. Amer. Rept. p. 120 (1853).

1. **Rhinochilus lecontei.**

Rhinocheilus lecontei, Baird & Girard, l. c.; Jan, Iconogr. Ophid. xlviii. t. 3. fig. 1; Bocourt, Miss. Sc. Mex., Rept. p. 602, t. 40. figg. 7, 7 *a–e*.

Rhinocheilus lecontei, var. *tessellatus*, Garman, Mem. Mus. Comp. Zool. viii. no. 3, p. 73.

Hab. NORTH AMERICA, Southern United States.—MEXICO, Coahuila (*Garman*).

Scales in twenty-three rows.

2. **Rhinochilus antonii.**

Rhinocheilus antonii, Dugès, Proc. Am. Phil. Soc. xxiii. p. 86.

Hab. MEXICO, Mazatlan on the Pacific coast.

Scales in seventeen rows. Ventrals 200; subcaudals 38 simple and 3 paired.

LEPTOCALAMUS.

Leptocalamus, Günth. Ann. & Mag. N. H. 1872, ix. p. 16.

The specimen on which this genus has been founded, *Leptocalamus torquatus*, is said to be from South America, and therefore cannot be included in this fauna. However, a second species has been described from Mexico, which Cope rightly supposes to belong to this genus.

1. **Leptocalamus unicolor.**

Geophis unicolor, Fisch. Abhandl. naturwiss. Ver. Brem. vii. p. 227, t. 15. figg. 1–3 (1880).

Enulius murinus, Bocourt, Miss. Sc. Mex., Rept. p. 537, t. 35. figg. 9, 9 *a–g* (nec Cope).

Leptocalamus unicolor, Cope, Proc. Am. Phil. Soc. xxii. p. 178 (1885), and Bull. U. S. Nat. Mus. no. 32, 1887, p. 84.

Geagras longicaudatus, Cope, Amer. Nat. 1884, p. 162.

Hab. MEXICO, Tehuantepec (*Sumichrast*).

Habit very long and slender; head small, narrow, not distinct from neck; eye small, pupil round. Rostral shield broad; posterior frontals about thrice the size of the anterior; vertical subtriangular, acutely pointed behind; supraoculars of moderate size; anterior ocular nearly as long as the two nasals together; two postoculars; seven upper labials, the third and fourth entering the orbit. Temporals 1 + 2. Scales in seventeen rows. Ventrals 203. Only one pair of chin-shields. Upper parts uniform brown, yellow below.

Enulius sumichrasti, Bocourt, Miss. Sc. Mex., Rept. p. 538, t. 31. figg. 6, 6 *a–c* (*Geagras sumichrasti*, Cope, Amer. Nat. 1884, p. 162), does not appear to be specifically distinct from *L. unicolor*.

STREPTOPHORUS.

Streptophorus, Dum. Mém. Ac. Sc. xxiii. p. 468 (1853).
Ninia, Baird & Girard, Cat. N.-Amer. Snakes, 1853, p. 49.

1. **Streptophorus atratus.**

Serpentula ceylonica, Seba, Thes. i. p. 20, t. 11.
Coluber atratus, Hallow. Proc. Ac. N. Sc. Phil. 1845, p. 245.
Streptophorus atratus, Cope, Proc. Ac. N. Sc. Phil. 1861, p. 76.
Ninia atrata, Cope, Journ. Ac. N. Sc. Phil. viii. p. 145 (1876).
Streptophorus sebæ, Dum. & Bibr. Erpét. vii. p. 515; Günth. Col. Sn. p. 16, and Ann. & Mag. N. H. 1872, ix. p. 18; Peters, MB. Ak. Wiss. Berl. 1859, p. 924; Jan, Arch. per la Zool. ii. p. 27 (1862), and Iconogr. Ophid. xii. t. 3. fig. 4, fig. 5 (var. *maculatus*), fig. 6 (var. *collaris*), xiii. t. 1. fig. 1 (var. *drozii*); Bocourt, Miss. Sc. Mex., Rept. p. 546, t. 32. fig. 7, t. 33. fig. 1; Cope, Proc. Ac. N. Sc. Phil. 1861, p. 76.
Ninia sebæ, Cope, Journ. Ac. N. Sc. Phil. viii. p. 145; Garman, Mem. Mus. Comp. Zool. viii. no. 3, p. 95 (1883).
Streptophorus drozii, Dum. & Bibr. Erpét. vii. p. 518; Günth. Col. Sn. p. 16.
Streptophorus lansbergi, Dum. & Bibr. Erpét. vii. p. 518; Jan, Arch. per la Zool. ii. p. 27 (1862), and Iconogr. Ophid. xiii. t. 1. fig. 2; Bocourt, Miss. Sc. Mex., Rept. p. 551, t. 32. figg. 9, 9 *a–c*.
Elapoidis fasciatus, Hallow. Journ. Ac. N. Sc. Phil. iii. p. 35, t. 4 (lower figure) (1855).
Streptophorus maculatus, Peters, MB. Ak. Wiss. Berl. 1861, p. 924; Bocourt, Miss. Sc. Mex., Rept. p. 548, t. 32. figg. 8, 8 *a–d*, t. 33. figg. 2–4.
Streptophorus labiosus, Bocourt, Miss. Sc. Mex., Rept. p. 550, t. 32. figg. 6, 6 *a–f*.

Head rather depressed, not very distinct from neck. Anterior frontals very small. Seven upper labials, the third and fourth entering the orbit (the third labial occasionally divided into two). Two postoculars. Temporals 1+2. Two pairs of chin-shields. Scales strongly keeled, in nineteen rows. Ventrals 135–153; anal entire. Upper parts of the head and neck black, with a yellow collar. The coloration of the body varies:

α. Var. *atratus, lansbergii.*

Hab. HONDURAS, Puerto Caballo (*Mus. Brit.*); COSTA RICA, San José (*Van Patten*).—COLOMBIA (*Mus. Brit.*); VENEZUELA (*Hallowell*), Caracas (*Mus. Brit.*); WESTERN ECUADOR (*Fraser*).

Lower parts uniform whitish; upper parts uniform blackish; collar sometimes very indistinct. Ventrals 152–157.

β. Var. *labiosus.*

Hab. GUATEMALA (*Bocourt*).

A narrow dark band or series of spots along the middle of the abdomen; upper parts uniform black; collar very indistinct. Ventrals 153.

One specimen in the British Museum.

γ. Var. *drozii, collaris.*

Hab. Mexico, Jalapa (*Höge*); Guatemala, Yzabal (*Salvin*); Honduras (*Mus. Brit.*); Nicaragua, Chontales (*Mus. Brit.*).

Lower parts uniform whitish; upper parts reddish, without spots; head and neck black, with a white collar. Ventrals 144–149.

δ. Var. *dorsalis.*

Hab. Mexico (*H. Finck*); British Honduras (*Colonial Exhibition*), Belize (*Bocourt*).

Lower parts uniform whitish; upper parts reddish, with a few small irregular black spots on the anterior part of the back; head and neck black, with a white collar. Ventrals 138–143.

ε. Var. *punctulatus.*

Hab. Guatemala, Pacific coast (*Salvin*).

Like var. δ, but the spots extend all over the back. Ventrals 138.

ζ. Var. *sebæ, fasciatus.*

Hab. Mexico (*Sallé*), Teapa in Tabasco (*H. H. Smith*); Guatemala (*Salvin*); British Honduras (*Mus. Brit. & Salvin*), Belize (*Mus. Brit.*).

Lower parts uniform white; head and neck black, with white collar. The spots on the upper parts of the body are transverse, narrow, generally forming anteriorly complete cross-bars, but the majority are broken in the median line of the back, the right half not corresponding in position to the left. Ventrals 135–150.

η. Var.

Hab. Mexico, Teapa in Tabasco (*H. H. Smith*).

Like var. ζ, but with the anterior black cross-bars as broad as the interspaces between them.

θ. Var. *maculatus.*

Hab. Nicaragua, Chontales (*Mus. Brit.*); Costa Rica, Irazu (*Rogers*), Cartago (*Mus. Brit.*).

Lower parts with small square black spots, generally irregularly arranged, but sometimes crowding together towards the median line of the abdomen, and forming a dark band. Upper parts with or without black transverse spots. The light collar generally indistinct. Ventrals 142–147.

2. **Streptophorus diadematus.**

Ninia diademata, Baird & Girard, Cat. Serp. N. Amer. 1853, p. 468, and in Pac. R. R. Exped., Rept. t. 27. fig. 37.

Streptophorus bifasciatus, Dum. & Bibr. Erpét. vii. p. 520; Günth. Col. Sn. p. 16; Cope, Proc. Ac. N. Sc. Phil. 1861 (1860), p. 77; Jan, Arch. per la Zool. ii. p. 26 (1862); Bocourt, Miss. Sc. Mex., Rept. p. 545, t. 32. figg. 10, 10 *a*, *b*.

Hab. MEXICO, Huatusco in Vera Cruz (*G. & S.*), Orizaba (*Sumichrast*), Jalapa (*Cope*), Teapa in Tabasco (*H. H. Smith*), Oaxaca (*Höge*).

Head rather depressed, not distinct from neck. Anterior frontals small. Six upper labials, the third and fourth entering the orbit, two postoculars. Temporals 1+2. Two pairs of chin-shields. Scales strongly keeled; in nineteen rows. Ventrals 149–160; anal entire. Upper parts black, with a yellow collar across the neck, lower parts white; each ventral scute with a subtriangular black spot, the spots forming a chain-like band along the middle of the abdomen.

CHERSODROMUS.

Chersodromus, Reinh. Vid. Meddel. naturh. Foren. Kjöbnh. 1860, p. 242.
Opisthiodon, Peters, MB. Ak. Wiss. Berl. 1861, pp. 460, 617.

This genus is closely allied to *Streptophorus*, from which it differs not only by having the two posterior frontals confluent into a single shield, but also by its dentition, the maxillary being very narrow and edentulous in front, dilated behind and armed with small teeth of equal size.

1. **Chersodromus liebmanni.**

Chersodromus liebmanni, Reinh. Vid. Meddel. naturh. Foren. Kjöbnh. 1860, p. 242; Jan, Arch. per la Zool. 1862, p. 25, and Iconogr. Ophid. xii. t. 3. fig. 2; Bocourt, Miss. Sc. Mex., Rept. p. 552, t. 34. figg. 8, 8 *a–e*.
Ninia liebmanni, Garman, Mem. Mus. Comp. Zool. viii. no. 3, p. 95 (1883).
Opisthiodon torquatus, Peters, MB. Ak. Wiss. Berl. pp. 461, 617.

Hab. MEXICO, Huatusco (*G. & S.*), Vera Cruz (*Smiths. Inst.*), Huanusco (*Mus. Berol.*)

Scales keeled, in seventeen rows; ventrals 130–134; anal entire; subcaudals 42. Seven upper labials, of which the third and fourth enter the orbit. Black above, whitish below, mottled with black, more especially on the hinder part of the body. A yellow band across the occipitals.

2. **Chersodromus nigricans.**

Chersodromus nigricans, Reinh. Vid. Meddel. naturh. Foren. Kjöbnh. 1860, p. 245.

Hab. ? CENTRAL AMERICA.

Uniform dark brown; ventral shields 131; subcaudals 41.

Probably not specifically distinct from the preceding. The locality whence the type was obtained is not recorded, but it was probably Central America.

CONTIA*.

Contia, Baird & Girard, N.-Amer. Snak. p. 110 (1853).

1. Contia episcopa.

Lamprosoma episcopum, Kennicott, U. S. & Mex. Bound. Survey, 1859, p. 22, t. 8. fig. 2 (bad).
Contia episcopa, Cope, Proc. Ac. N. Sc. Phil. xii. p. 251 (1861).
Homalosoma episcopum, Jan, Iconogr. Ophid. xiii. t. 4. fig. 2.

Hab. North America, Texas.—Mexico, Nuevo Leon (*W. Taylor*).

2. Contia calligaster.

Contia calligaster, Cope, Journ. Ac. N. Sc. Phil. viii. p. 146 (1875).

Hab. Costa Rica, Pico Blanco.

3. Contia pachyura.

Contia pachyura, Cope, Journ. Ac. N. Sc. Phil. viii. p. 145 (1875).

Hab. Costa Rica, Sipurio.

I have not seen a specimen of this or of the preceding species.

ABLABES.

Ablabes, sp., Dum. & Bibr. Erpét. vii. p. 304 (1854).
Trimetopon, Cope, Proc. Am. Phil. Soc. 1884, p. 177.

1. Ablabes gracilis.

Ablabes gracilis, Günth. Ann. & Mag, N. H. 1872, ix. p. 18, t. 3. fig. D.

Hab. Costa Rica, elevated country near Cartago.

Body and tail slender, subcylindrical; head narrow, not distinct from neck. A pair of narrow anterior frontals; posterior frontals confluent into one large shield. Rostral rounded. Nostrils small, between two nasals. One loreal; one anteocular and one postocular. Eye small. Scales smooth, with a single apical groove, in fifteen series. Anal and subcaudals double. The posterior maxillary teeth become gradually larger and are smooth. Rostral shield very broad and low; anterior frontals narrow, nearly the entire upper surface of the snout being occupied by the single posterior frontal. Vertical broad and long, five-sided, with the posterior angle produced and pointed; occipitals as long as the vertical and posterior frontal together; nasal shields small; loreal longer than deep; præocular narrow, not extending to the upper surface of the snout. Seven upper labials, the third and fourth entering the orbit. Temporals

* *Contia michoacanensis* (Cope, Proc. Am. Phil. Soc. xxii. p. 178) has been found by Boulenger to be an Opisthoglyph, and therefore will be inserted subsequently under the genus *Homalocranium.* It is figured on Tab. XXXVI. figg. B & C.

1+1+2. The first pair of lower labials form a suture together; two pairs of chin-shields, subequal in size. Ventrals 149; subcaudals 69. Upper parts nearly uniform blackish brown, the anterior and lateral scales somewhat lighter in the centre. An indistinct, rather narrow, light-coloured band across the neck, immediately behind the occipitals. Lower parts yellowish.

We have two specimens of this snake, the larger measuring 12 inches in length.

2. **Ablabes decipiens,** sp. n. (Tab. XXXVII. fig. A.)

Hab. Costa Rica, Irazu (*Rogers*).

Rather slender in habit, the tail being remarkably long and three-fourths of the length of the body. Head rather narrow and elongate; eye of moderate size. Scales in seventeen rows, without pores. Ventrals 131–151; anal divided; subcaudals 113. Vertical rather elongate, longer than the frontals, but shorter than the parietals. Eight upper labials, of which the fourth and fifth enter the eye; one præocular, sometimes with a second smaller one below; one or two postoculars. Temporals 1+1+2. The maxillary teeth are slender, widely set, eight or nine in number, the anterior considerably smaller than the middle and posterior, but the hindmost is not distinguished by increase of size or isolated position. Upper parts brownish black, with two narrow yellowish lines on each side, of which the lower, which runs along the outer series of scales, is the more distinct; the upper runs along the fifth series of scales. A broad yellow band crosses the temple and the posterior half of the occipitals; it is edged with black in front. Lower parts yellow, with the corners of the ventral shields blackish. In the females the ornamental markings are much fainter than in the male.

Total length 18 inches, tail (slightly incomplete) 8½ inches.

We have received three specimens of this snake, together with as many of a species of *Tachymenis*, to which it bears a most deceptive resemblance. Also to *Dromicus lateristriga*, *D. frenatus*, &c., it is so closely allied that, without examination of the dentition, it would be difficult to distinguish it.

The position of the yellow cross-band on the occipitals seems to be an important distinctive character when it is present.

3. **Ablabes sumichrasti.**

Masticophis, sp., Sumichrast, Bull. Soc. Zool. Fr. 1880, p. 183; La Naturaleza, 1882, p. 285.
Henicognathus sumichrasti, Bocourt, Miss. Sc. Mex., Rept. p. 628, t. 41. fig. 5.
Rhadinæa sumichrasti, Cope, Bull. U. S. Nat. Mus. no. 32, p. 79 (1887).

Hab. Mexico, Tehuantepec (*Sumichrast*).

4. **Ablabes serperaster.**

Rhadinæa serperaster, Cope, Proc. Ac. N. Sc. Phil. 1871, p. 212.

Hab. Costa Rica.

The teeth are equal, the scales smooth and poreless, the anal plate divided; nasals two, loreal one; one præ-, two postoculars. Scales in nineteen series. Upper labials eight, not elevated, the fourth and fifth entering the orbit. Temporals 1+2+3. Anterior frontals transverse, narrow ; postnasal larger than prænasal. Vertical wide; superciliary suture shorter than anterior, total length exceeding that of occipital suture. Loreal square; geneials subequal. Ventrals 164, subcaudals 78. Dark brown, with six longitudinal yellow or white lines, of which the first and second are brightest. The second dark band is wider than the first and vertebral; it and the third are partly divided by a faint white line. Another white line on each side is produced by a series of dark spots on the ends of the gastrosteges. Labial plates black, yellow-spotted. Head dark brown above, with a pale shade across frontal, and two just behind parietals. Chin and belly yellowish.

ELAPOCHRUS.

Elapochrus, Peters, MB. Ak. Wiss. Berl. 1860, p. 294.
Pliocercus, Cope, Proc. Ac. N. Sc. Phil. xii. p. 253 (1860).

1. **Elapochrus deppii.**

Elapochrus deppii, Peters, MB. Ak. Wiss. Berl. 1860 (June 7), p. 294, c. fig.
Pliocercus elapoides, Cope, Proc. Ac. N. Sc. Phil. 1860 (June 26), p. 253 ; Peters, MB. Ak. Wiss. Berl. 1869, p. 876 ; F. Müll. Verh. naturw. Ges. Basel, 1878, p. 596.
Liophis elapoides, Garman, Mem. Mus. Comp. Zool. viii. no. 3, p. 69 (1883) ; Bocourt, Miss. Sc. Mex., Rept. p. 635, t. 41. figg. 6, 6 *a–e*.
Liophis elapoides, var. *diastema,* Bocourt, l. c. p. 636, t. 41. fig. 8.
Liophis tricinctus, Jan, Iconogr. Ophid. xviii. t. 4. figs. 4–6.

Hab. MEXICO, Puebla (*Commiss. Geogr.*), Vera Cruz, Orizaba (*Sumichrast*), Mirador (*Smiths. Inst.*), Jalapa (*Smiths. Inst.*) ; GUATEMALA, Dueñas (*Salvin*).

2. **Elapochrus æqualis.** (Tab. XXXVI. fig. A.)

Pleiocercus æqualis, Salvin, P. Z. S. 1861, p. 227.
Pliocercus æqualis, var., F. Müll. Verh. naturw. Ges. Basel, 1878, p. 662, t. 2. fig. A.
Pliocercus sargii, Fischer, Wiegm. Arch. 1881, p. 225, t. 11. figs. 1–3.
Liophis elapoides, var. *æqualis,* Bocourt, Miss. Sc. Mex., Rept. p. 637, t. 41. figg. 7, 7 *a–e*.

Hab. MEXICO, Tezuitlan, Puebla, Tuxpan (*Commiss. Geogr.*) ; GUATEMALA, Vera Paz (*Salvin*), Coban (*Sarg*).

Body moderately slender, tail very long ; head depressed, anterior frontals one-third the size of posterior ; vertical five-sided, with a right posterior angle ; occipitals as long as vertical and posterior frontals together. Eight upper labials, the fourth and fifth entering the orbit ; loreal rhombic ; three præoculars, the lower one being a detached portion of the fourth labial ; two postoculars. Scales in seventeen rows, without pit; ventrals 131, anal divided; subcaudals $92+x$ (the tail is slightly mutilated). The dentition is distinctly diacranterian. Ground-colour reddish, with numerous black rings entirely encircling the body and nearly as broad as the interspaces, which again are spotted in an irregular fashion; the rings on the tail are closer together and more irregular. Head black, with a broad yellow band across the occipitals.

Total length 22 inches; length without the tail (which is mutilated) 13½ inches.

3. **Elapochrus dimidiatus.**

Pliocercus dimidiatus, Cope, Proc. Ac. N. Sc. Phil. 1865, p. 190; Journ. Ac. N. Sc. Phil. 1875, p. 138; Bull. U. S. Nat. Mus. no. 32, p. 79 (1887).

Hab. NICARAGUA (*Cope*); COSTA RICA (*Cope*).

4. **Elapochrus euryzona.**

Pliocercus euryzonus, Cope, Proc. Ac. N. Sc. Phil. 1862, p. 73, and 1871, p. 200.

Liophis (*Cosmiosophis*) *splendens*, Jan, Arch. per la Zool. 1863, p. 379, and Iconogr. Ophid. xviii. t. 5. fig. 1; Bouleng. Ann. & Mag. N. H. 1882, ix. p. 461.

Hab. PANAMA, Darien (*Smiths. Inst.*).—COLOMBIA; ECUADOR (*Whymper*).

HENICOGNATHUS.

Enicognathus, Dum. & Bibr. Erpét. vii. p. 328 (1854).

1. **Henicognathus annulatus.**

Enicognathus annulatus, Dum. & Bibr. Erpét. vii. p. 335, t. 80; Jan, Iconogr. Ophid. xvi. t. 4. fig. 3; Sumichrast, Bull. Soc. Zool. Fr. 1880, p. 181; Bocourt, Miss. Sc. Mex., Rept. p. 626, t. 40. figg. 6, 6 *a–e*.

Hab. MEXICO, Vera Cruz, Tabasco, Atitlan, Tehuantepec (*Sumichrast*); GUATEMALA, Vera Paz (*Salvin*).

LIOPHIS.

Liophis, Wagl. Nat. Syst. Amphib. p. 187 (1830).

1. **Liophis epinephelus.** (Tab. XXXVII. fig. B.)

Liophis epinephelus, Cope, Proc. Ac. N. Sc. Phil. 1862, p. 78, and Journ. Ac. N. Sc. Phil. 1857, p. 136.

Hab. COSTA RICA, Cartago (*Mus. Brit.*), San José (*Smiths. Inst.*), Irazu, Rio Sucio (*Rogers*); PANAMA, Chiriqui (*Champion*), Truando (*Smiths. Inst.*).

Scales in seventeen rows, some with, the majority without, pits; head rather depressed. Anterior frontals two-thirds the size of posterior; vertical longer than broad, with the lateral margins slightly convergent; occipitals shorter than the vertical and posterior frontals together; loreal short, one ante-ocular, not reaching the vertical, two postoculars; eight upper labials, the fourth and fifth entering the orbit; temporals

1+2. Anterior and posterior chin-shields subequal in length, the anterior in contact with five lower labials. Ventrals 139–149; anal divided; subcaudals 54–60. Dentition diacranterian. Greyish olive, shading into black on the posterior part of the body; anterior part of the body with black cross-bands, which soon become so broad as to be nearly confluent. The scales of the intervening spaces have a white margin, and as the interspaces on the middle and posterior parts of the body are exceedingly narrow, the white margins form irregular and interrupted transverse lines. A black stripe is generally visible along the side of the tail; abdomen irregularly chequered with black; lower part of the tail white.

An adult female is 24 inches long, the tail measuring 5½ inches.

2. **Liophis regalis.**

Diadophis regalis, Baird & Gir. N. Amer. Serp. pp. 115, 161, and Pacif. R. R. Report, Rept. t. 33. fig. 86; Garman, Mem. Mus. Comp. Zool. viii. no. 3, p. 73 (1883).

Diadophis punctatus, subsp. *regalis,* Cope, Bull. U. S. Nat. Mus. no. 32, p. 80.

Hab. MEXICO, Sonora (*B. & G.*), San Luis Potosi (*Garman*).

The dentition of this snake has not been described, and as I have not seen a specimen, I cannot be certain about its position in *Liophis.*

3. **Liophis lætus.**

Diadophis punctatus, var. *lætus,* Jan, Iconogr. Ophid. xv. t. 6. fig. 6; Bocourt, Miss. Sc. Mex., Rept. p. 622, t. 40. figg. 4, 4 *a–e.*

Liophis (*Diadophis*) *arnyi,* Günth. Ann. & Mag. N. H. 1868, i. p. 413 (not Kennicott).

Diadophis punctatus, var. *dugesii,* Villada, La Naturaleza, iii. p. 226, c. fig. (1875).

Diadophis punctatus, subsp. *arnyi,* Cope, Bull. U. S. Nat. Mus. no. 32, p. 80.

Hab. MEXICO, City of Mexico (*Doorman*), Mirador, Vera Cruz (*Cope*), Guanajuato (*Villada*), Zacualtipan (*Cope*).

I think that this snake should be, at least, specifically separated from *Diadophis punctatus,* having the dentition of a most marked diacranterian character, as is also well described by Villada. What the dentition of the more northern form, named *arnyi,* may be, I have no means of ascertaining; possibly it is the same as in the typical *D. punctatus,* in which the last tooth is scarcely larger than the preceding, though both are separated by a short space. However, *L. arnyi* has much fewer ventral shields (160); whilst *L. lætus* has 183–184, and according to Villada even 202.

CORONELLA.

Coronella, sp., Laur. Syn. Rept. p. 84 (1768).

We include in this genus only those species of Cope's *Rhadinæa* which possess a syncranterian or coryphodont dentition.

1. Coronella annulata. (Tab. XXXVIII.)

Coronella doliata, varr. A & C, Günth. Colubr. Sn. p. 41.
Coronella doliata, Salv. P. Z. S. 1861, p. 227.
Coronella doliata, varr. *formosa* et *conjuncta*, Jan, Iconogr. Ophid. xiv. t. 4. figg. B, C.
Coronella doliata, var. *conjuncta*, Bocourt, Miss. Sc. Mex., Rept. p. 611, t. 39. figg. 6, 6 *a–d* (Guanajuato).
Lampropeltis annulata, Kennicott, Proc. Ac. N. Sc. Phil. xii. pp. 257, 329 (Matamoros).
Lampropeltis micropholis, Cope, Proc. Ac. N. Sc. Phil. xii. p. 257 (Panama).
Lampropeltis polyzona, Cope, Proc. Ac. N. Sc. Phil. xii. p. 258 (Jalapa).
Ophibolus polyzonus (Cope), Sumichrast, Bull. Soc. Zool. Fr. 1880, p. 181.
Ophibolus triangulus, var. *zonatus*, Garman, Rept. & Batr. of N. Amer. p. 67.
Coronella formosa, Bocourt, Miss. Sc. Mex., Rept. p. 612, t. 39. figg. 3, 3 *a–e* (Mexico, Guatemala).
Coronella formosa, varr. *anomala* (Alta Vera Paz), *oligozona* (Tehuantepec, Guatemala), *polyzona* (Oaxaca), Bocourt, loc. cit. figg. 4, 4 *a–c*, 8, 8 *a–d*, 7, 7 *a–d*.
? *Ophibolus multistriatus* (Kennicott??), Cope, Bull. U. S. Nat. Mus. 1887, no. 32, p. 78.

Hab. Mexico, Tres Marias Is., Presidio, Mazatlan (*Forrer*), State of Guerrero (*H. H. Smith*), Mezquital del Oro in San Luis Potosi (*A. C. Buller*), Guanajuato (*Dugès*), Matamoros (*Kennicott*), city of Mexico (*Doorman*), Jalapa (*Smiths. Inst.*), Teapa in Tabasco (*H. H. Smith*), Oaxaca (*Bocourt*), Tehuantepec (*Sumichrast*), Yucatan; Guatemala, Dueñas, San Gerónimo (*Salvin*); Costa Rica, Irazu (*Rogers*); Panama (*Smiths. Inst.*), Chiriqui (*Champion*).

The colour-variations of Central-American specimens of this variable species may be arranged under the following categories; they cannot be brought into relation with the various localities where the specimens were collected:—

I. The pairs of black rings well defined.
  A. Scales of the red and yellow spaces with black tips.
    1. Some of the abdominal scutes within the red spaces with black spots.
      a. The yellow occipital collar incomplete.—Mexico.
      b. The yellow occipital collar complete, broad.—State of Guerrero, Amula.
    2. Abdominal scutes within the red spaces immaculate, or nearly so.
      a. The yellow occipital collar incomplete.—City of Mexico.
      b. The yellow occipital collar complete.—Huatusco, Teapa in Tabasco, Yucatan, Guatemala, Costa Rica, Chiriqui.
  B. Scales of the red and yellow spaces without black tips; yellow occipital collar complete.
    1. The black rings belonging to one pair separated from each other by 1–3 scales.—Mazatlan, Presidio, Mezquital del Oro.
    2. The black rings belonging to one pair separated from each other by 4–5 scales.—Tres Marias Is.

II. The black rings irregular in outline, the yellow colour which separates them being disposed in a transverse series of more or less confluent spots. Head black, with a broad occipital yellow collar; abdomen black.—Mexico. (Tab. XXXVIII. fig. B.)

III. The black rings well separated by the yellow rings, but encroaching so much upon the red as to be partly confluent, the red colour appearing on each side as a rounded spot.—Tehuantepec. (Fig. C.)

IV. The rings nearly obsolete, the presence of black rings being indicated on the tail and hind part of the trunk only.—Yucatan. (Fig. A.)

2. **Coronella mexicana.**

Ophibolus triangulus, var. *mexicanus*, Garman, Rept. & Batr. N. Amer. p. 66.

Hab. MEXICO, near San Luis Potosi.

3. **Coronella leonis,** sp. n. (Tab. XXXIX. fig. A.)

Hab. MEXICO, Nuevo Leon (*W. Taylor*).

Scales in twenty-three rows, without pit, smooth; head similar to that of *Coronella lævis*; anterior frontals not quite half the size of posterior; vertical five-sided, with the lateral margins convergent; one præocular not reaching the vertical; two postoculars; loreal longer than deep; seven upper labials, the third and fourth entering the orbit; temporals 2+3, of the two anterior only the upper one is in contact with the post-oculars. Ventrals 200; anal entire; subcaudals 50. Body pale olive-colour on the back, with twenty-seven salmon-coloured incompletely black-edged spots, some being of a transversely oval shape, but the majority presenting the appearance of being formed of two rounded portions. On the tail the spots lose their light centre and appear merely as brown spots. The lateral spots which are so conspicuous in most variations of *Coronella triangulum* are here nearly entirely absent. Vertical and each occipital with a black spot, red in the centre; abdomen with only a few blackish blotches irregularly scattered. A black band along the middle of the lower part of the tail.

The single specimen measures 23½ inches, the tail 3¾ inches.

This snake may be considered to be one of the aberrant forms of *Coronella triangulum.*

4. **Coronella godmani.** (Tab. XXXIX. fig. B.)

Dromicus godmanni, Günth. Ann. & Mag. N. Hist. 1865, xv. p. 94.

Rhadinæa godmani, Cope, Journ. Ac. N. Sc. Phil. viii. p. 139.

Henicognathus godmanii, Bocourt, Miss. Sc. Mex., Rept. p. 631, t. 40. figg. 5, 5 *a–d*.

Hab. GUATEMALA, Dueñas (*G. & S.*).

Scales smooth, without apical groove, in twenty-one series; ventrals 176; anal divided; subcaudals 88. Loreal longer than or as long as high; one præ-, two postoculars. Upper labials eight, the fourth and fifth entering the orbit. Temporals 1+2+3. Tail two-sevenths of the total length. Posterior chin-shields shorter than the anterior, which are in contact with four lower labials. Brownish grey, with a blackish vertebral band as broad as a scale; another similar band runs along the fifth outer series of scales.

Dark lines, besides, follow each of the other series of scales. Each upper labial with a bright-yellow, dark-edged spot; a pair of yellow spots on each side of the neck, at the commencement of the lateral band. Lower parts uniform whitish.

The maxillary is armed with thirteen or fourteen teeth, rather closely set; the anterior are small, but the teeth become gradually longer and stronger towards the middle, and still more towards the end of the jaw, there being no interspace between the hindmost and preceding teeth.

5. Coronella decorata.

Coronella decorata, Günth. Col. Sn. p. 35; Salv. P. Z. S. 1860, p. 455.

Rhadinæa decorata, Cope, Journ. Ac. N. Sc. Phil. 1875, p. 140.

? *Diadophis decoratus*, Garman, Mem. Mus. Comp. Zool. viii. no. 3, p. 71 (1883); Bull. Essex Inst. xix. p. 127 (1887).

Diadophis decoratus, Bocourt, Miss. Sc. Mex., Rept. p. 624, t. 40. figg. 3, 3 *a–e*.

Enicognathus vittatus (part.), Jan, Arch. per la Zool. ii. p. 271 (1863), and Iconogr. xvi. t. 2. fig. 2 (not fig. 3).

Hab. Mexico (*Sallé*); Guatemala, Volcan de Fuego (*Salvin*); Nicaragua (*Cope*); Costa Rica (*Cope*); Panama, Veragua (*Cope*).

6. Coronella ignita.

Dromicus ignitus, Cope, Proc. Ac. N. Sc. Phil. 1871, p. 201.

Rhadinæa ignita, Cope, Journ. Ac. N. Sc. Phil. viii. p. 140.

Hab. Panama, Darien.

7. Coronella vermiculaticeps.

Tæniophis vermiculaticeps, Cope, Proc. Ac. N. Sc. Phil. (1860) 1861, p. 249.

Rhadinæa vermiculaticeps, Cope, Journ. Ac. N. Sc. Phil. viii. p. 139.

Hab. Costa Rica (*Cope*); Panama, Veragua (*Cope*).

8. Coronella quinquelineata.

Rhadinæa quinquelineata, Cope, Proc. Am. Phil. Soc. xxiii. p. 277 (1886); and Bull. U. S. Nat. Mus. no. 32, p. 79 (1887).

Hab. Mexico, Zacualtipan (*Cope*), Tezuitlan in Puebla (*Commiss. Geogr.*).

9. Coronella fulviceps.

Rhadinæa fulviceps, Cope, Proc. Am. Phil. Soc. xxiii. p. 279 (1886) *.

Hab. Panama.

DROMICUS.

Dromicus, Coct. et Bibr. in Ramon de la Sagra's Hist. Cuba, Rept. p. 132 (1843).

Bocourt (Miss. Sc. Mex., Rept. p. 713) introduces into the Central-American fauna

* The dentition of this snake is not described, and hence its place in *Coronella* is doubtful.

Dromicus ater, and, on the authority of Jan, a variety named "*inconstans*," said to occur in Mexico. It appears to me very doubtful whether this truly West-Indian type extends into any part of the American Continent.

Also the occurrence in Mexico of *Dromicus pleii* (Dum. & Bibr.) (see Bocourt, Miss. Sc. Mex., Rept. p. 728) requires confirmation. The type of this species has a series of spots along each side of the abdomen ; it is a native of Martinique.

1. **Dromicus fugitivus.**

Coluber fugitivus, Donndorf, Zool. Beitr. iii. p. 206 (1798).
Dromicus fugitivus, Günth. Col. Sn. p. 132; Bocourt, Miss. Sc. Mex., Rept. p. 708, t. 50. figg. 4, 4 *a–e*.
Coluber cursor, Shaw, Gen. Zool. iii. pt. 2, p. 510 (1802).
Herpetodryas cursor, Schleg. Phys. Serp. ii. p. 199.
Dromicus cursor, Dum. & Bibr. vii. p. 650 ; Jan, Iconogr. xxiii. t. 5. figg. 1–3.

Hab. Guatemala (*Morelet*).—West Indies.

2. **Dromicus laureatus.** (Tab. XL. fig. A.)

Dromicus laureatus, Günth. Ann. & Mag. N. H. 1868, i. p. 419, t. 19. figs. E.
Rhadinœa loreata, Cope, Journ. Ac. N. Sc. Phil. viii. p. 140.

Hab. Mexico, Mexico city (*Doorman*), La Cumbre de los Arrastrados (*A. C. Buller*).

Scales smooth, in seventeen rows. Form of the head as in *Coronella lævis*; body and tail moderately slender; tail one-third of the total length. Eyes of moderate size. The rostral just reaches the upper surface of the head; anterior frontals scarcely half as large as posterior; vertical longer than the snout, but shorter than the occipitals, which are rounded behind ; loreal square ; one præorbital, reaching to the upper surface of the head, but not extending on to the vertical; two postoculars ; seven upper labials, the third and fourth entering the orbit; temporals 1+2+3, the anterior in contact with both postoculars; two pairs of chin-shields, nearly equal in length ; ventrals 163 ; anal bifid; subcaudals 95. A lead-coloured band, three scales broad, runs from the nape, along the median line of the back, to the end of the tail; flanks reddish, with a very indistinct greyish streak along the fourth outer series of scales; two yellow lines across the rostral, the lower running along the upper labials and across the neck, thus entirely encircling the head ; the upper runs along the canthus rostralis, and stops or terminates in the temporal region. Lower parts uniform yellowish. The dentition of this species is neither distinctly diacranterian nor syncranterian. The maxillary is armed with nine teeth, gradually becoming stronger, longer, and more widely set behind; the last is conspicuously the longest, but scarcely more distant from the penultimate than this is from the antepenultimate, although these three teeth are much more distant than the others are from each other.

3. **Dromicus clavatus.**

Dromicus clavatus, Peters, MB. Ak. Wiss. Berl. 1864, p. 388; Bocourt, Miss. Sc. Mex., Rept. p. 711, t. 45. figg. 2, 2 *a–e*.

Hab. Mexico.

4. **Dromicus omiltemanus,** sp. n. (Tab. XL. fig. B.)

Hab. Mexico, Omilteme in Guerrero 8000 feet (*H. H. Smith*).

Scales smooth, in seventeen rows, without pit; body and tail moderately slender, the latter not quite one-third of the total length; eye of moderate size. Anterior frontals about one-third of the posterior; vertical longer than the snout, and not much shorter than the occipitals, which are rounded behind; loreal square; one præorbital reaching to the upper surface of the head, but not in contact with the vertical; two postoculars; eight upper labials, the fourth and fifth entering the orbit; a minute portion of the third labial is separated, forming a rudimentary inferior præocular. Temporals 1+2+2; posterior chin-shields a little longer than anterior. Ventrals 154; anal divided; subcaudals 94. Upper parts olive-brown, bordered on each side by a well-defined yellowish line, which starts from the upper postocular, running along the fifth outer series of scales, and is lost nearly at the extremity of the tail. This yellowish line is accompanied by an olive-black band, two scales broad, which again is bordered by a yellowish line much more indistinct than the one described; the lowermost two series of scales are pinkish grey. Lower parts whitish, each ventral scute with a black dot on either side.

Total length 15½ inches, of which the tail takes 5 inches. Dentition distinctly diacranterian.

5. **Dromicus tæniatus.**

Enicognathus vittatus (part.), Jan, Arch. per la Zool. ii. p. 271 (1863) (number of ventral shields 129, evidently a mistake for 179); Iconogr. Ophid. xvi. t. 2. fig. 3 (and not fig. 2, which is *Calamaria vittata,* Rapp, MS., = *Coronella decorata,* Günth.).

Henicognathus vittatus, Bocourt, Miss. Sc. Mex., Rept. p. 630, t. 41. figg. 1, 1 *a–d*.

Dromicus tæniatus, Peters, MB. Ak. Wiss. Berl. 1863, p. 275.

Rhadinæa tæniata, Cope, Journ. Ac. N. Sc. Phil. viii. p. 140 (1875).

Hab. Mexico, City of Mexico (*Doorman*), La Cumbre de los Arrastrados (*A. C. Buller*).

6. **Dromicus fulvivittis.**

Rhadinæa fulvivittis, Cope, Journ. Ac. N. Sc. Phil. viii. p. 139 (1875).

Hab. Mexico, Orizaba in Vera Cruz (*Sumichrast*), Xautipa in Guerrero (*H. H. Smith*).

Distinguished from *D. tæniatus* chiefly by the total absence of ventral spots.

7. **Dromicus lachrymans** *.

Lygophis lachrymans, Cope, Proc. Am. Phil. Soc. xi. p. 154 (1871).
Rhadinæa lachrymans, Cope, Journ. Ac. N. Sc. Phil. viii. p. 140 (1875); and Bull. U. S. Nat. Mus. no. 32, p. 30.

Hab. CENTRAL MEXICO (*Cope*).

XENODON.

Xenodon, Boie, Isis, 1827, p. 541.

1. **Xenodon rhabdocephalus.**

Coluber rhabdocephalus (part.), Wied, Beitr. p. 355.
Coluber rhabdocephalus, Wied, Abbild. x. t. 4.
Xenodon rhabdocephalus, Schleg. Ess. t. 3. figg. 10, 11; Günth. Ann. & Mag. N. H. 1863, xii. p. 353, t. 5. fig. B; Jan, Iconogr. Ophid. xix. t. 4. fig. 1.

Hab. MEXICO, Amula in Guerrero (*H. H. Smith*); BRITISH HONDURAS (*Mus. Brit.*); GUATEMALA, San Gerónimo (*Salvin*).—SOUTH AMERICA, Guiana, Brazil, Argentine Republic.

2. **Xenodon angustirostris.**

Xenodon angustirostris, Peters, MB. Ak. Wiss. Berl. 1864, p. 390; Cope, Journ. Ac. N. Sc. Phil. viii. pp. 141, 157 (1875), and Bull. U. S. Nat. Mus. no. 32, p. 76 (1887).

Hab. GUATEMALA (*Cope*); COSTA RICA, Sipurio (*Gabb*); PANAMA (*Cope*), Veragua (*Peters*).

I have not seen an example of this species; but suppose that at least some of the specimens referred to it by Cope would have been named *X. rhabdocephalus* by myself †.

COLUBER.

Coluber, Günther, Col. Sn. p. 87 (1858).

1. **Coluber guttatus.**

Coluber guttatus, Linn. Syst. Nat. i. 1, p. 385; Günth. Cat. Col. Sn. p. 89.
Scotophis guttatus, emoryi, lætus, etc., Baird & Girard, Cat. N. Amer. Sn. pp. 75 etc.
Elaphis guttatus, Dum. & Bibr. Erpét. vii. p. 273; Jan, Iconogr. Ophid. xxi. t. 6.
Scotophis guttatus, Bocourt, Miss. Sc. Mex., Rept. p. 678, t. 46. figg. 4, 4 *a–e*.
Natrix emoryi, Cope, Bull. U. S. Nat. Mus. no. 32, p. 71.

---

* The information about this snake is very unsatisfactory. In one place it is described as having 78 subcaudals, whilst in another the number 60 is given as characteristic. The locality is at first stated to be unknown, but later it is said to be "Middle Mexico."

† Jan, Iconogr. Ophid. xix. t. 4. fig. 2, figures a *Xenodon bertholdi* from "Mexico." This is, without doubt, identical with *X. colubrinus*, Günth., a Brazilian species, the locality given by Jan resting upon an error.

Hab. North America, Eastern and South-eastern United States.—Mexico (*Mus. Brit.*), Chihuahua (*Cope*).

The specimens which, in my opinion, should be referred to this species vary in the number of longitudinal rows of scales, viz. from 25 to 29; all, however, have eight upper labials only.

2. **Coluber triaspis.**

Coluber triaspis, Cope, Proc. Ac. N. Sc. Phil. xviii. p. 128 (1866); Proc. Am. Phil. Soc. xviii. p. 271 (1879).

Coluber mutabilis, Cope, Proc. Am. Phil. Soc. xxii. p. 175 (1884).

Scotophis mutabilis, Bocourt, Miss. Sc. Mex., Rept. p. 680, t. 46. figg. 2, 2 *a–f.*

Hab. Mexico (*Sallé*), Guanajuato (*Dugès*), City of Mexico (*Doorman*), Tehuantepec (*Sumichrast*), Yucatan (*Boucard*); British Honduras, Belize (*Mus. Brit.; Cope*); Guatemala, Vera Paz (*Cope; Mus. Paris*); Costa Rica, San José (*Cope*).

This form passes into *C. guttatus*, but has generally smaller scales, viz. 31–35 series. Upper labials eight. Abdomen nearly uniform whitish.

3. **Coluber flavirufus.**

Coluber flavirufus, Cope, Proc. Ac. N. Sc. Phil. xviii. p. 319 (1866); Proc. Am. Phil. Soc. xxii. p. 175 (1884).

Natrix flavirufus, Cope, Bull. U. S. Nat. Mus. no. 32, p. 71.

Elaphis rodriguezi, Bocourt, Miss. Sc. Mex., Rept. p. 683, t. 46. figg. 1, 1 *a–e.*

Hab. Mexico (*Mus. Brit.*), Mirador, Tabasco, and Chiapas (*Cope*), Yucatan (*Schott*), Ruatan I. (*Gaumer*); Guatemala, Rio Polochic, Santa Rosa de Panzos (*Bocourt*).

Distinguished by nine (ten) upper labial shields. Scales in from 29 to 35 series. Possibly not a distinct species.

4. **Coluber chlorosoma,** sp. n. (Tab. XLI.)

Hab. Mexico, Atoyac in Vera Cruz, Amula in Guerrero (*H. H. Smith*), San Ramon 1500 feet (*A. C. Buller*).

Scales in thirty-three rows, smooth, only the median ones on the back showing faint traces of keels. Head rather narrow and elongate; eye of moderate size, with round pupil; anterior frontals half, or less than half, the size of the posterior. Vertical large, in contact with the præorbital. Loreal longer than deep, with a point at posterior angle. One præ-, two postoculars. Either nine upper labials, of which the fifth and sixth enter the orbit, or eight, the fourth and fifth being below the eye. The foremost series of temporals consists of three oblique, large, elongate shields, of which the two upper are in contact with the postoculars. Upper parts uniform greenish olive. On stretching the skin the base of each scale may be seen to be black, with a whitish spot on each side of the scale; lower parts uniform whitish. Ventrals 246–260; anal divided; subcaudals 111–113.

The largest of three specimens is 49 inches long, of which the tail measures 11. The smallest is only half the size, but also without a trace of spots.

At Atoyac this snake is eaten roasted by the poor country people.

SPILOTES.

Spilotes, Wagler, Nat. Syst. Amphib. p. 179 (1830).

1. Spilotes corais.

Coluber corais, Cuv. Règne Anim.; Schleg. Ess. t. 5. figg. 9, 10.

Spilotes corais, Dum. & Bibr. Erpét. vii. p. 222; Salv. P. Z. S. 1860, p. 456; 1861, p. 228; Günth. Cat. Col. Sn. p. 98; Cope, Bull. U. S. Nat. Mus. no. 32, p. 72.

Hab. CENTRAL AMERICA.—Eastern and Northern Tropical South America.

Var. 1. Tail black.

Spilotes melanurus, Dum. & Bibr. Erpét. vii. p. 224; Cope, Proc. Ac. N. Sc. Phil. xii. p. 564 (1861); Sumichrast, Arch. Sc. Bibl. univ. 1873, p. 259.

Spilotes corais, var. *melanurus*, Bocourt, Miss. Sc. Mex., Rept. p. 687, t. 44. figg. 1, 1 *a–e*.

Hab. NORTH AMERICA, Duval Co., Texas (*W. Taylor*).—MEXICO, Tampico (*Richardson*), Colima, Tuxpan, and Vera Cruz (*Cope*), Teapa in Tabasco (*H. H. Smith*), Oaxaca (*Sallé*), Tehuantepec (*Sumichrast*), Yucatan (*Mus. Brit.*); BRITISH HONDURAS, Stann Creek (*J. Robertson*); GUATEMALA, Dueñas, Lanquin, San Gerónimo (*Salvin*); COSTA RICA, Cartago (*Mus. Brit.*), San José and Talamanca (*Cope*); PANAMA (*Cope*).

Var. 2. Entirely black, or only the abdomen white.

Coluber obsoletus, Holbrook, N. Amer. Herpet. iii. p. 61, t. 12.

Georgia obsoleta, Baird & Gir. N. Amer. Rept. p. 158.

Spilotes erebennus, Cope, Proc. Ac. N. Sc. Phil. xii. p. 564 (1861).

Spilotes corais, subsp. *erebennus*, Cope, Journ. Ac. N. Sc. Phil. viii. p. 135 (1875).

Hab. NORTH AMERICA, Texas, Eagle Pass (*U. S. Nat. Mus.*).—MEXICO, Presidio, Tres Marias Is. (*Forrer*), Guanajuato (*Dugès*), Jalapa (*Commiss. Geogr.*), Atoyac in Vera Cruz (*H. H. Smith*).

2. Spilotes salvini. (Tab. XLII.)

? Seba, Thesaur. ii. t. 20. fig. 1.

? *Cerastes mexicanus*, Laur. Syn. Rept. p. 83 (1768).

Spilotes variabilis, var. A, Günth. Cat. Col. Sn. p. 99.

Spilotes pullatus, subsp. *auribundus*, Cope, Proc. Ac. N. Sc. Phil. xiii. p. 300 (1862).

Spilotes auribundus, Cope, Bull. U. S. Nat. Mus. no. 32, p. 71; Bocourt, Miss. Sc. Mex., Rept. p. 689, t. 44. figg. 5, 5 *a–f*.

Spilotes salvini, Günth. Ann. & Mag. N. H. 1862, ix. p. 125, t. 9. fig. 5 (head).

Hab. MEXICO, Huatusco * (*Mus. Brit.*), Vera Cruz, Tabasco (*Mus. Paris*), Atoyac (*H. H. Smith*), Tehuantepec (*Sumichrast*), Yucatan (*U. S. Nat. Mus.*); BRITISH HONDURAS, Belize (*Salvin*); GUATEMALA, Rio Motagua, Yzabal (*Salvin*), Atitlan (*Bocourt*); NICARAGUA (*Cope*).

Scales in nineteen rows, slightly keeled; one anterior and two posterior oculars; eight upper labials, the fourth and fifth of which enter the orbit, the sixth very small, triangular, not quite so large as the fourth, the seventh the largest; one anterior and generally two posterior temporals, the anterior not much larger than, and in contact with, the oculars; the lower posterior temporal, if present, is in contact with the seventh and eighth labials, but not with the front temporal.

Head rather thick; neck and tail slender; body compressed. Rostral rounded, rather small; anterior frontals half as large as the posterior; vertical five-sided, with the posterior angle slightly obtuse, not twice as long as broad; occipitals obtusely rounded behind, as long as the vertical; the anterior ocular not extending to, or only just reaching, the corner of the vertical; loreal square. Scales on the anterior part of the body much imbricate, in very oblique rows, those of the vertebral line of the anterior part of the body larger than the rest; each scale with two pits at the apex. Ventral scutes 215; anal entire; subcaudals 130. All the maxillary teeth of equal length.

Ground-colour yellow, with broad black cross-bands, which are much more regular and distinct on the posterior half of the body than on the anterior, where they are more confluent, and many scales within the bands remain entirely or partly yellow. Each scale of the yellow interspace has a black tip. Each shield of the head with black margins; occipitals black, with some symmetrical small yellow spots; suture of the eighth and seventh upper labials black. Lower parts yellow, many of the plates with black margins.

The largest specimen measures 77 inches, of which the tail takes 18 †.

3. **Spilotes pœcilonotus.** (Tab. XLIII.)

Spilotes pœcilonotus, Günth. Cat. Col. Sn. p. 100; Bocourt, Miss. Sc. Mex., Rept. p. 691, t. 43. figg. 4, 4 *a–e*.

? *Tropidodipsas lunulatus,* Cope, Proc. Ac. N. Sc. Phil. xii. p. 517 (1861) (an=juv. ?).

? *Phrynonax lunulatus,* Cope, ibid. xiv. p. 348, nota (1863).

? *Spilotes lunulatus,* Bocourt, Miss. Sc. Mex., Rept. p. 694, t. 42. figg. 1, 1 *a–e*.

Hab. MEXICO (*Mus. Brit.*), Orizaba (*Sumichrast*); GUATEMALA, Vera Paz, low forest (*Salvin*), Panzos, Rio Polochic (*Bocourt*); HONDURAS (*Mus. Brit.*).

* We have an adult and a young specimen from this locality; the former has *eighteen* rows of scales, the latter nineteen, which is the normal number.

† Cope includes in his list of Central-American Reptiles *Spilotes pullatus*; but the occurrence of this species in Central America seems to me to require further confirmation.

Body elongate, compressed; tail slender; head moderate, rounded in front, distinct from neck; eye large; mouth deeply cleft. Rostral shield much broader than high, very obtuse above, just reaching the surface of the head; anterior frontals short, much broader behind than in front; posterior ones much larger; vertical five-sided, very broad in front, with an obtuse angle behind, and convergent, very concave lateral edges; superciliaries very broad posteriorly; occipitals short, small, obtuse behind. Anterior ocular large, raised to the surface of the head, but not reaching the vertical; two posterior oculars; one loreal; two large nasals, nostril between; eight or nine upper labials; scales in twenty-one or twenty-three rows, those of the dorsal rows keeled; keels of the three medial rows excessively strong. Scales rather elongate, much imbricate, especially on the neck; ventral plates raised on the sides, 202–208; anal entire; subcaudals 126–132. Above uniform brown, except some scales of the medial row, which are yellow with black tip; head yellow and brown variegated; belly anteriorly uniform yellowish, ventral plates black-edged about the middle of body, posteriorly entirely black.

Length of cleft of mouth $1\frac{3}{8}$ inch, length of tail 18 inches; total length 5 feet 3 inches.

The number of upper labials, as well as their disposition, varies. One specimen has nine on both sides, the fourth, fifth, and sixth entering the orbit; a second shows the same arrangement on one side, but has eight labials only on the other, the fourth and fifth entering the orbit; the third specimen has eight labials on both sides, but on one the third, fourth, and fifth enter the orbit, on the other the fourth and fifth only.

If I am right in referring *T. lunulatus*, Cope, to this species, the young would have a more variegated coloration with darker cross-bands, which is, in fact, most likely the case.

4. **Spilotes argus.** (Tab. XLIV.)

Spilotes argus, Bocourt, Miss. Sc. Mex., Rept. p. 692, t. 48. figg. 10, 10 *a–f*.

Hab. MEXICO (*Mus. Brussels*), Atoyac in Vera Cruz (*H. H. Smith*).

Body and tail moderately elongate; head rounded in front; eye large, three-fifths of the length of the snout. Rostral shield rather broader than high, obtusely rounded above, just reaching the upper surface of the head. Anterior frontals half the size of posterior; vertical broad, with the lateral margin slightly concave, as long as the occipitals, which are obtusely rounded behind. Loreal longer than high; the single præocular large, raised to the surface of the head, but not reaching the vertical; two postoculars; ten upper labials, the fifth, sixth, and seventh, or the sixth and seventh only, entering the orbit. Temporals rather irregular, 2+2, but one only is in contact with the postoculars. Scales much imbricate, those of the median rows faintly keeled, in twenty-three series. Ventrals 213; anal entire; subcaudals $88+x$. The teeth of this species increase so slightly in size behind, that the dentition may well be described as isodont. The upper and lateral parts are black, each scale being provided with a yellow spot; on the scales near the head the yellow spot is small, but farther behind the black is reduced to the margin of the scale. On the posterior third of the body the yellow spots of some of the dorsal series, instead of being at the base of the scale, occupy the inner or outer half, thus leading to the formation of two black dorsal bands. Tail black, with four bands of yellow spots, each band composed of two series of spots; one band runs along the median line of the back of the tail, one along the meeting edges of the subcaudals, and one along each side. The scutes of the head are symmetrically spotted with yellow. Lower parts of the body yellow, the anterior ventral shields showing a blackish margin; this margin becomes gradually broader behind, the last shields being almost entirely black.

Our specimen is 56 inches long, the tail measuring 17 inches.

It differs in some respects from the one described by Bocourt. It has ten upper labial shields, of which the fifth, sixth, and seventh enter the orbit. The dorsal scales are also much more faintly keeled. The two black dorsal bands make their appearance only in the posterior third of the length of the body; and on the tail the yellow spots are arranged in four well-defined series, one running along the median line of the back, one on each side, and one along the middle of the lower surface. The coloration reminds one of that of *Coronella sayi*.

5. Spilotes chrysobronchus.

Spilotes chrysobronchus, Cope, Journ. Ac. N. Sc. Phil. viii. p. 136, t. 28. fig. 11 (head) (1875).
Spilotes chrysobranchus, Bocourt, Miss. Sc. Mex., Rept. p. 695, t. 48. figg. 9, 9 *a* (cop. from Cope).

Hab. Costa Rica (*Cope*).

Perhaps identical with *S. fasciatus*, Peters.

ZAMENIS.

Zamenis, Wagler, Nat. Syst. Amphib. p. 188 (1830).

1. Zamenis constrictor.

Coluber constrictor, Linn. Syst. Nat. ed. 12, i. 1, p. 385.
Bascanion constrictor, Baird & Girard, Cat. N. Amer. Rept. p. 93; Bocourt, Miss. Sc. Mex., Rept. p. 697, t. 48. fig. 5.
Coryphodon constrictor (part.), Dum. & Bibr. Erpét. vii. p. 183.
Zamenis constrictor, Boulenger, Cat. Sn. i. p. 387.
Coluber flaviventris, Say, in Long's Exped. Rock. Mount. ii. p. 185.
Bascanion flaviventris, Baird & Girard, Cat. N. Amer. Rept. p. 96; Bocourt, Miss. Sc. Mex., Rept. p. 697, t. 48. fig. 3, 3 *a*, *b*.

Hab. North America, United States.—Mexico (*Mus. Brit.*).

This species seems to be scarce in Northern Mexico, and the black variety does not extend even so far southwards.

2. Zamenis oaxacæ.

Coryphodon constrictor (part.), Dum. & Bibr. Erpét. vii. p. 184.
Coryphodon testaceus (part.), Günth. Cat. Col. Sn. p. 108.
Coryphodon oaxaca, Jan, Elench. sist. Ofid. p. 63.
Bascanion oaxaca, Bocourt, Miss. Sc. Mex., Rept. p. 701, t. 48. figg. 2, 2 *a–d*.
Zamenis oaxacæ, Boulenger, Cat. Sn. i. p. 386.

Hab. MEXICO, Oaxaca (*Mus. Paris*).

3. Zamenis rhombifer. (Tab. XLV.)

Coryphodon rhombifer, Günth. P. Z. S. 1860, p. 236.

Hab. PANAMA, Chiriqui (*Champion*).—COLOMBIA; E. PERU.

Body moderately slender; head rather short and high; eye large, two-thirds of the length of the snout; rostral shield rather broader than deep, reaching to the upper surface of the head. Anterior frontals one-half or one-third the size of posterior; vertical as long or nearly as long as occipitals. Nostril wide; loreal subquadrangular, with the posterior angle produced backwards; præocular single, concave, extending to the upper surface of the head, but not reaching the vertical; two postoculars. Temporals 2+2 (the two posterior may coalesce into a single shield). Normally nine upper labials, the fourth, fifth, and sixth entering the orbit. Scales keeled, in seventeen rows, those of the outer series smooth.

Ventrals 165; anal bifid; subcaudals *x*. (Esmeraldas.)
„ 150; „ „ 93. (Moyobamba.)
„ 157; „ „ 88. (Chiriqui.)

Upper parts brownish grey, with a series of large, irregularly rhombic, dark brown, black-edged spots along the back; they become fainter in large specimens, and only the black edges remain; sides of the body and of the ventral shields marbled with irregular black spots.

An adult specimen is 37 inches long, the tail measuring 12 inches.

We have figured the typical specimen from Esmeraldas, restoring its tail, and the specimen from Chiriqui, which is young.

4. Zamenis flavigularis.

Psammophis flavigularis, Hallowell, Proc. Ac. N. Sc. Phil. vi. p. 178 (1852); and in Sitgreave's Exped. Zuni R. p. 131, t. 11 (not good).
Coryphodon mento-varius, Dum. & Bibr. Erpét. vii. p. 187.
Bascanium mentovarium, Cope, Proc. Am. Phil. Soc. xviii. p. 271 (1879); Bull. U. S. Nat. Mus. no. 32. p. 71; Bocourt, Miss. Sc. Mex., Rept. p. 699, t. 45. figg. 6, 6 *a–d*.
Zamenis mentovarius, Boulenger, Cat. Sn. i. p. 389.
Bascanion suboculare, Cope, Proc. Ac. N. Sc. Phil. xviii. p. 319 (1866).
Bascanion lineatus, Bocourt, Miss. Sc. Mex., Rept. p. 700, t. 48. figg. 1, 1 *a–c*.
Zamenis lineatus, Boulenger, Cat. Sn. i. p. 388.

Hab. MEXICO, Colima (*U. S. Nat. Mus.*), Presidio and Tres Marias Is. (*Forrer*), Autlan in Jalisco, Guadalajara (*Buller*), Amula in Guerrero (*H. H. Smith*), Tehuantepec (*Sumichrast*); GUATEMALA, between Coban and Chisec (*Hague*), San Gerónimo, Vera Paz, Rio Chisoy (*Salvin*).

I do not consider the confluence of the two upper labial shields below the eye (as found in *Z. mentovarius*) to be of sufficient importance to justify distinction by a binomial term. Black spots at and below the tips of the scales are frequently very distinct (*Z. lineatus*, figured by Bocourt, op. cit. t. 48. fig. 1 *c*), but are also visible, though more faintly, in examples which would be referred to *Z. mentovarius*.

5. Zamenis flagelliformis.

Natrix flagelliformis, Laur. Syn. Rept. p. 79.
Masticophis flagelliformis, Baird & Girard, N. Amer. Rept. p. 98; Jan, Iconogr. Ophid. xx. t. 6.
Herpetodryas flagelliformis, Dum. & Bibr. Erpét. vii. p. 210; Günth. Cat. Col. Sn. p. 118.
Bascanium flagelliforme, Cope, Bull. U. S. Nat. Mus. no. 32, p. 71; and Proc. U. S. Nat. Mus. xiv. p. 625 (1892).
Zamenis flagelliformis, Boulenger, Cat. Sn. i. p. 389.
Herpetodryas psammophis (part.), Schleg. Phys. Serp. ii. p. 195; and Abbild. t. 44. figg. 21–24.

Hab. NORTH AMERICA, Southern States.—? MEXICO.

Cope mentions that he has seen specimens of this species from Guadalajara and Chihuahua. I myself have not been able to refer a Mexican specimen to it, if the specific characters are admitted to consist of a contracted end of the snout, occipitals narrowed behind, seventeen rows of scales, a moderate number of subcaudal scutes (not exceeding 112), &c.

6. Zamenis semilineatus. (Tab. XLVI. fig. A.)

Herpetodryas flavigularis, var., Günth. Cat. Col. Sn. p. 118*.
?*Bascanium semilineatum*, Cope, Proc. U. S. Nat. Mus. xiv. p. 626 (1892).
Zamenis tæniatus (part.), Boulenger, Cat. Sn. i. p. 390.

Hab. NORTH AMERICA, Arizona (*Cope*).—MEXICO, Presidio (*Forrer*).

Body and tail very slender, the tail being nearly one third of the total length. Head rather elongate, flat above, deeply concave in the loreal region. Vertical shield bell-shaped, with concave lateral edges; its anterior portion wider than a supra-ocular and in contact with the præorbital. Anterior frontals scarcely half the size of posterior. Eight supra-labials, the fourth and fifth entering the eye; loreal longer than deep; two præoculars, two postoculars; temporals 2+2+2; occipitals truncated behind; posterior chin-shields longer than anterior. Scales in seventeen rows. Ventrals 196–201; anal divided; subcaudals 132–134 (and more, *Cope*).

* Where nineteen series of scales have been counted by error.

A specimen from Presidio, fifty-six inches long, is light olive-coloured, greenish in the anterior fourth of the body; the lower part of the sides is ornamented with narrow continuous yellow and brownish lines. They are most distinct in the anterior part of the body and disappear in the posterior. One of these yellow lines runs along the meeting edges of the outermost and second series of scales, the second along the third and fourth series, and both have narrow black or brown edges. The upper labials are yellow with black upper margin; this black margin passes into the black upper edge of the outer yellow line. Lower parts yellow, with a few small black spots on the lower part of the head.

Twelve maxillary teeth, of which the two posterior are somewhat separated from, but scarcely larger than, the preceding.

A second specimen, which was probably obtained in North Mexico, scarcely differs from the one described, but the lower yellow band is somewhat broader and more distinct; and the scales of the back of the posterior part of the trunk are finely mottled with black.

This form seems to me to come nearest to *Bascanium semilineatum* of Cope, although the description of the colours does not apply in every respect. The "second" specimen was referred by me at first to *H. flavogularis*, which, however, seems to have a somewhat shorter tail, and has no longitudinal bands on the sides; even now I have some doubts as to the propriety of separating these two forms. Boulenger applies the name "*tæniatus*, Hallowell," to both these specimens; but, according to the unanimous testimony of other authors, *tæniatus* of Hallowell has only fifteen rows of scales, and ought to be kept distinct. Also *Leptophis lateralis* (Hallowell), from California, is, in my opinion, a very distinct species, characterized *inter alia* by a very peculiar physiognomy, and by the very large size of the anterior frontals.

7. **Zamenis ornatus.** (Tab. XLVI. fig. B.)

Masticophis ornatus, Baird & Girard, Cat. N. Amer. Rept. pp. 102, 159.
Bascanium ornatum, Cope, Proc. U. S. Nat. Mus. xiv. pp. 622, 629 (1892).
Zamenis tæniatus (part.), Boulenger, Cat. Sn. i. p. 391.

Hab. North America, Texas.—Mexico, Coahuila (*W. Taylor*).

This species is distinguished not only by its 15 rows of scales, but by an extremely narrow vertical shield, the anterior portion of which is narrower than a supraocular, and which is widely separate from the præocular.

8. **Zamenis melanolomus.**

Masticophis bilineatus *, Cope, Proc. Ac. N. Sc. Phil. xviii. p. 127 (1866) (nec Schleg.).

---

* Cope introduces also in his list of Central-American Reptiles (Bull. U. S. Nat. Mus. no. 32, 1887, p. 70) the true *Herpetodryas bilineatus* of Schlegel, figured by Jan, Iconogr. Ophid. xxxi. t. 4. fig. 3. But its origin

Masticophis melanolomus, Cope, Proc. Ac. N. Sc. Phil. xx. p. 134 (1868) *.

Hab. MEXICO, Yucatan (*Smiths. Inst. & Mus. Brit.*).

9. Zamenis pulcherrimus.

Masticophis pulcherrimus, Cope, Proc. Ac. N. Sc. Phil. xxvi. p. 65 (1874).

Drymobius pulcherrimus, Cope, Bull. U. S. Nat. Mus. no. 32, p. 70.

Drymobius (*Eudryas*) *pulcherrimus*, Bocourt, Miss. Sc. Mex., Rept. p. 725, t. 49. figg. 3, 3 *a–e*.

Zamenis pulcherrimus, Boulenger, Cat. Sn. i. p. 392.

Hab. MEXICO, Tehuantepec (*Sumichrast, Buller*); WEST NICARAGUA (*Cope*).

SALVADORA.

Salvadora, Baird & Girard, N. Amer. Rept. p. 104 (1853).

1. Salvadora grahamiæ.

Salvadora grahamiæ, Baird & Girard, N. Amer. Rept. pp. 104, 161; Jan, Iconogr. Ophid. i. t. 3. fig. 1; Günth. Ann. & Mag. N. H. 1863, xii. p. 349; Garman, Mem. Mus. Compar. Zool. Cambr. viii. pp. 38, 145 (1883); Bocourt, Miss. Sc. Mex., Rept. p. 661, t. 43. figg. 2, 2 *a–e*; Cope, Proc. U. S. Nat. Mus. 1892, p. 619.

Phimothyra grahamiæ et *bairdii*, Cope, Proc. Ac. N. Sc. Phil. xiii. p. 300 (1861), and xxxv. p. 14 (1883).

Zamenis grahami, Boulenger, Cat. Sn. i. p. 393.

Salvadora bairdii, Jan, Iconogr. Ophid. i. t. 3. fig. 2; Bocourt, Miss. Sc. Mex., Rept. p. 662, t. 43. figg. 3, 3 *a–d*; Dugès, La Naturaleza, i. p. 124 (1888); Cope, Proc. U. S. Nat. Mus. 1892, p. 619.

Phimothyra hexalepis, Cope, Proc. Ac. N. Sc. Phil. xviii. p. 304 (1866).

Hab. NORTH AMERICA, Southern United States.—MEXICO, Chihuahua (*U. S. Nat. Mus.*), Hacienda del Hobo (*Geddes*), Hacienda el Florencio in Zacatecas (*Buller*), Guanajuato (*Dugès*), Puebla (*Bocourt*), Amula in Guerrero 6000 feet (*H. H. Smith*), Orizaba (*Boucard*), Jalapa, Vera Cruz (*Mus. Philad. Acad.*), Tehuantepec (*Sumichrast*).

2. Salvadora mexicana.

Zamenis mexicanus, Dum. & Bibr. Erpét. vii. p. 695; Bocourt, Miss. Sc. Mex., Rept. p. 664, t. 46. fig. 5; Boulenger, Cat. Sn. i. p. 392.

Masticophis mexicanus, Jan, Iconogr. Ophid. xxii. t. 6. fig. 1.

Phimothyra mexicana, Cope, Proc. Ac. N. Sc. Phil. xvi. p. 167 (1864).

Lytorhynchus mexicanus, Cope, Proc. Am. Phil. Soc. xviii. p. 266 (1879).

Salvadora mexicana, Günth. Ann. & Mag. N. H. 1863, xii. p. 349; Cope, Bull. U. S. Nat. Mus. no. 32, p. 72.

---

is, in fact, unknown. The scutellation of the upperside of the head differs much from that of *Herpetodryas boddaertii*, to which, otherwise, it seems to have some resemblance.

* Cope, probably by some inadvertence, ascribes 15 scales to this snake, while our specimens have 17, and otherwise agree with Cope's description in every detail.

Hab. MEXICO, Izucar (*Boucard*), Guanajuato (*Dugès*), Colima (*U. S. Nat. Mus.*), Tierra Colorada in Guerrero (*H. H. Smith*).—CUBA, Cape Corrientes (*Mus. Paris*).

PITUOPHIS.

Pituophis, Holbrook, N. Amer. Herp. iv. p. 7 (1842).

1. **Pituophis deppii.**

Elaphis pleurostictus *, Dum. & Bibr. Erpét. vii. p. 244.
Pituophis pleurostictus, Bocourt, Miss. Sc. Mex., Rept. p. 666, t. 42. figg. 2, 2 *a–d*, 3, 3 *a–f*.
Elaphis deppei, Dum. & Bibr. Erpét. vii. p. 268.
Pituophis deppei, Jan, Iconogr. Ophid. xxii. t. 2. fig. 2.
Spilotes deppei, Cope, Bull. U. S. Nat. Mus. no. 32, p. 72 (1887).
Pituophis vertebralis, Günth. Cat. Col. Sn. p. 86 (nec Blainv.).
Arizona jani, Cope, Proc. Ac. N. Sc. Phil. xii. p. 369 (1861).

Hab. MEXICO, Hacienda del Castello in Guadalajara (*A. C. Buller*), Guanajuato (*Dugès*), city of Mexico (*Doorman*), Buenavista, San José Acateno in Vera Cruz (*Cope*), Tehuantepec (*Sumichrast*).

2. **Pituophis lineaticollis.** (Tab. XLVII.)

Arizona lineaticollis, Cope, Proc. Ac. N. Sc. Phil. xiii. p. 300 (1862).
Spilotes lineaticollis, Cope, Bull. U. S. Nat. Mus. no. 32, p. 72.

Hab. MEXICO, Southern Mexican plateau (*Sartorius*); GUATEMALA, Dueñas (*Salvin*).

Habit moderately slender, scarcely compressed. Head-shields normal; the anterior frontals small, not much more than one-fourth of the posterior. Rostral shield broader than high, not penetrating between the frontals; loreal longer than high; anterior ocular not reaching the vertical; three postoculars. Eight (or nine) upper labials, the fourth and fifth entering the orbit. Temporals scale-like. Scales in twenty-seven rows, those on the back keeled. Ventrals 244; anal entire; subcaudals 68.

Ground-colour brownish-olive, with black ornamental markings. Head without any spots; anterior part of the trunk with four black bands, which, posteriorly, break up into spots, the anterior ones on the back retaining the appearance of their twin composition, whilst the posterior and those on the tail are more or less simple cross-bands. The anterior part of the abdomen is unspotted, but on the posterior part the spots of the flanks encroach in an irregular fashion upon the ventral shields.

Our single specimen of this beautiful snake is 65 inches long, of which the tail measures 9½.

* I select the term *deppii* for this species in preference to *pleurostictus*, because the locality for the latter specimen has been erroneously given as "Monte Video," the difficulty of recognizing the species having been thus unduly increased.

3. **Pituophis elegans.**

Arizona elegans, Kennicott, U. S. Mex. Bound. Survey, ii. Rept. t. 13, p. 18; Bocourt, Miss. Sc. Mex., Rept. p. 676, t. 46. figg. 3, 3 *a, b.*

Rhinechis elegans, Cope, Proc. Am. Phil. Soc. xxiii. p. 284 (1886).

Hab. North America, Southern United States.—Mexico, Chihuahua (*Cope*).

4. **Pituophis melanoleuca.**

Coluber melanoleucus, Daud. Hist. Rept. vi. p. 409.

Pituophis melanoleucus, Holbr. N. Amer. Herpet. iv. t. 1, p. 7; Dum. & Bibr. Erpét. vii. p. 233; Günth. Cat. Col. Sn. p. 86.

Pituophis mexicanus, Dum. & Bibr. Erpét. vii. p. 236, t. 62; Günth. Cat. Col. Sn. p. 87; Jan, Iconogr. Ophid. xxii. t. 2; Bocourt, Miss. Sc. Mex., Rept. p. 674, t. 42. figg. 4, 4 *a–e*, t. 47. fig. 6.

Churchillia bellona, Baird & Girard, in Stansbury's Explorat., Rept. p. 350.

Pituophis bellona, Günth. Cat. Col. Sn. p. 87.

Pituophis sayi (Schleg.), Cope, Bull. U. S. Nat. Mus. no. 32, p. 72.

Hab. North America, United States.—Mexico, Nuevo Leon (*Taylor*), Colima and Chihuahua (*Cope*).

I conclude, from an examination of numerous specimens in the British Museum, that neither the number of labial shields nor that of series of scales can safely be used as characters to separate specifically southern from northern or western specimens.

DRYMOBIUS.

Drymobius, Fitzinger, Syst. Rept. p. 26 (1842).

1. **Drymobius margaritiferus.**

Herpetodryas margaritiferus, Schleg. Ess. ii. p. 184; and Abbild. t. 44. figg. 19, 20.

Drymobius margaritiferus, Fitz. Syst. Rept. p. 26; Cope, Bull. U. S. Nat. Mus. no. 32, p. 70; Bocourt, Miss. Sc. Mex., Rept. p. 716, t. 49. figg. 2, 2 *a–d.*

Leptophis margaritiferus, Dum. & Bibr. Erpét. vii. p. 539.

Dromicus margaritiferus, Günth. Cat. Col. Sn. p. 126; Salv. P. Z. S. 1860, p. 456.

Thamnosophis margaritiferus, Jan, Iconogr. Ophid. xxxi. t. 6. fig. 3.

Hab. Mexico, Colima (*Xantus*), Presidio (*Forrer*), Zacualtipan, Tuxpan, Misantla, Coatzacoalcos (*Cope*), Santo Domingo de Guzman (*A. C. Buller*), Omilteme in Guerrero, Atoyac in Vera Cruz, Teapa in Tabasco (*H. H. Smith*), Tehuantepec (*Sumichrast*); British Honduras, Belize (*Mus. Brit.*); Guatemala, Dueñas and Lanquin (*Salvin*); Costa Rica (*Gabb*); Panama (*Leconte*).

2. **Drymobius boddærtii.**

Coluber boddærtii, Seetzen, in Meyer's Zool. Archiv, ii. p. 59 (1796).

Herpetodryas boddærtii, Schleg. Ess. ii. p. 185; Dum. & Bibr. Erpét. vii. p. 210; Günth. Cat. Col. Sn. p. 115; Salv. P. Z. S. 1860, p. 456; Jan, Iconogr. Ophid. xlix. t. 1. fig. 1.

Drymobius (*Eudryas*) *boddærti* (Cope), Bocourt, Miss. Sc. Mex., Rept. p. 720, t. 51. figg. 5, 5 *a–c*.
Dromicus pleei, Jan, Iconogr. Ophid. xxiv. t. 5. fig. 1 (nec D. & B.).
Drymobius (*Eudryas*) *pleei*, Bocourt, Miss. Sc. Mex., Rept. p. 728, t. 51. figg. 3, 3 *a–d*.
Herpetodryas rappii, Salv. P. Z. S. 1860, p. 456 (nec Günth.).
Herpetodryas lævis, Fischer, Wiegm. Archiv, 1881, 1. p. 227, t. 11. figg. 4–6.
Drymobius (*Eudryas*) *lævis* (Cope), Bocourt, Miss. Sc. Mex., Rept. p. 722, t. 51. figg. 6, 6 *a–d*.
Drymobius (*Eudryas*) *dorsalis*, Bocourt, loc. cit. p. 724, t. 51. figg. 2, 2 *a–d*.
Coryphodon alternatus, Bocourt, Bull. Soc. Philom. 1884, p. 133.

Hab. Mexico, Tres Marias Is. (*Forrer*), Jalapa, Tuxpan, and Actopam (*Cope*), Zacualpan (*H. Finck*), Hacienda del Hobo (*P. Geddes*), Vera Cruz, Huatuzco (*H. H. Smith*); Guatemala, Dueñas, low forests of Vera Paz, and Lake Peten (*Salvin*), Coban (*Sarg*); Nicaragua (*Cope*); Costa Rica (*Smiths. Inst.*); Panama (*Cope*), Veragua (*Cope*).—Tropical South America; West Indies.

As in the majority of isodont Colubrine Snakes, so also in this species the posterior teeth are frequently somewhat larger than the middle ones; and this increase in size may be in some individuals sufficiently great to lead to the description of the dentition as coryphodont. After examining a considerable number of specimens, however, I come to the conclusion not to separate such individuals, particularly since I have compared specimens from the northernmost as well as the southernmost range, among them such as have absolutely the same style of isodont dentition.

As to the two young specimens from Dueñas which were referred to *Herpetodryas rappii* by Mr. Salvin, I believe them now to be the young of *H. boddærtii*, distinct from the typical Venezuelan *H. rappii*. The fact is, that what I considered (Cat. Col. Sn. p. 117) a specific peculiarity of *H. rappii*, viz. the remarkable difference of coloration between young and old, is a character common to the majority of these American snakes, be they referred to *Zamenis*, *Coryphodon*, *Drymobius*, *Herpetodryas*, &c. The young are ornamented with a series of quadrangular dark cross-bands on the back, with which similar spots on the side alternate. If a knowledge of the locality of such a young snake does not assist in the determination, it is most difficult, and perhaps impossible, to assign it to its proper species. It is a remarkable fact, supported by other analogous instances among vertebrates, that this juvenile pattern of coloration may in some forms persist throughout the various stages of growth. Thus in specimens from Western Ecuador which I have referred as a variety to *H. rappii* (P. Z. S. 1859, p. 412) the quincunciate pattern of coloration remains unchanged with age, but the tints are less intense in old than in young individuals.

3. **Drymobius cæruleus.**

Dromicus cæruleus, Fischer, Jahrb. Hamb. Wiss. Anstalt, ii. p. 103, t. 4. fig. 7 (1885).
? *Drymobius* (*Eudryas*) *cæruleus*, Bocourt, Miss. Sc. Mex., Rept. p. 727, t. 51. figg. 4, 4 *a–d*.

Hab. Guatemala, Coban (*Sarg*), Vera Paz (*Mus. Paris*).

4. **Drymobius dendrophis.**

Herpetodryas dendrophis, Schleg. Phys. Serp. ii. p. 196; and Abbild. t. 44. figg. 25–28; Günth. Ann. & Mag. N. H. 1863, xii. p. 358, t. 6. fig. B.

Dendrophidium dendrophis (Fitz.), Cope, Proc. Am. Phil. Soc. 1886, p. 278.

Dendrophidion dendrophis, Bocourt, Miss. Sc. Mex., Rept. p. 730, t. 49. figg. 4, 4 *a–d*; and Le Naturaliste, 1889, p. 46, figg. 1–4.

Drymobius dendrophis, Cope, Bull. U. S. Nat. Mus. no. 32, p. 70.

Herpetodryas poitei, Dum. & Bibr. Erpét. vii. p. 208.

Herpetodryas brunneus, Salv. P. Z. S. 1860, p. 456 (nec Günth.).

Herpetodryas nuchalis, Peters, MB. Ak. Wiss. Berl. 1863, p. 285.

Dendrophidium melanotropis, Cope, Journ. Ac. N. Sc. Phil. viii. p. 134, t. 26. fig. 1 (1875).

Elaphis melanotropis, Cope, Bull. U. S. Nat. Mus. no. 32, p. 71.

Dendrophidium chloroticum, Cope, Proc. Am. Phil. Soc. xxiii. p. 278 (1886).

Hab. Guatemala, Vera Paz, between Coban and Lanquin (*Salvin*), Peten (*Morelet*), Coban (*Cope*); Mexico (*Sallé*); Nicaragua (*Cope*); Costa Rica (*Gabb*).—South America, Cayenne, Venezuela, ? Ecuador.

This snake varies considerably in the length of the tail, as may be seen from the following table:—

A.	Specimen from Vera Paz, ♂.	161 Ventrals.	Anal divided.	105 subcaudals.
B.	" " " " ♀.	139 "	" entire.	140 "
C.	" " " "	160 "	" divided.	117 "
D.	" " " "	163 "	" "	153 "
E.	" " Guatemala, ♂.	154 "	" "	119 "

Shields of the head regular, the anterior frontals about two-fifths the size of the posterior; vertical as long as the suture between the occipitals, with the lateral margins convergent posteriorly. Loreal as high as long or longer than high; one præocular raised to the surface of the crown, but not reaching the vertical. Two, sometimes three, postoculars. Nine upper labials, the fourth, fifth, and sixth entering the orbit, or ten upper labials, the fifth, sixth, and seventh being below the eye. Eye large, its longitudinal diameter being equal to its distance from the nostril, and in young specimens it is proportionally even larger. Scales more or less strongly keeled; sometimes one, sometimes more, of the outer series of scales being smooth; in seventeen rows. Maxillary teeth gradually increasing in size posteriorly, the posterior being conspicuously larger and stronger than the middle ones; thirty-two or thirty-four in number.

The ground-colour may be brownish-olive or even blackish. Young specimens have the body ornamented with irregular, blackish-edged, narrow cross-bands, more conspicuous on the anterior than on the posterior part of the body; sometimes these bands are modified into narrow blackish bars enclosing lighter-coloured ocelli. With age this ornamental colour generally disappears entirely or traces of it are visible only on the anterior part of the body. But in all specimens the ground-colour of the sides encroaches largely upon the ventral shields. One female specimen from Vera Paz is nearly black, with the head and tail of a lighter (in life probably red) colour.

An adult male is 49 inches long, of which the tail measures 16½. An adult female measures 53 inches, the tail being 20.

Herpetodryas brunneus, from Guayaquil, is closely allied to *D. dendrophis*, but distinguished by a smaller eye. This character would be less conspicuous in young specimens.

HERPETODRYAS.

Herpetodryas, Boie, Isis, 1827.

1. Herpetodryas fuscus.

Coluber fuscus, Linn. Mus. Ad. Fried. p. 32, t. 17. fig. 1.
Herpetodryas fuscus, Dum. & Bibr. Erpét. vii. p. 209; Cope, Bull. U. S. Nat. Mus. no. 32, p. 70.
Herpetodryas carinatus, var., Schleg. Ess. ii. p. 175, t. 7. figg. 3, 4.

Hab. PANAMA, Darien (*Cope*).—TROPICAL SOUTH AMERICA, east of the Andes.

2. Herpetodryas carinatus.

Coluber carinatus, Linn. Syst. Nat. i. 1, p. 384.
Herpetodryas carinatus (part.), Schleg. Ess. ii. p. 175, t. 7. figg. 5, 6.
Herpetodryas carinatus, Dum. & Bibr. Erpét. vii. p. 207; Cope, Journ. Ac. N. Sc. Phil. viii. p. 135 (1875).

Hab. COSTA RICA (*Gabb*), San José (*Van Patten*); PANAMA, Chiriqui (*Champion*), Darien (*Cope*).—TROPICAL SOUTH AMERICA, east of the Andes.

The carination of the median dorsal scales of the specimen from Chiriqui is so faint that the specimen might be referred to *H. fuscus*.

3. Herpetodryas grandisquamis.

Spilotes grandisquamis, Peters, MB. Ak. Wiss. Berl. 1868, p. 451.
Herpetodryas grandisquamis, Cope, Journ. Ac. N. Sc. Phil. viii. p. 135 (1875).

Hab. COSTA RICA.

4. Herpetodryas melas.

Herpetodryas melas, Cope, Proc. Am. Phil. Soc. xxiii. p. 278 (1886).

Hab. NICARAGUA.

LEPTOPHIS.

Leptophis, Bell, Zool. Journ. ii. p. 328 (1826).

1. **Leptophis mexicanus.**

Leptophis mexicana, Dum. & Bibr. Erpét. vii. p. 536; Cope, Journ. Ac. N. Sc. Phil. viii. p. 133 (1875); Jan, Iconogr. Ophid. xlix. t. 6. fig. 3.

Ahætulla mexicana, Günth. Cat. Col. Sn. p. 154; Salv. P. Z. S. 1860, p. 457.

Hapsidophrys mexicanus, Cope, Proc. Am. Phil. Soc. xxii. p. 382 (1885).

Hab. Mexico, Tampico, Amula, Teapa (*H. H. Smith*), Atoyac (*Richardson*), Jalisco, north of the Rio de Santiago, Jicaltepec (*Commiss. Geogr.*), Vera Cruz, Tehuantepec (*Sumichrast*), Santo Domingo de Guzman (*A. C. Buller*), Yucatan (*Mus. Brit.*); British Honduras, Stann Creek (*Rev. J. Robertson*), Belize (*Mus. Brit.*); Guatemala, Lanquin (*Salvin*); Nicaragua (*Cope*); Costa Rica (*Van Patten*).

2. **Leptophis occidentalis.**

Ahætulla occidentalis, Günth. P. Z. S. 1859, p. 412.

Leptophis occidentalis, Cope, Bull. U. S. Nat. Mus. no. 32, p. 69.

Hab. Panama, Darien and Panama (*Cope*).—Western Ecuador (*Fraser*).

3. **Leptophis modestus.** (Tab. XLVIII.)

Ahætulla modesta, Günth. Ann. & Mag. N. H. 1872, ix. p. 26, t. 6. fig. C (head).

Philothamnus modestus, Cope, Bull. U. S. Nat. Mus. no. 32, p. 69.

Hab. Guatemala, Rio Chisoy, below the town of Cubulco (*Salvin*).

Scales in fifteen rows, very strongly keeled, except those in the outermost series. Ventrals 171; anal bifid; subcaudals 171. Snout depressed, not pointed. Rostral shield just reaching to the upper surface of the crown; anterior frontals subtruncated in front, about half the size of the posterior. Vertical not much longer than broad, rather shorter than the snout or than the occipitals, which are truncated behind. Loreal longer than deep. Anteocular extending to the upper surface of the crown, but not reaching the vertical; two small and short postoculars. Eight upper labials, the fourth and fifth of which enter the orbit. Temporals $1+\frac{1}{1.1}$. Eye rather smaller than in the other species of this genus, a little more than one-half of the length of the snout. Uniform greenish-olive above, light green below. A narrow black band from the eye along the suture between the labials and temporals.

The posterior three or four teeth are larger than the preceding, but not separated from them by an interspace.

The single specimen known is 52 inches long, the tail measuring 22.

4. Leptophis depressirostris.

Philothamnus depressirostris, Cope, Proc. Ac. N. Sc. Phil. 1861, p. 557.

Hab. Panama, Cocuyas de Veraguas (*Mus. Acad. Phil.*).

5. Leptophis saturatus.

Leptophis saturatus, Cope, Journ. Ac. N. Sc. Phil. viii. p. 133, t. 28. fig. 10 (1875) (head).
Hapsidophrys saturatus, Cope, Bull. U. S. Nat. Mus. no. 32, p. 69.

Hab. Nicaragua (*Cope*); Costa Rica (*Gabb*).

I have not seen a specimen of this species.

6. Leptophis diplotropis.

Ahætulla diplotropis, Günth. Ann. & Mag. N. H. 1872, ix. p. 25, t. 6. fig. A.

Hab. Mexico, Tres Marias Is. and Presidio in Western Mexico (*Forrer*), Amula in Guerrero (*H. H. Smith*), Tehuantepec (*Sumichrast*), Santo Domingo de Guzman, and La Laguna in the Sierra de Alica (*A. C. Buller*).

7. Leptophis præstans.

Thrasops præstans, Cope, Proc. Ac. N. Sc. Phil. 1868, p. 309.
Leptophis præstans, Cope, Journ. Ac. N. Sc. Phil. viii. p. 133 (1875).
Thrasops (*Ahætulla*) *sargii,* Fisch. Wiegm. Archiv, 1881, p. 229, t. 11. figg. 7–9.

Hab. Mexico, Amula in Guerrero, Teapa in Tabasco (*H. H. Smith*); British Honduras, Stann Creek (*Rev. J. Robertson*); Guatemala, Peten (*Cope*), Coban (*Sarg*); Costa Rica (*Gabb*).

The number of postoculars varies from 2 to 4, and that of the upper labials from 8 to 9; generally only the fifth, seventh, or ninth middle series of scales are keeled, the keels (structure) passing into mere black lines (colour).

8. Leptophis bilineatus.

Diplotropis bilineata, Günth. Ann. & Mag. N. H. 1872, ix. p. 24, t. 6. fig. B.
Leptophis bilineatus, Cope, Bull. U. S. Nat. Mus. no. 32, p. 69.

Hab. Mexico, Tehuantepec (*Sumichrast*); Nicaragua (*Cope*); Costa Rica (*Mus. Brit.*).

9. Leptophis æruginosus.

Leptophis æruginosus, Cope, Journ. Ac. N. Sc. Phil. viii. p. 132, t. 28. fig. 9 (1875).

Hab. Costa Rica (*Gabb*).

Founded upon a young specimen, which may possibly be the young of the preceding species.

SYNCHALINUS.

Synchalinus, Cope, Proc. Am. Phil. Soc. 1893, p. 345.

1. **Synchalinus corallioides.**

Synchalinus corallioides, Cope, l. c.

Hab. Costa Rica, Buenos Aires.

TROPIDONOTUS.

Tropidonotus, Kuhl, Bull. Sc. Nat. ii. p. 81 (1824).

1. **Tropidonotus ordinatus.**

Coluber ordinatus, Linn. Syst. Nat. i. p. 379.

Tropidonotus ordinatus, Boie, Isis, 1827, p. 535; Günth. Cat. Col. Sn. p. 73; Salv. P. Z. S. 1860, p. 455; 1861, p. 228; Bouleng. Cat. Sn. i. p. 205.

Coluber sirtalis, Linn. Syst. Nat. i. p. 383.

Coluber eques, Reuss, Mus. Senckenb. i. p. 152, t. 8. fig. 2 (1834).

Coluber bipunctatus, Latr. Rept. iv. p. 85.

Tropidonotus bipunctatus, Schleg. Ess. ii. p. 320; Dum. & Bibr. Erpét. vii. p. 582.

Eutænia macrostemma, Kennicott, Proc. Ac. N. Sc. Phil. 1860, p. 331; Cope, Bull. U. S. Nat. Mus. no. 32, p. 74.

Eutænia megalops, Kennicott, Proc. Ac. N. Sc. Phil. 1860, p. 330; Cope, Bull. U. S. Nat. Mus. no. 32, p. 74.

Eutænia cyrtopsis, Kennicott, Proc. Ac. N. Sc. Phil. 1860, p. 333; Cope, Bull. U. S. Nat. Mus. no. 32, p. 73; Bocourt, Miss. Sc. Mex., Rept. p. 774, t. 55. figg. 5.

Thamnophis scalaris, Cope, Proc. Ac. N. Sc. Phil. 1860, p. 369.

Eutænia scalaris, Cope, Proc. Ac. N. Sc. Phil. 1866, p. 204.

Tropidonotus scalaris, Bouleng. Cat. Sn. i. p. 205.

Tropidonotus collaris, Jan, Arch. per la Zool. iii. p. 209 (1865); and Iconogr. Ophid. xxv. t. 5. fig. 2.

Eutænia collaris, Cope, Proc. Am. Phil. Soc. xxii. p. 173 (1884); Bull. U. S. Nat. Mus. no. 32, p. 73; Dugès, La Naturaleza, i. p. 127, t. 13. fig. 16 (1888).

Eutænia flavilabris, Cope, Proc. Ac. N. Sc. Phil. 1866, p. 306; Bull. U. S. Nat. Mus. no. 32, p. 73.

Eutænia sumichrasti, Cope, Proc. Ac. N. Sc. Phil. 1866, p. 306; Proc. Am. Phil. Soc. xxii. p. 174 (1884); Bull. U. S. Nat. Mus. no. 32, p. 73.

Eutænia cyrtopsis, var. *sumichrasti*, Bocourt, Miss. Sc. Mex., Rept. p. 775, t. 57. fig. 3.

Eutænia phenax, Cope, Proc. Ac. N. Sc. Phil. 1868, p. 134; Bull. U. S. Nat. Mus. no. 32, p. 73.

Eutænia cyrtopsis, var. *phenax*, Bocourt, Miss. Sc. Mex., Rept. p. 778.

Entænia pulchrilatus, Cope, Proc. Am. Phil. Soc. xxii. p. 164 (1884); Dugès, La Naturaleza, i. p. 128 (1888); Cope, Bull. U. S. Nat. Mus. no. 32, p. 73.

Eutænia pulchrilata (!), Bocourt, Miss. Sc. Mex., Rept. p. 767, t. 57. figg. 4.

Eutænia insigniarum, Cope, Proc. Am. Phil. Soc. xxii. p. 172 (1884); Dugès, La Naturaleza, i. p. 129 (1888); Ives, Proc. Ac. N. Sc. Phil. 1892, p. 462.

Eutænia cyrtopsis, var. *fulvus*, Bocourt, Miss. Sc. Mex., Rept. p. 777, t. 57. figg. 2.

Hab. North America.—Mexico, throughout the country; Guatemala, Dueñas, San Gerónimo, Rio Chisoy (*Salvin*), Alta Vera Paz and Coban (*Bocourt*).

A species as variable as it is common. The variations observable in Central-American specimens cannot be localized, and seem to occur indiscriminately in any locality in which the species abounds.

2. Tropidonotus saurita.

Coluber saurita, Linn. Syst. Nat. i. p. 385.

Tropidonotus saurita, Boie, Isis, 1827, p. 535; Schleg. Ess. ii. p. 321; Dum. & Bibr. Erpét. vii. p. 585; Günth. Cat. Col. Sn. p. 72; Jan, Iconogr. Ophid. xxvi. t. 2. fig. 1; Bouleng. Cat. Sn. i. p. 212.

Thamnophis saurita, Bocourt, Miss. Sc. Mex., Rept. p. 754, t. 55. fig. 1, t. 56. fig. 1.

Coluber proximus, Say, in Long's Exped. Rocky Mount. i. p. 187.

Eutænia proxima, Baird & Gir. N.-Amer. Rept. p. 24; Cope, Bull. U. S. Nat. Mus. no. 32, p. 74.

Thamnophis saurita, var. *proxima*, Bocourt, Miss. Sc. Mex., Rept. p. 757, t. 56. figg. 2.

Eutænia rutiloris, Cope, Proc. Am. Phil. Soc. xxii. p. 388 (1885).

Thamnophis saurita, var. *faireyi*, Bocourt, Miss. Sc. Mex., Rept. p. 758, t. 56. figg. 3.

Hab. North America, east of the Rocky Mountains.—Mexico, Matamoros (*Smithson. Inst.*), Guerrero (*H. H. Smith*), Jalapa (*Höge*), Orizaba and Tuxpan in Vera Cruz (*U. S. Nat. Mus.*), Tehuantepec (*Sumichrast*), Cozumel I. (*Gaumer*); British Honduras (*Mus. Brit.*); Guatemala (*Morelet*).

3. Tropidonotus chrysocephalus.

Eutænia chrysocephala, Cope, Proc. Am. Phil. Soc. xxii. p. 173 (1884); Bocourt, Miss. Sc. Mex., Rept. p. 762, t. 57. figg. 1.

Tropidonotus chrysocephalus, Bouleng. Cat. Sn. i. p. 203.

Hab. Mexico, Omilteme and Amula, alt. 6000–8000 ft., both in Guerrero (*H. H. Smith*), Orizaba in Vera Cruz (*Sumichrast*).

4. Tropidonotus præocularis.

Eutænia præocularis, Bocourt, Miss. Sc. Mex., Rept. p. 770, t. 56. figg. 7.

Hab. British Honduras, Belize.

Known from two specimens, and distinguished by nineteen rows of strongly keeled scales, eight upper labials, three præoculars, &c.

5. Tropidonotus multimaculatus.

Atomarchus multimaculatus, Cope, Amer. Nat. 1883, p. 1309.

Eutænia multimaculata, Cope, Proc. Am. Phil. Soc. xxiii. p. 284 (1886).

Hab. Mexico, Chihuahua city (*U. S. Nat. Mus.*).

I have not seen a specimen of this species.

6. **Tropidonotus scaliger.**

Tropidonotus scaliger, Jan, Arch. per la Zool. iii. p. 214 (1865); Bouleng. Cat. Sn. i. p. 203.
Eutænia scalaris, Dugès, La Naturaleza, i. p. 129, t. 13. fig. 17 (1888); Bocourt, Miss. Sc. Mex., Rept. p. 768, t. 56. figg. 6 (excl. synonymy).

Hab. MEXICO, Guanajuato (*Dugès*), Mexico city (*Doorman*).

7. **Tropidonotus godmani,** sp. n.

Hab. MEXICO, Omilteme and Amula in Guerrero (*H. H. Smith*).

Scales in seventeen rows, strongly keeled, with the exception of those of the outermost series, which are short, rhombic, smooth or very faintly keeled. Head moderate, distinct from neck. The diameter of the eye is equal to its distance from the nostril. Anterior frontals subtriangular, truncated in front. Vertical very little shorter than occipitals; one præocular, which just reaches the upper surface of the head; three postoculars. Seven upper labials, of which the third and fourth enter the orbit; five lower labials are in contact with the anterior chin-shields, which are shorter than the posterior. Temporals 1+2. Ventrals 145, 146, 151; subcaudals 76, 76, 81.

Brown above, with a yellowish vertebral line, one scale broad; a greenish lateral band occupies the sixth and seventh rows of scales; a few small black markings occupy the sides of the body, but are generally covered by the overlapping neighbouring scales. Lower parts greenish, sometimes finely marbled with blackish. Upper labials whitish, with the posterior margin black.

Seven examples, showing a remarkable agreement in structure and colour, were collected at Omilteme and Amula. One of the largest specimens is 24 inches long, of which the tail takes 6. The stomach of one contained young mice.

8. **Tropidonotus variabilis.**

Hemigenius variabilis, Dugès, Proc. Am. Phil. Soc. 1888, p. 182 (cum fig.); La Naturaleza, 1890, p. 403, t. 27. fig. 14; Bocourt, Miss. Sc. Mex., Rept. p. 741, t. 53. figg. 5.

Hab. MEXICO, Guanajuato and Guadalajara (*Dugès*).

This species is not isodont, but has the dentition of a *Tropidonotus*.

9. **Tropidonotus melanogaster.**

Tropidonotus melanogaster, Peters, MB. Ak. Wiss. Berl. 1864, p. 389; Günth. Zool. Record, 1865, p. 154; Bouleng. Cat. Sn. i. p. 225.
Eutænia melanogaster, Cope, Bull. U. S. Nat. Mus. no. 32, p. 73; Bocourt, Miss. Sc. Mex., Rept. p. 779, t. 53. fig. 4, t. 55. fig. 3.
Tropidonotus mesomelanus, Jan, Arch. per la Zool. iii. p. 230 (1865); Iconogr. Ophid. xxvii. t. 5. fig. 3, and t. 6. fig. 2.
Regina mesomelæna, Cope, Bull. U. S. Nat. Mus. no. 32, p. 74; Dugès, La Naturaleza, i. p. 130 (1888).
Tropidonotus baronis-mülleri, Troschel, in Müll. Reise Mex. iii. p. 610.

Hab. MEXICO, La Laguna 6500 ft. and Colonia Brizuela in Jalisco (*A. C. Buller*), Guanajuato (*Dugès*), Mexico city (*Geddes*), Vera Cruz (*Commiss. Geogr.*), Tehuantepec (*Sumichrast*).

10. Tropidonotus validus.

Regina valida, Kennicott, Proc. Ac. N. Sc. Phil. 1860, p. 334; Cope, Bull. U. S. Nat. Mus. no. 32, p. 74.

Tropidonotus validus, Cope, Proc. Ac. N. Sc. Phil. 1860, p. 342; Bouleng. Cat. Sn. i. p. 237.

Tropidonotus quadriserialis, Fisch. Verh. naturf. Ver. Hamburg, iii. p. 82, t. 1. fig. 2 (1879) (copied by Bocourt, Miss. Sc. Mex., Rept. p. 752, t. 54. figg. 6).

Hab. NORTH AMERICA, Utah, Arizona, Lower California.—MEXICO, Presidio (*Forrer*), Mazatlan (*Fischer*), Colima (*Xantus*), San Blas (*Dugès*).

11. Tropidonotus grahami.

Regina grahamii, Baird & Gir. N.-Amer. Rept. p. 47.

Tropidonotus grahamii, Günth. Cat. Col. Sn. p. 78; Bouleng. Cat. Sn. i. p. 240.

Tropidonotus leberis, var. *grahami*, Jan, Iconogr. Ophid. xxvii. t. 5. fig. 1.

Hab. NORTH AMERICA, Michigan to Texas.—? MEXICO.

The occurrence of this snake in Mexico is doubtful, the specimens in the British Museum, labelled with that locality, having been obtained from a dealer.

12. Tropidonotus sipedon.

Coluber sipedon, Linn. Syst. Nat. ed. 1768, i. p. 219; ed. 1766, i. p. 379.

Tropidonotus sipedon, Holbr. N.-Amer. Herpet. iv. p. 29, t. 6.

Coluber fasciatus, Linn. Syst. Nat. ed. 1766, i. p. 378.

Tropidonotus fasciatus, Boie, Isis, 1827, p. 535; Holbr. N.-Amer. Herpet. iv. p. 25, t. 5; Günth. Cat. Col. Sn. p. 76; Dum. & Bibr. Erpét. vii. p. 566; Jan, Iconogr. Ophid. xxvi. t. 3; Bouleng. Cat. Sn. i. p. 242.

Tropidonotus rhombifer, Hallow. Proc. Ac. N. Sc. Phil. 1852, p. 177; Jan, Iconogr. Ophid. xxvi. t. 4. fig. 2; Cope, Bull. U. S. Nat. Mus. no. 32, p. 73; Bocourt, Miss. Sc. Mex., Rept. p. 751, t. 54. figg. 5.

Tropidonotus pogonias (part.), Dum. & Bibr. Erpét. vii. p. 574.

Hab. NORTH AMERICA, east of the Rocky Mountains.—MEXICO, Tampico (*Richardson*), Misantla (*Commiss. Geogr.*); COSTA RICA, Cartago (*Mus. Brit.*).

TRETANORHINUS.

Tretanorhinus, Dum. & Bibr. Erpét. vii. p. 348 (1854).

1. Tretanorhinus nigroluteus.

Tretanorhinus nigroluteus, Cope, Proc. Ac. N. Sc. Phil. 1861, p. 298; 1868, p. 309; Proc. Am. Phil. Soc. xxii. p. 183 (1885); Günth. Ann. & Mag. N. H. 1872, ix. p. 27; Bocourt, Le Nat. 1891, p. 122; Bouleng. Cat. Sn. i. p. 282.

Regina nigrolutea, Cope, Bull. U. S. Nat. Mus. no. 32, p. 74.
Helicops agassizii, Jan, Arch. per la Zool. iii. p. 254 (1865) ; Iconogr. Ophid. xxviii. t. 2. fig. 1.
Helicops bifrenatus, Bocourt, Bull. Soc. Philom. viii. p. 134 (1884).
? *Tretanorhinus lateralis*, Bocourt, Le Nat. 1891, p. 122.

Hab. Nicaragua (*Cope*); Panama (*Mus. Brit.*), Colon (*Mus. Paris*).

2. Tretanorhinus dimidiatus.

? *Tropidonotus dimidiatus*, Boie, Isis, 1827, p. 535.
Tropidonotus dimidiatus, Cope, Proc. Ac. N. Sc. Phil. 1862, p. 297.
Regina dimidiata, Cope, Bull. U. S. Nat. Mus. no. 32, p. 74.

Hab. Mexico, Jalapa (*Cope*).

This is an altogether dubious species ; and without the examination and comparison of the typical specimens described by Boie and Cope no opinion can be formed as to the correctness of Cope's identification, or Boulenger's view, according to which Boie's snake is the Indian *Atretium schistosum*. Unfortunately, it is not known in what Museum Boie's specimen is preserved, if it be still in existence ; the example described by Cope was contained in the Museum of the Philadelphia Academy, but it cannot now be found there.

3. Tretanorhinus mocquardi.

Tretanorhinus mocquardii, Bocourt, Le Nat. 1891, p. 122.

Hab. Central America.

I have not seen a specimen of this species.

HELICOPS.

Helicops, Wagler, Nat. Syst. Amphib. p. 170 (1830).

1. Helicops septemvittatus.

Calopisma septemvittatum, Fisch. Verh. naturw. Ver. Hamburg, iii. p. 84, t. 1. fig. 3 (1879).
Limnophis septemvittatus, Cope, Bull. U. S. Nat. Mus. no. 32, p. 75.
Helicops septemvittatus, Bouleng. Cat. Sn. i. p. 275.

Hab. Mexico.

Unknown to me.

ISCHNOGNATHUS.

Ischnognathus, Dum. & Bibr. Mém. Ac. Sc. xxiii. p. 468 (1853).
Storeria, Baird & Girard, Cat. N. Amer. Rept. 1853, p. 135.

1. Ischnognathus dekayi.

Tropidonotus dekayi, Holbr. N.-Amer. Herp. iv. p. 53, t. 14.
Ischnognathus dekayi (part.), Dum. & Bibr. Erpét. vii. p. 507.
Ischnognathus dekayi, Günth. Cat. Col. Sn. p. 81; Salv. P. Z. S. 1860, p. 456; 1861, p. 228; Bouleng. Cat. Sn. i. p. 286.
Storeria dekayi (Holbrook), Cope, Bull. U. S. Nat. Mus. no. 32, p. 75; Bocourt, Miss. Sc. Mex., Rept. p. 742, t. 53. figg. 1, 2.
Storeria tropica, Cope, Proc. Am. Phil. Soc. xxii. p. 175 (1885).
Storeria dekayi, var. *anomala*, Dugès, Proc. U. S. Nat. Mus. xi. p. 9, cum fig. (1888); La Naturaleza, 1890, p. 401.

Hab. NORTH AMERICA.—MEXICO, Guanajuato (*Dugès*), Vera Cruz (*Cope*), Orizaba (*Godman*, *Dugès*); GUATEMALA, Coban, San Gerónimo (*Salvin*), Peten (*Hague*).

2. Ischnognathus occipito-maculatus.

Coluber occipito-maculatus, Storer, Rep. Fish. & Rept. Mass. p. 230.
Storeria occipito-maculata, Baird & Gir. N.-Amer. Rept. p. 137; Cope, Bull. U. S. Nat. Mus. no. 32, p. 75; Bocourt, Miss. Sc. Mex., Rept. p. 745, t. 53. figg. 6.
Ischnognathus occipito-maculatus, Günth. Cat. Col. Sn. p. 81; Jan, Iconogr. Ophid. xxx. t. 1. fig. 2.
Ischnognathus dekayi (part.), Dum. & Bibr. Erpét. vii. p. 507.

Hab. NORTH AMERICA.—MEXICO, Zacualtipan in Hidalgo (*Cope*).

The occurrence of this species in Mexico rests at present solely on Cope's authority.

3. Ischnognathus storerioides.

Tropidoclonium storerioides, Cope, Proc. Ac. N. Sc. Phil. 1865, p. 190.
Storeria storerioides, Garman, N.-Amer. Rept. p. 29.
Natrix storerioides, Cope, Proc. U. S. Nat. Mus. xi. p. 391 (1889).
Ischnognathus storerioides, Bouleng. Cat. Sn. i. p. 288.

Hab. MEXICO, Ciudad in Durango (*Forrer*), Mexico city (*Doorman*), Jalisco to 8500 feet (*A. C. Buller*).

A young specimen from Ciudad in Durango does not show any sign of a division of the nasal shield; probably with the growth of the snake such a division would have taken place.

4. Ischnognathus copii.

Adelophis copei (Dugès), Cope, Proc. Am. Phil. Soc. xviii. p. 265 (1879); La Naturaleza, i. p. 18, t. 3. fig. 1 (1887) (copied in Bocourt, Miss. Sc. Mex., Rept. p. 739, t. 53. figg. 3).
Ischnognathus copii, Bouleng. Cat. Sn. i. p. 289.

Hab. MEXICO, Guadalajara.

I do not know this snake from autopsy, and entertain some doubts as to its pertinence to *Ischnognathus*.

HYDROMORPHUS.

Hydromorphus, Peters, MB. Ak. Wiss. Berl. 1859, p. 276.

1. **Hydromorphus concolor.**

Hydromorphus concolor, Peters, loc. cit. p. 277, fig. 3 (head and dentition).

Hab. Costa Rica (*Hoffmann*).

HYPSIGLENA.

Hypsiglena, Cope, Proc. Ac. N. Sc. Phil. 1861 (1860), p. 246.
Pseudodipsas, Peters, MB. Ak. Wiss. Berl. 1861 (1860), p. 251.

1. **Hypsiglena torquata.**

Leptodeira torquata, Günth. Ann. & Mag. N. H. (3) v. p. 170, t. 10. fig. A (Febr. 1860).
Hypsiglena torquata, Cope, Bull. U. S. Nat. Mus. no. 32, p. 78.
Hypsiglena ochrorhynchus, Cope, Proc. Ac. N. Sc. Phil. xii. p. 246 (June 1860).
Comastes quincunciatus, Jan, Iconogr. Ophid. xxxviii. t. 1. fig. 1.
Hypsiglena chlorophæa, Cope, Proc. Ac. N. Sc. Phil. xii. p. 246.
Hypsiglena texana, Stejneger, U. S. Dep. Agric. N. Amer. Faun. no. 7. pt. 2, p. 205.

Hab. North America, California, Texas (*Taylor*).—Mexico, Chihuahua and Guaymas (*Cope*), Presidio and Ventanas in Durango (*Forrer*), Colima (*Xantus*), Guanajuato (*Dugès*), Jalisco (*coll. Salv. & Godm.*), Mezquital del Oro (*Buller*), Zacatecas (*Cameron*); Nicaragua (*Mus. Brit.*), Island of Laguna (*Mus. Liverp.*).

Specimens of this species vary in the coloration of the neck, the two principal types of variation having been described as *L. torquata* and *H. ochrorhynchus*. We have, however, another specimen from Jalisco, without distinct white collar, which shows only 19 (not 21) series of scales, only 7 (not 8) upper labials, and which lacks the lower small præocular. Specimens from Zacatecas and Mezquital have also 19 (20) scales and 7 upper labials, but the collar is distinct; the lower præocular may be present or absent in the same specimen. Although these specimens differ somewhat in the shape of their head, I think that a larger series will prove their specific identity.

2. **Hypsiglena discolor.** (Tab. XLIX. fig. A.)

Leptodeira discolor, Günth. P. Z. S. 1860, p. 317.
Hypsiglena discolor, Cope, Bull. U. S. Nat. Mus. no. 32, p. 78.

Hab. Mexico, Oaxaca (*Sallé*).

The head is rather broad and depressed, the snout rounded; the eye is of moderate size, its vertical diameter being about one-third the width between the eyes; the trunk is rounded and, like the tail, somewhat slender. The rostral shield reaches just to the upper surface of the snout; the frontals are nearly square; the anterior pair are one-third the size of the posterior, which are slightly bent downwards to the side of the head; the vertical is pentagonal, longer than broad; the occipitals are rounded posteriorly. Nostril situated between two nasals; loreal quadrangular; one anterior and two posterior oculars; seven or eight upper labial shields, the third and fourth or fourth and fifth entering the orbit. There is one elongate temporal shield in contact with both the oculars; the other temporals, five in number, are scale-like. The median lower labial is triangular and rather small; nine lower labials, the first of which is in contact with its fellow behind the median shield. There are two pairs of chin-shields of nearly equal size. Scales in nineteen rows, smooth, rhombic, those of the sides similar to those on the back. The number of the ventral plates varies between 182 and 179, that of the caudal between 88 and 87; anal divided.

The ground-colour of the upper parts is dirty white; the upper part of the head is brown; there is a whitish collar behind the occipitals. Fifty-one or fifty-four black bands cross the trunk and extend on to the edge of the belly; they are broader than the interspaces between, and become interrupted and spot-like on the tail.

All the lower parts are uniform whitish.

	inches.	lines.
Total length	21	1
Length of the head	0	7
Greatest width of the head	0	$5\frac{1}{3}$
Length of the trunk	14	6
„ „ tail	6	0

3. **Hypsiglena latifasciata,** sp. n. (Tab. XLIX. fig. B.)

Hab. Southern Mexico (*Mus. Godman & Salvin*).

Scales in twenty-one series. Head broad, depressed, distinct from neck. Pupil vertical. Anterior frontals two-thirds of posterior; vertical much longer than broad, as long as occipitals. Eight upper labials, the third, fourth, and fifth entering the orbit; loreal higher than long; one anterior and two posterior oculars; temporals $1+2+3$; two pairs of chin-shields, the posterior the shorter; trunk with alternate black and whitish cross-bands, the black ones being about four times as broad as the whitish, seven on the trunk and three on the tail. Lower parts uniform whitish; the upper part of the head backwards to the occipitals and the sides of the head are mottled with brown, whilst the whole occipital region is occupied by a nearly uniform yellowish patch, bordered behind by a black nuchal band and on the sides by a brownish band running from the eye backwards. Ventrals 186; anal divided; subcaudals 74.

A single specimen, $13\frac{3}{4}$ inches long, the tail measuring 3 inches.

4. **Hypsiglena ornata.**

Comastes ornatus, Bocourt, Bull. Soc. Philom. 1884, p. 141.
Hypsiglena ornata, Cope, Bull. U. S. Nat. Mus. no. 32, p. 78.

Hab. PANAMA, Isthmus of Darien.

PETALOGNATHUS.

Petalognathus, Dum. & Bibr. Erpét. vii. p. 463 (1854).

1. **Petalognathus nebulatus.**

Coluber nebulatus, L. Syst. Nat. i. p. 383.
Dipsas nebulata, Schleg. Ess. ii. p. 275, t. 11. figs. 14, 15.
Petalognathus nebulatus, Dum. & Bibr. Erpét. vii. p. 464.
Leptognathus nebulatus, Günth. Cat. Col. Sn. p. 177; Jan, Iconogr. Ophid. xxxvii. t. 5. fig. 3; Cope, Bull. U. S. Nat. Mus. no. 32, p. 66.

Hab. MEXICO, Mexico city (*Doorman*), Vera Cruz (*Commiss. Geogr.*), Teapa (*H. H. Smith*); GUATEMALA, Coban (*Sarg*); NICARAGUA (*Smiths. Inst.*); COSTA RICA (*Gabb*).—SOUTH AMERICA, Guiana, Venezuela, Western Ecuador.

TROPIDODIPSAS.

Tropidodipsas, Günth. Cat. Col. Sn. p. 180 (1858).

1. **Tropidodipsas fasciata.** (Tab. L. fig. B.)

Tropidodipsas fasciata (part.), Günth. Cat. Col. Sn. p. 181.
Leptognathus fasciata, Cope, Proc. Ac. N. Sc. Phil. 1868, p. 109; Bull. U. S. Nat. Mus. no. 32, p. 67.
Leptognathus (*Tropidodipsas*) *subannulatus*, Müll. Verh. naturf. Ges. Basel, viii. p. 274, t. 1. fig. 5 (1887).

Hab. MEXICO, Jicaltepec, Vera Cruz (*Cope*).

The specimens which I described in 1858 under the above name have been referred by later authors to different species; one of these, *T. philippii*, must be admitted to be distinct, having fifteen, instead of seventeen series of scales. Much less certain is the specific distinctness of *T. dumerilii* and its synonyms.

The specimens which I retain under the original name of *T. fasciata* have seventeen rows of scales, a longer snout, larger eye, loreal distinct from the two præoculars, and, especially, a pair of posterior chin-shields, which are nearly as long as the anterior. The latter character is the most important, and offers the evidence that *L. dumerilii* of Jan is identical with *T. sartorii* and not with *T. fasciata*.

According to Cope, this snake inhabits the "tierra caliente." Our figure is taken from the type in the British Museum (spec. *a*).

2. **Tropidodipsas dumerilii.** (Tab. L. fig. A.)

Tropidodipsas fasciata, spec. *c*, Günth. Cat. Col. Sn. p. 181.
Leptognathus dumerilii, Jan, El. System. 1863, p. 101 (March).
Tropidodipsas sartorii, Cope, Proc. Ac. N. Sc. Phil. 1863, p. 100 (April).
Leptognathus sartorii, Cope, Proc. Ac. N. Sc. Phil. 1868, p. 109; Bull. U. S. Nat. Mus. no. 32.
Geophis annulatus, Peters, MB. Ak. Wiss. Berl. 1870, p. 643, t. 1. fig. 2.
Galedon annularis, Jan, Iconogr. Ophid. xxxvi. t. 5. fig. 1.
Leptognathus sexscutatus, Bocourt, Bull. Soc. Philom. 1884, p. 137.
Leptognathus leucostomus, Bocourt, Bull. Soc. Philom. 1884, p. 138.
Leptognathus (*Tropidodipsas*) *bernoullii*, Müll. Verh. naturf. Ges. Basel, viii. p. 272, t. 1. f. 3 (1887).
Leptognathus (*Tropidodipsas*) *cuculliceps*, Müll. Verh. naturf. Ges. Basel, viii. p. 273, t. 1. fig. 4 (1887).

Hab. MEXICO, Vera Cruz, Mirador (*Sartorius*), Orizaba (*Mus. Brit.*), Yucatan (*Mus. Brit.* and *Mus. Paris*); GUATEMALA, Coban (*Sarg*), Atitlan (*Commiss. Geogr.*), Vera Paz, Hacienda de Chitalon (*Bernoulli*).

This form has also seventeen rows of scales, but a very short snout, and consequently the posterior chin-shields are reduced to a pair of scale-like scutes, much shorter than the anterior. The number of upper labials varies between six and eight, and the scutellation between the eye and nostril is subject to great variation, which has led to the creation of several of the specific names enumerated. The keels of the scales are very faint, and may disappear altogether in preserved specimens.

The coloration of the specimen figured is that of a living example which was received by the Zoological Society in 1884, and lived for a short time in the Gardens. The (in spirit) white, black-tipped scales of this beautiful snake are coral-red during life; none of them were yellow.

3. **Tropidodipsas philippii.**

Tropidodipsas fasciata, spec. *e*, Günth. Cat. Col. Sn. p. 181.
Leptognathus philippii, Jan, Iconogr. Ophid. xxxvii. t. 5. fig. 1.
Leptognathus albocinctus, Fisch. Jahrb. Hamburg. Wiss. Anst. ii. p. 107, t. 4. fig. 9 (head) (1885).

Hab. MEXICO, Mazatlan (*Mus. Hamb.*).

4. **Tropidodipsas anthracops.**

Leptognathus anthracops, Cope, Proc. Ac. N. Sc. Phil. 1868, pp. 108, 136; Bull. U. S. Nat. Mus. no. 32, p. 66.

Hab. WEST NICARAGUA (*Cope*).

LEPTOGNATHUS.

Leptognathus, Dum. & Bibr. Erpét. vii. p. 473 (1854).
Anholodon, Dum. & Bibr. loc. cit. p. 1165 (1854).

1. Leptognathus mikanii.

Dipsas mikanii, Schleg. Ess. ii. p. 277.
Anholodon mikanii, Dum. & Bibr. Erpét. vii. p. 1165.
Leptognathus mikanii, Günth. Cat. Col. Sn. p. 178; Ann. & Mag. N. H. 1872, ix. p. 29; Cope, Proc. Ac. N. Sc. Phil. 1868, p. 135; Jan, Iconogr. Ophid. xxxvii. t. 6. fig. 3.
Leptognathus oreas, Cope, Proc. Ac. N. Sc. Phil. 1868, pp. 108, 109.

Hab. MEXICO, Tehuantepec (*Sumichrast*).—SOUTH AMERICA, Western Ecuador, Brazil.

The following notes are taken from the single Mexican specimen in the British Museum, $12\frac{1}{4}$ inches long, which represents a prettily-coloured variety of this species:—

Posterior frontals large, not entering the orbit. Vertical as broad as long, with an obtuse angle behind. Loreal entering the orbit; another well-developed anteocular above it; two postoculars. Eight upper labials, the fourth and fifth entering the orbit. Temporals $2+3\,(2)+3$. Three pairs of chin-shields subequal in size, as broad as long; a pair of lower labials form a suture together in front of the chin-shields. Ventrals 188; anal entire; subcaudals ca. 85. Yellowish, with numerous narrow black cross-bands, 44 on the trunk and 23 on the tail, as broad as the interspaces of the ground-colour; each more or less completely divided into two by a yellow transverse line, which is broader within the anterior black bands than within the posterior. The bands do not extend on the belly, which is chequered with black. Upper parts of the head black, finely mottled with yellow.

2. Leptognathus annulatus. (Tab. XLIX. fig. C.)

Leptognathus annulatus, Günth. Ann. & Mag. N. H. 1872, ix. p. 30.

Hab. COSTA RICA, Cartago (*Mus. Brit.*).

Scales smooth, in fifteen rows, the vertebral scales being enlarged, hexagonal. Habit slender; neck very thin; head broad and short. Eye of moderate size, with vertical pupil. Anterior frontals short and small; posterior frontals large, extending down on the sides of the snout, and forming the antero-superior part of the orbit. Vertical with nearly parallel outer edges, and with a right angle behind, shorter than the occipitals. Loreal broadly entering the orbit; a small separate præocular below. Two postoculars. Seven or eight upper labials, the fourth and fifth (or fifth and sixth) entering the orbit. Temporals $1+2+3$. The first pair of lower labials not in contact with each other. Four pairs of chin-shields, the anterior pair the smallest, the second the largest, much longer than broad. Ventrals 164; anal entire; subcaudals 113. Upper parts light brownish, powdered with darker; lower parts yellowish, mottled with brown.

Body and tail encircled by black rings, which are shorter than the head, but wider than the interspaces; there are about forty of these rings on the trunk. Head irregularly spotted with brown.

The single specimen known is 17½ inches long, the tail measuring 6 inches.

3. Leptognathus argus.

Leptognathus argus, Cope, Journ. Ac. N. Sc. Phil. viii. p. 130 (1875).

Hab. Costa Rica, Sipurio (*Cope*).

4. Leptognathus pictiventris.

Leptognathus pictiventris, Cope, Journ. Ac. N. Sc. Phil. viii. p. 130 (1875).

Hab. Costa Rica.

5. Leptognathus articulatus.

Dipsas brevis, Cope, Proc. Ac. N. Sc. Phil. 1860, p. 266 (nec Dum. & Bibr.).
Leptognathus articulata, Cope, Proc. Ac. N. Sc. Phil. 1868, pp. 107, 135.

Hab. Costa Rica; Panama, Veragua.

6. Leptognathus viguieri.

Leptognathus viguieri, Bocourt, Bull. Soc. Philom. viii. p. 136 (1884).

Hab. Panama, Isthmus of Darien.

I have not seen a specimen of this or of the preceding species; to judge from the descriptions, these snakes must be closely related to each other.

7. Leptognathus brevis.

Leptognathus brevis, Dum. & Bibr. Erpét. vii. p. 476.

Hab. Mexico (*Mus. Paris*).

8. Leptognathus brevifacies.

Tropidodipsas brevifacies, Cope, Proc. Ac. N. Sc. Phil. 1866, p. 127.
Leptognathus brevifacies, Cope, ibid. 1868, p. 108.
Dipsadomorus fasciatus, Bocourt, Bull. Soc. Philom. viii. p. 135 (1884).
Leptognathus torquatus, Cope, Proc. Am. Phil. Soc. xxii. p. 172 (1885).

Hab. Mexico, Yucatan (*Smiths. Inst.* and *Mus. Paris*).

9. Leptognathus semicinctus.

Leptognathus semicinctus, Bocourt, Bull. Soc. Philom. viii. p. 139 (1884).

Hab. Mexico; Guatemala, Alta Vera Paz (*Bocourt*).

MESOPELTIS.

Mesopeltis, Cope, Proc. Ac. N. Sc. Phil. 1866, p. 318.
Asthenognathus, Bocourt, Bull. Soc. Philom. viii. p. 141 (1884).

1. **Mesopeltis sanniolus.**

Mesopeltis sanniolus, Cope, Proc. Ac. N. Sc. Phil. 1866, p. 318.

Hab. Mexico, Yucatan (*Schott*).

2. **Mesopeltis dimidiatus.** (Tab. LI. figg. A, B.)

Leptognathus dimidiatus, Günth. Ann. & Mag. N. H. 1872, ix. p. 31.
Petalognathus multifasciatus, Bocourt, Bull. Soc. Philom. viii. p. 139 (1884).
Mesopeltis multifasciatus, Cope, Bull. U. S. Nat. Mus. no. 32, p. 67.
Leptognathus (*Asthenognathus*) *grandoculis,* Müll. Verh. naturf. Ges. Basel, viii. p. 271, t. 1. fig. 2 (1887).

Hab. Mexico (*Mus. Brit.*); Guatemala, Vera Paz (*Morelet*), Mazatenango (*Bernoulli*); Nicaragua, Matagalpa 3250 feet (*Rothschuh*).

Scales smooth, in fifteen rows, those of the vertebral series not enlarged. Body much compressed, neck slender, head broad and short. Eye rather large, with vertical pupil. Anterior frontals short and small; posterior frontals large, extending down on the sides of the snout, and forming the antero-superior part of the orbit. Vertical with nearly parallel outer edges, and with a right angle behind, shorter than the occipitals. Loreal confluent with the single præocular; two postoculars. Eight upper labials, the fifth and sixth entering the orbit; (seven upper labials, the fourth, fifth, and sixth entering the eye, *Müller*). Temporals 1+2+3. Three pairs of chin-shields, the anterior the largest, crescent-shaped, much longer than broad; the middle shorter, but still longer than broad; the posterior of about the same size as the middle, and divergent. An azygos scale-like shield between the front chin-shields and the minute median labial. The first pair of lower labials are not joined together by a suture, being separated by the azygos shield; the five following lower labials are in contact with the anterior chin-shields. Ventrals 186–193; anal entire; subcaudals 98–126. Body and tail with broad blackish-brown rings separated by whitish interspaces much narrower than the rings; there are twenty-five black rings on the trunk and sixteen on the tail. The white interspaces are again each subdivided by a narrow black transverse line. Upper parts of the head black, with small whitish spots irregularly placed; a pair of large whitish spots on the neck, forming a kind of collar. Anterior chin-shields black. In the larger specimen (24 inches) the blackish bands are purplish-brown, with broad blackish margins, so broad as to almost entirely occupy the interspaces, at least on the back; the remainder of each whitish interspace is finely powdered with purplish. A black spot on the posterior chin-shields.

I have seen two specimens—the type, which is 17 inches long, the tail measuring 5 inches, and a larger one from Nicaragua, 24 inches long.

Addendum to page 107.

HENICOGNATHUS.

The following new species has been received quite recently, since the publication of the part containing the account of *Henicognathus*, and may find an appropriate place at the end of the Aglyphodont Colubrines.

2. **Henicognathus venustissimus,** sp. n. (Tab. LI. fig. C.)

Hab. Nicaragua, Matagalpa, Hacienda Rosa de Jerico 3250 feet (*Rothschuh*).

Form of the body and head as in *Coronella annulata*, but with a stronger and longer tail. Frontals together a little shorter than the vertical, which is nearly as long as an occipital. Loreal rhombic; one præocular, just reaching the upper surface of the head; two postoculars, both in contact with the occipital. Temporals 2+2. Nine upper labials, the fourth, fifth, and sixth entering the orbit. Chin-shields nearly of the same length. Scales in 17 rows. Ventrals 145; anal divided. Maxillary teeth of equal size, small, 43 in number. Coral-red or reddish-olive, with more or less irregular, broad, black, white-edged rings. The black rings encroach but slightly upon the abdomen. Head and neck black, with a broad white interocular band; the anterior two black body-bands are complete, but the following (eight or nine in number) are broken in the middle of the back; they are complete again on the tail. Each scale of the red spaces with a black tip. Lower parts whitish, with some blackish spots, irregularly crowded towards the sides of the abdomen. Each subcaudal with a large black spot.

I have seen two specimens, both with mutilated tails; the larger measures 17 inches, without tail.

Fam. COLUBRIDÆ GLYPHODONTES.

HOMALOCRANIUM.

Homalocranion, Dum. Mém. Ac. Sc. xxiii. p. 490 (1853).
Tantilla, Baird & Gir. Cat. N.-Amer. Rept. p. 132 (1853).
Lioninia, Hallowell, Proc. Ac. N. Sc. Phil. 1861, p. 484.
Microdromus, Günth. Ann. & Mag. N. H. 1872, ix. p. 17.

Synopsis of the Central-American species *.

I. Seven upper labial shields.
  A. Two nasal shields.
    a. Scales in thirteen series.
      1. *miniatum.*
    b. Scales in fifteen series.
      α. The first pair of lower labial shields in contact with each other.
        2. *atriceps*; (3. *melanocephalum*); 4. *jani*; 5. *boulengeri*; 6. *bocourti*; 7. *armillatum*; 8. *breve*; 9. *michoacanense*; 10. *annulatum.*
      β. The median lower labial shield is in contact with the anterior chin-shields.
        11. *deppii*; 12. *tæniatum*; 13. *mœstum*; 14. *schistosum*; 15. *reticulatum*; 16. *canulum.*
  B. One nasal shield.
        17. *mexicanum*; 18. *virgatum.*
II. Six upper labial shields.
        19. *bimaculatum*; 20. *rubrum.*

---

* The following species are recorded from Texas, and in all probability occur south of the Rio Grande. I introduce them here for purposes of comparison with their allies:—

Homalocranium planiceps.

Coluber planiceps, Blainv. Nouv. Ann. Mus. Hist. Nat. iv. p. 294, t. 27. fig. 3 (1835).
Homalocranion planiceps, Dum. & Bibr. Erpét. vii. p. 857; Jan, Iconogr. Ophid. xv. t. 2. fig. 2; Bocourt, Miss. Sc. Mex., Rept. p. 581, t. 36. figg. 7, 7 *a–d.*

Hab. NORTH AMERICA, Texas, Duval Co. (*W. Taylor*), California (*Botta*).

Head rather narrow, depressed. Eye small, not quite one third the length of the snout. Vertical shield rather broad, with an obtuse angle in front and an acute one behind. Seven upper labials; posterior nasal scarcely in contact with the anteocular; one ante- and two postoculars; temporals 1+1. The first pair of lower labials form a suture together. Scales in fifteen rows, without apical groove. Ventrals 153, 134 (*Blainville*), 140 (*Bocourt*); subcaudals 45. Upper parts light olive, each scale with a brownish dot at the base, lower parts white; head black above, the black extending over some portion of the middle of the neck.

This species was hitherto known from a single specimen only in the Paris Museum. The specimen in the British Museum is 10 inches long, of which the tail takes $1\frac{3}{4}$.

Although Jan's and Bocourt's accounts are taken from the same specimen, the type in the Paris Museum, they do not agree in important details. Jan figures the first pair of lower labials as distinctly separate;

1. Homalocranium miniatum.

Tantilla miniata, Cope, Proc. Ac. N. Sc. Phil. 1863, p. 100; Journ. Ac. N. Sc. Phil. viii. p. 144 (1876).

Hab. Mexico, Mirador, near Vera Cruz (*Cope*).

Vertical shield scarcely longer than broad. Seven upper labials; posterior nasal in contact with the anteocular; one ante- and two postoculars. Temporals two large, two small. The median lower labial in contact with the chin-shields. Scales in thirteen rows. Blackish brown above, a yellow collar two scales wide across the neck and touching the occipitals; lower parts salmon-coloured.

2. Homalocranium atriceps, sp. n. (Tab. LII. fig. B.)

Hab. Mexico, Nuevo Leon (*W. Taylor*).

Head narrow, flat, and depressed; eye small, not quite one-third the length of the snout. Vertical shield of moderate width, with a very obtuse angle in front, and a very pointed one behind. Seven upper labials, of which the hindmost is rather higher than the sixth; posterior nasal generally in contact with the anteocular; one ante- and two postoculars; temporals 1+1. The first pair of lower labials are in contact with each other. Scales in fifteen rows; ventrals 127; subcaudals 58. Upper parts uniform light reddish olive; upper part of the head blackish, bordered behind by a

---

Bocourt describes them in the same manner, but figures them as joining in the median line! The safest way, therefore, will be to rely upon the original figure given by Blainville, which agrees with that of Bocourt and with the arrangement of these shields in the specimen in the British Museum.

Homalocranium gracile.

Tantilla gracilis, Baird & Gir. Cat. N.-Amer. Rept., Serp. p. 132 (1853); Cope, Journ. Ac. N. Sc. Phil. viii. p. 143 (1876); Garman, Mem. Mus. Comp. Zool. viii. no. 3, p. 87, t. 6. fig. 3 (1883).

Homalocranion gracile, Jan, Arch. per la Zool. ii. p. 50 (1862); Iconogr. Ophid. xv. t. 2. fig. 1.

? *Homalocranion gracilis*, Bocourt, Miss. Sc. Mex., Rept. p. 579, t. 36. figg. 5, 5 *a–d*.

Tantilla hallowellii, Cope, Proc. Ac. N. Sc. Phil. 1860, p. 77.

Hab. North America, Texas, Duval Co. (*W. Taylor*), Indianola (*B. & G.*).

Head short, flat, with the snout rather produced; eye small, nearly one-fourth of the length of the snout. Vertical shield very broad, with a very obtuse angle in front and a pointed one behind; six upper labials; the posterior nasal separated from or in contact with the anteocular; one ante- and one postocular; temporals 1+1. Median lower labial generally in contact with the chin-shields. Scales in fifteen rows, without apical groove. Ventrals 114 or 115. Anal divided; subcaudals 46 to 48. Upper parts uniform very light brownish; head greyish above; lower parts white.

Bocourt possibly describes and figures a different species, in which the first pair of labials are in contact with each other, separating the mental from the chin-shields. The British Museum possesses a similar specimen from Texas.

narrow, straight, white, transverse collar, crossing the neck behind the occipitals. No white spots on the head. Lower parts white.

Two specimens, 7 inches long, of which the tail takes nearly 2.

3. Homalocranium melanocephalum.

Coluber melanocephalus, Linn. Mus. Ad. Frid. p. 24, t. 15. fig. 2; Syst. Nat. i. p. 378 [the species described by Daudin under this name is a compound of different forms; and that by Kuhl (Beytr. p. 87) belongs to an Indian genus]; Boie, Isis, xx. p. 526 (1827).

Natrix melanocephalus, Merr. Tent. p. 110.

Elaps melanocephalus, Wagl. in Spix, Serp. Brasil. p. 8, t. 2 *b*. fig. 1.

Cloelia melanocephala, Wagl. Syst. Amph. p. 187.

Calamaria melanocephala, Schleg. Ess. ii. p. 38, t. 1. fig. 30.

Homalocranion melanocephalum, Dum. & Bibr. Erpét. vii. p. 859; Jan, Iconogr. Ophid. xv. t. 2. figg. 4, 5; Bocourt, Miss. Sc. Mex., Rept. p. 588, t. 37. figg. 4, 4 *a–d*.

Tantilla melanocephala, Cope, Proc. Ac. N. Sc. Phil. 1871, p. 205; Journ. Ac. N. Sc. Phil. viii. p. 144 (1876).

Cloelia dorsata, Wagl. Syst. Amph. p. 187.

Tantilla armillata, Cope, Journ. Ac. N. Sc. Phil. viii. p. 143 (1876).

Hab. Tropical America.

Head flat, depressed; eye about half as long as the snout. Vertical shield rather elongate, longer than broad. Seven upper labials, one ante- and two postoculars; temporals 1+1. The first pair of lower labials form a suture together. Scales in fifteen rows, without apical groove. Ventrals 137–162; anal divided; subcaudals 44–62. Upper parts brownish, with a black dorsal line, which sometimes is very indistinct, and may be entirely absent; sometimes with a dark line along each side. Head and neck black above, with symmetrical whitish markings, which, however, are by no means constant; sometimes the snout is white, and a pair of occipital spots are present, which may expand into a collar; the black coloration of the neck is usually margined with white behind. A white spot behind the eye is nearly always present. Lower parts whitish.

The synonymy and description given above embrace individuals collected from a wide geographical range, and more or less slightly varying from each other. Schlegel, Duméril and Bibron, myself, and Jan included the whole under one specific name—*melanocephalum*; and in a general work in which the whole of the variations have to come under consideration it is, perhaps, more convenient to treat of them in the form of varieties of one specific type, while in the account of a more restricted fauna a strictly binominal method may be adopted, if the variations can be duly characterized. I propose, therefore, to arrange first the materials before me of *Homalocranium melanocephalum* (in the wider sense) in a synoptical table, and afterwards separately to describe the Central-American forms:—

I. Posterior nasal in contact with the anteorbital.

A. Neck with a broad black cross-band bordered behind by a complete white cross-band, which may be broken up into spots. A black stripe running along the median line of the back is rarely absent. Ventrals 136–145; subcaudals 48–67.—*Hab.* South America, Berbice, Pará, Moyobamba, Parana. = *Coluber melanocephalus*, L., Wagl., Jan (fig. 5), Bocourt.

B. No broad black nuchal band; no black dorsal stripe.

1. Ventrals 136–147; subcaudals 49; vertical not dilated.—*Hab.* Guatemala, Nicaragua. = var. *jani* (? Jan, fig. 4).
2. Ventrals 161, subcaudals 47. Vertical much dilated, almost as broad as long.—*Hab.* Mexico, Vera Cruz. = var. *boulengeri*, m.
3. Ventrals 172–176; subcaudals 55. Vertical longer than broad, much broader than supraoculars.—*Hab.* Mexico, Guanajuato. = var. *bocourti*, m. (*H. coronatum*, Bocourt, nec Baird & Gir.).

II. Posterior nasal not in contact with the anteorbital. A broad black band crosses the neck, bordered with white behind; in old specimens it may become indistinct.

A. Median line of the back with a black stripe.

1. A similar stripe along the side of the body. Ventrals 148–161; subcaudals 52–62.—*Hab.* Western Ecuador, Quito. = var. *fraseri*, m.
2. No dark stripe along the side of the body. Ventrals 159–171; subcaudals 44–52.—*Hab.* Guatemala, Costa Rica. = var. *armillatum*, Cope.

B. No black dorsal stripe. Ventrals 149–154; subcaudals 30–32.—*Hab.* Brazil, Pernambuco. = var. *pernambucense*, m.

4. **Homalocranium jani**, sp. n. (Tab. LII. fig. D.)

? *Homalocranion melanocephalum*, Jan, Iconogr. Ophid. xv. t. 2. fig. 4.

Hab. Guatemala (*F. D. G.*); Nicaragua, Matagalpa 3250 feet (*Rothschuh*).

Head flat, depressed; eye less than one-half of the length of the snout. Vertical shield moderately broad, longer than broad, with a very obtuse angle in front; one ante- and two postoculars; posterior nasal in contact with the anteocular. Seven upper labials; temporals 1+1. The first pair of lower labials form a suture together. Scales in fifteen rows. Ventrals 136–147; anal divided; subcaudals 49. Upper parts brownish or greyish, without a black dorsal line. The Guatemalan specimen shows a light line along each side of the back. Coloration of the side of the head as in *H. melanocephalum*, but the neck is without a black broad band; a narrow whitish more or less complete collar crossing the occiput. Lower parts white.

The larger of the two specimens is 9½ inches long.

5. **Homalocranium boulengeri**, sp. n. (Tab. LII. fig. F.)

Hab. Mexico, Huatusco in Vera Cruz (*F. D. G.*).

Head flat, broad, short, depressed; eye small, not quite half as long as the snout. Vertical shield broad, almost as broad as long, with an obtuse angle in front and an acute one behind. Seven upper labials; posterior nasal in contact with the anteocular; one ante- and two postoculars; temporals 1+1. The first pair of lower labials form a suture together. Scales in fifteen rows, without apical groove. Ventrals 161; anal divided; subcaudals 46. Upper parts light brownish (perhaps reddish in life); upper part of the head and neck black, with a white collar, two scales broad and including the extremities of the occipital shields. Snout black; a white spot in front and one behind the eye.

The specimen is 9½ inches long, the tail measuring 1⅔.

6. **Homalocranium bocourti**, sp. n.

Homalocranion coronatum, Bocourt, Miss. Sc. Mex., Rept. p. 589, t. 37. figg. 5, 5 *a–d* (nec Baird & Girard *).

Hab. MEXICO, Guanajuato (*Dugès*).

Vertical shield moderately broad, considerably longer than broad, much broader than the supraocular. Seven upper labials; posterior nasal in contact with the anteocular; one ante- and two postoculars; temporals 1+1. The first pair of lower labials form a suture together. Scales in fifteen rows. Ventrals 172–176; anal divided; subcaudals 55. Upper parts light reddish; upper part of the head brown, with a narrow white collar two scales broad and including the extremities of the occipital shields. Lower parts whitish.

Total length 9½ inches, of which the tail takes 2.

7. **Homalocranium armillatum.** (Tab. LII. fig. C.)

Tantilla armillata, Cope, Journ. Ac. N. Sc. Phil. viii. p. 143 (1876).

Hab. GUATEMALA, Dueñas (*Salvin*); COSTA RICA, Cartago (*Mus. Brit.*).

Head flat, much depressed; eye less than one-half of the length of the snout. Vertical shield moderately broad, longer than broad, with an obtuse angle in front; occipitals elongate, as long as the distance between the end of the snout and the end of the vertical shield; one ante- and two postoculars; posterior nasal separated from the anteocular by the posterior frontal or nearly so. Seven upper labials; temporals 1+1+2. The first pair of lower labials form a suture together. Scales in fifteen rows. Ventrals 159, 171; anal divided; subcaudals 44, 52. Upper parts brownish or reddish, with a black vertebral line; upper parts of the head and the neck brown; behind the brown neck a narrow white collar, which is cut into two by the black vertebral line. Each occipital with a small white spot at the end; sides of the head coloured as in *H. melanocephalum* (s. str.). Lower parts white.

One specimen is 8¾ inches long, the tail measuring 1¾; the other is somewhat longer.

* Who describe their species as having a hexagonal vertical, acute anteriorly and posteriorly.

8. **Homalocranium breve,** sp. n.

Hab. British Honduras (*Salvin*).

Head narrow, depressed, somewhat contracted in front; eye small, one-third of the length of the snout. Vertical shield rather broad, with a very obtuse angle in front and a pointed one behind. Seven upper labials, of which the hindmost is much higher than the sixth; posterior nasal in contact with the anteocular; one ante- and two postoculars; temporals 1+1. The first pair of lower labials are in contact with each other. Scales in fifteen rows. Ventrals 111. Upper parts uniform light greyish-olive, somewhat darker on the head; no collar or white spots. Lower parts white.

One specimen, 5 inches long without tail, which is injured.

9. **Homalocranium michoacanense.** (Tab. XXXVI. figg. B, C.)

Contia michoacanensis, Cope, Proc. Am. Phil. Soc. xxii. p. 178 (1885)*.

Hab. Mexico, Michoacan (*Dugès*), Mezquital del Oro (*A. C. Buller*).

Body moderately slender; tail rather short; head depressed, elapiform. Rostral shield much wider than high; anterior frontals two-thirds the size of posterior; vertical five-sided, with the posterior angle pointed; supraoculars large, pointed in front; occipitals as long as the vertical and posterior frontals together; nasal undivided above, but divided below the nostril, in contact with the first and second upper labials. Loreal nearly square; one præ- and two postoculars. Temporals 1+2. Seven upper labials, the third and fourth entering the orbit, the sixth much larger than the seventh; anterior chin-shields several times larger than the posterior, in contact with four lower labials. Scales in fifteen rows; ventrals 161–162, anal divided, subcaudals 50.

Red, with black cross-bands, which are arranged in pairs, the rings of each pair being separated by a yellow interspace; the rings do not extend to the abdomen. Head yellow, with a broad black patch across the vertical, supraoculars, and anterior part of the occipitals. The ring of the anterior pair forms a broad collar, interrupted in the median line below. In one specimen the trunk shows thirteen pairs of black bands and three on the tail. In the second the bands of the posterior half of the trunk are more crowded and numerous, and nearly equidistant, without yellow between them. In Cope's specimen the bands of each pair seem to have been confluent.

Total length 11 inches, of which the tail measures 2.

10. **Homalocranium annulatum.**

Tantilla annulata, Böttger, Zool. Anzeig. 1892, p. 419.

Hab. Nicaragua (*Mus. Lübeck*).

Vertical shield longer than broad; posterior nasal in contact with the præocular;

* See *antea*, p. 104 nota.

no loreal; one præ- and two postoculars. Temporals 1+1. Seven upper labials, the seventh higher than the sixth. The first pair of lower labial shields in contact with each other. Eye half as large as the space between eye and end of the snout. Scales in fifteen rows. Ventrals 149; anal divided; subcaudals 54. Head black; snout to the middle of the posterior frontals yellow; the fifth supralabial yellow; two narrow yellow collars, each broadly edged with black, the anterior crossing the extremity of the occipital shields. The anterior third of the trunk with alternating yellow, black-edged, transverse spots, extending from the vertebral line to the ventrals; these spots become indistinct on the remainder of the body. Lower parts uniform yellow.

Known from a single specimen, 19 inches long.

11. Homalocranium deppii.

Homalocranion deppii, Bocourt, Miss. Sc. Mex., Rept. p. 584, t. 36. figg. 11, 11 *a–d*.

Hab. SOUTHERN MEXICO (*F. D. G.*).

Head flat, depressed; eye of moderate size, at least half as long as the snout. Vertical shield broader than long, six-sided, with an obtuse angle in front and an acute one behind. Seven upper labials; posterior nasal broadly in contact with the anteocular; one ante- and two postoculars; temporals 1+1. Mental slightly in contact with the chin-shields. Scales in fifteen rows. Ventrals 152; anal divided; subcaudals 55. Upper parts brownish-olive, the head darker; back with five parallel brown lines, half a scale broad—one running along the median line, one on either side along the fifth, and the other along the third outer series of scales. A narrow yellowish collar, interrupted in the middle by the median brown line. Lower parts whitish.

Our specimen is 10 inches long, of which the tail takes $2\frac{2}{3}$.

12. Homalocranium tæniatum.

Homalocranion tæniatum, Bocourt, Miss. Sc. Mex., Rept. p. 587, t. 37. figg. 3, 3 *a–e*.
? *Homalocranium trivittatum,* Müll. Verh. naturf. Ges. Basel, vii. p. 678 (1882).

Hab. GUATEMALA (*Bocourt*); HONDURAS, Bonacca I. (*Mus. Brit.*).

Vertical shield longer than broad, with an obtuse angle in front. Seven upper labials; posterior nasal in contact with the anteocular; one ante- and two postoculars; temporals 1+1. The median lower labial is slightly in contact with the chin-shields. Scales in fifteen rows. Ventrals 149, 163; anal divided; subcaudals 68, 69. Brownish, with three light longitudinal bands, which are only half as broad as the space between them. A yellowish collar, two scales broad, crosses the neck immediately behind the head, and is not connected with the longitudinal bands.

I should, without hesitation, refer *H. trivittatum* as a synonym to this species, but for the *six* upper labials ascribed to it by Müller; however, as this author mentions a seventh labial, which he calls a "postlabial," it is probable that the two snakes in reality do not differ in this respect.

13. **Homalocranium mœstum.** (Tab. LII. fig. E.)

Homalocranium mœstum, Günth. Ann. & Mag. N. H. 1863, xii. p. 352.
Homalocranion mœstum, Bocourt, Miss. Sc. Mex., Rept. p. 583, t. 36. figg. 9, 9 *a–d.*

Hab. Mexico, Yucatan (*Cope*); Guatemala, Peten (*Salvin*).

Anterior frontals very broad, with the lateral portion produced backwards, about half as large as the posterior frontals. Vertical six-sided, much longer than broad, with an obtuse angle in front and an acute one behind; occipitals rounded behind, as long as the vertical and posterior frontals together; posterior nasal low, elongate, in contact with the single præocular; two postoculars. Upper labials seven, the third and fourth entering the orbit, the last being the largest and highest. The median lower labial is in contact with the chin-shields, which are twice as long as broad; then follow two or three other pairs of small scale-like chin-shields. Scales smooth, in fifteen rows. Ventrals 150. Entirely deep black, with a broad yellow collar extending over the hind part of the occipitals laterally to the eyes, and below over the whole chin and throat.

14. **Homalocranium schistosum.**

Homalocranion schistosum, Bocourt, Miss. Sc. Mex., Rept. p. 584, t. 36. figg. 10, 10 *a–e.*

Hab. Mexico (*Boucard*); Guatemala (*F. D. G. & O. S.*), Alta Vera Paz (*Bocourt*).

Head very short, flat, and depressed; eye small, not quite half as long as the snout. Vertical shield broad, almost as broad as long, with an obtuse angle in front and a slightly acute one behind. Seven upper labials, of which the hindmost is elevated; posterior nasal in contact with the anteocular; one ante- and two postoculars; temporals 1+1. The first pair of lower labials are widely separated from each other. Scales in fifteen rows, without apical groove. Ventrals 135, 121 (*Bocourt*); subcaudals 38. Upper parts greyish-brown, with an indistinct narrow lighter collar across the end of the occipitals. No white spots on the head. Lower parts white.

Our specimen is 8½ inches long, of which the tail takes 1½.

15. **Homalocranium reticulatum.**

Tantilla reticulata, Cope, Proc. Ac. N. Sc. Phil. (1860) 1861, p. 77; Journ. Ac. N. Sc. Phil. viii. p. 144 (1876).

Hab. Panama, Cocuyas de Veraguas (*Mitchell*).

This snake, of which I have not seen a specimen, is described as follows:—

"Vertical plate broad, slightly angular in front, projecting posteriorly for half its length between the occipitals. Occipitals and both pair of frontals rather broad. Rostral broad, visible from above. . . . . Two postoculars, upper one in contact posteriorly with the occipital, the lower touching one temporal. A second temporal equal to the first, and a third very small one behind it. Superior labials seven, last largest, third and fourth entering the orbit, both low. Four geneials, anterior in contact with inferior rostral. Scales in fifteen rows, last one slightly larger. Ventrals 148; caudals 67 pairs. Total length 10 inches 3 lines; tail 3 inches.

"Colour above chestnut-brown, much darker posteriorly, extending upon the tips of the ventrals. Anteriorly the scales are edged with darker, presenting a reticulated appearance. Central dorsal row of scales lighter, forming a pale vitta, disappearing on the tail. Third and fourth rows on each side also lighter, forming indistinct bands. A collar of the same pale yellow-brown crosses the ends of the occipitals. Cephalic plates clouded and edged with darker; a deep brown mark extending from the occipitals to the mouth across the yellowish labials. Beneath pale yellow, deepening posteriorly."

16. Homalocranium canulum.

Tantilla vermiformis, Cope, Proc. Ac. N. Sc. Phil. 1867, p. 126 (nec Hallowell).

Tantilla canula, Cope, Journ. Ac. N. Sc. Phil. viii. p. 144 (1876); Proc. U. S. Nat. Mus. xiv. p. 598 (1892).

Hab. MEXICO, Yucatan (*Schott*).

Anterior and posterior frontals narrow, transverse; vertical small, longer than suture to rostral, angular in front. Supraoculars rather large; occipitals longer than vertical. Seven upper labials; posterior nasal in contact with the præocular; two postoculars. Temporals 1+1. The first pair of lower labials are separated from each other. Scales in fifteen rows. Ventrals 110; subcaudals 37. Lead-coloured, darker above.

17. Homalocranium mexicanum.

Elapomorphus mexicanus, Günth. Ann. & Mag. N. H. 1862, ix. p. 57, t. 9. fig. 1.

Hab. MEXICO (*Sallé*).

Nasal simple. Head flat, depressed, short; eye small, but rather more than half as long as the snout. Anterior frontals broad, but very short, their longitudinal diameter being only one-fourth of that of the posterior; vertical six-sided, of moderate length; occipitals not much longer than vertical. Posterior frontal in contact with the second labial; one anterior ocular, not reaching to the upper surface of the head; two post-oculars. Seven upper labials, the third and fourth of which enter the orbit, the seventh being the largest. A rather large temporal shield in contact with the postoculars; a small one behind between the seventh labial and the occipital. The median lower labial is exceedingly small; the first pair of lower labials are narrow, transverse, and form a suture together; two pairs of chin-shields, of nearly equal size. Scales in fifteen rows, without apical groove. Ventral shields 158; anal bifid; subcaudals 52. Brownish-olive, with three blackish longitudinal bands, viz. one, almost linear, along the vertebral series of scales; the two others along the sides, each composed of two blackish lines—one line running along the middle of the second outer series, the other along that of the third. A pair of small yellowish spots on the neck; a yellowish band across the front part of the snout; lips with a black spot below the eye. Lower parts uniform yellowish.

Length of head 4 lines, of trunk 10 inches, of tail 2 inches 9 lines.

18. **Homalocranium virgatum.** (Tab. LII. fig. A.)

Microdromus virgatus, Günth. Ann. & Mag. N. H. 1872, ix. p. 17, t. 4. fig. B.
Homalocranion virgatum, Bocourt, Miss. Sc. Mex., Rept. p. 585, t. 36. figg. 4, 4 *a, b* (after Günther).
? *Homalocranion sexfasciatum,* Fischer, Abhandl. naturw. Ver. Bremen, vii. p. 225, t. 14. figg. 8–10 (1882).

Hab. Costa Rica, Cartago.

Nasal generally undivided. Rostral shield just reaching to the upper surface of the snout; anterior frontals scarcely half the size of the posterior, narrow; vertical five-sided, longer than broad; occipitals as long as the vertical and frontals together, rounded behind. Nasal not, or but slightly, in contact with the anteocular. One ante-, two postoculars. Seven upper labials, the third and fourth entering the orbit, the hindmost the largest. Temporals 1+1. The first pair of lower labials not, or but slightly, in contact with each other. Anterior chin-shields much larger than the scale-like posterior. Ventrals 180; subcaudals 71. Upper parts greyish, with a white collar; a pair of brown bands edged with black, and two scales broad, run along the back from the collar to about the middle of the tail. A similar band along each side of the body, and sometimes a narrow blackish line along each side of the abdomen. Lower parts uniform white. Upper labials white, with a black spot below the eye and on the rostral shield.

In this species several of the characters which have been used for specific, and even generic, distinction, break down. The nasal shield is generally simple, and therefore this species ought to be referred to *Elapomorphus*, which formerly was distinguished from *Homalocranium* by its undivided nasal. But in two of our specimens this shield is distinctly divided. *Elapomorphus* seems to differ from *Homalocranium* by the smaller number of maxillary teeth. All our specimens of *H. virgatum*, although varying in the extent of the nasal and mental shields, have exactly the same coloration.

19. **Homalocranium bimaculatum.**

Tantilla bimaculata, Cope, Journ. Ac. N. Sc. Phil. viii. p. 148 (1876).
Homalocranion bimaculatum, Bocourt, Miss. Sc. Mex., Rept. p. 580, t. 36. figg. 6, 6 *a–d*.

Hab. Mexico, San Blas (*Forrer*), Mazatlan (*Cope*).

Head very short; eye small, scarcely one-third the length of the snout. Vertical shield broad, six-sided, with an obtuse angle in front. Six upper labials; posterior nasal in contact with the anteocular; one ante- and one postocular. Temporals 1+1.

Median lower labial in contact with the chin-shields. Scales in fifteen rows, without apical groove. Ventrals 127; anal divided; subcaudals 27. Upper parts of a light brownish colour, with three parallel, brown, punctated lines—one along the middle series of scales, the other along the meeting-edges of the third and fourth outer series; these lines start from the brown patch which covers the upperside of the head. An oval white spot on and behind each occipital, on each side of the brown vertebral line. Lower parts white.

Our largest specimen is $7\frac{2}{3}$ inches long, of which the tail takes 1.

Tantilla calamarina, from various parts of Mexico (Cope, Proc. Ac. N. Sc. Phil. 1866, p. 320, and Journ. Ac. N. Sc. Phil. viii. p. 143), has been distinguished on account of the separation of the anteocular from the nasal and supraocular—a not very reliable character, if unsupported by other distinctive features.

20. Homalocranium rubrum.

Tantilla rubra, Cope, Journ. Ac. N. Sc. Phil. viii. p. 144 (1876); Sumichrast, Bull. Soc. Zool. Fr. v. p. 181 (1880).

Homalocranion rubrum, Bocourt, Miss. Sc. Mex., Rept. p. 590, t. 37. figg. 6, 6 *a–d*.

Hab. MEXICO, Guanajuato (*Dugès*), Orizaba (*Mus. Brit.*), Tehuantepec (*Sumichrast*).

Head flat, depressed; eye small, not quite half as long as the snout. Vertical shield of moderate width, considerably longer than broad. Seven upper labials; posterior nasal in contact with, or slightly separated from, the anteocular; one ante- and two postoculars. Temporals 1+1. The first pair of lower labials form a suture together. Scales in fifteen rows, without apical groove. Ventrals 161–179; anal divided; subcaudals 68–72. Upper parts uniform red; upper parts of the head and neck black, with a sharply-defined white collar, three scales broad, and generally including the extremities of the occipital shields. Extremity of the snout generally white; a white spot behind the eye is rarely absent.

The following snakes I do not know from autopsy, nor am I able, from the descriptions given, to assign to them their proper place in the system:—

Lioninia vermiformis, Hallow. Proc. Ac. N. Sc. Phil. 1861, p. 484; *Tantilla vermiformis*, Cope, ibid. 1862, p. 74, and Journ. Ac. N. Sc. Phil. viii. p. 145 (1876).—*Hab.* NICARAGUA.

Symphimus leucostomus, Cope, Proc. Am. Phil. Soc. xi. p. 150 (1869).—*Hab.* MEXICO, Chihuitan, Oaxaca.

SCOLECOPHIS.

Scolecophis, Fitz. Syst. Rept. 1843, p. 25 ; Cope, Proc. Ac. N. Sc. Phil. 1862, p. 74.

Distinguished from *Homalocranium* by the presence of a loreal shield.

1. Scolecophis atrocinctus.

Calamaria atrocincta, Schleg. Ess. ii. p. 47.

Homalocranion atrocinctum, Dum. & Bibr. Erpét. vii. p. 864 ; Salv. P. Z. S. 1860, p. 454, and 1861, p. 227.

Homalocranion (*Platycranion*) *atrocinctum,* Jan, El. sist. Ofidi, p. 40 ; Iconogr. Ophid. xv. t. 2. fig. 7.

Tantilla atrocincta, Cope, Journ. Ac. N. Sc. Phil. viii. p. 145 (1876).

Scolecophis atrocinctus, Bocourt, Miss. Sc. Mex., Rept. p. 577, t. 37. figg. 2, 2 *a–e*.

Elaps zonatus, Hallow. Journ. Ac. N. Sc. Phil. 1855, p. 35.

Hab. Mexico, Toluca (*Cope*) ; Guatemala, Dueñas (*Salvin*), Vera Paz (*Bocourt*) ; Honduras (*Hallowell*).

PROCINURA.

Procinura, Cope, Proc. Am. Phil. Soc. xviii. p. 262 (1880).

" Dentition opisthoglyph. Form that of *Elapomorphus.* Two nasal, one loreal, and one preocular plates ; anterior and posterior frontals distinct. Scales smooth, excepting those of the posterior dorsal and caudal regions, where they are keeled, those of the latter so much so as to be tubercular. Anal plate double.

" This genus is near to *Scolecophis,* Cope, but the peculiar tubercular carination of the tail distinguishes it."

1. Procinura æmula.

Procinura æmula, Cope, Proc. Am. Phil. Soc. xviii. p. 262.

Hab. Mexico, Southern Chihuahua, Batopilas (*Cope*).

Scales in fifteen rows. Ventrals 148 ; subcaudals 41. Loreal square ; one ante- and two postoculars ; seven upper labials, the third and fourth entering the orbit ; temporals 1 + 2. Body surrounded by broad black rings, which have a broad yellow border and are separated by red interspaces of twice their width. Each red scale has a central black spot.

The only known species inhabits a rocky, mountainous region.

Although this curious snake is in its colour an almost perfect repetition of *Elaps fulvius,* it is not known that the two species inhabit the same region.

POGONASPIS.

Pogonaspis, Cope, Proc. Ac. N. Sc. Phil. 1894, p. 204.

Described as differing from *Homalocranium* by the presence of a single pair of chin-shields.

1. **Pogonaspis ruficeps.**

Pogonaspis ruficeps, Cope, Proc. Ac. N. Sc. Phil. 1894, p. 204.

Hab. Costa Rica, Jimenez (*Alfaro*).

Posterior frontals in contact with the second labial. Vertical and occipitals elongate, the former angulate in front and pointed behind. One ante-, two postoculars. Upper labials seven, the third and fourth entering the orbit, the fifth and sixth lower than the large seventh. Temporals 1+1. The lower labials of the first pair in contact with each other. Scales smooth *. Ventrals 146; anal divided. Light brown above, yellow below. Median dorsal row of scales deep brown; a narrow yellow line on the adjacent borders of the third and fourth rows of scales, which is bounded above by a single dark brown row, and below by several rows with dark brown centres. All the colours less conspicuous posteriorly. Upper surface of the head light reddish-brown, with a pale spot at the extremity of each occipital. A black spot below the eye. (*Cope.*)

Note.—Another Central-American Snake described by Mr. Cope is *Enulius murinus*, Proc. Am. Phil. Soc. xi. p. 558 (1871), from Nicaragua. As originally described, this genus would seem to be well characterized and easily recognizable. But although Mr. Cope has referred to this genus on several subsequent occasions, his most recent observation (Proc. Ac. N. Sc. Phil. 1894, p. 205) is calculated to excite some doubt as to the accuracy of the original description. He had described it as having a long, strong, grooved tooth, whilst recently he associates with it *Leptocalamus torquatus* (Günth.), an aglyphous snake, of which I have now examined three specimens, one from Nicaragua. Under these circumstances, and until by a renewed examination of the type the glyphodont dentition of *Enulius* is clearly established, one cannot help thinking that Bocourt was right after all in recognizing Cope's *Enulius murinus* in the aglyphous snake described and figured by him.

CERCOCALAMUS.

Cercocalamus, Günth. Ann. & Mag. Nat. Hist. 1863, xi. p. 21.

Differs from *Homalocranium* by having undivided subcaudal scutes.

1. **Cercocalamus collaris.**

Cercocalamus collaris, Günth. Ann. & Mag. Nat. Hist. 1863, xi. p. 21, t. 3. fig. A.

Hab. Central America.

Rostral shield much broader than high, not recurved backwards; anterior frontals half as large as the posterior, which are bent downwards on the side; vertical not quite

* The author has omitted to state the number of rows.

twice as long as broad, hexagonal, with an obtuse angle in front, and with an acute one behind. Occipitals tapering behind, somewhat longer than vertical. The single nasal is elongate, touching the præorbital, which is not raised to the upper surface of the head; one postorbital. Seven upper labials, the third and fourth of which enter the orbit; the fifth upper labial is in immediate contact with the occipital, separating the anterior temporal from the postorbital. There is a large, elongate, posterior, temporal shield (probably formed by two confluent shields) on the side of each occipital. Six lower labials; the first pair not forming a suture together behind the median labial shield. Two pairs of elongate chin-shields. Ventrals 142; subcaudals 46. Upper parts brownish-olive; a broad black collar across the neck; a very narrow, indistinct, blackish vertebral line on the anterior part of the trunk.

The typical specimen is still the only one known of this snake. It is 9 inches long, the tail measuring 2.

STENORHINA.

Stenorhina, Dum. Mém. Ac. Sc. xxiii. p. 490 (1853); Dum. & Bibr. Erpét. vii. p. 865; Günth. Cat. Col. Sn. p. 246.

Microphis, Hallow. Proc. Ac. N. Sc. Phil. vii. p. 97 (1854).

Bergenia, Steindachner, Reise Novara, Rept. p. 92 (1869).

Body cylindrical, calamariform; tail of proportionate length; head small, flat, not distinct from neck. Rostral rounded, not prominent. Two pairs of frontals, the anterior confluent with the anterior nasal. Posterior nasal confluent with, or distinct from, a loreal. Nostril between the two shields; one anterior, two posterior oculars. Scales smooth, in seventeen rows.

1. Stenorhina degenhardti.

Calamaria degenhardtii, Berthold, Neue Rept. von Neu-Granada, Götting. 1846, p. 8, t. 1. figg. 3, 4.

Stenorhina degenhardtii, Peters, MB. Ak. Wiss. Berl. 1861, p. 461; Jan, Arch. per la Zool. ii. p. 63; Bocourt, Miss. Sc. Mex., Rept. p. 594, t. 37. figg. 7, 7 *a–e*; Cope, Journ. Ac. N. Sc. Phil. viii. p. 142 (1876).

Stenorhina degenhardtii, var. *quinquelineata,* Jan, Iconogr. Ophid. xlviii. t. 2. fig. 6.

Stenorhina degenhardtii, var. *ocellata,* Jan, Iconogr. Ophid. xlviii. t. 2. fig. 5.

Stenorhina degenhardtii, var. *apiata,* Cope, Journ. Ac. N. Sc. Phil. viii. p. 142; Bocourt, Miss. Scient. Mex., Rept. p. 597, t. 37. figg. 10, 10 *a*; Sumichrast, Bull. Soc. Zool. Fr. 1880, p. 180.

Stenorhina degenhardtii, var. *kennicottiana,* Bocourt, Miss. Scient. Mex., Rept. p. 595, t. 37. figg. 9, 9 *a, b*.

Stenorhina ventralis, Dum. & Bibr. Erpét. vii. p. 867; Günth. Cat. Col. Sn. p. 246; Salv. P. Z. S. 1861, p. 228.

Stenorhina freminvillei, Dum. & Bibr. Erpét. vii. p. 868, t. 70; Günth. Cat. Col. Sn. p. 246; Garman, Mem. Mus. Comp. Zool. viii. no. 3, p. 85 (1883); Bocourt, Miss. Sc. Mex., Rept. p. 596, t. 37. figg. 8, 8 *a*.

Stenorhina kennicottiana, Cope, Proc. Ac. N. Sc. Phil. 1861, p. 242.

Microphis quinquelineatus, Hallow. Journ. Ac. N. Sc. Phil. 1855, p. 33, t. 4. fig. 1 (not t. 3, as stated in text).

Stenorhina quinquelineata, Cope, Proc. Ac. N. Sc. Phil. 1861, p. 243; Bocourt, Miss. Sc. Mex., Rept. p. 597, t. 37. figg. 11, 11 *a–c*.

Stenorhina lactea, Cope, Proc. Ac. Nat. Sc. Phil. 1862, p. 303.

Bergenia mexicana, Steindachner, Reise Novara, Rept. p. 93.

Seven upper labial shields, the third and fourth entering the orbit. Loreal exceptionally present. Occipitals short, scarcely longer than vertical. Coloration variable:—

Var. α. Upper parts nearly uniform brown, lower lighter-coloured, mottled with brown. Ventrals 151.

Hab. MEXICO (*Mus. Brit.*); NICARAGUA, Chontales Mines (*Rix*); COSTA RICA, Cartago (*Mus. Brit.*).

Var. β. Like var. *α*, with or without a few black transverse spots, and with a blackish median line or band along the abdomen and tail; ventrals 149–156: *degenhardtii*, Berth. = *ventralis*, D. & B.

Hab. MEXICO, Atoyac in Vera Cruz (*H. H. Smith*), Vera Cruz, Mirador (*Cope*), Tehuantepec (*Sumichrast*); GUATEMALA, Pacific Coast (*Salvin*), Coban (*D. & B.*); COSTA RICA (*Cope*).—COLOMBIA, Popayan (*Berthold*).

Var. γ. Back with transverse ocellated dark-edged spots; lower parts marbled with brown: *ocellata*, Jan.

Hab. GUATEMALA, Vera Paz, Low Forest (*F. D. G. & O. S.*).—VENEZUELA, Puerto Cabello (*Jan*).

Var. δ. Brown, with numerous narrow, more or less irregular dark brown or black bands; ventrals 155: *kennicottiana*, Cope.

Hab. ISTHMUS OF PANAMA.

Var. ε. Nearly uniform yellowish-olive; ventrals 165: *freminvillii*, D. & B.

Hab. MEXICO, Tehuantepec, Tapana (*Sumichrast*), Yucatan (*Mus. Brit.*); HONDURAS (*Cope*).

Var. ζ. Yellowish-olive, with a short black stripe behind the eye, and initial traces of a black vertebral line: *apiata* and *lactea*, Cope.

Hab. MEXICO, Putla (*Mus. Brit.*), Tehuantepec (*Sumichrast*); SALVADOR, La Union (*Dow*); COSTA RICA, Cartago (*Mus. Brit.*).

Var. η. Whitish, back with three complete black dorsal lines, and with a narrow incomplete, undulated line on each side.

Hab. MEXICO (*Mus. Brit.*).

Var. θ. One complete black dorsal line, and a pair of narrow undulated lines on each side: *freminvillii*, var., D. & B.

Hab. GUATEMALA, Peten (*D. & B.*), Plain of Zacapa (*Salvin*).

Var. ι. Five complete black lines on the back and sides; ventrals 170: *quinquelineata*, Hallow.

Hab. MEXICO, Jalapa (*Höge*); HONDURAS (*Hallowell*).

TACHYMENIS.

Tachymenis, Wiegm. Act. Ac. Cæs. Leop.-Carol. xvii. p. 251 (1834).
Coronella, sp., Günth. Cat. Col. Sn. p. 36 (1858).
Coniophanes, Cope, Proc. Ac. N. Sc. Phil. 1860, p. 248.
Glaphyrophis, Jan, Prodr. in Arch. per la Zool. 1863, p. 94.
Erythrolamprus, sp., Cope, Bull. U. S. Nat. Mus. no. 32, 1887, p. 77.

Body cylindrical, coronelliform; tail of proportionate length; head of proportionate shape, distinct from neck. Rostral rounded, not prominent. Two pairs of frontals; two nasals. Scales smooth. Grooved tooth of moderate size.

1. Tachymenis piceivittis.

Coniophanes piceivittis, Cope, Proc. Am. Phil. Soc. 1869 (July), p. 149; Bocourt, Miss. Sc. Mex., Rept. p. 656, t. 41. figg. 2, 2 *a–e*; Sumichrast, Bull. Soc. Zool. Fr. v. p. 181 (1880).
Tachymenis piceivittis, Günth. Ann. & Mag. N. H. 1872, ix. p. 20.
Tachymenis tæniata, Peters, MB. Ak. Wiss. Berl. 1869 (December), p. 876.

Hab. MEXICO, Tehuantepec (*Sumichrast*), Yucatan (*Mus. Brit.*).

Scales in twenty-five rows. Upper labials eight, the fourth and fifth entering the orbit. A small second præocular below the large one. Upper parts and upper half of the sides black, with a narrow white band on each side of the back, commencing from the canthus rostralis and ending on the tail. Lower parts and lower half of the sides white.

The resemblance in coloration of this snake to the aglyphodont *Dromicus fulvivittis* (*anteà* p. 113) is so extraordinary that without examination of the teeth and number of scales the two forms would be readily mistaken for each other. It would be of interest to know whether the two live together in the same districts. At present we know the aglyphodont form only from Guerrero and Orizaba.

2. Tachymenis bipunctata.

Coronella bipunctata, Günth. Cat. Col. Sn. p. 36.

Coniophanes bipunctatus, Cope, Proc. Ac. N. Sc. Phil. 1866, p. 128; Bocourt, Miss. Sc. Mex., Rept. p. 653, t. 40. figg. 8, 8 *a–c*.

Erythrolamprus bipunctatus, Cope, Bull. U. S. Nat. Mus. no. 32, 1887, p. 77.

Glaphyrophis pictus, Jan, Arch. per la Zool. ii. p. 304; Iconogr. Ophid. xviii. t. 5. fig. 4 (see Günth. Zool. Record, 1866, p. 126).

Hab. MEXICO, Orizaba, Vera Cruz, and Tehuantepec (*Sumichrast*); BRITISH HONDURAS, Belize (*Godman*), Stann Creek (*J. Robertson*).

Scales in twenty-one rows; upper labials eight, the fourth and fifth entering the orbit. Anterior frontals square, truncated in front; posterior together larger than the vertical. Along each side of the abdomen and lower part of the tail a perfectly regular series of round, deep-black spots or (in old specimens) ocelli.

3. Tachymenis fissidens.

Coronella fissidens, Günth. Cat. Col. Sn. p. 36 (1858).

Coniophanes fissidens, Cope, Proc. Ac. N. Sc. Phil. 1860, p. 248; Journ. Ac. N. Sc. Phil. viii. p. 138 (1875); Bocourt, Miss. Sc. Mex., Rept. p. 650, t. 41. figg. 3, 3 *a–d*; Sumichrast, Bull. Soc. Zool. Fr. v. p. 181 (1880).

Tachymenis fissidens, Garman, Rept. N. Amer. 1883, p. 60.

Erythrolamprus fissidens, Cope, Bull. U. S. Nat. Mus. no. 32, 1887, p. 77.

Tæniophis imperialis, Baird, U. S. & Mex. Bound. Surv., Rept. p. 23, t. 19. fig. 1 (1859).

Erythrolamprus imperialis, Cope, Bull. U. S. Nat. Mus. no. 32, 1887, p. 77.

Coniophanes punctigularis, Cope, Proc. Ac. N. Sc. Phil. 1860, p. 248; Journ. Ac. N. Sc. Phil. viii. p. 137 (1875).

Erythrolamprus punctigularis, Cope, Bull. U. S. Nat. Mus. no. 32, 1887, p. 77.

Coniophanes proterops, Cope, Proc. Ac. N. Sc. Phil. 1860, p. 249; Journ. Ac. N. Sc. Phil. viii. p. 138 (1875); Bocourt, Miss. Sc. Mex., Rept. p. 654.

Erythrolamprus proterops, Cope, Bull. U. S. Nat. Mus. no. 32, 1887, p. 77.

Glaphyrophis lateralis, Jan, Arch. per la Zool. ii. p. 304; Iconogr. Ophid. xviii. t. 5. figg. 3, 3 *b* (see Günth. Zool. Record, 1866, p. 126).

Dromicus chitalonensis, Müll. Verh. naturf. Ges. Basel, vi. p. 407 (1878).

Erythrolamprus violaceus, Cope, Bull. U. S. Nat. Mus. no. 32, 1887, p. 77 *.

Hab. NORTH AMERICA, Texas.—MEXICO, Tamaulipas (*Smithson. Inst.*), Tampico (*H. H. Smith*), Hacienda del Hobo (*Geddes*), Playa Vicente (*Sallé*), Jalapa (*Cope*), Vera Cruz, Jicaltepec, Mirador, Arroyo del Potrero (*Cope*), Orizaba (*Sumichrast*), Teapa in Tabasco (*H. H. Smith*), Tehuantepec (*Sumichrast*), Chiapas (*Cope*); BRITISH

* This nominal species is introduced by Mr. Cope seemingly on the authority of Hallowell; however, on referring to the original description in Proc. Ac. N. Sc. Phil. 1860, p. 485, it proves to stand under the heading "*Coniophanes fissidens,* nobis."

HONDURAS, Belize (*Godman*), Stann Creek (*J. Robertson*); GUATEMALA, Vera Paz, low forest (*Salvin*), Coban (*Smithson. Inst.*), Pacific coast (*Salvin*), Chitalon, near Mazatenango (*Müller*); NICARAGUA, Matagalpa 3250 feet (*Rothschuh*); COSTA RICA (*Gabb*).

Scales in nineteen or twenty-one rows. Upper labials eight (very rarely seven), the fourth and fifth entering the orbit; one præocular. Back brownish or greyish-olive in various shades, sides darker, the two colours sharply defined from each other, and sometimes separated by a whitish or white band-like shade or band. Back with or without a brown or black vertebral line or band. The various colours may assume a distribution in real bands running from the neck to the tip of the tail. The upper part of the head is encircled by a more or less distinct or complete yellow line, which runs from the upper part of the temple along the canthus, and across the rostral to the other side, and may be confluent with or separated from a pair of yellow spots occupying the neck. These spots may be isolated or form the commencement of a lateral band. Upper labials longitudinally traversed by a yellowish band. Lower parts whitish, with irregularly scattered black dots, more constant on the throat than farther behind; rarely crowded together on the sides of the ventral scutes.

Feeds on frogs.

4. **Tachymenis lateritia.**

Coniophanes lateritius, Cope, Proc. Ac. N. Sc. Phil. 1862, p. 524.
Tachymenis melanocephala, Peters, MB. Ak. Wiss. Berl. 1869 (1870), p. 876.

Hab. MEXICO, Guadalajara.

Scales in nineteen rows. Upper labials seven, the third and fourth entering the orbit; vertical nearly as broad as long. Body bright vermilion, powdered with brown; yellow below. Head and neck, to the distance of ten scales, black. Throat black-spotted.

5. **Tachymenis grammophrys.**

Erythrolamprus grammophrys, Dugès, Proc. Am. Phil. Soc. 1888, p. 181, fig.; La Naturaleza, ser. 2, i. p. 402, t. 27. fig. 13 (1890).

Hab. MEXICO, Tengohecho in Michoacan.

Scales in seventeen rows. Seven upper labials, of which the third and fourth enter the orbit; one præocular; temporal in contact with postoculars; ventrals 175. Back brown, sides reddish, not banded; lower parts yellowish, without spots. An interrupted white line along each side of the upper surface of the head, commencing on the canthus. Upper labials longitudinally traversed by a white line.

6. **Tachymenis decipiens**, sp. n. (Tab. LIII. fig. A.)

Hab. Costa Rica, Irazu (*Rogers*).

Scales in seventeen rows. Upper labials seven, the third and fourth entering the orbit; one præocular; temporal not in contact with postoculars; ventrals 165–173. Upper parts dark greyish-brown; two yellow lines run along each side of the body and tail. Lower parts uniform yellowish.

Rather slender in habit, the tail being long, about one-half of the length of the body. Snout rather short; eye of moderate size. Anterior frontals very short and small. Vertical shorter than the occipitals, but as long as or longer than the frontals; one præocular; one or two postoculars. Temporals 1+2, the anterior widely separated from the postoculars, the sixth labial forming a suture with the occipital. Upper parts dark greyish-brown, with two narrow yellowish lines on each side, of which the lower runs along the outer series of scales; the upper takes its origin from a spot on each side of the neck, and is continued as a series of spots, of which one occupies the antero-lateral angle of the occipital, and the others the frontal shields. Upper labials brownish-black in the upper, and yellowish in the lower half.

The largest specimen is 19 inches long without the tail, which is mutilated.

As mentioned already on p. 105, this snake bears the closest resemblance to the aglyphodont *Ablabes decipiens.* Mr. Boulenger considers the latter to be synonymous with *Liophis lateristriga* (Berthold)=*Urotheca lateristriga*, Cat. Sn. ii. p. 181. The resemblance of the two snakes is certainly very great indeed, but the dental character of the genus *Urotheca*—" maxillary teeth 11 to 14, increasing in size posteriorly, followed after a short interspace by two enlarged ones"—does not apply to the types of *A. decipiens* from Costa Rica, as far as the hindmost teeth are concerned; these teeth are not enlarged, and not separated from the preceding by a larger interspace. I cannot describe such a dentition otherwise than isodont. But Mr. Boulenger is quite right in referring another specimen, from Ecuador, to *Urotheca*, which possesses a distinctly diacranterian dentition.

Thus, in spite of the great outward similarity of *Urotheca lateristriga* and *Ablabes decipiens*, I am inclined to regard them as distinct; and in this I am confirmed by the number of præoculars, which in the former is increased and may be as high as three, while *A. decipiens* has one only, sometimes with an additional small subocular. Also the yellow cross-band on the occipitals, which is so characteristic of *A. decipiens*, is absent in *U. lateristriga*.

Since we know of the existence in Central America of two extremely similar snakes, one with an isodont (*A. decipiens*), the other (*Tachymenis decipiens*) with a glyphodont dentition, the presence of a third (*Urotheca lateristriga*) in Ecuador and Colombia with diacranterian teeth is not surprising.

HYDROCALAMUS.

Hydrocalamus, Cope, Proc. Am. Phil. Soc. xxii. p. 177 (1885).

1. Hydrocalamus quinquevittatus.

Homalopsis quinquevittatus, Dum. & Bibr. Erpét. vii. p. 975.
Calopisma quinquevittatum, Jan, Iconogr. Ophid. xxx. t. 2.
Hydrocalamus quinquevittatus, Cope, Proc. Am. Phil. Soc. xxii. p. 176 (1885).
Hydrops lubricus, Cope, Proc. Ac. N. Sc. Phil. 1871, p. 217.

Hab. MEXICO, Coatzacoalcos R., Vera Cruz; GUATEMALA, Peten (*Smithson. Inst.*; *Mus. Paris*).

In spite of the extraordinary similarity of this snake to *Tachymenis bipunctata*, I think it advisable to retain it in a distinct genus. Its physiognomy is entirely distinct from the Coronelline type, and is that of a Freshwater Snake—a fact sufficiently recognized by previous authors who have associated this species with Freshwater genera. This feature consists of a short and depressed snout, small eye, and small frontal shields (the anterior of which are pointed, the posterior together being about equal in extent to the vertical). The occipitals are narrow and elongate. The nasal seems to be undivided.

TOMODON.

Tomodon, sp.*, Dum. & Bibr. Erpét. vii. p. 932 (1854).
Manolepis, Cope, Proc. Am. Phil. Soc. xxii. p. 176 (1885).

Differing from *Tachymenis* in possessing a large and long grooved tooth. Nasal single; anal divided.

1. Tomodon nasutus.

Tomodon nasutus, Cope, Proc. Ac. N. Sc. Phil. 1864, p. 166; Sumichrast, Arch. Bibl. Univ. 1873, p. 262, and Bull. Soc. Zool. Fr. v. p. 182 (1880).
Monolepis nasutus, Cope, Proc. Am. Phil. Soc. xxii. p. 176 (1885).

Hab. MEXICO, Colima (*Xantus*), Tehuantepec (*Sumichrast*).

Scales in nineteen series; upper labials eight. Snout projecting, obliquely truncated. Brownish, with a blackish-brown band from the neck to the extremity of the tail.

I have not seen a specimen of this snake.

CONOPHIS.

Conophis, Peters, MB. Ak. Wiss. Berl. 1860, p. 519.
Tomodon, sp., Dum. & Bibr. Erpét. vii. p. 932.

---

* *Tomodon strigatus*, Günth., described from a specimen labelled "India," then believed to come from Nicaragua, seems to be, in fact, a South-American snake. (See Boulenger, Ann. & Mag. N. H. 1886, xviii. p. 437.)

Differing from *Tachymenis* in possessing a large and long grooved tooth. Nasal as well as anal divided.

1. **Conophis lineatus.**

Tomodon lineatum, descr. pars, Dum. & Bibr. Erpét. vii. p. 936.

Eudrome à flancs linéolés, Dum. & Bibr. loc. cit. t. 73.

Psammophis lineatus, Günth. Cat. Col. Sn. p. 135.

Tomodon lineatus, Salv. P. Z. S. 1860, p. 455; Jan, Iconogr. Ophid. xix. t. 6. fig. 3.

Conophis lineatus, Cope, Journ. Ac. N. Sc. Phil. viii. p. 137 (1875).

Conophis concolor, Cope, Proc. Ac. N. Sc. Phil. 1866, p. 318.

Conophis pulcher, Cope, Proc. Ac. N. Sc. Phil. (1868) 1869, p. 308; Bocourt, Miss. Sc. Mex., Rept. p. 646, t. 28. figg. 6, 6 *a–g*.

Hab. MEXICO, Yucatan (*Mus. Brit.*); GUATEMALA, Dueñas (*Salvin*), Vera Paz and Chiapas (*Smithson. Inst.*); NICARAGUA (*Cope*); COSTA RICA, Cartago (*Mus. Brit.*), San José (*van Patten*).

Scales in nineteen rows; upper labials eight. Body longitudinally banded with black; three of the bands take their origin from the snout, breaking up on the body into continuous lines or series of spots; sometimes the vertebral band dissolved into two pairs of black lines, sometimes an additional band along the side. The bands may disappear altogether, with the exception of traces on the head.

Duméril and Bibron have evidently confounded in their description forms with seven and eight labials; but as they have figured a specimen with eight labials, I have taken this as the type of their species. The arrangement of the bands, particularly the width of the vertebral band, is subject to infinite variation.

2. **Conophis vittatus.**

Tomodon lineatum, descr. pars, Dum. & Bibr. Erpét. vii. p. 936.

Conophis lineatus, Bocourt, Miss. Sc. Mex., Rept. p. 643, t. 38. figg. 5, 5 *a–e*.

Conophis vittatus, Peters, MB. Ak. Wiss. Berl. 1860, p. 519; Bocourt, Miss. Sc. Mex., Rept. p. 644, t. 38. figg. 7, 7 *a–e*.

Conophis sumichrastii, Cope, Journ. Ac. N. Sc. Phil. viii. p. 137 (1875); Sumichrast, Bull. Soc. Zool. Fr. v. p. 182 (1880).

Hab. MEXICO, Tepetlapa in Guerrero (*H. H. Smith*), Guadalajara (*Cope*), Santo Domingo de Guzman (*A. C. Buller*), Oaxaca (*Mus. Paris*), Tehuantepec (*Sumichrast*).

Scales in nineteen rows; upper labials seven. Body longitudinally banded with black, three bands taking their origin from the snout; in younger specimens the bands are uniformly black, whilst in older ones the vertebral, and sometimes the lateral band becomes lighter along the middle, the black colour being reduced to more or less narrow margins. The black bands may appear as parallel straight lines four in number.

PHILODRYAS.

Philodryas, Wagl. Nat. Syst. Amph. p. 185 (1830) ; Günth. Cat. Col. Sn. p. 123.
Dryophylax, Dum. & Bibr. Erpét. vii. p. 1103.

1. Philodryas putnami.

Dromicus putnami, Jan, Iconogr. Ophid. xxiv. t. 6. fig. 3.

Hab. Mexico, Jalisco (*Jan*), La Cumbre de los Arrastrados, alt. 8500 feet (*A. C. Buller*).

This snake proves to be opisthoglyph.

ERYTHROLAMPRUS.

Erythrolamprus, Boie, Isis, xix. p. 981 (1826).

Body of moderate dimensions, subcylindrical. Head distinct from neck; eye of moderate size, with the pupil round. Two nasals; a loreal. Scales smooth, in fifteen rows. Anal divided. Posterior maxillary tooth grooved.

1. Erythrolamprus æsculapii.

Coluber æsculapii, Linn. Mus. Ad. Fred. p. 29, t. 11. fig. 2.
Erythrolamprus æsculapii, Boie, Isis, xx. p. 538 (1827) ; Dum. & Bibr. Erpét. vii. p. 845.
Coluber venustissimus, Wied, Reise Brasil. ii. p. 75 ; and Abbild. Nat. Bras., Rept. t. 39.
Erythrolamprus venustissimus, Boie, Isis, xix. p. 981 (1826) ; Dum. & Bibr. Erpét. vii. p. 851, t. 74 ; Günth. Cat. Col. Sn. p. 47 ; Cope, Journ. Ac. N. Sc. Phil. viii. p. 141 (1875) ; Bocourt, Miss. Sc. Mex., Rept. p. 658, t. 38. figg. 4, 4 *a–e*.
Coronella venustissima, Schleg. Ess. ii. p. 53.
Erythrolamprus intricatus, Dum. & Bibr. Erpét. vii. p. 855.

Hab. ? Mexico (*Boucard*); Nicaragua, Matagalpa (*Rothschuh*), Chontales Mines (*Rix*); Costa Rica, Irazu (*Rogers*); Panama, Chiriqui (*Champion*).—Tropical South America.

One ante-, two postoculars. Red, with black rings, the rings either single or subdivided, or in pairs.

I have suspected for some time that, in species normally opisthoglyphous, specimens may occur in which the groove of the posterior maxillary tooth is obsolete or entirely absent. Such is undoubtedly the case in the three specimens of *Erythrolamprus* from Nicaragua which I have examined. These specimens were collected in two distinct localities by different collectors.

OXYRRHOPUS.

Oxyrrhopus, Wagl. Syst. Amph. p. 185 (1830).
Cloelia sp. et *Erythrolamprus* sp., Wagl. Syst. Amph. p. 187 (1830).
Brachyruton et *Oxyrrhopus,* Dum. & Bibr. Erpét. vii. pp. 1002, 1011 (1854).
Oxyrhopus, Günth. Cat. Col. Sn. p. 188 (1858).

1. Oxyrrhopus plumbeus.

Coluber plumbeus, Wied, Reise Brasil. i. p. 95, and Abbild. xiii. t. 6; Schleg. Ess. ii. p. 152, t. 6. figg. 3, 4.

Brachyruton plumbeum, Dum. & Bibr. Erpét. vii. p. 1004; Jan, Iconogr. Ophid. xxxv. t. 1. fig. 3.

Oxyrhopus plumbeus, Günth. Cat. Col. Sn. p. 189; Cope, Journ. Ac. N. Sc. Phil. viii. p. 131 (1875), and Bull. U. S. Nat. Mus. no. 32, p. 76.

Hab. Mexico, Tehuantepec (*Cope*); Guatemala (*Mus. Brit.*); Costa Rica (*Gabb*).—Tropical South America.

2. Oxyrrhopus cloelia.

Coluber cloelia, Daud. Rept. vi. p. 330, t. 78.

Brachyruton cloelia, Dum. & Bibr. Erpét. vii. p. 1007; Jan, Iconogr. Ophid. xxxv. t. 1. fig. 1.

Oxyrhopus cloelia, Günth. Cat. Col. Sn. p. 189; Cope, Proc. Am. Phil. Soc. xvii. p. 92 (1878); Bull. U. S. Nat. Mus. no. 32, p. 76 (1887); Sumichrast, Bull. Soc. Zool. Fr. v. p. 184 (1880).

Scolecophis scytalinus, Cope, Proc. Ac. N. Sc. Phil. 1866, p. 320.

Hab. Mexico, Tehuantepec (*Sumichrast*), Yucatan (*Schott*); Panama, Veraguas and Isthmus of Darien (*Cope*).—Tropical South America.

3. Oxyrrhopus occipitoluteus.

Brachyruton occipitoluteum, Dum. & Bibr. Erpét. vii. p. 1009; Jan, Iconogr. Ophid. xxxv. t. 1. fig. 2.

Oxyrhopus occipitoluteus, Cope, Bull. U. S. Nat. Mus. no. 32, p. 76 (1887).

Hab. Panama, Darien (*Cope*).

Not seen by me.

4. Oxyrrhopus petolarius.

Coluber petolarius, Linn. Mus. Ad. Fred. t. 9. fig. 2.

Oxyrhopus petolarius, Wagl. Syst. Amph. p. 185; Günth. Cat. Col. Sn. p. 190; Dum. & Bibr. Erpét. vii. p. 1033; Cope, Journ. Ac. N. Sc. Phil. viii. p. 132.

Lycodon petolarius, Schleg. Ess. ii. p. 122.

Oxyrhopus subpunctatus, Dum. & Bibr. Erpét. vii. p. 1016.

Oxyrhopus rhombifer, Dum. & Bibr. Erpét. vii. p. 1018; Jan, Iconogr. Ophid. xxxv. t. 5. fig. 2.

Oxyrhopus multifasciatus, Dum. & Bibr. Erpét. vii. p. 1019; Jan, Iconogr. Ophid. xxxv. t. 6. fig. 2.

Oxyrhopus bipræocularis, Dum. & Bibr. Erpét. vii. p. 1030.

Hab. Guatemala, Vera Paz, low forest (*Salvin*); Nicaragua, Chontales Mines (*Crowfoot*), Matagalpa, alt. 3250 feet (*Rothschuh*); Costa Rica (*Salvin*), Sipurio (*Cope*).—Tropical South America.

One of the characters by which Duméril and Bibron distinguish their *O. petolarius* is the extent of the præocular, which is said to be in contact with the vertical. This is really the case in the majority of specimens, but it is a by no means constant character; not only in South-American, but also in the few Central-American specimens which I have examined, and which evidently belong to the same species, the extent of the præocular varies.

As to *Oxyrhopus doliatus* of Dum. & Bibr., this species is characterized by black rings completely encircling the body. I have never seen an example with this coloration, and the specimens described by me under that name (Cat. Col. Sn. p. 192) belong to a widely different species, viz. *Coluber scolopax* of Klein or *Lycognathus scolopax* of Dum. & Bibr.

Scytale coronatum (Schn.) is introduced by Cope into the Central-American fauna on the authority of Sternberg, who brought a specimen from Panama (Bull. U. S. Nat. Mus. no. 32, p. 76).

LEPTODIRA.

Sibon *, Fitzinger.
Leptodeira (Fitzinger, n. n.), Günth. Cat. Col. Sn. p. 165 (1858).

Synopsis of the Central-American species.

I. Scales in 19 series.	
A. Occipital not in contact with the labials.	
1. Eight upper labials	1. *nigrofasciata.*
2. Seven upper labials	2. *pacifica.*
B. Occipital in contact with the sixth labial	3. *mystacina.*
II. Scales in 21 series.	
A. One præocular reaching the vertical; ventrals 194	4. *annulata.*
B. Two præoculars, the upper reaching the vertical.	
1. Collar not divided by a black line. Ventrals 189	5. *affinis.*
2. [Collar divided by a black line. Ventrals 192–211	9. *polysticta.*]
C. Præocular not reaching the vertical.	
1. Ventrals 180–193	6. *yucatanensis.*
2. Ventrals 165–166	7. *splendida.*
III. Scales in 23 series.	
A. Chin-shields subequal in length.	
1. A white or whitish collar, not interrupted in the middle; ventrals 168–176	8. *personata.*
2. Nape divided in the middle by a median dark line.	
a. Ventrals 192–211	9. *polysticta.*
b. Ventrals 165–170	10. *ocellata.*
B. Posterior chin-shields much longer than anterior; ventrals 188	11. *frenata.*
IV. Scales in 25 series	12. *rhombifera.*

* Linnæus named a snake *Coluber sibon*, which, however, does not seem to enter the present genus. Moreover, Fitzinger's definition of *Sibon* is extremely vague, and I prefer to retain the name which I chose for this genus in 1858.

1. Leptodira nigrofasciata.

Leptodira nigrofasciata, Günth. Ann. & Mag. N. H. 1868, i. p. 425.

Hab. NICARAGUA (*Seemann*).

Scales in nineteen rows. Anterior frontals nearly square, two-thirds the size of posterior; vertical as long as the snout, and a little shorter than the occipitals; loreal as high as long. One præocular, not quite reaching to the vertical; a minute lower præocular on one side of the head; one postocular; anterior temporal separating entirely the occipital from the upper labials. Eight upper labials, the third, fourth, and fifth entering the orbit. Temporals 1+2+3. Ventrals 174; subcaudals 74.

Body with sixteen broad black cross-bands, which are from three to four times broader than the interspaces of the whitish ground-colour, and confluent on the posterior part of the trunk. Head black above, separated by a white collar from the first black band; tail with about eight black bands. Lower parts whitish.

One specimen, 14 inches long, of which the tail takes 3½.

2. Leptodira pacifica.

Leptodira pacifica, Cope, Proc. Ac. N. Sc. Phil. (1868) 1869, p. 310.

Hab. MEXICO, Mazatlan (*Cope*), Presidio (*Forrer*).

Scales in nineteen rows. Anterior frontals nearly square, one-third the size of posterior; occipitals elongate, truncate behind; loreal as high as long. Two præ-, two postoculars, the upper much larger than the lower. Temporals 1+2, the anterior entirely separating the occipital from the labials. Seven upper labials, the third and fourth entering the orbit. Ventrals 164; subcaudals 61.

Bright rosy-brown, with four or more series of blackish-brown very small spots, of which the median pair may form short cross-bars by their union. A whitish spot on the nape, longitudinally divided by a black bar, and a number of small spots on the upperside of the head. Uniform cream-coloured below.

3. Leptodira mystacina.

Leptodira mystacina, Cope, Proc. Am. Phil. Soc. xi. p. 151 (1869) *.

Hab. MEXICO, Tehuantepec (*Sumichrast*).

Scales in nineteen rows. Anterior frontals nearly square, one-half the size of posterior; vertical as long as the snout, and a little shorter than the occipitals; loreal as high as long. Two præ-, and two postoculars, the lower small in both sets; the upper præocular in contact with the vertical. Temporals 1+2+3; the foremost temporal is separated from the postocular either by a minute intercalated scute, or by the sixth labial, which touches the occipital. Eight upper labials, the (third), fourth, and fifth entering the orbit. Ventrals 193–4, 187 (*Cope*).

* In Proc. U. S. Nat. Mus. xiv. 1892, Mr. Cope unites this form with *Leptodira nigrofasciata* (p. 678).

Trunk with about thirteen broad black cross-bands, twice as broad as the interspaces of the whitish ground-colour, more or less confluent or alternating on the hinder part of the trunk. Upper part of the head mottled with brown, separated by a pure white collar from the first black cross-band. A brown streak from the eye to the angle of the mouth. Lower parts white.

4. **Leptodira annulata.**

Coluber annulatus, Linn. Mus. Ad. Fred. p. 34, t. 8. fig. 2; Syst. Nat. i. p. 386.

Leptodeira annulata, part., Günth. Cat. Col. Sn. p. 166; Cope, Bull. U. S. Nat. Mus. no. 32, p. 67 (1887) (*Sibon annulatum*, part.).

Dipsas septentrionalis, Kennicott, U. S. Mex. Bound. Surv., Rept. p. 16, t. 8. fig. 1.

Hab. North America, Texas.—Northern Mexico, Matamoros (*Kennicott*).

Scales in twenty-one rows. Anterior frontals nearly square, half the size of posterior; vertical as long as the snout, and shorter than the occipitals; loreal about as high as long. One præocular, reaching the vertical; two postoculars. Eight upper labials, the third, fourth, and fifth entering the orbit. Temporals 1+2+3, the foremost in contact with the postoculars. Ventrals 194.

Whitish, with about twenty-four subrhombic transverse black spots on the back, broader than the intervening spaces of the ground-colour. A whitish unspotted collar. Upper part of the head with brown spots.

5. **Leptodira affinis,** sp. n.

Hab. Central America (*Mus. Brit.*); exact locality unknown.

Scales in twenty-one rows. Anterior frontals about one-third of posterior; vertical rather shorter than occipital and in contact with the upper præocular; loreal longer than deep. Præoculars two, the lower small; two postoculars. Eight upper labials, of which the fourth and fifth enter the orbit. Temporals 1+2+3. Posterior chin-shields longer than anterior. Ventrals 189.

Ground-colour reddish-olive. A series of transverse or irregular, rounded, brown spots edged with darker runs along the back; they descend some distance down the sides, and are about thirty-seven in number on the trunk. Another series of smaller and very numerous spots runs along the lower part of the side. The first dorsal spot is large and subquadrangular, truncated in front, leaving a collar not divided by a brown median line. Upperside of the head mottled with brown; a brown temporal band. Lower parts yellowish.

A single female specimen is 21 inches long; the tail, which is somewhat mutilated, would have measured about $2\frac{1}{2}$ inches.

6. **Leptodira yucatanensis.**

Leptodira annulata, var., Cope, Proc. Ac. N. Sc. Phil. 1866, p. 127.
Sibon annulatum, subsp. *yucatanensis*, Cope, Bull. U. S. Nat. Mus. no. 32, p. 67.

Hab. MEXICO, Yucatan (*Schott*), Cozumel I. (*Gaumer*).

Scales in twenty-one rows. Anterior frontals nearly square; loreal square. Two (one, *Cope*) præoculars, the upper not reaching the vertical; two postoculars. Eight upper labials; eye over the third, fourth, and fifth. Head broad; neck constricted. Ventrals 180–193.

Grey, with about twenty-five rounded brown spots edged with black, extending downwards on the sides, and sometimes partly confluent. A black band from the eye, across the angle of the mouth, to the first cross-band. A light collar, not divided by a band.

7. **Leptodira splendida,** sp. n. (Tab. LIII. fig. B.)

Hab. MEXICO, Izucar (*Mus. Brit.*).

Scales in twenty-one rows. Anterior frontals square, one-third of posterior; vertical scarcely as long as the snout, and shorter than the occipitals; loreal longer than high. Three præoculars, of which the upper is the largest, but does not reach the vertical; two postoculars, the lower small. Eight upper labials, the fourth and fifth entering the orbit. Temporals 1+2+3. Ventrals 165–166; subcaudals 76.

Body above with broad subquadrangular black transverse spots, about twenty on the trunk; the spots do not extend downwards to the abdominal scutes, and generally some of them coalesce so as to form a short zigzag band; they are separated by yellow or brownish-yellow dorsal spots, the sides of the body being grey. The collar is of the same yellow colour, but longitudinally divided by a black line. A black stripe from the eye to the first black cross-band. Upper parts of the head mottled with brown. Lower parts uniform yellowish.

Two specimens, of which the larger measures 26 inches, the tail measuring $6\frac{1}{2}$.

8. **Leptodira personata.** (Tab. LIV. figg. A and B, ad. and young.)

Leptodeira annulata, part., Günth. Cat. Col. Sn. p. 166.
Leptodira personata, Cope, Proc. Ac. N. Sc. Phil. (1868) 1869, p. 310.

Hab. MEXICO, Presidio (*Forrer*), Mazatlan (*Bischoff*), Santo Domingo de Guzman and Hacienda Santa Gertrudis (*A. C. Buller*), City of Mexico (*Geddes*), Jalapa (*Höge*).

Scales in twenty-three rows. Anterior frontals as long as wide, narrowed in front, about one-half of posterior. Vertical as long as, or shorter than, the snout, and shorter than the occipitals; loreal as long as high. Præoculars (one) two, of which the lower is very small, the upper reaches nearly the vertical; two postoculars. Eight upper labials, the (third) fourth and fifth entering the orbit. Temporals 1+2+3. Anterior and posterior chin-shields subequal in length. Ventrals 168–176.

A white or whitish collar, not divided in the middle by a blackish longitudinal streak. Ground-colour lighter than the rhombic spots on the back, which on the trunk are about twenty-five in number. In young specimens the ground-colour is almost white and the spots are nearly black, but with age the contrast of colours is less striking. The upper part of the head, again, is black or brown. Lower parts whitish.

A common species.

9. **Leptodira polysticta,** sp. n. (Tab. LV. fig. A.)

Leptodeira annulata, part., Günth. Cat. Col. Sn. p. 166.

Hab. Mexico, Jalapa (*Höge*), Oaxaca (*Sallé*), Yucatan (*Mus. Brit.*); British Honduras, Belize (*Godman*); Honduras (*Dyson*); Panama (*Mus. Brit.*).

Scales in twenty-three rows; in twenty-one in a specimen from Yucatan. Anterior frontals one-half or more of the size of the posterior. Vertical as long as, or shorter than, the snout, and shorter than the occipitals; loreal as long as high. Præoculars two or three, of which the upper is the largest, and reaches or nearly reaches the vertical; two postoculars. Eight upper labials, of which the fourth and fifth enter the orbit. Temporals 1+2+3. Anterior and posterior chin-shields subequal in length. Ventrals 192–211.

A whitish collar is not always distinct, but whether present or not, a brown or black longitudinal line divides the neck. Back with a series of numerous small transverse bars, not extending to the ventrals; a series of very small spots along each side of the body.

10. **Leptodira ocellata,** sp. n. (Tab. LV. fig. B.)

Hab. Nicaragua, Chontales Mines (*Rix*); Costa Rica (*Mus. Brit.*).

Scales in twenty-three rows. Anterior frontals about one-third of posterior; vertical rather shorter than occipital; loreal large, square. Præoculars two, the lower very small, the upper reaching the vertical or nearly so; sometimes a detached portion of a labial forms a third præocular; two postoculars. Eight upper labials, of which the fourth and fifth enter the orbit. Temporals 1+2+3. Anterior and posterior chin-shields equal in length. Ventrals 165–170.

Ground-colour reddish- or brownish-olive; a series of dark brown transverse or irregularly-rounded spots along the back, which do not descend to the sides, and are about thirty-eight in number on the trunk. Sides with a series of round brown spots, the spots being opposite to the interspaces of the ground-colour and accompanied by smaller irregular spots. Neck without a distinct collar, and longitudinally divided in the middle by a brown narrow line proceeding from the first dorsal spot.

Upperside of the head mottled with brown; a brown temporal band. Lower parts yellowish.

The largest of four specimens is 22 inches long, the tail measuring $5\frac{1}{2}$.

11. **Leptodira frenata.**

Sibon frenatum, Cope, Proc. U. S. Nat. Mus. ix. p. 184 (1887).

Hab. MEXICO, Jalapa.

Scales in twenty-three rows. Body rather slender; head very distinct and depressed. Loreal nearly square. Præoculars two, the upper not reaching the vertical; two postoculars. Nine upper labials; eye resting on the fourth and fifth, and separated from the third by the small inferior præocular. Temporals 1+2+3. Posterior chin-shields much longer than anterior. Ventrals 188.

The coloration is described as follows:—Above black, below white. At distances of from six to nine scales, narrow cross-bands of one scale in width rise from the abdominal border-colour, and meet or terminate in alternating positions; on or near the middle line of the back these bands are more or less grey, sometimes darker in the middle. The top of the head is grey, densely mottled with blackish, having a concentric space of light grey between a black spot behind the head-shields and the beginning of the black of the superior surfaces. A broad black band passes downwards and posteriorly from the eye, and, crossing the angle of the mouth, covers the side of the neck and unites with the black of the following regions.

The typical specimen seems to have been very young, only 12 inches long.

12. **Leptodira rhombifera.** (Tab. LIV. fig. C.)

Leptodira rhombifera, Günth. Ann. & Mag. N. H. 1872, ix. p. 32.

Hab. GUATEMALA, Rio Chisoy near Cubulco (*Salvin*).

Scales in twenty-five series. Anterior frontals very small; vertical as long as snout; loreal longer than deep. Præocular single, barely reaching the vertical; two postoculars. Eight upper labials, of which the fourth and fifth enter the orbit. Temporals 1+2+3. Chin-shields rather elongate, equal in length. Body rather slender; head broad and depressed. Ventrals 170.

Brownish. Trunk with about twenty-six large, subrhombic, dark brown spots edged with black; yellowish cross-bands, brightest on the median line, separate these rhombic spots from one another. Upper part of the head brown, powdered with black. A black stripe with a yellowish margin on each side connects the crown of the head with the first rhombic spot. Abdomen yellowish; subcaudals powdered with brown.

One specimen, 23 inches long, the tail measuring 5.

TRIMORPHODON.

Dipsadomorphus, sp., Fitz. Syst. Rept. p. 27; Günth. Cat. Col. Sn. p. 174.
Trimorphodon, Cope, Proc. Ac. N. Sc. Phil. (1861) 1862, p. 297.

Upon a material consisting of fourteen Mexican specimens, Mr. Cope established, beside the Dumerilian *Dipsas biscutatus*, and the extralimital *Lycodon lyrophanes* from California, five other species as follows:—

Trimorphodon tau, Cope, Proc. Am. Phil. Soc. 1869, p. 152, from Tehuantepec.

Trimorphodon upsilon, Cope. Proc. Am. Phil. Soc. 1869, p. 152, from Chihuahua, Guanajuato, Zacualtipan, and Guadalajara.

Trimorphodon collaris, Cope, Journ. Ac. Nat. Sc. Phil. viii. p. 131 (1875), from Orizaba.

Trimorphodon wilkinsonii, Cope, Proc. Am. Phil. Soc. 1886, p. 285, from Chihuahua.

Trimorphodon lambda, Cope, Proc. Am. Phil. Soc. 1886, p. 286, from Guaymas.

Mr. Cope has given several synopses of these species (the most recent in Proc. Am. Phil. Soc. 1886, p. 286) showing the characters by which he has arranged his specimens, and which I have attempted to apply to the material before me (sixteen specimens). However, I have found these characters to break down completely, so that scarcely one of my specimens would correspond with any of Mr. Cope's descriptions, and the majority would have had to be described under new names. Therefore I am rather inclined to think that these snakes are subject to much individual variation in scales, scutellation, and colours. The latter, especially, change with growth.

Among our sixteen specimens I distinguish, beside *T. biscutatus*, only one other form, for which I retain one of the Copean names—*T. upsilon*—as the one which has been used by Cope for specimens from various localities.

1. **Trimorphodon biscutatus.**

Dipsas biscutata, Dum. & Bibr. Erpét. vii. p. 1153.
Dipsadomorphus biscutatus, Günth. Cat. Col. Sn. p. 176; Salv. P. Z. S. 1860, p. 457; 1861, p. 228.
Dipsas biscutata, var. *latifascia*, Peters, MB. Ak. Wiss. Berl. 1869, p. 877.
Trimorphodon biscutatus, Cope, Proc. Am. Phil. Soc. 1886, p. 286.
Trimorphodon major, Cope, Proc. Am. Phil. Soc. 1869, p. 153; Sumichrast, Bibl. Univers. 1873, p. 253.
? *Trimorphodon lambda*, Cope, Proc. Am. Phil. Soc. 1886, p. 286.

Hab. Mexico, Presidio, Mazatlan (*Forrer*), Colima (*Xantus*), Puebla (*Peters*), Guadalajara (*Cope*), Oaxaca (*Sallé*), Tehuantepec (*Sumichrast*); Guatemala, Lanquin (*Salvin*); Nicaragua (*Dow*); Panama (*Dow*).

More or less regular concentric brown chevron-bands across the head and neck. Scales in 24–27 series. Upper labials generally nine.

According to Cope, this snake is known by the name of "culebra casera" in Tehuantepec.

2. Trimorphodon upsilon.

Trimorphodon upsilon, Cope, Proc. Am. Phil. Soc. 1869, p. 152.
Eteirodipsas biscutata, Jan, Iconogr. Ophid. xxxix. t. 1. fig. 3.

Hab. MEXICO, Chihuahua, Guadalajara, Zacualtipan (*Cope*), Ventanas in Durango (*Forrer*), Guanajuato (*Dugès*), La Cumbre de los Arrastrados (*A. C. Buller*), Jalapa (*Höge*).

An uninterrupted white or whitish collar across the neck. Scales in 23–24 series. Upper labials generally eight.

DIPSAS *.

Dipsas, sp., Boie, Isis, 1827, p. 549.

1. Dipsas cenchoa.

Coluber cenchoa, Linn. Syst. Nat. i. p. 389.
Dipsas cenchoa, Wied, Beitr. Reise Brasil. i. p. 396; Günth. Cat. Col. Sn. p. 174; Salv. P. Z. S. 1860, p. 457; Cope, Bull. U. S. Nat. Mus. no. 32, p. 68.
Imantodes cenchoa, Dum. & Bibr. Erpét. vii. p. 1065.
Dipsas weigelii (Fitz.), Schleg. Ess. ii. p. 278, t. 11. figg. 19, 20.
Himantodes leucomelas, Cope, Proc. Ac. N. Sc. Phil. 1861, p. 296.

Hab. MEXICO, Jalapa and Huatusco in Vera Cruz (*Godman*), Teapa in Tabasco (*H. H. Smith*); GUATEMALA, Coban (*Salvin*); COSTA RICA, Irazu (*Rogers*); PANAMA (*Mus. Brit.*), Chiriqui (*Champion*).—TROPICAL SOUTH AMERICA.

The specimens referred by me to this species vary in their scutellation and coloration. Males seem to have larger eyes than females. The præorbitals, variable in number, may or may not touch the vertical; the upper labials are eight or nine in number; frequently two temporals are in contact with the postoculars, but almost as frequently these two anterior temporals coalesce into a single shield. Thus the great width of the median dorsal scales remains the principal distinctive character of this species.

2. Dipsas gemmistrata.

Dipsas cenchoa, Cope, Proc. Ac. N. Sc. Phil. 1860, p. 264 (nec Linn.).
Himantodes cenchoa, var. *elegans,* Jan, Iconogr. Ophid. xxxviii. t. 2.
Himantodes gemmistratus, Cope, Proc. Ac. N. Sc. Phil. 1861, p. 296.
Dipsas gemmistrata, Cope, Bull. U. S. Nat. Mus. no. 32, p. 68.

---

* I strongly suspect that *Leptognathus stratissima,* Cope, Proc. Am. Phil. Soc. xxiii. p. 280 (1885), from Panama, will prove to be a species of this genus.

Hab. MEXICO, Toluca, Guadalajara (*Cope*), Tehuantepec (*Sumichrast*); GUATEMALA, Pacific Coast (*Salvin*); SALVADOR (*Dow*); COSTA RICA, Cartago (*Mus. Brit.*).

The scales of the median dorsal series are not broader than long or even longer than broad. In the scutellation of the head this species varies as much as *Dipsas cenchoa*, but the variations do not coincide with the geographical distribution of the species. In the type-specimens the spots on the body numbered about forty-two, and were connected by a median dorsal vitta.

A variety is stated to have the very numerous dorsal spots extending entirely across the back and sides.

In our specimens the spots are likewise exceedingly numerous, disposed in three series—one of larger ones of rhombic shape along the back, the two others running along the lower part of the sides. The spots have a light centre, scarcely darker than the ground-colour, and dark brown edges.

3. **Dipsas splendida,** sp. n. (Tab. LVI. fig. A.)

Hab. MEXICO, Yucatan (*Mus. Brit.*).

The scales of the median dorsal series are enlarged, but longer than broad; scales in seventeen series. Habit less slender than *Dipsas cenchoa*; head small; occipitals but little longer than vertical; loreal longer than deep; one anterior ocular, not reaching the vertical; two postoculars; eight upper labials, the third, fourth, and fifth entering the orbit; temporals 1+2+3; posterior chin-shields about as long as anterior; ventrals 223. Ground-colour of the body brownish grey; body with about thirty-five brownish-black rhombic spots, which are produced in the median line forward and backward so as to nearly touch the preceding and succeeding; laterally each spot is continued downwards to the ventral scutes; each spot has very clear edges of white; the ground-colour forms an uninterrupted collar on the neck.

One female specimen, 28 inches long, of which the tail takes 7.

4. **Dipsas tenuissima.**

Himantodes tenuissimus, Cope, Proc. Ac. N. Sc. Phil. (1866) 1867, p. 317.

Hab. MEXICO, Hacienda Santa Gertrudis (*A. C. Buller*), Yucatan (*Mus. Brit.*).

The scales of the median dorsal series are small like the others, which are arranged in seventeen series; anteoculars two or one, postoculars two; upper labials eight, the fourth and fifth, and sometimes the third, entering the orbit. Temporals 1 or 2+3. Body exceedingly slender and compressed. Ventrals 250; subcaudals 157. Whitish, with about fifty transverse black light-edged spots on the body, which approach closely on the median line; about thirty-five spots on the tail.

5. **Dipsas gracillima,** sp. n. (Tab. LVI. fig. B.)

Hab. SOUTHERN MEXICO (*Godman*).

Body very slender and compressed. Scales in seventeen rows, those of the vertebral line not enlarged; ventrals 244; subcaudals about 110.

Head moderately broad; loreal a little longer than deep; one anteocular, which does not reach the vertical; two postoculars; eight upper labials, the third, fourth, and fifth entering the orbit; temporals 1+2+3.

Light brownish, with about seventy-four brown cross-bands edged with darker, and about twice as broad as the interspaces between them. The spots on the tail are more indistinct and irregular.

The length of the entire specimen is 21½ inches, of which the tail takes 5½.

DRYIOPHIS.

Dryiophis, sp. (Boie), Fitzinger, N. Class. Rept. 1826, pp. 29, 30, 60.
Tragops et *Oxybelis*, Duméril & Bibron, Erpét. vii. pp. 813, 822.

1. **Dryiophis acuminata.**

Coluber acuminatus, Wied, Abbild. Lief. 14, t. 1.
Dryiophis acuminata, Günth. Cat. Col. Sn. p. 156; Cope, Proc. Ac. N. Sc. Phil. 1861, p. 555; Cope, Bull. U. S. Nat. Mus. no. 32, p. 68.
Dryinus æneus, Wagl. in Spix, Serp. Brasil. p. 12, t. 3.
Oxybelis æneus, Wagl. Syst. Amph. p. 183; Dum. & Bibr. Erpét. vii. p. 819.
Dryinus auratus, Bell, Zool. Journ. 1825, p. 329, t. 12.
Dryiophis aurata, Schleg. Ess. ii. p. 255, t. 10. figg. 16–18.
Dryiophis vittatus, Girard in U. S. Naval Astron. Exped., Rept. p. 211, t. 36.

Hab. MEXICO, Tres Marias Is., Presidio (*Forrer*), Guaymas, Tuxpan, and Tabasco (*Cope*), Sarabia and Santo de Domingo de Guzman (*A. C. Buller*), Tepetlapa in Guerrero (*H. H. Smith*), Coatzocoalcos R. (*U. S. Nat. Mus.*), Tehuantepec (*Sumichrast*), Yucatan (*Mus. Brit.*), Bonacca I. (*Gaumer*); GUATEMALA, Rio Motagua (*Salvin*); HONDURAS (*Mus. Brit.*), Ruatan I. (*Gaumer*); SALVADOR (*Dow*); NICARAGUA (*U. S. Nat. Mus.*); COSTA RICA, Irazu (*Rogers*); PANAMA, Veraguas and Taboga I. (*U. S. Nat. Mus.*).—SOUTH AMERICA to Southern Brazil.

Specimens of this widely spread and common species vary particularly in the extent of the præocular, which frequently reaches the vertical, and in the comparative length of the vertical and occipital shields. It ranges from the Mexican State of Sonora to Southern Brazil.

2. **Dryiophis fulgida.**

Coluber fulgidus, Daud. Hist. Nat. Rept. vi. p. 532, t. 80.
Dryiophis fulgidus, Boie, Isis, 1827, p. 546; Wagl. Ic. t. 10; Günth. Cat. Col. Sn. p. 158; Cope, Bull. U. S. Nat. Mus. no. 32, p. 68; Ives, Proc. Ac. N. Sc. Phil. (1891) 1892, p. 460.
Oxybelis fulgidus, Dum. & Bibr. Erpét. vii. p. 817.
Dryiophis catesbyi, Schleg. Ess. ii. p. 252; Abbild. t. 14. figg. 3–6.

Hab. Mexico, Tehuantepec (*Sumichrast*), Yucatan (*Mus. Brit.*); British Honduras (*Mus. Brit.*); Salvador (*Dow*); Panama (*Dow*).—South America to Brazil.

The number of upper labial shields varies between nine and ten, the fourth, fifth, and sixth, or the fifth, sixth, and seventh entering the orbit.

3. **Dryiophis brevirostris.**

Dryiophis brevirostris, Cope, Proc. Ac. N. Sc. Phil. 1860, p. 555; Journ. Ac. N. Sc. Phil. viii. p. 132, t. 26. fig. 2 (1875).

Hab. Nicaragua, Chontales Mines (*Rix*); Costa Rica (*Smithson. Inst.*); Panama, Veraguas (*Cope*).

Addendum to page 140.

The following snake, which the British Museum has received quite recently, is interesting in several respects, so that it appears desirable to insert its description at the earliest opportunity.

NEOPAREAS, gen. nov.

Body slender, strongly compressed; tail slender; head broad, deep, and short; snout short; pupil of the eye vertical. Scales smooth, in fifteen rows, those of the vertebral series not enlarged; anal entire; subcaudals two-rowed. Upper labials narrow; chin-shields not divided in the middle, gradually passing into the ventral shields. Maxillary teeth small, non-grooved.

1. **Neopareas bicolor,** sp. n. (Tab. LVI. fig. C.)

Hab. Nicaragua, Chontales Mines (*Rix*).

Vertical shield broad and large, nearly as large as an occipital; anterior frontals much broader than long, one-third of posterior. A large square shield occupies the space between the nasal and the eye; a small anteocular above it which does not reach the vertical; three, sometimes four, postoculars; ten or eleven upper labials, of which the fourth, fifth, sixth, and seventh, or the third, fourth, fifth, and sixth enter the orbit; temporals $1+2+3+3$. The median lower labial is much broader than long; the lower labials of the first pair coalesce, forming one shield, which stretches from one side to the other behind the median labial. There are three undivided gular shields along the middle; the first is nearly round, the second sub-semicircular, the third of nearly the same shape as, but narrower than, the first ventral. Ventrals 195; subcaudals 129.

Ground-colour deep black, with white rings completely encircling the body—the first round the neck, fourteen round the trunk, and twelve round the tail; the white rings are only half as broad as the interspaces of the black ground-colour.

I know this extraordinary species from one specimen only, 21½ inches long, of which the tail takes 8.

Fam. NOTHOPSIDÆ.

NOTHOPSIS.

Nothopsis, Cope, Proc. Ac. N. Sc. Phil. 1871, p. 201.

1. Nothopsis rugosus.

Nothopsis rugosus, Cope, Proc. Ac. N. Sc. Phil. 1871, p. 201, t. 17. figg. 1–7.

Hab. PANAMA, Isthmus of Darien.

Fam. BOIDÆ.

LOXOCEMUS.

Loxocemus, Cope, Proc. Ac. N. Sc. Phil. 1861, pp. 76, 304; Günth. Zool. Rec. 1864, p. 123.
Plastoseryx, Jan, Wiegm. Arch. 1862, p. 242.

1. Loxocemus bicolor.

Loxocemus bicolor, Cope, Proc. Ac. N. Sc. Phil. 1861, p. 76; Günth. Ann. & Mag. N. Hist. 1862, ix. p. 55; Bocourt, Miss. Sc. Mex., Rept. p. 515, t. 30. figg. 5, 5 *a–c*; Boul. Cat. Sn. i. p. 74.
Plastoseryx bronni, Jan, Iconogr. Ophid. iii. t. 1.
Loxocemus bronni, Bocourt, Miss. Sc. Mex., Rept. p. 516, t. 30. figg. 6, 6 *a*, *b*.
Loxocemus sumichrasti, Bocourt, Journ. Zool. v. 1876, p. 344; Sumichrast, Bull. Soc. Zool. Fr. 1880, p. 180.

Hab. MEXICO, Colima (*Xantus*), Tehuantepec (*Sumichrast*); GUATEMALA (*Dow*).

EPICRATES.

Epicrates, Wagler, Syst. Amphib. p. 168 (1830).

1. **Epicrates cenchris.**

Boa cenchria, Linn. Syst. Nat. i. p. 374; Wied, Abbild. Naturg. Bras., Rept. t. 18; Schleg. Ess. ii. p. 385, t. 14. figg. 3–5.

Epicrates cenchria, Gray, Viper. Sn. p. 96; Cope, Journ. Ac. N. Sc. Phil. viii. p. 130 (1875).

Epicrates cenchris, Dum. & Bibr. Erpét. vii. p. 555; Jan, Iconogr. Ophid. viii. t. 1; Boul. Cat. Sn. i. p. 94.

Epicrates maurus, Gray, Viper. Sn. p. 94.

Cliftia fusca, Gray, Viper. Sn. p. 99.

Epicarsius cupreus, Fisch. Abhandl. naturf. Ver. Hamburg, iii. 1856, p. 96, t. 2. fig. 1.

Epicrates cupreus, Jan, Iconogr. Ophid. vii. t. 1.

Hab. Costa Rica (*Mus. Hamburg*); Panama, Chagres River (*Mus. Brit.*).—Tropical South America.

The variation in the number of rows of scales has a very great range, viz. from 39 to 51.

CORALLUS.

Corallus, Daudin, Hist. Rept. v. p. 256 (1803); Gray, Viper. Sn. p. 97.

1. **Corallus cooki.**

Corallus cookii, Gray, Zool. Misc. p. 42 (1842); Boul. Cat. Sn. i. p. 99.

Corallus hortulanus (part.), Gray, Viper. Sn. p. 97.

Diphosoma ruschenbergerii, Cope, Journ. Ac. N. Sc. Phil. viii. p. 129 (1876); Bocourt, Miss. Sc. Mex., Rept. p. 527.

Hab. Panama (*Ruschenberger*).—Colombia; Venezuela; Windward Is.

Like other Boid snakes, *Corallus hortulanus* and allied specific forms vary in an extraordinary manner as to the number of scales. If *C. cooki* were not separated from *C. hortulanus*, this species would include specimens with scales varying from 39 to 59, the increase in numbers being due to the small size of the scales on the sides of certain individuals. So far as the evidence from our specimens goes, the lesser number of rows of scales (39–47) is found in the snakes from the West Indies, Venezuela, Colombia, and Panama, whilst the larger number (53–59) occurs in specimens from Guiana and Brazil. There is a specimen from the Schomburgk collection in the British Museum, said to be from "British Guiana," having the scales in 43 series only. But as the collection which Sir R. Schomburgk gave to the British Museum contained not a few Reptiles from other localities, besides British Guiana, this specimen cannot be used as an argument against the above proposition. Under these circumstances I think it advisable to keep the large-scaled form separate, as one with a distinct geographical range. I attach no weight to other characters which have been adduced as indicating the specific distinctness of the two forms.

2. **Corallus annulatus.**

Xiphosoma annulatum, Cope, Journ. Ac. N. Sc. Phil. viii. p. 129, t. 28. fig. 6 (1876).
Corallus annulatus, Boul. Cat. Sn. i. p. 102.

Hab. Costa Rica (*Gabb*).

UNGALIA.

Ungalia, Gray, Zool. Misc. p. 46 (1842).

1. **Ungalia moreleti.**

Tropidophis moreletii, Bocourt, Bull. Soc. Philom. ix. p. 113 (1885).
Ungalia moreletii, Bocourt, Miss. Sc. Mex., Rept. t. 42. figg. 5, 5 *a, b.*

Hab. Guatemala, Vera Paz (*Mus. Paris*).

I have not seen an example of this snake.

UNGALIOPHIS.

Peropodum * g. n., Müller, Verh. nat. Ges. Basel, vi. p. 652 (1878).
Peropodum, Bocourt, Miss. Sc. Mex., Rept. p. 522.
Ungaliophis, Müller, Verh. nat. Ges. Basel, vii. p. 141 (1882).
Ungualiophis, Cope, Bull. U. S. Nat. Mus. no. 32, p. 56.

1. **Ungaliophis continentalis.**

Peropodum sp. n., Müll. Verh. nat. Ges. Basel, vi. p. 652, t. 1 (1878).
Peropodum guatemalensis, Bocourt, Miss. Sc. Mex., Rept. p. 523, t. 31. figg. 5, 5 *a, b* (cop. Müller).
Ungaliophis continentalis, Müll. Verh. nat. Ges. Basel, vii. p. 142 (1882).

Hab. Guatemala, Retalhuleu (*Bernoulli*).

BOA.

Boa, sp., Linnæus, Syst. Nat. i. p. 373 (1760).
Boa, Wagler, Syst. Amph. p. 169 (1830).

1. **Boa imperator.**

Boa constrictor (part.), Linn. Syst. Nat. i. p. 373 (1766); Schleg. Ess. ii. p. 373.
Boa constrictor, Günth. P. Z. S. 1860, p. 233; Salv. P. Z. S. 1861, p. 228.
Boa constrictor, var. *isthmica,* Garman, N.-Amer. Rept. p. 9.

---

* *i. e.,* new genus of the family Peropodes.

Boa imperator, Daud. Rept. v. p. 150; Dum. & Bibr. Erpét. vi. p. 519; Gray, Viper. Sn. p. 101; Peters, MB. Ak. Wiss. Berl. 1869, p. 875; Cope, Journ. Ac. N. Sc. Phil. viii. p. 130; Jan, Iconogr. Ophid. vi. t. 1; Bocourt, Miss. Sc. Mex., Rept. p. 519, t. 30. figg. 8, 8 *a–c*; Garman, N.-Amer. Rept. p. 8; Cope, Bull. U. S. Nat. Mus. no. 32, p. 65; Boul. Cat. Sn. i. p. 119; Sumichrast, Bull. Soc. Zool. Fr. 1880, p. 179.

Boa eques, Eyd. & Soul. Voy. Bonite, Zool. Rept. p. 144, t. 4; Dum. & Bibr. Erpét. vii. p. 521; Gray, Viper. Sn. p. 101; Jan, Iconogr. Ophid. vi. t. 2.

Hab. MEXICO, Ventanas in Durango (*Forrer*), Colima (*Xantus*), Atoyac in Vera Cruz (*H. H. Smith*), Vera Cruz (*Cope*), Tehuantepec (*Sumichrast*), Yucatan (*Schott*); BRITISH HONDURAS (*Godman, Dyson*); GUATEMALA, Vera Paz (*Owen*); NICARAGUA (*Cope*); COSTA RICA (*Godman*).—ECUADOR (*Fraser*).

To describe this form as *Boa constrictor*, var., or even simply as *Boa constrictor*, would be as justifiable as to distinguish it under a separate name. The case is perfectly analogous to that of *Corallus hortulanus*, Central-American specimens having fewer series of scales (viz. 65–79) than the *Boa constrictor* of Tropical America south of the Isthmus of Panama and east of the Andes (85–95). But the larger-scaled form extends farther southwards than *Corallus cooki*, viz. into Western Ecuador and Peru. According to Sumichrast, it is common on the Isthmus of Tehuantepec. Mr. Champion observed it at Bugaba, in the State of Panama, but he did not procure specimens.

2. **Boa mexicana.**

Boa mexicana (Rapp), Jan, Iconogr. Ophid. p. 82; Bocourt, Miss. Sc. Mex., Rept. p. 521, t. 30. figg. 9, 9 *a*, *b*; Cope, Proc. Am. Phil. Soc. xxii. p. 382 (1885).

Boa diviniloquax, var. *mexicana*, Jan, Iconogr. Ophid. v. t. 4.

Hab. MEXICO, Jicaltepec (*Cope*).

Scales in 55 rows.

Fam. ELAPIDÆ.

ELAPS.

Elaps (Schneider), sp., Wagler, Syst. Amph. p. 193.

Elaps, Günther, P. Z. S. 1859, p. 81.

These snakes are known in Mexico by the name "coralillas."

1. **Elaps fulvius.**

Coluber fulvius, Linn. Syst. Nat. i. p. 381.

Elaps fulvius, Dum. & Bibr. Erpét. vii. p. 1215; Günth. Cat. Col. Sn. p. 235, and Ann. & Mag. N. H. 1859, iv. p. 169; Jan, Iconogr. Ophid. xlii. t. 2. fig. 2; Cope, Bull. U. S. Nat. Mus. no. 32, p. 86.

Elaps corallinus, Cuv. Règne Anim. t. 35. figg. 1, 1 *a, b*; Wied, Nov. Act. 1820, p. 108, t. 4, and Abbild. vi. t. 4; Schleg. Ess. ii. p. 440, t. 16. figg. 1–5; Dum. & Bibr. Erpét. vii. p. 1207; Günth. Cat. Col. Sn. p. 233, and Ann. & Mag. N. H. 1859, iv. p. 169; Salv. P. Z. S. 1860, p. 457, and 1861, p. 228; Jan. Iconogr. Ophid. xli. t. 6. figg. 1–3; A. Dugès, La Naturaleza, vii. p. 200 (1885).

Elaps circinalis, Dum. & Bibr. Erpét. vii. p. 1210; Cope, Journ. Ac. N. Sc. Phil. viii. p. 148 (1875).

Elaps diastema, Dum. & Bibr. Erpét. vii. p. 1222.

Elaps diastema, var. *michoacanensis,* A. Dugès, La Naturaleza, (2) i. p. 487, t. 32 (1891).

? *Elaps epistema,* Dum. & Bibr. Erpét. vii. p. 1222.

Elaps nigrocinctus, Girard, Proc. Ac. N. Sc. Phil. 1854, p. 226; U. S. Naval Astron. Exped. ii. Zool. p. 210, t. 35; Cope, Journ. Ac. N. Sc. Phil. viii. p. 148 (1875); Bull. U. S. Nat. Mus. no. 32, p. 86.

Elaps divaricatus, Hallow. Journ. Ac. N. Sc. Phil. iii. p. 36 (1855).

Elaps fitzingeri, Jan, Rev. et Mag. Zool. 1858, pp. 516, 521.

Elaps fulvius, var. *fitzingeri,* Jan, Iconogr. Ophid. xlii. t. 2. fig. 3.

Elaps ornatissimus, Jan, Iconogr. Ophid. xlii. t. 1. fig. 1; Cope, Journ. Ac. N. Sc. Phil. 1875, p. 148; Bull. U. S. Nat. Mus. no. 32, p. 87.

Elaps apiatus, Jan, Rev. et Mag. Zool. 1858, p. 522; Cope, Bull. U. S. Nat. Mus. no. 32, p. 87.

Elaps affinis, Jan, Rev. et Mag. Zool. 1858, p. 525.

Elaps fulvius, var. *affinis,* Jan, Iconogr. Ophid. xlii. t. 1. fig. 2.

? *Elaps euryxanthus,* Kennicott, Proc. Ac. N. Sc. Phil. 1860, p. 337.

Elaps corallinus, var. *crebripunctatus,* Peters, MB. Ak. Wiss. Berl. 1869, p. 877.

Hab. Texas to Southern Brazil.

I arrange the Central-American specimens under three groups, which, however, pass into each other, and within which the variation also is not inconsiderable:—

I. The black bands broad or of moderate width, and with yellow borders and black spots on the red interspaces generally in northern specimens; without yellow borders or black spots generally in southern examples.—*Hab.* North America, Texas.—Mexico, Northern parts (*Mus. Brit.*), City of Mexico (*Doorman*), Guanajuato (*Dugès*); British Honduras, Belize (*Godman & Salvin; Rev. J. Gegg*); Guatemala, Dueñas (*Salvin*), Vera Paz, low forest (*Godman & Salvin*); Costa Rica (*Godman & Salvin*); Panama, Chiriqui (*Champion*).

II. The black bands narrow.

A. The black bands close together.—*Hab.* Mexico, Vera Cruz (*Jan, Cope*); British Honduras (*Mus. Brit.*), Stann Creek (*Rev. J. Robertson*), Belize (*Parsons*); Guatemala, Yzabal (*Salvin*); Costa Rica (*Van Patten*).

B. The black bands more or less widely distant from one another.—*Hab.* Mexico (*Fink, Sallé*), Rio Santiago in Jalisco (*Godman & Salvin*), Michoacan (*Dugès*), Omilteme in Guerrero, Atoyac in Vera Cruz, Teapa in Tabasco (*H. H. Smith*), Tehuantepec (*Sumichrast*), Yucatan (*Schott*); British Honduras (*Colonial Exhibit.*); Salvador (*Dow*); Nicaragua (*U. S. Nat. Mus.*); Costa Rica (*U. S. Nat. Mus.*); Panama (*U. S. Nat. Mus.*).

2. Elaps bernardi.

Elaps bernardi, Cope, Bull. U. S. Nat. Mus. no. 32, 1887, p. 87.

Hab. MEXICO, Zacualtipan in Hidalgo (*Bernard*).

I have not seen a specimen of this snake.

3. Elaps aglæope.

Elaps aglæope, Cope, Proc. Ac. N. Sc. Phil. 1859, p. 344; ? Sumichrast, Bull. Soc. Zool. Fr. 1880, p. 184.

Elaps fulvius (*aglæope*), Salv. P. Z. S. 1860, p. 458.

Elaps corallinus, var., A. Dugès, La Naturaleza, vii. p. 201 (1885).

Hab. ? MEXICO, Tehuantepec (*Sumichrast*); GUATEMALA, Lanquin in Vera Paz (*Salvin*), Peten (*Berendt*); HONDURAS (*Cope*).

This may be regarded as another of the numerous variations of *E. fulvius*, in which the black bands are very narrow and numerous, alternating with incomplete bands, which generally are merely transverse series of black spots. I keep it distinct, as the passage from this extreme form to the typical *E. fulvius* is much less complete than between the other variations.

4. Elaps ephippifer.

Elaps aglæope, Cope, Proc. Am. Phil. Soc. 1869, p. 162 (nec 1859).

Elaps ephippifer, Cope, Proc. Am. Phil. Soc. 1886, p. 281.

Hab. MEXICO, Pacific side of the Isthmus of Tehuantepec (*Sumichrast*).

Not seen by me; possibly a colour-variation of *E. fulvius*.

5. Elaps multifasciatus.

? *Elaps mipartitus*, Dum. & Bibr. Erpét. vii. p. 1220.

Elaps mipartitus, Günth. Ann. & Mag. N. H. 1859, iv. p. 172.

Elaps semipartitus, Jan, Iconogr. Ophid. xlii. t. 2. fig. 1.

Elaps multifasciatus, Jan, Rev. et Mag. Zool. 1858, p. 521; Cope, Proc. Ac. N. Sc. Phil. 1871, p. 209.

Hab. NICARAGUA, Chontales (*Mus. Brit.*); COSTA RICA (*Van Patten*); PANAMA (*Mus. Brit.*).—COLOMBIA, Bogota; VENEZUELA, Caracas.

The snakes referable to the names quoted above are extremely similar to one another as far as coloration is concerned, but they differ in the shape of the body, or number of ventral shields, in an extraordinary degree, although the extremes seem to be bridged over by intermediate forms. It is unfortunate that Duméril and Bibron have omitted to give the number of ventrals for their *E. mipartitus*.

I find the following variations recorded by authors, or in the specimens at my disposal:—

Specimen from	Caracas (*Jan*).	278 ventrals.
„	Bogota (*Mus. Brit.*).	277 „
„	Costa Rica (*Cope*).	268 „
„	Panama (*Mus. Brit.*).	261 „
„	Chontales (*Mus. Brit.*).	239 „
„	Venezuela (*Mus. Brit.*).	239 „
„	No locality (*Mus. Brit.*).	233 „
„	Caracas (*Mus. Brit.*).	210 „
„	Venezuela (*Mus. Brit.*).	204 „

6. **Elaps ruatanus,** sp. n. (Tab. LVII. fig. B.)

Hab. Honduras, Ruatan I. (*G. F. Gaumer*).

Habit rather stout for an *Elaps*; ventral scutes 184–204. Vertical shield narrow, sometimes twice as long as broad. Trunk with about forty-five complete black rings, as broad on the abdomen as on the back, and separated by white (reddish) interspaces about as wide as the black rings. The black rings, however, are not always of the same width; in some individuals broader black rings (four scales wide) alternate with narrower ones (two or three scales wide). The head is of the same bright colour as the lighter interspaces on the trunk, with the exception of the upper part of the snout and a narrow fringe of the anterior lower labials. The first black ring is posterior to the occipitals, and complete. Tail with four black rings and black tip.

Of this form I have eight specimens before me, agreeing with each other in the general pattern of coloration, and showing only a slight variation of the number of ventral scutes; I count 184–186 in six, and 200–204 in two of the specimens. Therefore this appears to be a distinguishable local form.

7. **Elaps lemniscatus.**

Coluber lemniscatus, Linn. Mus. Ad. Fried. p. 34, t. 14. fig. 1.
Elaps lemniscatus, Schneid. Hist. Amph. p. 291; Dum. & Bibr. Erpét. vii. p. 1217.
Elaps marcgravii, Wied, Abbild. Naturg. Bras., Rept. t. 42; Dum. & Bibr. Erpét. vii. p. 1209.
Elaps marcgravii, var. *laticollaris*, Peters, MB. Ak. Wiss. Berl. 1869, p. 877.

Hab. Mexico, Puebla (*Peters*).—Tropical South America.

8. **Elaps decoratus.**

Elaps decoratus, Jan, Rev. et Mag. Zool. x. p. 525 (1858); Iconogr. Ophid. xlii. t. 6. fig. 4; Günth. P. Z. S. 1859, p. 85, t. 18. fig. A.

Hab. ? Mexico (*Jan*).—Brazil (*Mus. Brit.*).

9. **Elaps elegans.**

Elaps elegans, Jan, Rev. et Mag. Zool. x. p. 524 (1858); Iconogr. Ophid. xlii. t. 5. fig. 2.

Hab. MEXICO, Huatuzco in Vera Cruz (*Godman*), Teapa in Tabasco (*H. H. Smith*); GUATEMALA, Vera Paz, low forest (*Salvin*).

Fam. HYDROPHIDÆ.

PELAMIS.

Pelamis, sp., Daudin, Hist. Nat. Rept. vii. p. 357 (1802).

1. **Pelamis bicolor.**

Hydrus bicolor, Schneid. Hist. Amph. p. 242.

Pelamis bicolor, Daud. Hist. Nat. Rept. vii. p. 366; Fischer, Abhandl. naturw. Ver. Hamburg, 1856, p. 61; Cope, Proc. Ac. N. Sc. Phil. 1859, p. 347; Journ. Ac. N. Sc. Phil. viii. p. 148 (1875); Günth. Trans. Zool. Soc. vi. p. 400.

Hab. WEST COAST OF MEXICO (*Mus. Berol.*); COSTA RICA, Pacific coast (*Champion, Van Patten*); BAY OF PANAMA (*Seemann, Dow, Salvin, Sallé*).—INDO-PACIFIC OCEAN.

This is the species of sea-snake mentioned by Seemann [Voyage 'Herald,' i. p. 265 (1848)]. It has been seen by M. Sallé and Mr. Salvin * in considerable numbers near the Bay of Panama, and more recently by Mr. Champion off Punta Arenas, Costa Rica.

Fam. CROTALIDÆ.

ANCISTRODON.

Agkistrodon, Beauvois, Trans. Am. Phil. Soc. 1799, p. 381.

Ancistrodon, Cope, Proc. Ac. N. Sc. Phil. xi. p. 336 (1860).

Cenchris, Daudin, Hist. Nat. Rept. v. p. 356 (1802).

1. **Ancistrodon bilineatus.** (Tab. LVIII. figg. A & B, ad. et juv.)

Ancistrodon bilineatus, Günth. Ann. & Mag. N. H. 1863, xii. p. 364; Cope, Proc. Ac. N. Sc. Phil. 1865, p. 191; Bull. U. S. Nat. Mus. no. 32, p. 89; Müller, Verh. naturf. Ges. Basel, 1878, p. 404; Sumichrast, Bull. Soc. Zool. Fr. 1880, p. 185.

"*Cantil*" in Tehuantepec.

Hab. MEXICO, Tres Marias Is., Presidio (*Forrer*), Colima and Guadalajara (*Xantus*), Tehuantepec (*Sumichrast*), Yucatan (*Mus. Brit.*); BRITISH HONDURAS, Belize (*Mus. Brit.*); GUATEMALA, Pacific coast (*Salvin*); NICARAGUA (*Cope*).

* *Cf.* Günth. P. Z. S. 1858, p. 381.

Ground-colour brownish-olive, brown or black. Young and light-coloured individuals have broad dark cross-bands, which are irregularly edged with white spots. The white spots are retained after the bands have disappeared, and are more numerous on the lower part of the sides and abdomen than on the back. A white or yellow line runs from the rostral along the canthus rostralis and the superciliary edge to behind the angle of the mouth; a white or yellow band along the upper labials. Rostral with a vertical yellow or white band, continued on the chin.

Scales keeled, in twenty-three series. Ventrals 137; subcaudals 65, of which the posterior are double. Upper labials eight.

BOTHROPS.

Bothrops, sp., Wagler, Syst. Amphib. p. 174 (1830).
Bothrops et *Ophryacus,* Cope, Bull. U. S. Nat. Mus. no. 32, 1887, p. 88.

Two subgenera:—

I. Scales of the superciliary margin erect. (Ophryacus.) . . . . 1. *B. undulatus.*
II. A broad superciliary scute. (Bothrops.) . . . . 2. *B. atrox.*

1. Bothrops undulatus.

Atropus undulatus, Jan, Rev. et Mag. Zool. 1859, p. 157.
Ophryacus undulatus, Cope, Bull. U. S. Nat. Mus. no. 32, p. 88.

Hab. Mexico, Omilteme in Guerrero (*H. H. Smith*), Orizaba (*Mus. Brit.*), Vera Cruz, Actopam (*Com. Geogr.*), Oaxaca (*Sallé*).

With one exception, the specimens I have examined are young, and of a dark brown coloration with the black dorsal markings mentioned by Jan. A row of whitish spots runs along the lower part of the sides. This is also the coloration of a young specimen from Omilteme. However, an adult female from the same locality, 23 in. long, is of a yellowish-green colour, the scales being densely punctulated with black, and only outline traces of the black rhombic dorsal markings remaining. Lower parts yellowish, the posterior abdominal and the caudal scutes punctulated with black.

2. Bothrops atrox.

Coluber atrox, Linn. Mus. Ad. Fried. i. p. 33, t. 22. fig. 2.
Bothrops atrox, Wagl. Syst. Amph. p. 174; Dum. & Bibr. Erpét. vii. p. 1507; Jan, Iconogr. Ophid. xlvii. t. 2; Cope, Journ. Ac. N. Sc. Phil. viii. p. 151 (1875).
Trigonocephalus atrox, Schleg. Ess. ii. p. 535, t. 19. figg. 5, 6.
Craspedocephalus atrox, Fitz. N. Class. Rept. p. 61; Gray, Cat. Viper. Sn. p. 6.

Hab. Mexico, Orizaba (*Sumichrast*), Mirador and Tuxpan (*Cope*), Atoyac in Vera Cruz, Teapa in Tabasco (*H. H. Smith*), Tehuantepec (*Sumichrast*); British Honduras, Stann Creek (*Rev. J. Robertson*); Guatemala, Pacific Coast and Vera Paz, low forest (*Salvin*), Coban (*Sarg*); Nicaragua, Chontales (*Rix*), Matagalpa (*Rothschuh*), San Juan (*Xantus*); Costa Rica (*Gabb*); Panama (*Cope*), Chiriqui (*Champion*).—Tropical South America.

LACHESIS.

Lachesis, Daudin, Hist. Nat. Rept. v. p. 349 (1802).

1. Lachesis mutus.

Crotalus mutus, Linn. Syst. Nat. i. p. 373.
Lachesis mutus, Daud. Hist. Nat. Rept. v. p. 351.
Lachesis stenophrys, Cope, Journ. Ac. N. Sc. Phil. viii. p. 152 (1875).

Hab. Costa Rica, Sipurio (*Gabb*); Panama (*Mus. Brit.*).—Tropical South America.

BOTHRIECHIS.

Bothriechis, Peters, MB. Ak. Wiss. Berl. 1859 (March), p. 278.
Teleuraspis, Cope, Proc. Ac. N. Sc. Phil. 1859 (Dec.), p. 338.
Thamnocenchris, Salvin, P. Z. S. 1860, p. 459.
Bothriopsis, Peters, MB. Ak. Wiss. Berl. 1861, p. 359.
Bothriechis, Teleuraspis, Bothriopsis, et *Porthidium*, Cope, Proc. Ac. N. Sc. Phil. 1871, p. 206.

The Central-American species may be arranged in the following subgeneric groups:—

I. Hinder part of the body and tail compressed, prehensile: arboreal species.
  A. Upper surface of the head uniformly scaly.
    1. A series of small, more or less erect scales between the superciliary scute and the eye. (Teleuraspis.) . . . . 1. *B. schlegeli.*
    2. No small scales between the superciliary scute and the eye. (Bothriechis.)
      a. The second labial entering the fossa . . . . 2. *B. nigroviridis.* 3. *B. lateralis.*
      b. The second labial not entering the fossa . . . . 4. *B. bicolor.*
  B. Fore part of the head with scutes above. (Thamnocenchris.) . . . 5. *B. aurifera.*
II. Hind part of the body and tail non-prehensile: terrestrial species.
  A. Upper surface of the head uniformly scaly.
    1. Snout pointed . . . . 6. *B. lansbergii.*
    2. Snout obtuse.
      a. Superciliary scute broad . . . . 7. *B. godmani.* 8. *B. ophryomegas.*
      b. Superciliary scute narrow . . . . 9. *B. nummifera.* *B. quadriscutata.*
  B. Upper surface of the head partly covered with scutes. (Bothriopsis.) 10. *B. scutigera.*

1. Bothriechis schlegeli.

Trigonocephalus schlegelii, Berthold, Abhandl. Ges. Wiss. Göttingen, iii. p. 13, t. 1. figg. 5, 6 (1847).

Teleuraspis schlegelii, Cope, Proc. Ac. N. Sc. Phil. 1859, p. 338; Journ. Ac. N. Sc. Phil. viii. pp. 149, 157, t. 27. fig. 2 (1875).

Bothrops schlegelii, Jan, Iconogr. Ophid. xxvii. t. 6. fig. 2.

Lachesis nitida, Günth. P. Z. S. 1859, p. 414, t. 20. fig. C.

Bothrops (*Teleuraspis*) *nigroadspersus*, Steind. SB. Ak. Wiss. Wien, lxii. p. 348, t. 8 (1870).

Hab. GUATEMALA (*Godman*); NICARAGUA, Chontales Mines (*Rix*), Hacienda Rosa de Jerico, alt. 3250 feet (*Rothschuh*); COSTA RICA (*Salvin*); PANAMA, Chiriqui (*Champion*), Veraguas and Panama (*Cope*).—SOUTH AMERICA to Ecuador.

Specimens of this species vary considerably in the number of scales (19–25), the development of keels and spinosity of the scales of the head, and in coloration. One specimen from Chontales is an albino, being almost entirely of a lemon-yellow colour: an anomaly which occurs also in the West-African Viperine Snake, *Atheris chloroëchis.*

2. Bothriechis nigroviridis.

Bothriechis nigroviridis, Peters, MB. Ak. Wiss. Berl. 1859, p. 278; Cope, Journ. Ac. N. Sc. Phil. viii. p. 150 (1875).

Hab. COSTA RICA, Volcan de Barba (*Hoffmann*), Pico Blanco (*Gabb*), Rio Sucio and Volcan de Irazu (*Rogers*).

Feeds on frogs.

3. Bothriechis lateralis.

Bothrops bilineatus, Peters, MB. Ak. Wiss. Berl. 1859, p. 278 (nec Wied).

Bothriechis lateralis, Peters, MB. Ak. Wiss. Berl. 1862, p. 674; Cope, Bull. U. S. Nat. Mus. no. 32, p. 88.

Hab. COSTA RICA (*Hoffmann*, *Salvin*).

4. Bothriechis bicolor.

Bothrops bicolor, Bocourt, Ann. Sc. Nat. (5) x. p. 201 (1868).

Bothrops (*Bothriechis*) *bernoullii*, Müll. Verh. naturf. Ges. Basel, vi. p. 399, t. 3. fig. A (head) (1878).

Hab. WESTERN GUATEMALA, forest of San Agustin, in the Department of Solola (*Bocourt*), Cuesta de Atitlan (*Müller*).

5. Bothriechis aurifera.

Thamnocenchris aurifer, Salv. P. Z. S. 1860, p. 459, t. 32.

Bothriechis aurifer, Cope, Proc. Ac. N. Sc. Phil. 1871, p. 205.

Hab. GUATEMALA, Coban (*Salvin*).

6. **Bothriechis lansbergii.**

Trigonocephalus lansbergii, Schleg. Mag. Zool. 1841, Rept. t. 1, c. fig. med.

Bothrops lansbergii, Günth. Ann. & Mag. N. H. 1863, xii. p. 350; Jan, Iconogr. Ophid. xxvii. t. 4. fig. 1.

Porthidium lansbergii, Cope, Proc. Ac. N. Sc. Phil. 1871, p. 207.

Teleuraspis castelnaui, var. *brachystoma,* Cope, Proc. Ac. N. Sc. Phil. 1859, p. 339.

Bothriechis brachystoma, Cope, Proc. Ac. N. Sc. Phil. 1861, p. 295.

Bothriopsis brachystoma, Cope, Proc. Ac. N. Sc. Phil. 1871, p. 208; Sumichrast, Bull. Soc. Zool. Fr. 1880, p. 185.

Bothrops nasutus, Bocourt, Ann. Sc. Nat. (5) x. p. 202 (1868).

Porthidium nasutum, Cope, Proc. Am. Phil. Soc. 1879, p. 271.

Bothriopsis proboscideus, Cope, Journ. Ac. N. Sc. Phil. viii. p. 150, t. 27. fig. 3, t. 28. fig. 13 (1875).

"*Chatilla*" in Tehuantepec.

Hab. MEXICO, Tehuantepec * (*Sumichrast*), Yucatan (*Mus. Brit.*); GUATEMALA, Vera Paz, low forest (*Salvin*), Panzos (*Bocourt*), Peten (*Berendt*); NICARAGUA, Hacienda Rosa de Jerico, alt. 3250 feet (*Rothschuh*), Chontales (*Mus. Brit.*); COSTA RICA, Limon (*Zeledon*); PANAMA, Chiriqui (*Champion*).—COLOMBIA; VENEZUELA.

From an examination of the series of specimens before me, I have come to the conclusion that the presence or absence of a suture in the nasal scute cannot be used as a taxonomic character in this species.

7. **Bothriechis godmani.** (Tab. LVII. fig. A.)

Bothriechis godmanni, Günth. Ann. & Mag. N. H. 1863, xii. p. 364, t. 6. fig. G (upperside of head).

Bothriopsis godmanii, Cope, Proc. Ac. N. Sc. Phil. 1871, p. 208; Müller, Verh. naturf. Ges. Basel, vi. p. 402, t. 3. fig. B (head) (1878).

Bothrops brammianus, Bocourt, Ann. Sc. Nat. (5) x. p. 201 (1868).

Bothriechis trianguligera †, Fischer, Osterprogramm d. Akad. Gymnas. Hamburg, 1883, p. 13.

Hab. GUATEMALA, Totonicapam (*Salvin*), Mazatenango (*Bernoulli*), ? San Lucas, alt. 1558 metres (*Bocourt*), Volcan de Fuego (*Godman & Salvin*).

Scales in twenty-one series, strongly keeled, except those in the outer row, which are smooth. Scales on the upper surface of the head faintly keeled and of unequal size, three of them being as large as the largest scales of the trunk; superciliaries well developed, without small scales along the orbital margin; canthus rostralis angular, covered with small shields, the hindmost of which enters the orbit. Rostral shield triangular, erect, not prominent or elevated; nasals small, separated by a suture. Nine

* A specimen from Tehuantepec seems to come nearest to Cope's description of his *Bothriechis brachystoma.*

† I am indebted to the kindness of Prof. Kraepelin and Dr. Pfeffer, who have given me the opportunity of examining the type of Fischer's description.

upper labials; the second does not enter into, and the third is almost immediately below, the facial pit; the fifth is the largest, below the eye, from which it is separated by two series of very small scales; the posterior gradually decrease in size. Ten lower labials. Ventrals 142; subcaudals 28, the tail being short, very thin, and *not prehensile.* Upper parts uniform brown; a black band runs from the eye to behind the angle of the mouth; labial shields yellow; a series of rounded, black, light-edged spots along the side of the anterior part of the trunk. Lower parts yellowish, marbled with blackish in the posterior half of the body; an irregular series of subquadrangular black spots along each side of the anterior half of the belly.

The largest specimen, a female, measures 17½ inches, the tail occupying 1⅝.

8. **Bothriechis ophryomegas.**

Bothrops ophryomegas, Bocourt, Ann. Sc. Nat. (5) x. p. 201 (1868).

Hab. GUATEMALA, Escuintla (*Bocourt*).

Not seen by me.

9. **Bothriechis nummifera.**

Atropos mummifer, Rüpp. Verz. Senckenberg. Mus., Rept. p. 21.

Teleuraspis nummifer, Cope, Proc. Ac. N. Sc. Phil. 1859, p. 339; 1860, p. 345; Günth. Ann. & Mag. N. H. 1863, xi. p. 25, t. 3. fig. C (head).

Bothrops nummifer, Jan, Iconogr. Ophid. xlvii. t. 5. fig. 2.

Bothriechis nummifera, var. *notata,* Fisch. in Wiegm. Arch. 1880, p. 222, t. 8. figg. 10–12.

Atropos mexicanus, Dum. & Bibr. Erpét. vii. p. 1521, t. 83 bis.

Bothriechis mexicanus, Cope, Proc. Ac. N. Sc. Phil. 1861, p. 294.

Bothrops affinis, Bocourt, Ann. Sc. Nat. (5) x. p. 201 (1868).

Bothriopsis affinis, Cope, Proc. Ac. N. Sc. Phil. 1871, p. 209.

Hab. MEXICO, common as far north as Tuxpan (*Cope*), Huatuzco in Vera Cruz (*Godman*), Teapa in Tabasco (*H. H. Smith*); GUATEMALA, Lower Vera Paz (*Godman & Salvin*), San Agustin, alt. 610 metres (*Bocourt*), Coban (*Sarg*); NICARAGUA, Hacienda Rosa de Jerico (*Rothschuh*); COSTA RICA (*Cope*).

This species has been split into two, on the strength of a few minute scales being sometimes intercalated between the rostral and nasal. It was therefore of some interest to clear up this matter by an examination of embryos taken from the oviduct of the same mother. The mother possesses those supplementary scales, but her offspring were found to vary in this respect.

10. **Bothriechis scutigera.**

Bothriechis scutigera, Fisch. in Wiegm. Arch. 1880, p. 218, t. 8. figg. 8, 9 (head).

Hab. GUATEMALA (*Fischer*).

CROTALOPHORUS.

Crotalophorus (Linnæus), Gray, Ann. Phil. 1825, p. 205.

1. Crotalophorus miliarius.

Crotalus miliarius, Linn. Syst. Nat. i. p. 372.
Caudisona rava, Cope, Proc. Ac. N. Sc. Phil. 1865, p. 191.
Crotalophorus ravus, Cope, Proc. Am. Phil. Soc. 1885, p. 382.
Crotalus miliarius, var. *ravus*, Garman, Mem. Mus. Comp. Zool. viii. p. 120 (1884).

Hab. MEXICO, Tableland of Southern Mexico (*Cope*), Puebla (*Com. Geogr.*).

CROTALUS.

Crotalus, sp., Linnæus, Syst. Nat. i. p. 372 (1766).

1. Crotalus triseriatus.

Uropsophus triseriatus, Wagl. Nat. Syst. Amphib. p. 176.
Crotalus triseriatus, Cope, Proc. Am. Phil. Soc. xxii. p. 179 (1885); Bull. U. S. Nat. Mus. no. 32, p. 90.
Crotalus lugubris, A. Dugès, La Naturaleza, iv. p. 25 (1877) (nec Jan).

Hab. MEXICO, Guanajuato (*A. Dugès*), Toluca, Zacualtipan, and Jalapa (*Cope*), Orizaba (*Mus. Brit.*).

2. Crotalus polystictus.

Caudisona polysticta, Cope, Proc. Ac. N. Sc. Phil. 1865, p. 191.
Crotalus polystictus, Cope, Proc. Am. Phil. Soc. 1884, p. 179; Bull. U. S. Nat. Mus. no. 32, p. 89; A. Dugès, La Naturaleza, (2) i. p. 134, t. 12. fig. 12 (head) (1888).
Crotalus lugubris, Jan, Prodrom. p. 31; Iconogr. Ophid. xlvi. t. 3. fig. 3.
Crotalus jimenezii, A. Dugès, La Naturaleza, iv. p. 23, t. 1. figg. 18–20 (1877).

Hab. MEXICO, Colima and Guanajuato (*A. Dugès*), Valley of Mexico (*Herrera*).

Probably identical with the preceding.

3. Crotalus intermedius.

Crotalus intermedius, Fisch. Abh. naturw. Ver. Bremen, vii. p. 230, t. 14. figg. 1–4 (head) (1882).

Hab. MEXICO (*Sallé*).

4. Crotalus omiltemanus, sp. n. (Tab. LVIII. fig. C.)

Hab. MEXICO, Omilteme in Guerrero (*H. H. Smith*).

This species belongs to the group *Uropsophus*, and is distinguished from the other allied forms by the small number of labial shields and large number of ventral scutes.

Body rather more elongate than in the allied species, the ventrals being 178–185 in number. Scales in twenty-one series, those of the back strongly keeled, the three outer series smooth. Head triangular, depressed; upper surface of the snout in front of the supraciliaries covered with enlarged scutes, varying in size and number; crown of the head covered with small smooth scales. Two nasal shields, the anterior much larger than the posterior and encroaching upon the anterior and upper surface of the snout. Rostral shield reverted above. Nine upper labials separated from the eye by a single series of scales.

Dark greyish-black, with a dorsal series of larger, more or less rounded black spots, and two or three lateral series of smaller ones; lower parts blackish; an oblique black band from the eye to the lower part of the neck; it is white-edged above and below, the lower edge forming a wide band along the upper labial shields.

Of the two specimens the larger is 21 inches long; it possesses a strongly compressed five-jointed rattle, which has a groove along each side.

5. **Crotalus palmeri.**

? *Crotalus tigris*, Kennicott, in Mex. Bound. Surv., Rept. p. 14, t. 4; Baird, in Pac. R. R. Exped. Report, x. t. 35. figg. 1 *a*–1 *g* (see also Stejneger, N. Amer. Faun. no. 7, p. 214).

Crotalus tigris, var., or *Crotalus palmeri*, Garman, Bull. Essex Inst. xix. p. 124 (1888).

Hab. North America, ? California and Nevada (*Death Valley Exped.*), ? Deserts of Gila and Colorado (*Kennicott*).—Mexico, Monclova in Coahuila (*Garman*).

6. **Crotalus pallidus,** sp. n. (Tab. LIX. fig. B.)

Hab. Mexico, City of Mexico (*Doorman*).

Scales strongly keeled, in twenty-five series, the two or three outer ones being smooth; none of the scales are striated. As to the scutellation of the head, a small pair of internasal scutes may be noticed as well as a larger scute occupying the canthus rostralis between the internasal and the supraciliary; the space surrounded by these scutes, as well as the rest of the head, is covered by small faintly keeled scales. Nasal divided, the posterior portion being much smaller than the anterior. Upper labials twelve or thirteen, *separated from the eye by two series of scales only.* Ventrals 149.

Upper parts pale brownish, with a dorsal series of about thirty-six rounded or sub-rhombic small spots, which are but faintly visible; lower parts uniform whitish.

A single specimen, 15½ inches long; it has a strongly compressed five-jointed rattle, grooved along the middle of the side.

7. **Crotalus salvini,** sp. n. (Tab. LIX. fig. A.)

Crotalus molossus, Garman, Bull. Essex Inst. xix. p. 123 (nec Baird & Gir.) (1888).

Hab. MEXICO, Huamantla, alt. 8000 feet (*Rymer Jones*), Alvarez Mts., near San Luis Potosi (*Garman*).

Habit stout; 173 ventral scutes; scales very strongly keeled, with a groove on each side of the high keel, in twenty-five rows, the scales of the four outer series being smooth. The upper surface of the head is covered with small scutes symmetrically arranged and passing posteriorly into scales; these scutes are concave in the middle, and have the margins turned upwards in a peculiar manner; four of them, which are small, take the place of the anterior frontals, the four succeeding that of the posterior frontals; the place of the vertical is occupied by two pairs of scutes with an unpaired median; finally, the remaining scale-like scutes are distinctly so arranged as to indicate the initial formation of a pair of occipitals. All the scutes and scales of the upper part of the head are finely striated. Sides of the head deep, the upper labial shields being separated from the orbit by four series of scales. Rostral as wide as deep. Nostril large; the anterior nasal large and quadrangular, the posterior being a narrow rim round the nostril. Thirteen small upper labials, smaller than the scales covering the temple.

Greyish-olive, with thirty-two subquadrangular dark brown spots along the back, two series of smaller spots along the side; upper surface of the head black anteriorly, an oblique deep brown spot from the eye towards the angle of the mouth. Tail banded. Lower parts uniform whitish.

A single female specimen is 30 inches long, and had ten embryos in its oviduct; its rattle is strongly compressed, with a lateral groove, and consists of nine joints.

Beside other points of disagreement, a comparison of the figure of the head, given by Baird and Girard in Pac. R. R. Report, x. t. 24. fig. 5, of *Crotalus molossus*, will prove the distinctness of Salvin's Rattlesnake from the latter species.

8. **Crotalus horridus.**

? *Crotalus horridus*, Linn. Syst. Nat. i. p. 372.

Crotalus horridus, Daud. Hist. Rept. v. p. 311, t. 69. fig. 1; Schleg. Ess. ii. p. 561, t. 20. figg. 12–14; Dum. & Bibr. Erpét. vii. p. 1472, t. 84 bis, fig. 2; Jan, Iconogr. Ophid. xlvi. t. 3. fig. 1.

Crotalus durissus, Cope, Sumichrast et al.

" *Vibora de cascabel*," Mexico.

Hab. MEXICO, southern parts (*Godman & Salvin*), Omilteme in Guerrero (*H. H. Smith*), Tehuantepec (*Sumichrast*); BRITISH HONDURAS (*Mus. Brit.*).—TROPICAL AMERICA.

9. **Crotalus basiliscus.**

Caudisoma basilisca, Cope, Proc. Ac. N. Sc. Phil. 1864, p. 166.

Crotalus basiliscus, Cope, Bull. U. S. Nat. Mus. no. 32, p. 89; Proc. Am. Phil. Soc. 1884, p. 180; A. Dugès, La Naturaleza, (2) i. p. 133 (1888).

Crotalus rhombifer, A. Dugès, La Naturaleza, iv. p. 22 (nec Latr.) (1877).

Hab. MEXICO, Presidio (*Forrer*), Colima (*Xantus*), Mezquital del Oro, Zacatecas (*A. C. Buller*), City of Mexico (*Doorman*), San Luis Potosi, Guanajuato, Guadalajara, Silao, Zamora (*Dugès*), Puebla (*Com. Geogr.*).

10. **Crotalus adamanteus.**

Crotalus adamanteus, Beauv. Trans. Am. Phil. Soc. iv. p. 368 (1824); Baird & Gir. Cat. N. Amer. Rept. p. 3; Cope, Bull. U. S. Nat. Mus. no. 32, p. 90.

Crotalus atrox, Baird & Gir. Cat. N. Amer. Rept. pp. 5, 156; Garman, Bull. Essex Inst. xix. 1888, p. 113.

Crotalus adamanteus, var. *atrox,* Garman, Mem. Mus. Comp. Zool. viii. p. 113 (1883).

Hab. NORTH AMERICA, Southern United States.—MEXICO, City of Chihuahua (*Wilkinson*), San Pedro and Monclova (*Garman*).

BIOLOGIA CENTRALI-AMERICANA.

ZOOLOGIA.

Class BATRACHIA.

Division BATRACHIA SALIENTIA.

Fam. RANIDÆ.

RANA.

Rana, sp., Linn. Syst. Nat. i. p. 354; Bouleng. Batr. Sal. p. 7.
Ranula, Peters, MB. Ak. Wiss. Berlin, 1859, p. 402.
Pohlia, Steind. Reise der Novara, Amphib. p. 15.
Trypheropsis, Cope, Proc. Ac. N. Sc. Philad. 1868, p. 117.

Tongue deeply notched behind; vomerine teeth. Fingers free; toes webbed, without adhesive disks; outer metatarsals with a distinctly developed web between them. Omosternum and sternum with a bony style.

1. **Rana montezumæ.**

Rana montezumæ, Baird, Proc. Ac. N. Sc. Philad. 1855, p. 61; Girard, U. S. Mex. Bound. Surv., Rept. p. 27, t. 36. figg. 1–6; Brocchi, Miss. Sc. Mex., Batr. p. 14, t. 4. figg. 2, 2 *a*, *b*; Cope, Bull. U.S. Nat. Mus. no. 32, p. 19; Bouleng. Batr. Sal. p. 35; Dugès, La Naturaleza, 2 ser. i. p. 137 (1888).
Rana adtrita, Troschel, in Müller's Reisen in den Verein. Staat. etc. iii., Wirbelth. p. 82 (1865).

Hab. Mexico, city of Mexico (*Doorman*), Guanajuato (*Dugès*), Lake of Texcoco (*Godman*), Tehuantepec (*Sumichrast*), Tabasco (*Berendt*); Guatemala (*Cope*).

Form robust; snout rather short and wide, in front and on the sides sloping towards the labial margin, with very indistinct canthus rostralis; loreal region not concave. Tympanum as large as the eye. Vomerine teeth in two short transverse straight groups, on a level with the hind margin of the choanæ. Toes not enlarged at the tips, broadly webbed, the web extending to the last phalanx of the fourth toe. A short, narrow, inner metatarsal tubercle. The first finger very little longer than the second.

Skin with tubercles, minute on the back and small on the sides; a narrow, more or less indistinct, glandular fold on each side of the back. The hind limb, without tarsus, as long as, or rather shorter than, the body; fourth toe (from metatarsal joint) five-ninths of the length of the body. Brownish, with rounded darker ocellated spots, and generally with greyish or whitish dots. Lower parts and thighs punctulated with white; upper parts of the legs with irregular dark cross-bars. Also the tympanum is dotted like the skin round it. Sometimes uniform brown.

Male with a large external vocal sac between the angle of the mouth and the humerus.

Length of body of adult female 110 millim., of adult male 70 millim.

The peculiar coloration of this species is so well described by Troschel that I cannot doubt that his *Rana adtrita* is based upon a specimen of *R. montezumæ*. The only divergence in his description is that the tympanum is said by him to be smaller than the eye-opening.

2. **Rana halecina.**

Rana halecina (Kalm, Iter Amer. iii. p. 46 *); Holbrook, N.-Amer. Herpet. i. p. 89, t. 13; Dum. & Bibr. Erpét. Gén. viii. p. 352; Hallow. Proc. Ac. N. Sc. Philad. 1856, p. 141; Günth. Batr. Sal. p. 13; Cooper, U. S. Explor. & Surv. xii. t. 29. fig. 1; Salvin, P. Z. S. 1860, p. 460; Brocchi, Miss. Sc. Mex., Batr. p. 10; Bouleng. Batr. Sal. p. 41; Cope, Bull. U. S. Nat. Mus. no. 32, p. 19; Dugès, La Naturaleza, 2 ser. i. p. 138 (1888).

Rana halecina austricola, Cope, Proc. Am. Phil. Soc. xxiii. p. 517 (1886).

Rana pipiens, Gm. Syst. Nat. iii. p. 1052.

Rana lecontei, Baird & Gir. Proc. Ac. N. Sc. Philad. 1853, p. 301; Günth. Batr. Sal. p. 15; Brocchi, Bull. Soc. Philom. i. p. 179, and Miss. Sc. Mex., Batr. p. 14, t. 4. figg. 1, 1 *a–c*; Bouleng. Batr. Sal. p. 42.

Rana utricularia, Harlan, Sillim. Journ. x. p. 59; Journ. Ac. N. Sc. Philad. v. p. 337; Bouleng. Batr. Sal. p. 40.

Rana berlandieri, Baird, U. S. Mex. Bound. Surv., Rept. p. 27, t. 36. figg. 7–10.

Rana virescens (Kalm), Garman, Bull. Ess. Inst. xvi. p. 41; Cope, Bull. U. S. Nat. Mus. no. 34, p. 397 (1889); Werner, Jahresb. naturw. Ver. Magdeburg, 1894, p. 128.

Hab. NORTH AMERICA, United States.—MEXICO, Monterey in Tamaulipas (*Cope*), Presidio de Mazatlan (*Forrer*), Guanajuato (*Dugès*), Amula (*H. H. Smith*), Guadalajara, Talpa, and La Cumbre de los Arrastrados (8500 feet), in Jalisco (*Buller*), Jalapa, Vera Cruz (*O. S. & F. D. G.*); GUATEMALA, Coban, Dueñas, Lanquin (*Salvin*); COSTA RICA, Bebedero (*Underwood*), San José and La Palma Lagoon to alt. 1500 metres (*Mus. Nacional de Costa Rica*).

* According to Garman (Bull. Ess. Inst. xvi. p. 41), Kalm did not use the specific term "*halecina*," but commences his specific diagnosis with the word "*virescens*." I am not inclined to assist in substituting the latter term for one so generally adopted hitherto.

Head of moderate size, with the snout rather pointed. Tympanum nearly the size of the eye. Vomerine teeth between and not touching the choanæ, in two short more or less transverse series. The first finger extends beyond the second. Toes with the web distinctly emarginate and not extending to the tips. Only one (the inner) metatarsal tubercle, which is very small. Hind limb, without tarsus, as long as, fourth toe (from metatarsal joint) three-fifths of the length of, the body. Back with a narrow glandular fold on each side. Olive, or greenish or brownish above, with oval or rounded blackish spots edged with lighter; legs with cross-bars.

Male with small internal vocal sacs which may be accompanied by a dilatation of the outer skin.

Many specimens exhibit what is termed an external vocal sac: this is shown by an ovate space of the skin of greater or less extent above the shoulder-joint being dilated, thin, and rugose; it admits of a greater distension of the vocal sac proper, which is present in all male specimens. A number of adult male individuals show no such modification of the outer skin, and for them the specific name "*halecina*" has been retained, while the others have been distinguished as *Rana utricularia*. I cannot discover any other constant distinctive character; and observing that male specimens from the same locality and of nearly the same size may or may not have this modification of the outer skin, I am disinclined to admit it as a specific distinction.

Brocchi (Miss. Sc. Mex., Batr. p. 12, t. 3. figg. 1, 1 *a–c*) describes and figures a *Rana macroglossa* from Guatemala. Until the typical specimen is re-examined and redescribed, it is not possible to form an opinion as to this frog. The figure certainly does not assist. According to the description this frog would be characterized by having a tympanum of about half the extent of the eye, but in the figure it is represented of the same size as in *R. halecina*. The two inner fingers, stated to be unequal in length, are represented as equal in the special figure (1 *a*). No vomerine teeth whatever are shown in fig. 1 *c*, and the tongue, "légèrement échancrée en avant," is figured as pointed in front, &c.

3. **Rana forreri.** (Tab. LX. fig. A.)

Rana forreri, Bouleng. Ann. & Mag. N. H. 1883, xi. p. 343.

Hab. Mexico, Presidio de Mazatlan and Ciudad in Durango (*Forrer*).

Head of moderate size, with the snout rather pointed; canthus rostralis rather obtuse, the loreal region slanting outwards and slightly concave. Vomerine teeth in two short, scarcely oblique series, between the choanæ *. The first finger is very little longer than the second when both are adpressed to each other. Toes with the web

* In the typical example they extend a little behind the choanæ; but this is not always the case, as in two of the specimens from Ciudad, which, however, are referred to *R. halecina* in the Natural History Museum.

distinctly emarginate and not extending to the tips. Only one metatarsal tubercle, which is narrow and rather short. Hind limb, without tarsus, as long as, fourth toe (from metatarsal joint) four-sevenths of the length of, the body. Interorbital space narrow, not quite half the width of the upper eyelid. Tympanum nearly as large as the eye. Back with a narrow glandular fold on each side, and, like the sides, with numerous tubercles and short glandular tracts. Greenish-olive, with numerous large black spots, irregular in shape, and edged with light green. Legs with cross-bars. Lower parts whitish.

The type from Presidio is an adult female, 70 millim. long; but three other examples, collected by Forrer at Ciudad, clearly belong to the same species. Of these latter, two are adult females, and one a half-grown male with the inner vocal sacs but little developed. In a perfectly mature condition this specimen would have had external vocal sacs, of which already traces are visible.

4. **Rana omiltemana,** sp. n. (Tab. LXI. fig. A.)

Hab. Mexico, Omilteme in Guerrero 8000 feet (*H. H. Smith*).

Head of moderate size, with the snout rather pointed. Tympanum as large as, and close to, the eye. Vomerine teeth in two very oblique (sometimes almost longitudinal) series, which extend backwards beyond the level of the hind margin of the choanæ. First finger extending beyond the second. Toes with the web distinctly emarginate and not reaching the tips; only one (the inner) metatarsal tubercle, which is narrow and short. Hind limb, without tarsus, as long as, fourth toe (from metatarsal joint) two-thirds of the length of, the body. Back with a narrow glandular fold on each side; upper eyelid not granulated. Greyish-brown, with rather small rounded or subquadrangular dark spots on the back; small black dots scattered between the spots, a blackish line along the canthus rostralis. Hind limbs with cross-bars; præanal region reddish. Lower parts whitish; throat sometimes mottled with black.

I have seen five adult males. Their vocal sac is small, but there is generally an outer vocal sac developed between the tympanum and shoulder-joint.

Distance between snout and vent	60 millim.
Diameter of tympanum	5·5 ,,
Length of hind limb to tibio-tarsal joint	60 ,,
Distance from heel to end of foot	51 ,,
,, of metatarsal joint from tip of fourth toe . . .	37 ,,

R. omiltemana shows resemblances to several other species, e. g. *Rana halecina*, *R. sylvatica*, but perhaps to no other more than to *R. clamata*, from which it differs, *inter alia*, by a relatively shorter second finger and by its smaller size.

5. **Rana bonaccana,** sp. n. (Tab. LX. fig. B.)

Hab. Mexico, Island of Bonacca, off the coast of Yucatan (*Gaumer*).

Form rather robust; head large; snout moderately pointed, with very distinct canthus rostralis, and high, concave loreal region. Nostril nearer to the end of the snout than to the eye. Tympanum as large, or nearly as large, as the eye, and equal to the distance of the eye from the nostril. Vomerine teeth in two oblique rows, placed between the hinder part of the choanæ, not close together. First finger but little longer than the second when both are adpressed to each other. Toes broadly webbed, only the last phalanx of the fourth being free. Only one narrow metatarsal tubercle. Fingers and toes with rounded (not dilated or truncated) ends. Hind limb, without tarsus, rather shorter than, fourth toe (from metatarsal joint) half as long as, the body. A very narrow glandular fold on each side of the back; skin of the upper parts and sides covered with minute asperities, of the lower parts smooth. Greyish, irregularly marbled with brown and black, the spots on the legs being more or less transverse. Lower parts whitish, densely and finely marbled with brown; throat blackish-brown; no black spot in the temporal region, but one of oblong shape on the front part of the humerus; and a black stripe below the canthus rostralis.

Several specimens agree in form and coloration; the measurements of an adult female are:—

Distance between snout and vent	100	millim.
Diameter of tympanum	9	,,
Length of hind limb to tibio-tarsal joint	89	,,
Distance from heel to end of foot	75	,,
,, of metatarsal joint from end of fourth toe	51	,,

Neither *R. clamitans* nor *R. draytonii* has so angular a canthus rostralis or so broad a forehead as this species.

6. **Rana maculata.**

Rana maculata, Brocchi, Miss. Sc. Mex., Batr. p. 13, t. 3. figg. 2, 2 *a–c*.

Hab. Guatemala, Totonicapam (*Bocourt*).

Form robust; head of moderate size; snout short and moderately pointed; canthus rostralis distinct. Tympanum two-thirds the size of the eye. Vomerine teeth in two oblique groups between the hinder part of the choanæ. The two inner fingers nearly of the same length. Toes broadly webbed, the membrane extending to the penultimate phalanx of the fourth toe. A small inner metatarsal tubercle. Skin smooth (dorsal glandular folds apparently but little developed). Upper parts greenish-olive, with rather small, rounded black spots; hind limbs banded. Temple with a black band from the eye to behind the tympanum.

Werner (Jahresb. naturw. Ver. Magdeburg, 1894, p. 128) places this frog among the synonyms of *R. halecina*; but if it is correctly described (and description and figure agree tolerably well), it must be a distinct species, apparently of the *R. sylvatica* type.

7. **Rana pustulosa.** (Tab. LXI. fig. B.)

Rana pustulosa, Bouleng. Ann. & Mag. N. H. 1883, xi. p. 343.

Hab. MEXICO, Ventanas in Durango (*Forrer*).

Form of the body moderately robust; head of moderate size, with broad and rounded snout; canthus rostralis angular; loreal region high and concave. Tympanum half the size of the eye. Vomerine teeth in two short oblique series extending to behind the level from the hind margin of the choanæ. First finger extending beyond the second. Toes with very broad web, reaching the last phalanx of the fourth toe. Tips of the toes slightly swollen. An inner small metatarsal tubercle. Hind limb, without tarsus, shorter than, fourth toe (from metatarsal joint) half as long as, the body. Back with a narrow glandular fold; skin of the upper parts covered with small pustules. Greyish-brown, with indistinct blackish spots; limbs indistinctly barred. Lower parts dull yellow, throat marbled with blackish. An indistinct light line runs along the upper lip and side of the abdomen.

The measurements of an adult female are:—

Distance between snout and vent	110 millim.
Diameter of tympanum	6 ,,
Length of hind limb to tibio-tarsal joint	96 ,,
,, fourth toe from metatarsal joint	56 ,,

Some small frogs collected by Dr. A. C. Buller at the Hacienda Santa Gertrudis, Jalisco, may possibly be the young of this species; their snout is, however, more pointed; the skin shows only minute asperities; the web reaches only the penultimate joint. A whitish line runs along the glandular fold, and another, broader and more distinct, along the upper lip and side of the abdomen.

8. **Rana palmipes.**

Rana palmipes, Spix, N. Sp. Test., Ran. p. 29, t. 5. fig. 1; Peters, MB. Ak. Wiss. Berl. 1872, p. 205; 1873, p. 622.

Ranula gollmeri, Peters, l. c. 1859, p. 402 (from a young, apparently not well-preserved specimen).

Rana affinis, Peters, l. c. 1859, p. 403; 1871, p. 402.

Ranula affinis, Cope, Proc. Ac. N. Sc. Philad. 1866, p. 129; 1868, p. 117.

Pohlia palmipes, Steind. Reise der Novara, Amph. p. 15, t. 1. fig. 5; Günth. Zool. Rec. iv. p. 147.

Rana vaillanti, Brocchi, Miss. Sc. Mex., Batr. p. 11, t. 2. figg. 1, 1 *a–c*.

Rana palmipes, part., Bouleng. Batr. Sal. p. 48.

The synonymy as given above is that compiled by Peters, Cope, and Boulenger; but I am not satisfied that the different names and descriptions really refer to the same species. The difficulty is not likely to be solved soon—first, because the typical specimens are evidently in an indifferent state of preservation, and, secondly, because no collection contains sufficient material from the various localities of Tropical America to permit of a satisfactory examination. Central-American specimens differ very considerably in the extent of the web between the toes, in the size of the tympanum, and in coloration; as far as the material before me goes, I distinguish among them at least two species, for one of which I retain the name of *R. palmipes*, but whether or not this is the species so-called by Spix, I am unable to say. In South-American specimens similar differences occur and have been noticed by Cope, who described a *Ranula brevipalmata* and *R. nigrilatus* as distinct species. They may possibly be identical with Central-American examples, but without direct comparison of specimens from the northern and southern countries it is impossible to arrive at a definite conclusion.

Hab. MEXICO (*Sallé*); GUATEMALA, Vera Paz (*Salvin*).—Northern parts of Tropical South America.

Form of the body somewhat slender; head of moderate size, with the snout rather long; canthus rostralis angular; loreal region high and concave. Tympanum two-thirds the size of the eye. Vomerine teeth in two short groups between the hind margins of the choanæ. First finger slightly extending beyond the second. Toes with small terminal disks and with the web reaching nearly to the last phalanx of the fourth toe. A small inner metatarsal tubercle. Hind limb without tarsus shorter than, fourth toe (from metatarsal joint) half as long as, the body. Back with a narrow glandular fold; skin smooth. Brownish (in spirits), without or with indistinct marblings; legs barred; thighs finely mottled behind; lower parts whitish, thighs sometimes spotted below. Upper lip with a whitish line.

Measurements of an adult (?) female:—

Distance between snout and vent	58 millim.
Diameter of tympanum	5·5 „
Length of hind limb to tibio-tarsal joint	50 „
„ fourth toe from metatarsal joint	30 „

9. **Rana melanosoma,** sp. n. (Tab. LXIII. fig. B.)

Rana palmipes, part., Bouleng. Batr. Sal. p. 48.

Hab. GUATEMALA, Dueñas (*Salvin*) *; NICARAGUA, Hacienda Rosa de Jericho, 3250 feet (*Rothschild*).

Form of the body and size of the head moderate; snout moderately long, rounded,

* Two specimens from Lanquin (*Salvin*) are not in a sufficiently good condition to allow me to give an opinion as to their species.

and rather sloping in front; canthus rostralis obtusely angular; loreal region moderately high and slightly concave. Tympanum half the size of the eye. Vomerine teeth in two short, slightly oblique groups between the choanæ, and rather more remote from the choanæ than from each other. First finger subequal in length to second. Toes with small terminal disks, and with the web reaching to the distal end of the penultimate phalanx, or nearly so. A very small inner metatarsal tubercle. Hind limb without tarsus longer than, fourth toe (from metatarsal joint) rather less than half as long as, the body. Back with a narrow and low glandular fold on each side; skin smooth. Upper parts very dark coloured (nearly black in spirits); legs obscurely cross-barred. Lips, throat, and hind part of the thighs marbled black and white; female nearly uniform whitish below.

Male with vocal sacs, but without corresponding external enlargement of the integument.

Measurements of an adult male:—

Distance between snout and vent	65	millim.
Longitudinal diameter of eye	8	„
„ „ tympanum	4·5	„
Length of hind limb to tibio-tarsal joint	72	„
„ fourth toe from metatarsal joint	35	„

10. **Rana godmani**, sp. n. (Tab. LXIII. figg. A, A′.)

Hab. Costa Rica, Rio Sucio (*Underwood*).

Form of the head and body resembling that of *Hylorana*. Snout moderately long, rather pointed, flat above, with distinct canthus rostralis, and with the sides rather high and subvertical. Tympanum nearly the size of the eye (females). Vomerine teeth in two oblique series between the choanæ. First finger longer than the second. Toes with the ends swollen, ends of the fingers less so. Web reaching to the proximal end of the penultimate phalanx, which, however, shows a narrow fringe in its whole length. A small inner metatarsal tubercle. Hind limb without tarsus shorter than the body, fourth toe (from metatarsal joint) nearly one-half the length of the body. Body smooth, with the exception of a rather broad and prominent glandular fold along each side of the back. Back greenish, with numerous or few black specks and spots. A black stripe runs from the end of the snout along the lower edge of the canthus rostralis and of the glandular fold. An oblique black stripe descends from the upper corner of the tympanum towards the shoulder-joint. Upper lip with a more or less distinct whitish stripe. Hind limbs without cross-bands and with the concealed parts unspotted. Lower parts whitish.

	♀.	♀.	♀.
Distance between snout and vent	79 millim.	70 millim.	53 millim.
Longitudinal diameter of eye	6 „	7 „	5 „
„ „ tympanum . . .	5 „	5 „	4 „
Length of entire hind limb	126 „	115 „	86 „
„ hind limb to heel	70 „	63 „	46 „
„ fourth toe from metatarsal joint .	40 „	37 „	26 „

I have examined three specimens of this fine species, which seem to be females.

11. **Rana cæruleopunctata.**

Rana cæruleopunctata, Steind. Verh. zool.-bot. Ges. Wien, 1864, p. 264, t. 15. fig. 1 (details incorrect, see p. 265); Bouleng. Batr. Sal. p. 50.

Ranula cæruleopunctata, Cope, Proc. Ac. N. Sc. Philad. 1866, p. 130.

Ranula chrysoprasina, Cope, l. c.; Bull. U. S. Nat. Mus. no. 32, p. 19 (1887).

Trypheropsis chrysoprasinus, Cope, l. c. 1868, p. 117; Journ. Ac. N. Sc. Philad. viii. p. 114, t. 23. fig. 12 (1876).

Hylarana chrysoprasina, Brocchi, Miss. Sc. Mex., Batr. p. 65.

Rana chrysoprasina, Bouleng. Batr. Sal. p. 49.

Hab. Nicaragua (*Cope*); Costa Rica, La Palma, alt. 1500 metres (*Mus. Nacional de Costa Rica*).

Body slender; head moderate; snout produced, pointed, overreaching the mouth, with the canthus rostralis and anterior edge sharply angular. Loreal region high, vertical, slightly concave. Tympanum about as large as the eye. Vomerine teeth when developed in two oblique series between the hind margins of the choanæ. First finger very little longer than the second. Toes with small terminal disks, and with the webs emarginate, leaving the last two phalanges of the fourth toe free. Two small metatarsal tubercles, the outer very small, at the base of the fourth toe. Hind limb, without tarsus, as long as the body; fourth toe (from metatarsal joint) half as long. Back with a very thin and scarcely distinct lateral glandular fold; skin smooth, on the hinder part of the back indistinctly granulated. Upper parts dark olive, hind limbs barred; a bluish-white band from the tip of the snout along the upper lip to the shoulder; a fine whitish line along the canthus rostralis and glandular fold; lower parts whitish, with or without fine blackish marbling; a brown band along the front of the upper arm. Hinder part of the thighs black, with a few large golden spots.

Male without vocal sacs.

Measurements of an adult male:—

Distance between snout and vent	43 millim.
Length of hind limb to tibio-tarsal joint	43 „
„ fourth toe to metatarsal joint	22 „

LÆVIRANA.

Levirana, Cope, Proc. Am. Phil. Soc. 1894, p. 197.

Differs from *Rana* by the absence of vomerine teeth.

1. **Lævirana vibicaria.**

Levirana vibicaria, Cope, Proc. Am. Phil. Soc. 1894, p. 197.

Hab. COSTA RICA, Irazu Range and Isla Nueva (*G. K. Cherrie*).

Form robust; snout short, wide; hind limbs rather short, the heel reaching to the front of the eye. Canthus rostralis strongly marked, angular near the nostril, sloping towards the apex, which does not project beyond the border of the lip. Loreal region slightly concave. Tympanum rather more than half the width of the eye. Vomerine teeth none; choanæ rather small. Tongue deeply emarginate behind. Toes with slightly enlarged terminal disks, supported by T-shaped phalanges. First finger a little shorter than second. Toes half-webbed. One flat, oval, metatarsal tubercle. Skin smooth; a glandular fold on each side of the back. Olive-grey above, side of the head and a band on each side of the back black. Two rows of small black spots on each side of the vertebral column. Yellowish below; concealed surface of femur reddish, unspotted; upper surface of femur with a black stripe on the distal half. Lips unspotted.

Length of head and body	65	millim.
„ hind limb	101	„
„ hind foot	49	„
„ tarsus (astragalo-calcaneum)	16	„

Fam. **HYLAPLESIIDÆ.**

HYLAPLESIA.

Hylaplesia sp., Boie, Isis, 1827, p. 294.
Dendrobates, Wagl. Syst. Amphib. p. 202.
Hylaplesia, Günth. Batr. Sal. p. 124.

Synopsis of Central-American Species.

First finger not extending as far as second; the tarso-metatarsal joint not reaching the end of the snout . . . . 1. *tinctoria,* Schn.

First finger reaching slightly beyond the second; tarso-metatarsal joint reaching far beyond the end of the snout . . . . 2. *trivittata,* Spix.

First and second fingers of equal length; the tarso-metatarsal joint reaching the end of the snout; size of tympanum about equal to that of a disk . . . . 3. *typographa,* Kef.

First finger longer than the second; the tarso-tibial joint reaching nearly to the end of the snout; a white band along the upper lip and the lower part of the side . . . . 4. *talamancæ,* Cope.

1. **Hylaplesia tinctoria.**

Calamita tinctorius, Schneid. Hist. Amph. p. 175.

Hylaplesia tinctoria, Boie, Isis, 1827, p. 294; Günth. Batr. Sal. p. 125; Brocchi, Miss. Sc. Mex., Batr. p. 89, t. 11. figg. 2, 2 *a–d.*

Dendrobates tinctorius, Wagl. Syst. Amph. p. 202; Dum. & Bibr. Erpét. Gén. viii. p. 252, t. 90. fig. 1; Steind. Verh. zool.-bot. Ges. Wien, 1864, p. 260, t. 13. fig. 1, t. 15. fig. 2; Bouleng. Batr. Sal. p. 142; Cope, Bull. U. S. Nat. Mus. no. 32, p. 19.

Dendrobates histrionicus, Berthold, Gött. Nachr. 1845, p. 43; Gött. Abhandl. p. 15, t. 1. fig. 8.

Phyllobates auratus, Girard, U. S. Astron. Exped. ii. p. 209; Proc. Ac. N. Sc. Philad. vii. p. 226 (1854).

Hylaplesia aurata, Cope, Proc. Ac. N. Sc. Philad. 1863, p. 49.

Dendrobates lugubris, Schmidt, Denkschr. Wien. Ak. 1858, p. 250; Bouleng. Batr. Sal. p. 145.

Phyllobates chocoensis, Posada Arango, Rev. et Mag. Zool. 1872, t. 27.

Dendrobates tinctorius, subsp. *vittatus,* Cope, Proc. Am. Phil. Soc. xxxi. p. 340 (1894).

Hab. Nicaragua (*Bransford*); Costa Rica, La Palma, Talamanca, Buenos Ayres (*Mus. Nacional de Costa Rica*), Las Delicias, Santa Clara (*Biolley*); Panama (*Pfeiffer*), Veraguas (*Warczewicz*), Taboga I. (*U. S. Astron. Exped.*), Isthmus of Darien (*Mus. Brit.*).

Extremely variable in coloration: uniform black; greyish above and black below; black with white or grey spots; greyish with black spots; black with white transverse or longitudinal bands and spots.

2. **Hylaplesia trivittata.**

Hyla trivittata, Spix, Spec. Nov. p. 35, t. 9. fig. 1.

Hyla nigerrima, Spix, l. c. p. 36, t. 9. fig. 2.

Dendrobates nigerrimus, Wagl. Syst. Amph. p. 202; Steind. Verh. zool.-bot. Ges. Wien, 1864, p. 257, t. 13. fig. 2.

Dendrobates trivittatus, Wagl. l. c.; Peters, MB. Ak. Wiss. Berlin, 1872, p. 213; Bouleng. Batr. Sal. p. 144.

Hylaplesia picta, Tschudi, Batr. p. 71.

Dendrobates obscurus, Dum. & Bibr. Erpét. Gén. viii. p. 655.

Dendrobates trivittatus, var. *maculatus,* Peters, MB. Ak. Wiss. Berlin, 1873, pp. 617, 618.

Hab. Panama, Chiriqui (*Peters*).—Tropical America.

Sometimes uniform black, generally with white bands and spots variously disposed.

3. **Hylaplesia typographa.**

Dendrobates typographus, Keferst. Gött. Nachr. 1867, p. 360; Wiegm. Arch. 1868, p. 298, t. 9. fig. 7; Cope, Journ. Ac. N. Sc. Philad. viii. p. 102 (1875).

Dendrobates ignitus, Cope, Proc. Ac. N. Sc. Philad. 1874, p. 68.

Hab. Nicaragua, Chontales Mines (*Rix*); Costa Rica (*Mus. Nacional de Costa Rica*).

4. **Hylaplesia talamancæ.**

Dendrobates talamancæ, Cope, Journ. Ac. N. Sc. Philad. viii. p. 102, t. 23. fig. 6 (1875).

Hab. COSTA RICA (*Mus. Nacional de Costa Rica*), Cold Harbour, on the East Coast (*Gabb*).

NOTE.—I hesitate to introduce into this Fauna a frog which Cope described as *Phyllobates truncatus* (Proc. Ac. N. Sc. Philad. 1860, p. 372) and later on as *Hylaplesia truncata* (ibid. 1863, p. 42). He was uncertain as to its origin, and although he gives "Panama" as locality in Bull. U. S. Nat. Mus. no. 32, p. 19, he does not add an authority for this statement. However, I reproduce here what is known of this frog; it is omitted in Boulenger's Catalogue:—

Skin smooth, above and below. Tongue small, linear, entire. First and second fingers equal in length; disks one-fourth the size of the tympanum; anterior limbs more than half the length of posterior. Front and canthus rostralis convex; snout concave, truncate, elevated. Tympanum less than half the size of the eye. A transverse bony ridge on the occiput. Coloration a dark ferruginous maroon above and below. A pale curved line on each side, beneath this another which unites with its fellow on the anterior part of the abdomen. From this point of junction a medial band takes its rise and bifurcates behind. A pair of parallel lines on the throat, which unite anteriorly, following the curve of the mandible. Limbs sparsely and irregularly faintly spotted.

"A certain serrulation of the alveolar ridge of the maxillary bone greatly resembles dentition."

Fam. BRACHYCEPHALIDÆ.

ATELOPUS.

Atelopus, Dum. & Bibr. Erpét. Gén. viii. p. 660; Philippi, Wiegm. Arch. 1894, p. 214; Bouleng. Ann. Mag. N. H. 1894, xiv. p. 374.

Phryniscus sp., Dum. & Bibr. Erpét. Gén. viii. p. 722 (nec Wiegm.).

Phrynidium, Martens, Nomencl. Rept. Mus. Berol. p. 40 (1856).

Hylæmorphus, Schmidt, Denksch. Wien. Ak. 1858, p. 255.

Phirix, Schmidt, l. c. p. 256.

1. **Atelopus varius.** (Tab. LXII. figg. A, B, C.)

Atelopus varius, Stannius, Handb. Zoot., Amphib. 1856, p. 16 (nom. nud.); Keferst. Wiegm. Arch. 1868, p. 292, t. 8. figg. 33, 34; Cope, Journ. Ac. N. Sc. Philad. viii. p. 102 (1875); Brocchi, Miss. Sc. Mex., Batr. p. 96, t. 11. figg. 1, 1 *a–d.*

Phrynidium varium, Martens, Nomencl. Mus. Zool. Berol. p. 40.

Phryniscus varius, Günth. Batr. Sal. p. 45; Bouleng. Batr. Sal. p. 152.

Atelopus varius, var. *maculatus,* Peters, MB. Ak. Wiss. Berlin, 1873, p. 618.

Hylæmorphus dumerilii, Schmidt, Denksch. Wien. Ak. 1858, p. 255, t. 3. figg. 23, 24.

Phryniscus dumerilii, Brocchi, Miss. Sc. Mex., Batr. p. 98.

Hab. GUATEMALA (*Godman*); COSTA RICA, Bebedero, Escazu (*Underwood*), Izaru (*Rogers*), San Marcos, Tivas (alt. 1000 met.), Santa Clara (alt. 400 met.) (*Mus. Nacional de Costa Rica*); PANAMA, Veraguas, Chiriqui (*Mus. Berol.*).

Coloration very variable. The ground-colour is black, the ornamental colours pink, yellow, and green. The latter are distributed in various patterns, in the form of spots, or longitudinal or transverse bands, irregularly and more rarely symmetrically. The upper or lower parts may be black, almost uniform, or pink with scarcely a trace of black. In some of the most aberrant colour-variations the ground-colour is greyish, and the yellow bands are edged with black; in others the colours assume a reticulated pattern; and, finally, specimens occur of a brownish-olive, mottled and dotted with darker, the abdomen in these being yellow, with some irregular black linear figures.

Extremely common and widely distributed in Costa Rica.

Fam. ENGYSTOMATIDÆ.

ENGYSTOMA.

Engystoma, Fitz. N. Class. Rept. p. 65; Günth. Batr. Sal. p. 51; Bouleng. Batr. Sal. p. 160.

Tongue elliptic, free and entire behind; a transverse ridge of the mucosa between the choanæ. Vomerine teeth none. Tympanum hidden; ostia pharyngea narrow. Fingers and toes free, without disks. Outer metatarsals united. Precoracoids and omosternum absent; sternum cartilaginous. Skin smooth.

1. Engystoma ustum.

Engystoma rugosum, Günth. Batr. Sal. p. 52 (nec Dum. et Bibr.).
Engystoma carolinense, part., Günth. Batr. Sal. p. 51; Brocchi, Miss. Sc. Mex., Batr. p. 93.
Engystoma ustum, Cope, Proc. Ac. N. Sc. Philad. 1866, p. 131; 1871, p. 216; Brocchi, Miss. Sc. Mex., Batr. p. 94, t. 10. fig. 1; Bouleng. Batr. Sal. p. 161.
Systoma ustum, Cope, Journ. Ac. N. Sc. Philad. vi. p. 194 (1867).
Engystoma mexicanum, Peters, MB. Ak. Wiss. Berlin, 1869, p. 881.

Hab. Mexico, Presidio de Mazatlan (*Forrer*), Agua Delgada, Guadalajara (*Buller*), Guanajuato (*Dugès*), Cordova (*Sallé*), Vera Cruz and Tehuantepec (*Sumichrast*), Coatzocoalcos (*Streets*).

Metatarsus with two tubercles.

2. Engystoma pictiventre.

Engystoma pictiventre, Cope, Proc. Am. Phil. Soc. 1886, p. 272.

Hab. Nicaragua (*Bransford*).

One small metatarsal tubercle. Snout twice as long as the eye, and projecting well beyond the mouth. No fold across the occiput; skin smooth. First finger shorter than second; first toe very short, second * considerably longer than fifth; fourth elongate. Olivaceous-brown above. A black band with a very narrow pale upper border extends from the end of the snout to the lower part of the groin. No inguinal spot. Black below, with white spots, those on the abdomen very large, the others smaller. (After Cope.)

Length of body	22·5 millim.
" hind leg (measured from groin)	29 "
" hind foot	14 "
" astragalar portion of hind foot	4·5 "

3. **Engystoma elegans.** (Tab. LXII. fig. D.)

Engystoma carolinense, part., Günth. Batr. Sal. p. 51.
Engystoma elegans, Bouleng. Batr. Sal. p. 162.

Hab. Mexico, Cordova (*Sallé*).

Snout rather pointed, projecting beyond the mouth, with rather obtuse canthus rostralis and sloping sides. First finger shorter than second; a single small metatarsal tubercle; third and fifth toes equal in length. The length of the body equals that of the hind limb to the root of the fourth toe. Skin smooth; a fold across the occiput. Light brown, back almost entirely covered with a large subtriangular patch, pointed towards the head, and with the base behind; sides of the head and neck dark brown. Lower parts marbled with brown, the brown colour forming a network; groin with a dark brown spot; thigh, calf, and tarsus with a single cross-band each, edged with darker.

Distance of vent from snout	32 millim.
Total length of hind limb (measured from vent)	43 "
Length of foot with tarsus	21 "

GLOSSOSTOMA, gen. nov.

Resembling *Engystoma*. Tongue very large, elliptic, free and entire behind; no transverse ridge of the mucosa between the choanæ. Vomerine teeth none. Tympanum hidden below the skin; ostia pharyngea narrow. Fingers and toes without disks, the latter webbed. Outer metatarsals united. Sternal apparatus as in *Engystoma*. Skin quite smooth everywhere, a cross-fold behind the superciliaries.

1. **Glossostoma aterrimum,** sp. n. (Tab. LXIV. fig. A.)

Hab. Costa Rica (*Underwood*).

Resembling an *Engystoma* in general habit. Snout depressed, rather pointed, of

* Cope may have intended to write "third."

moderate length; the entire hind leg not much longer than the body; eye not very small, its diameter equal to its distance from the nostril. First finger shorter than second, third considerably longer than fourth. Toes one-third webbed, without subarticular tubercles; a single very small and flat metatarsal tubercle; no tarsal fold. Skin very smooth and thick. Uniform deep black; minute scattered granules with bluish points on each side of the vent.

Length of the body	55 millim.
„ hind limb (from vent)	65 „
Distance between vent and heel	35 „

HYPOPACHUS.

Hypopachus, Keferstein, Gött. Nachr. 1867, p. 352; Bouleng. Batr. Sal. p. 159.

Differing from *Engystoma* by the presence of very small precoracoids.

1. Hypopachus variolosus.

Engystoma variolosum, Cope, Proc. Ac. N. Sc. Philad. 1866, p. 131; Proc. Am. Phil. Soc. 1869, p. 166.

Systoma variolosum, Cope, Journ. Ac. N. Sc. Philad. vi. p. 194 (1867).

Hypopachus variolosus, Cope, Journ. Ac. N. Sc. Philad. viii. p. 101 (1875); Brocchi, Miss. Sc. Mex., Batr. p. 92, t. 10. fig. 2; Bouleng. Batr. Sal. p. 159; Cope, Bull. U. S. Nat. Mus. no. 32, p. 18 (1887).

Hypopachus seebachii, Keferst. Gött. Nachr. 1867, p. 352; Wiegm. Arch. 1868, p. 293, t. 9. figg. 1, 2.

Hypopachus oxyrrhinus, Bouleng. Ann. & Mag. N. H. 1883, xi. p. 344; Zool. Record, 1889, p. 21.

Hypopachus cuneus, Cope, Bull. U. S. Nat. Mus. no. 34, p. 388 (1889).

Hab. North America, Texas, Duval Co. (*Taylor*).—Mexico, Presidio de Mazatlan (*Forrer*), Guadalajara (*Dugès*); Costa Rica (*Seebach, coll. Salvin-Godman*), Bebedero (*Underwood*), San José (*Biolley*).

Two very large metatarsal tubercles, oval and compressed. Abdomen brownish, with rounded whitish spots.

2. Hypopachus inguinalis.

Hypopachus inguinalis, Cope, Proc. Am. Phil. Soc. 1869, p. 166; Brocchi, Miss. Sc. Mex., Batr. p. 91, t. 10. figg. 3, 4; Bouleng. Batr. Sal. p. 160.

Hypopachus variolosus, subsp., Cope, Bull. U. S. Nat. Mus. no. 32, p. 18.

Hab. Guatemala, Coban (*Salvin*).—Ecuador (*Fraser*).

Two prominent blunt metatarsal tubercles. Abdomen whitish, marbled with black.

Fam. CYSTIGNATHIDÆ.

LEPTODACTYLUS.

Leptodactylus, Fitz. Class. Rept. p. 38.
Cystignathus, Wagl. Syst. Rept. p. 202.
Tarsopterus, Reinh. & Lütk. Vid. Meddel. 1861, p. 177.
Gnathophysa, Cope, Nat. Hist. Rev. 1865, p. 112.
Entomoglossus, Peters, MB. Ak. Wiss. Berlin, 1870, p. 647.

Fingers and toes free, without disks. Vomerine teeth present. Tongue entire or slightly notched behind. Tympanum distinct. Outer metatarsals united. Sternum with a bony style.

1. **Leptodactylus pentadactylus,** Laur.

For the synonymy, see Peters, MB. Ak. Wiss. Berlin, 1872, p. 197; and Bouleng. Batr. Sal. p. 241.

Hab. Eastern Costa Rica (*Cope*: Bull. U. S. Nat. Mus. no. 32, p. 18), Las Delicias, Santa Clara (alt. 400 met.).—Tropical South America.

Vomerine teeth in two arched series behind the choanæ. Snout rounded, with the canthus rostralis rather obtuse. Tympanum two-thirds or more the size of the eye. Toes not fringed. Two small metatarsal tubercles, the outer one being often absent in young specimens. Skin smooth, generally with large glandular patches on the sides; a glandular fold on each side of the back; abdominal disk more or less indistinct. Coloration subject to great variation: generally greenish-brown above, with transverse rows of black spots; lips without white stripe. Hind limbs cross-banded; thigh marbled with blackish and yellow behind. Male with a strong conical spine on the inner side of the first digit, and with another bi- or tricuspid one on each side of the breast.

Common and widely distributed as this species is in Tropical South America, it does not seem to extend northwards beyond Costa Rica, where it is perhaps scarce or local. I have seen only one very young specimen (body 36 millim. long) from that country; it is peculiarly coloured, very like an example of the same size from Ecuador in the Natural History Museum. The lower parts, sides of the body, and limbs are deep black; lower parts of the body and limbs with small round milky-white spots. Upper parts of the body purplish-grey; interorbital space and anterior part of the back each with one, the snout with two, black cross-bands. Two black longitudinal bands start from the eye, the upper one running along the side of the back to the middle of the pelvis, the second obliquely descending above the tympanum towards the abdomen; canthus rostralis with a black stripe; three black triangular spots on the upper lip. Upper parts of the limbs not banded.

2. Leptodactylus albilabris.

Cystignathus albilabris, Günth. Ann. & Mag. N. H. 1859, iv. p. 217.
Cystignathus labialis, Cope, Proc. Am. Phil. Soc. xvii. p. 90; xviii. p. 269.
Leptodactylus caliginosus, Brocchi, Miss. Sc. Mex., Batr. p. 87, t. 5. fig. 1 (nec Girard).
Leptodactylus labialis, Brocchi, ibid. p. 20.
Leptodactylus albilabris, Bouleng. Batr. Sal. p. 245, t. 16. fig. 4.

Hab. MEXICO, Cordova and Tehuantepec (*Sumichrast*).—ANTILLES, St. Thomas, Sta. Cruz.

Vomerine teeth in two transverse series behind the choanæ. Snout rather pointed and projecting beyond the mouth, with the canthus rather obtuse. Tympanum more than one-half the size of the eye. Toes not fringed; thighs short. Two metatarsal tubercles; a tarsal fold. Skin smooth or with a few small tubercles, a glandular fold on each side of the back; a ventral disk formed by transverse folds. Brownish or blackish, with darker markings; a blackish stripe along the canthus, and a white one along the upper lip. Limbs cross-banded; hinder side of the thigh with a series of white spots.

3. Leptodactylus fragilis.

Cystignathus fragilis, Brocchi, Bull. Soc. Philom. i. p. 182 (1877).
Cystignathus gracilis, Cope, Proc. Am. Phil. Soc. 1869, p. 162; xviii. 1879, p. 269; Bull. U. S. Nat. Mus. no. 32, p. 17 (1887) (nec Dum. & Bibr.).
Leptodactylus fragilis, Brocchi, Miss. Sc. Mex., Batr. p. 19, t. 5. fig. 2; Bouleng. Batr. Sal. p. 241.

Hab. MEXICO, Tehuantepec and Vera Cruz (*Sumichrast*).

Vomerine teeth in two transverse (arched ?) series behind the choanæ. Tympanum nearly as large as the eye. Two metatarsal tubercles. Toes not fringed. Sides with a glandular fold and with tubercles. Brownish above, with black spots.

4. Leptodactylus microtis.

Cystignathus microtis, Cope, Proc. Am. Phil. Soc. xviii. p. 265 (1879).
Leptodactylus microtis, Bouleng. Batr. Sal. p. 244; Cope, Bull. U. S. Nat. Mus. no. 32, p. 17 (1887).

Hab. MEXICO, ? Guanajuato (*Dugès*).

Vomerine teeth in two short transverse series behind the choanæ; snout not elongate, without canthus. Eye not large, and tympanum only two-fifths of its diameter. Limbs stout, especially the femur; toes not fringed. Skin without glandular folds on the sides. Dark brown, with a blackish, pale-edged triangle between the eyes, the apex pointing backwards. No light stripe on the upper lip. (After Cope.)

Length of body	28 millim.
„ hind leg	38 „
„ hind foot	20 „

5. Leptodactylus perlævis.

Cystignathus perlævis, Cope, Proc. Am. Phil. Soc. xviii. p. 269 (1879).

Hab. MEXICO, Western Tehuantepec, Tapana (*Sumichrast*).

Vomerine teeth in two short transverse series, well separated from each other, and extending but little beyond the inner border of the choanæ. Snout projecting a little beyond the mouth, with distinct canthus. Tympanum more than one-half of orbit. Toes not fringed. Skin particularly smooth and shining. Greyish, sides speckled with whitish; an interorbital dark spot; upper lip marbled. Posterior face of femur dark, with white specks. Hind limbs obscurely cross-banded above. (After Cope.)

Length of body	38 millim.
„ hind limb	55 „
„ hind foot	28 „

6. Leptodactylus caliginosus.

Leptodactylus caliginosus, Girard, Proc. Ac. N. Sc. Philad. 1853, p. 422; Bouleng. Batr. Sal. p. 247; and Zool. Record for 1887, Rept. p. 31.

Cystignathus melanonotus, Hallowell, Proc. Ac. N. Sc. Philad. 1860, p. 480; Cope, Proc. Am. Phil. Soc. xviii. p. 269 (1879).

Cystignathus echinatus, Brocchi, Bull. Soc. Philom. i. p. 181 (1877).

Leptodactylus validus, Garman, Bull. Essex Inst. xix. p. 14 (1887).

Leptodactylus echinatus, Brocchi, Miss. Sc. Mex., Batr. p. 18, t. 5. fig. 4.

Leptodactylus melanonotus, Brocchi, ibid. p. 20; Cope, Bull. U. S. Nat. Mus. no. 32, p. 18 (1887).

Hab. MEXICO, Mazatlan, Misantla, Tehuantepec; NICARAGUA; COSTA RICA.—SOUTH AMERICA, Colombia, Ecuador, Demerara, Brazil, Paraguay; ANTILLES.

Vomerine teeth in two slightly arched or straight series behind the choanæ. Snout rounded, moderately long. Tympanum two-thirds the size of the eye. Toes fringed; two metatarsal tubercles. Skin smooth, with a few small flat tubercles, without longitudinal folds. Brownish, with indistinct darker markings; a large angular spot on the occiput. Hinder side of the thigh marbled with darker. Male with two spinous tubercles on the inner side of the first finger.

An examination of specimens from various localities reveals an extraordinary amount of variation in the comparative length of the hind limbs, in the extent of the transverse series of vomerine teeth, and in the coloration of the abdomen—variations which generally in this and allied genera are used for specific distinction. The extremes of these variations are, however, connected by intermediate forms; and therefore I agree with Boulenger, who, at any rate at present, leaves them under the same specific name.

BORBOROCŒTES.

Borborocœtes, Bell, Voy. 'Beagle,' Rept. p. 30.
Borborocœtes, sp., Bouleng. Batr. Sal. p. 252.

Fingers and toes free, without disks. Vomerine teeth present. Tongue entire or slightly notched behind. Tympanum distinct. Outer metatarsals united. Sternum without bony style.

1. Borborocœtes mexicanus.

Borborocœtes mexicanus, Bouleng. P. Z. S. 1898, p. 481, t. 39. fig. 2.

Hab. MEXICO, Hacienda el Florencio, Zacatecas (*Buller*).

Habit raniform. Snout rounded, a little longer than the eye, with obtuse canthus. Tympanum two-thirds the diameter of the eye. Vomerine teeth in two rounded groups behind the choanæ. First finger longer than second; subarticular tubercles prominent; two metatarsal tubercles, of which the inner is unusually large and compressed. Skin smooth above and below. Greyish above, with darker markings; a dark cross-bar between the eyes; a dark lumbar spot; a dark streak on each side of the head and body not extending to the groin; lips and limbs cross-banded. White below.

Length of body	36 millim.
„ hind limb	63 „
Distance between snout and vent	36 „

SYRRHAPHUS.

Syrrhophus *, Cope, Am. Nat. 1878, p. 253.
Syrrhopus, Bouleng. P. Z. S. 1888, p. 206.
Hypodictyon, Cope, Proc. Am. Phil. Soc. 1885, p. 383.
Malachylodes, Cope, Proc. Am. Phil. Soc. 1879, p. 264.
Phyllobates, part., Cope; Bouleng. Batr. Sal. p. 194.

Toes free; disks, if developed, generally small. Outer metatarsals united. Vomerine teeth absent. Tongue entire or very little emarginate behind. Pupil horizontal; tympanum distinct. Sternum without bony style.

1. Syrrhaphus verruculatus.

Phyllobates verruculatus, Peters, MB. Ak. Wiss. Berlin, 1870, p. 650.
Syrrhopus verruculatus, Bouleng. P. Z. S. 1888, p. 206.

Hab. MEXICO, Huanusco.

Abdomen granular; upper parts with small warts. Snout with rounded canthus and

* The etymology of this name is somewhat obscure; it is probably derived from συρραφεὺς (one who joins together), suggested by the united outer metatarsals; the name, therefore, should be *Syrrhaphus*.

vertical loreal region. Tympanum two-thirds the diameter of the eye. Disks of fingers moderate, of toes small; first finger shorter than second; two metatarsal tubercles. Yellowish-brown above, with black spots; loreal and temporal regions blackish-brown. Brownish-yellow below, marbled with black. (After Peters.)

Length of body	20 millim.
„ hind limb	31 „
„ hind foot	14 „

2. **Syrrhaphus omiltemanus,** sp. n.

Hab. Mexico, Omilteme in Guerrero 8000 feet (*H. H. Smith*).

Similar in general habit to *Hylodes rhodopis*. Abdomen coarsely granular, upper parts with minute tubercles. Snout of moderate width, acuminate in front, with angular canthus and subvertical loreal region. Tympanum three-fourths the size of the eye. Disks of fingers and toes exceedingly small; first and second fingers subequal in length. Fingers with large swollen subarticular tubercles, those of the toes not continued on the metatarsals; one metatarsal tubercle. Upper parts brownish; a blackish band runs along the canthus rostralis and, behind the eye, above the tympanum towards the side of the body, where it widens and disappears. A small black spot on each side of the back, behind the shoulder. Hind limbs with short, narrow cross-bands which do not encroach upon the unspotted concealed surfaces of the limbs. Whitish below.

Length of body	21 millim.
„ hind limb	33 „
Distance between vent and heel	18 „

3. **Syrrhaphus ridens.**

Phyllobates ridens, Cope, Proc. Ac. N. Sc. Philad. 1866, p. 131; Journ. Ac. N. Sc. Philad. viii. p. 155 (1875).

Hypodictyon ridens, Cope, Proc. Am. Phil. Soc. 1885, p. 383.

Hab. Nicaragua, Rio San Juan (*Cope*); Panama (*Cope*).

Abdomen closely areolate. Snout projecting beyond the mouth; canthus rostralis distinct. Tympanum one-eighth the size of the eye. Fingers and toes long, with well-marked disks. Heel reaching the end of the snout, when brought forwards. Greyish-brick-red, with grey bars on the lips, hind limbs, and on the interorbital space. A black spot on the tympanum. Lower parts pale brownish. (After Cope.)

4. **Syrrhaphus verrucipes.**

Syrrhophus verrucipes, Cope, Proc. Am. Phil. Soc. viii. p. 383 (1885).

Hab. Mexico, Zacualtipan in Hidalgo (*Cope*).

Skin smooth everywhere, except on the sides, which are tubercular-areolate. Head moderately wide, with vertical loreal region and obtuse canthus rostralis. Tympanum more than two-thirds the diameter of the eye. Toes with small disks, those of the fingers a little larger; two small metatarsal tubercles; subtubercular tubercles large. Grey above, with numerous black spots; limbs with black cross-bars, which continue on the external digit of each foot. Upper surface of femur with close white spots on a dark ground; posterior surface dark, with a few small white spots. White below. (After Cope.)

Length of body	26 millim.
„ hind limb	39 „
„ hind foot	18 „
„ tarsus	8 „

5. **Syrrhaphus leprus.**

Syrrhophus leprus, Cope, Proc. Am. Phil. Soc. xviii. p. 268 (1879).

Hab. Mexico, Santa Efigenia, Tehuantepec (*Sumichrast*).

Skin smooth everywhere. Head wide, snout broadly acuminate, with angular canthus and vertical sides. Tympanum one-third the size of the eye. Digital dilatations small. Brown, dotted with moderately large grey spots; lower parts pale pinkish-brown. (After Cope.)

Length of body	24 millim.
„ hind limb	35 „
„ hind foot	10 „

Found in woods. The dorsal spots are yellow in life.

6. **Syrrhaphus guttulatus.**

Malachylodes guttulatus, Cope, Proc. Am. Phil. Soc. xviii. p. 264 (1879).
Syrrhopus guttulatus, Bouleng. P. Z. S. 1888, p. 206.

Hab. Mexico, Guanajuato (*Dugès*).

Skin smooth, except some small tubercles on the eyelids and a trace of areolation on the posterior part of the sides and abdomen. Head flat and rather wide, with obtuse canthus rostralis. Digital dilatations small. Brown, with numerous small, faint, pale spots; limbs with rufous bands. Uniform yellowish below. (After Cope.)

Length of body	22 millim.
„ hind limb	30 „
„ hind foot	14 „

7. Syrrhaphus cystignathoides.

Phyllobates cystignathoides, Cope, Proc. Am. Phil. Soc. xvii. p. 89 (1877).
Syrrhophus cystignathoides, Cope, ibid. xviii. p. 268 (1879).

Hab. Mexico, Potrero near Cordova, Vera Cruz.

Skin entirely smooth. Snout rather narrow, not projecting beyond the mouth, without canthus. (Disks of fingers and toes moderate.) Tympanum half the diameter of the orbit. Heel reaching to the middle of the eye when brought forwards; two metatarsal tubercles. Brownish; back with large blackish spots, sides marbled with lighter and darker; limbs obscurely cross-banded. (After Cope.)

Length of body	22·5 millim.
„ hind limb	33 „
„ hind foot	15·5 „
„ tarsus	5·5 „

8. Syrrhaphus hylæformis.

Phyllobates hylæformis, Cope, Journ. Ac. N. Sc. Philad. viii. p. 107, t. 23. fig. 8 (1875).
Syrrhopus hylæformis, Bouleng. P. Z. S. 1888, p. 206.

Hab. Costa Rica, Pico Blanco, 7000 feet.

Smooth above and below. Head rather broad, with the snout depressed; canthus rostralis obtuse. Tympanum one-sixth the size of the eye. Fingers and toes with rather large disks *. Rich brown, with a red vertebral stripe. Femora light brown in front and behind. Lower parts rosy or orange.

Length of body (♂)	27 millim.
„ hind limb	37 „
„ hind foot	18 „

PALUDICOLA.

Paludicola, Wagl. Syst. Amphib. p. 206.
Leiuperus, Dum. et Bibr. Erpét. Gén. viii. p. 420.
Paludicola sp., Bouleng. Batr. Sal. p. 229.

Fingers and toes free, without dilatations. Vomerine teeth none. Tongue entire or indistinctly notched behind. Tympanum distinct or (generally) hidden. Outer metatarsals united. Sternum with a bony style.

1. Paludicola nitida.

Leiuperus nitidus, Peters, MB. Ak. Wiss. Berlin, 1869, p. 878.

* Very small in the figure.

Hab. MEXICO *.

General habit slender. Snout as long as the eye, with rounded canthus. Tympanum very indistinct, about one-third the size of the eye. Skin smooth, only a few minute granules on the back. First finger a little shorter than second; subarticular tubercles prominent, extending over the metatarsals. Greenish, marbled with darker; a dark triangular spot between the eyes; loin with an oval black spot marbled with green; limbs with brown bands. Lower parts brownish, with obscure lighter spots. (After Peters.)

Length of body	20 millim.
„ fore limb	14 „
„ hind limb	26 „
„ foot	12 „

2. **Paludicola (?) mexicana.**

Leiuperus mexicanus, Brocchi, Bull. Soc. Philom. i. p. 184 (1877); Miss. Sc. Mex., Batr. p. 21, t. 5. figg. 3, 3 *a–c*.

Paludicola (?) *mexicana,* Bouleng. Batr. Sal. p. 237.

Hab. MEXICO.

Habit stout, head large and broad. Snout truncated, with angular canthus. Tympanum distinct, half the size of the eye. Toes with a rudimentary web at the base; two metatarsal tubercles. Brownish, with some dark marbling on the upper parts; white beneath, throat speckled with grey. (After Brocchi.)

TOMODACTYLUS, gen. nov.

Fingers and toes free, with small truncate dilatations. Vomerine teeth none. Tongue entire or indistinctly notched behind. Tympanum distinct. Outer metatarsals united. A lumbar gland. Sternum with a bony style.

These frogs differ by the presence of digital disks from the species described under *Paludicola*, *Leiuperus*, and other generic names: a character which, in my opinion, should carry generic distinction.

1. **Tomodactylus amulæ,** sp. n. (Tab. LXIV. fig. C.)

Hab. MEXICO, Amula in Guerrero (*H. H. Smith*).

General habit rather slender. Snout moderate, with sharp canthus and subvertical sides; diameter of the eye a little less than its distance from the end of the snout.

* Boulenger (Batr. Sal. p. 237) gives "Puebla" as the locality where this frog was collected, misunderstanding the title of Peters's paper. Peters merely gives "Puebla" as the address of Hr. Berkenbusch, the donor of the collection, expressing his regret that "any notice of the special localities of the species sent is wanting."

Tympanum vertically oval, about one-sixth of the size of the orbit. Upper parts with scattered small tubercles; abdomen granular; a large lumbar gland. Vomerine teeth none. Fore limb slender, hind limbs short. Fingers and toes quite free, with small, but distinct, truncated disks. Subarticular tubercles very prominent, those of the toes continued along the metatarsals; two metatarsal tubercles. First and second fingers subequal in length. Brownish; snout in front of a line intersecting the interorbital space whitish. Lumbar gland partly brown, partly black, with two or three white spots. Thighs broadly marbled with brown and whitish. White below.

Length of body	23 millim.
„ fore limb	16 „
„ hind limb	31 „
Distance between vent and heel	17 „

One adult (?) and two young specimens.

LIOHYLA.

Leiyla, Keferst. Gött. Nachr. 1868, p. 330.
Hylodes sp., Bouleng. Batr. Sal. p. 198.
Lithodytes sp., Cope, Bull. U. S. Nat. Mus. no. 32, p. 16 (1887).
Liohyla, Cope, Proc. Am. Phil. Soc. xxxi. 1894, p. 335.

Toes with a very narrow web and with small thin disks. Outer metatarsals united. Vomerine teeth; tongue free and emarginate behind. Pupil horizontal; tympanum distinct. Sternum without bony style.

I retain this genus only because it is convenient to lessen the great number of species described as *Hylodes.*

1. **Liohyla guentheri.** (Tab. LXVI. figg. A, A'.)

Leiyla guentherii, Keferst. Gött. Nachr. 1868, p. 330; or Wiegm. Arch. 1868, p. 296, t. 9. figg. 4, 5.
Lithodytes guentherii, Cope, Bull. U. S. Nat. Mus. no. 32, p. 16 (part.).
Hylodes palmatus, Bouleng. Batr. Sal. p. 201 (part.).

Hab. Costa Rica (*Underwood, Seebach*), Rio Ceibo, Buenos Aires (*Pittier*).

Rather slender in general form. Snout broad, somewhat projecting beyond the mouth, with more or less obtuse canthus and sloping, concave sides; eye scarcely equal to its distance from the nostril. Tympanum round, with the vertical diameter but little longer than the longitudinal, half the area of the orbit (in females). Vomerine teeth in two prominent transverse series, well separated from each other, behind, but within, the level of the choanæ. Upper parts of the body and of the limbs smooth, sometimes with several flat tubercles; sides slightly tubercular; a glandular fold above the tympanum; abdomen smooth, without folds or disk. Length of the body less than the distance between vent and heel. Disks of the fingers large, truncated, much larger than those of the toes; first finger rather longer than second; fingers free; toes one-fourth or nearly one-third webbed. Metatarsals without pads; a single metatarsal tubercle; a tarsal fold. Brown above, the upper part of the snout of a lighter colour; sides of the body, anterior portion of the hind limbs and the posterior portion of the calves marbled with black and yellow; hinder part of the thighs with small yellow rounded spots. Abdomen white, uniform or with scattered small brownish-black spots.

	Ad. fem.	Hgr fem.
Length of body	79 millim.	40 millim.
Distance between vent and heel	88 „	43 „
Length of tarsus	21 „	12 „
„ foot	43 „	22 „

I hesitate to identify this frog with *L. rugulosa*, as the form of its snout and the extent of the web are differently described and represented by Keferstein. *Hylodes raniformis* from Colombia (Bouleng. Ann. & Mag. N. H. 1896, xvii. p. 19) is much more nearly allied to, and perhaps identical with, it. The extent of the web between the toes is about the same in both *H. guentheri* and *H. raniformis*.

The specimen from the Rio Ceibo agrees singularly well with the figure given by Keferstein, only the tuberosities on the back are much smaller and indistinct; it is quite intermediate between Keferstein's example and the types of *H. raniformis*.

2. **Liohyla rugulosa.**

Liyla rugulosa, Cope, Proc. Am. Phil. Soc. 1869, p. 160; xxxi. p. 335 (1894).
Hylodes rugosus, Brocchi, Miss. Sc. Mex., Batr. p. 54.
Hylodes palmatus, Bouleng. Batr. Sal. p. 201 (part.).

Hab. Mexico, Omilteme in Guerrero (*H. H. Smith*), Hacienda Santa Gertrudis, Jalisco (*Buller*), Tehuantepec (*Sumichrast*); ? Guatemala, Dueñas (*Salvin*); Nicaragua, Hacienda Rosa de Jericho 3250 feet (*Rothschuh*).

Habit not slender. Snout rather obtuse, with the canthus rostralis not very sharp, and with sloping sides. Eye equal to its distance from the nostril. Tympanum round, two-fifths or three-fifths the size of the eye in females. Upper parts, and still more the sides, with small tubercles or short folds of the skin: an adult female has the sides of the abdomen extensively granular. Length of the body (distance of vent from snout) equal to the distance between vent and heel. Disks of fingers and toes small. The two inner fingers equal in length. Toes barely one-fourth webbed. Two metatarsal tubercles, of which the outer is very small and may be entirely absent; a tarsal fold. Brownish-olive or dark brown, marbled with darker colour, generally a lighter interorbital cross-band; hinder part of the thighs spotted with white. Lower parts white, throat often marbled with brown.

Immature specimens may have a narrow median yellow line from the snout to the vent, or a broader whitish band.

	Ad. fem. (Tehuantepec.)	Male.
Distance of vent from snout	45 millim.	33 millim.
„ „ heel	45 „	33 „
Length of tarsus and foot	31 „	23 „

3. Liohyla bocourti.

? *Hylodes bocourti*, Brocchi, Miss. Sc. Mex., Batr. p. 50, t. 16. figg. 2, 2 *a–c*.

Hab. ? Guatemala, Coban (*Brocchi*); Costa Rica, Rio Sucio (*Rogers*).

Habit not slender. Head broad. Snout obtuse, with the canthus rostralis somewhat indistinct, and with the sides sloping. Eye equal to its distance from the nostril. Tympanum round, small, about one-fourth of the size of the eye. Vomerine teeth in two prominent linear groups, close together, behind the level of the choanæ. Tongue slightly emarginate behind. Skin smooth. Length of the body nearly equal to the distance of the vent from the heel. Disks of fingers and toes large. The two inner fingers subequal. Toes barely one-fourth webbed. One metatarsal tubercle; a tarsal fold (?). Reddish-olive, with a lighter interorbital cross-band. A few brown vertical bands on the upper jaw; a brown band below the canthus rostralis, another from behind the eye over the tympanum, and another on the shoulder. Hind limbs with a few faint cross-bands; hinder side of the thighs not spotted. Lower parts whitish.

Distance between vent and snout	51 millim.
„ „ heel	53 „
Length of tarsus and foot	38 „

Of this species I have only one, rather desiccated, specimen before me; and the type, to judge from the figure, does not appear to have been in a much better state of preservation. However, that figure leaves much to be desired in several respects; I strongly suspect it to be unreliable. The separate figure (2 *c*) of the snout is narrower and more pointed than is shown in the figure of the entire animal. The tongue, described as without posterior notch, is figured as possessing one, and the vomerine teeth are represented as more widely apart than I have seen in any of these frogs.

4. Liohyla fleischmanni. (Tab. LXIV. figg. B, ♂ ♀.)

Hylodes fleischmanni, Boettger, Kat. Batr. Samml. Senckenb. Mus. 1892, p. 27.

Hab. Costa Rica, La Palma, San José, Iscazu (*Pittier de Fahéga, Underwood, Mus. Brit.*).

Head rather slender. Snout not obtuse, with the canthus rostralis not very sharp, and with sloping sides. Eye equal to its distance from the nostril. Tympanum round: in males two-thirds of, or equal to, the size of the eye; in females smaller, sometimes equal to only one-third of the eye. Upper parts and sides to a greater or lesser degree tubercular, sometimes nearly smooth. Length of the body less than the distance between vent and heel. Disks of fingers and toes moderately developed. *The two inner fingers equal in length.* Toes barely one-fourth webbed. One metatarsal tubercle; a tarsal fold. Bluish-black above, with very indistinct markings; or dark greenish with black marbling, the snout in front of an interorbital line being of a lighter colour. Cross-bands on the thighs, and spots on their hinder sides, rather indistinct. Whitish below; throat generally marbled with brownish. Groin and lower part of the thighs generally rose-coloured, particularly in the slate-coloured individuals. Young and old sometimes with a median white dorsal line.

	Ad. ♂.	Ad. ♀.	Ad. ♀.
Distance between vent and snout .	45 millim.	73 millim.	65 millim.
" " heel . .	49 "	81 "	73 "
Length of tarsus and foot	34 "	59 "	52 "
Diameter of tympanum	4 "	4 "	4 "

This frog cannot be referred to, or confounded with, *Liohyla guentheri*, having a much narrower web between the hind toes, besides other distinctive characters. According to Keferstein, the web of the species described by him leaves free one phalanx of the first and second toes, two on the third and fifth, three on the fourth. In *L. fleischmanni* the number of free phalanges is one more on each of the toes.

It seems to be the most common species at La Palma and in the neighbourhood of San José.

I am indebted to Professor Boettger for direct comparison of some of our specimens with the types in the Senckenberg Museum.

5. **Liohyla pittieri,** sp. n. (Tab. LXV. figg. B, B′.)

Hab. Costa Rica (*Underwood*), Boruca (*Pittier de Fahéga*).

Closely allied and very similar to *L. fleischmanni*. Habit rather robust, the length of the body being a little less than the distance between vent and heel. Snout not projecting beyond the mouth, of moderate length, with obtuse canthus rostralis and sloping sides. Tympanum round, three-fourths the size of the eye in the male, and about one-half in the female. Vomerine teeth as in *L. fleischmanni*. Male with some short tubercular folds and small tubercles on the side of the back, female nearly smooth. Abdomen smooth, without folds. Disks of the fingers small, of the toes moderately developed. *First finger longer than the second.* Toes barely one-fourth webbed. One metatarsal tubercle; a tarsal fold. *The hinder side of the thighs black, with well-defined small round lemon-yellow spots*; throat brownish, minutely speckled with yellow; abdomen whitish. Male brownish, faintly marbled with darker colour, sides marbled with yellowish; a narrow yellow vertebral line and a light interorbital cross-band. Female nearly uniform brownish-black above; sides marbled with yellowish.

	♀.	♂.
Length of the body	60 millim.	51 millim.
Distance between vent and heel	62 „	52 „
Length of hind limb (from vent) . . .	104 „	87 „

I have long hesitated before I ventured to distinguish this frog under a distinct name; however, the receipt of a second specimen agreeing in every essential distinctive character with the first, removed every doubt about the species being distinct from *L. fleischmanni*. The male has the testicles well developed, yet lacks vocal sacs. These organs are present in much smaller specimens of *L. fleischmanni*.

6. **Liohyla engytympanum**, sp. n. (Tab. LXV. fig. A.)

Hab. Costa Rica, Monte Redondo (*Underwood, in Mus. Brit.*).

In general habit and form of the head similar to *L. rugulosa*. Snout rather obtuse, with the canthus rostralis distinct, and with sloping sides; eye scarcely equal to its distance from the nostril. *Tympanum an irregular oval, deeper than long, barely half the area of the orbit* (in adult female). Vomerine teeth in two short prominent transverse series, slightly separated from each other, behind but within the level of the choanæ; choanæ rather wider than the eustachian openings. Upper and lateral parts of the body, supra-orbital region, and upper parts of the hind limbs with tubercles and granules; a glandular fold above the tympanum; abdomen smooth, with indistinct granules on the sides; transverse and longitudinal folds indicate the presence of an abdominal disk. Length of the body a little less than the distance between vent and heel. Disks of fingers and toes subequal, rounded, moderately developed; first finger rather longer than the second; a rudimentary web between the three inner fingers. Metatarsals without pads, except at the base of the inner toe; outer metatarsal tubercle scarcely visible; a tarsal fold. Toes one-fourth webbed. Greyish, with a yellow vertebral and gular line; an indistinct dark interorbital cross-band. Upper lip and flanks with blackish spots. Upper parts of the limbs obscurely banded; hind part of the femur rather finely and conspicuously marbled with black and white. Lower parts white.

One adult female.

Length of the body	59 millim.
Distance between vent and heel	62 „
Length of foot	47 „

7. Liohyla ranoides.

Lithodytes ranoides, Cope, Proc. Am. Phil. Soc. xxiii. p. 275 (1886).
Liohyla ranoides, Cope, Proc. Am. Phil. Soc. xxxi. p. 335 (1893).

Hab. NICARAGUA (*Cope*); COSTA RICA, Buenos Aires, Terraba (*Cope*).

Habit not slender. Head relatively rather long. Snout acuminate, somewhat projecting, with the canthus rostralis distinct, and with the sides subvertical. Eye equal to its distance from the nostril. Tympanum oval, its vertical much longer than its horizontal diameter, scarcely one-third the size of the eye. Skin nearly smooth, but a pair of feeble folds form an obscure pattern on the scapular regions. The length of the body seems to be nearly equal to the distance between vent and heel. Disks of fingers and toes truncate. Toes nearly one-third webbed. Two small metatarsal tubercles, of which the outer is obsolete. Coloration variable, greyish to dark brown, with or without a white median dorsal line or band, and sometimes with a light interorbital cross-band. Upper parts more or less distinctly marbled with darker colour; lower parts whitish. Limbs dark cross-banded; the cross-bands of the tibia extend on the skin that covers the flexors of the foot so as to be seen from below. (After Cope.)

This species was originally described from four Nicaraguan specimens, of which the largest measured 26 millim. Although Cope, at a later period, recognized the same species in some Costa Rican frogs, I am unable to find it among the numerous examples sent by Pittier and Underwood. Its chief characters seem to be the acuminate snout, short hind limbs, webbed toes, and the peculiar small, oval tympanum, which Cope ascribes to all his specimens, and which therefore cannot be regarded as an individual or accidental modification. In this respect *L. ranoides* resembles *L. engytympanum*; but the latter has a shorter obtuse snout and longer hind limbs.

8. Liohyla berkenbuschi.

Hylodes berkenbuschii, Peters, MB. Ak. Wiss. Berlin, 1869, p. 879.

Hab. MEXICO (*Peters*).

Snout acuminate, with distinct canthus rostralis. Tympanum large (sex unknown), its diameter two-thirds of that of the eye; choanæ narrower than the eustachian tubes. Vomerine teeth on two small approximate transverse prominences behind the choanæ. Sides of the body with some oblong glandular pads and *with a strong fold from the shoulder to the thigh.* Abdominal disk smooth, a cross-fold on the chest. First finger a little shorter than the second. The basal phalanges of the toes are united by a web extending as narrow fringes to the disks. Disks of fingers and toes well developed; one metatarsal tubercle; a tarsal fold. Upper parts brownish, with darker spots, of which one, triangular in shape, between the eyes, is preceded by the lighter-coloured snout; upper lip and limbs cross-banded; sides with black spots, marbled with yellow;

hinder side of the thighs with small black and light spots. Lower parts brownish, marbled with darker. (After Peters.)

Length of body	43 millim.
" hind limb	70 "
" foot	32 "

HYLODES.

Hylodes, Fitz. Syst. Rept. p. 38; Dum. et Bibr. Erpét. Gén. viii. p. 619; Günth. Batr. Sal. p. 91; Cope, Journ. Ac. N. Sc. Philad. vi. p. 96 (1866).
Batrachyla, Bell, Voy. Beagle, Rept. p. 43.
Craugastor, Cope, Proc. Ac. N. Sc. Philad. 1862, p. 153.
Strabomantis, Peters, MB. Ak. Wiss. Berlin, 1863, p. 405.
Lithodytes sp., Cope, Journ. Ac. N. Sc. Philad. vi. p. 96 (1866); Bull. U.S. Nat. Mus. no. 32, p. 16 (1887).
Epirhexis, Cope, Journ. Ac. N. Sc. Philad. vi. p. 96 (1866).
Limnophys, Espada, Jorn. Ac. Sc. Lisb. iii. p. 60 (1870).
Hylodes sp., Bouleng. Batr. Sal. p. 198.

Toes free; disks, if developed, generally small. Outer metatarsals united. Vomerine teeth; tongue entire or very little emarginate behind. Pupil horizontal; tympanum generally distinct. Sternum without bony style.

The species of this genus are very imperfectly known, and, owing to their great number and great variability, they are most difficult of specific differentiation as well as of recognition from the descriptions extant. The following scheme may in some measure facilitate the determination of Central-American specimens:—

I. Abdomen granular.
A. Crown of the head flat.
1. Disks of the fingers and toes very small.
a. Metatarsal tubercles small.
1. *sallæi*; 2. *plicatus*; 3. *polyptychus*; 4. *underwoodi.*
b. Inner metatarsal tubercle large.
5. *calcitrans.*
2. Disks of the fingers or toes developed.
a. Tympanum distinct.
6. *platyrhynchus.*
b. Tympanum very small, indistinct.
7. *cerasinus.*
B. Edges of the fronto-parietals crest-like.
8. *biporcatus.*

II. Abdomen smooth.
A. Crown of the head flat.
1. Disks of the fingers and toes very small.
a. Head of moderate size.
9. *euryglossus*; 10. *rhodopis*; 11. *rugosus*; 12. *venustus*; 13. *florulentus*; 14. *augusti.*
b. Head very large.
15. *laticeps.*
2. Disks of the fingers or toes developed.
16. *melanostictus*; 17. *brocchii*; 18. *nubilus*; 19. *alfredi*; 20. *bransfordi*; 21. *diastema*; 22. *lanciformis.*
B. Crown of the head with raised osseous ridges.
23. *petriculus*; 24. *megacephalus*; 25. *gulosus.*

1. Hylodes sallæi.

Hylodes sallæi, Günth. P. Z. S. 1868, p. 487, t. 38. fig. 3.
Hylodes rhodopis, part., Bouleng. Batr. Sal. i. p. 203.
? *Hylodes habenatus,* Dugès, La Naturaleza, (2) ii. p. 479 (1897).

Hab. Mexico, Jalapa, Vera Cruz (*Sallé*); Guatemala, Vera Paz (*Salvin*).

Posterior half of the abdomen granular. No fold across chest. Habit moderate. Head not wider than the trunk; snout somewhat acuminate, with distinct canthus rostralis and with the sides subvertical. Tympanum about half the area of the eye (in both sexes). Vomerine teeth in two short groups, well separated from each other, behind the level of the choanæ. Skin of the upper parts smooth, either entirely or with several exceedingly fine linear folds—one from the eye above the tympanum to the shoulder, another oblique from the eye towards the centre of the back, and one or two on each side of the hinder half of the back. Disks of the fingers and toes very small; first finger a little longer than the second. Two small metatarsal tubercles; series of subarticular tubercles continued on the metatarsals; no tarsal fold. The length of the body is almost equal to the distance between vent and heel. Greyish-olive; hind limbs obscurely cross-banded; an indistinct dark band between the eyes. A black band runs from the nostril through the eye above and round the tympanum; vent black; generally one single small black spot or a pair in the centre of the back; lower parts white.

A small species.

	Type.	♀.	♀.
Length of body	28 millim.	30 millim.	36 millim.
Distance between vent and heel . .	27 „	29 „	37 „

This species has been identified by Cope (Proc. Am. Phil. Soc. xi. 1869, p. 160) with his *Lithodytes rhodopis*, and in this he has been followed by Boulenger, Fischer, and others. No doubt the two forms are very closely allied, but in *H. sallæi* the greater part of the abdomen is granular, while in all the notes by Cope this part of the skin is represented as smooth. There is, of course, the possibility that in consequence of the imperfect preservation the granulation has disappeared or become indistinct in Cope's specimens, as I myself have noticed in some examples, particularly in the one obtained by Salvin in Vera Paz; however, closer examination under a lens reveals the areolated condition of the skin.

If these specimens be referred to the smooth-bellied *L. rhodopis*, the structure of the skin of the abdomen, used by Cope for the generic discrimination of *Lithodytes* and *Hylodes*, and by Boulenger for division of the species into groups, would cease to be a help even for specific distinction.

2. **Hylodes plicatus**, sp. n. (Tab. LXVI. fig. B.)

Hab. MEXICO, Jalapa (*Mus. Brit.*).

A very distinct fold across the chest, separating the smooth throat from the granular abdomen. Habit rather slender; snout somewhat acuminate, with sharp canthus rostralis and subvertical sides. Tympanum as large as the eye (in a male). Vomerine teeth in two small groups, well separated from each other, behind, but within the level of, the choanæ. Skin of the upper parts smooth, with the exception of two pairs of exceedingly fine, curved, dorsal folds; another fold above the tympanum. Disks of the fingers and toes very small; first finger a little longer than the second. Two small metatarsal tubercles; series of subarticular tubercles continued on the metatarsus; no tarsal fold. The length of the body is almost equal to the distance between vent and heel. Light brownish-olive above, with a dark interorbital band; hind limbs cross-barred. A black band runs from the nostril through the eye to above and round the tympanum; vent black. Lower parts white.

Length of body	19·5 millim.
Distance between vent and heel	20·5 ,,

3. **Hylodes polyptychus.** (Tab. LXVI. fig. C.)

Hylodes polyptychus, Cope, Proc. Am. Phil. Soc. xxiii. p. 276 (1886).
Hylodes steinegerianus, Cope, Proc. Am. Phil. Soc. 1894, p. 338.

Hab. NICARAGUA, Hacienda Rosa de Jericho, alt. 3250 feet (*Rothschuh*); COSTA RICA, La Palma (*Underwood*).

Habit moderate, the length of the body being nearly equal to the distance between vent and heel. Snout rather short, with distinct canthus rostralis and subvertical sides. Tympanum nearly the size of the eye in males, but smaller in females. Vomerine teeth in two short oblique series, well separated from each other, starting from the inner posterior angle of the choanæ. Back with prominent, linear, tubercular longitudinal folds or series of tubercles (sometimes as many as eight of these folds may be counted); skin between the folds, on the sides of the body, on the superciliaries, and on the non-concealed portions of the limbs, tubercular; abdomen granulated. Disks of the fingers absent, of the toes small. First and second fingers subequal in length; two metatarsal tubercles; a tarsal fold; subarticular tubercles continued on the metatarsals. Back between the two most prominent folds darker than the sides, on which the light colour forms an oblique band. Upper lip and limbs cross-barred; often a light line along the hinder side of the thigh. Lower parts whitish, with or without brown marbling.

	♀ (Nicaragua).	♂ (Costa Rica).	♀ (Costa Rica).
Length of body	26 millim.	24 millim.	31 millim.
Distance between vent and heel . .	23 „	25 „	31 „

In specimens from Costa Rica the disks of the toes are rather more developed than in Nicaraguan individuals; but from a direct comparison of specimens from these localities, I come to the conclusion that the characters on which Cope has based *H. steinegerianus* do not hold good for specific distinction.

4. Hylodes underwoodi. (Tab. LXVI. fig. D.)

Hylodes underwoodi, Bouleng. Ann. & Mag. N. H. 1896, xviii. p. 340.

Hab. Costa Rica, La Palma, alt. 1600 metres (*Underwood*).

Habit moderate; snout moderately obtuse, with obtuse canthus and slightly sloping, concave sides. Vomerine teeth in two oblique groups behind the choanæ, well separated from each other. Tympanum not quite the size of the eye, in adult females conspicuously smaller. Fingers and toes slender, with prominent subarticular tubercles which are continued on the metatarsals. Disks very small, quite inconspicuous on the fingers. First and second fingers equal in length. Two metatarsal tubercles. The distance between vent and heel equals the length of the body. Upper parts rough with small tubercles; a pair of)(-like folds on the anterior half of the back, widely distant from each other; other similar folds on the posterior half are less distinct or constant; abdomen granular *. Upper parts brownish-black, with a bluish-white chevron across the anterior third of the back; or lighter-coloured, with darker markings, such as a dark triangular spot between the eyes, a lumbar spot, and cross-bars on the lips and limbs. Hinder side of the limbs nearly uniform brownish, with some minute marbling.

Length of body	30 millim.
Distance between vent and heel	30 „

* Mr. Boulenger describes it as smooth, but the granulation is distinct enough, and may be seen even in soft, badly preserved examples.

5. **Hylodes calcitrans,** sp. n. (Tab. LXVII. fig. B.)

Hab. Mexico, Omilteme in Guerrero (*H. H. Smith*), Jalisco (*Richardson*).

Body rather stout, its length exceeding the distance between vent and heel; snout rather broad, but pointed in front, with sloping sides and distinct, curved canthus. Tympanum two-thirds the size of the eye. Vomerine teeth in two short oblique groups behind, and within the level of, the choanæ. Upper parts smooth, abdomen granular. Tips of the fingers and toes scarcely dilated. First and second fingers subequal. Subarticular tubercles well developed, but not continued on the metatarsals. *Two metatarsal tubercles, of which the inner is unusually large*, projecting, nearly as large as the tympanum. Toes short. No tarsal fold. Coloration variable. Upper part of the snout light-coloured, separated by a dark interorbital band from the colour of the back, which is generally brownish-grey marbled with a darker tint. A blackish band along the canthus rostralis sometimes extends over the whole side of the head, and may be continued on the side of the body. Limbs with narrow cross-bars, which sometimes encroach upon the concealed hinder surface of the hind limb. The darker spots on the back are sometimes confluent, and form a pair of longitudinal bands. Lower parts whitish, throat brownish.

Length of body	36 millim.
,, hind limb from vent	58 ,,
Distance between vent and heel	33·5 ,,
Length of tarsus	9 ,,
,, foot	16 ,,

Numerous examples show all the unusual development of the inner metatarsal tubercle, but in specimens from Jalisco the granulation of the abdomen is less conspicuous, possibly owing to their imperfect state of preservation.

6. **Hylodes platyrhynchus,** sp. n. (Tab. LXVII. fig. A.)

Hab. Costa Rica (*Mus. Nac. de Costa Rica*).

Head large, broad, its width being increased by the loreal region, which, instead of being subvertical, slopes outwards in a very oblique plane; canthus rostralis distinct. Posterior half of the abdomen granular and finely areolated. Eyes obliquely directed forwards. Tympanum one-third the size of the eye (in a female). Vomerine teeth in two long, straight, subcontinuous series behind the choanæ, and laterally extending to the level of the inner margins of the choanæ. Upper parts nearly smooth, with the exception of a linear fold above and round the tympanum, and another from the eye curving towards the middle of the side of the back; some small tubercles on the hinder half of the superciliary region. Disks of the fingers well developed, truncate, larger than those of the toes. First finger a little shorter than second. One metatarsal tubercle; metatarsals without subarticular tubercles; no tarsal fold. The length of the body is rather less than the distance between vent and heel. Greyish-olive; upper lip and hind limbs cross-barred, the bars extending over the hinder part of the thigh. A black band along the canthus rostralis, and over and round the tympanum; some other blackish band-like spots follow the linear folds; a dark interorbital band. Lower parts white.

Length of body (♀)	56 millim.
Distance between vent and heel	59 „

I have seen only one specimen of this species; it is nearly allied to *H. brocchii*, but differs in the form of its snout, the long vomerine series, and the areolated abdomen.

7. Hylodes cerasinus.

? *Hylodes cerasinus*, Cope, Journ. Ac. N. Sc. Philad. viii. p. 112, t. 26. fig. 4 (1876).

Hab. Costa Rica, La Palma (*Underwood*), ? Pico Blanco (*Cope*).

Habit moderate; head moderately wide and long; snout a little longer than the orbit, with distinct canthus and sloping sides. Tympanum very small, indistinct; tongue broad, subtruncate behind. Vomerine teeth in two small groups, behind the very narrow choanæ. Skin smooth above, abdomen granulated. Disks of the fingers rather large, truncated, of toes well developed. First finger shorter than second. The distance between vent and heel is less than the length of the body. One metatarsal tubercle; the subarticular tubercles continued below the metatarsals. Rose-coloured along the middle of the back, sides and an interorbital cross-band slate-coloured; lower parts whitish; upper lip and hind limbs indistinctly cross-banded; hinder side of thigh marbled with brown and whitish.

Length of body	24 millim.
Distance between vent and heel	23 „
Length of hind limb	41 „

In several points Cope's description does not well apply to the single example in the Natural History Museum, from which my diagnosis is taken, and unfortunately the figure given by him is quite useless. Indeed, I should have been inclined to refer that specimen to *Hylodes cruentus* (Peters, MB. Ak. Wiss. Berlin, 1873, p. 609), from Chiriqui, but for the fact that Peters describes the tympanum as distinct, as a vertical oval, barely one-third the size of the eye.

8. Hylodes biporcatus.

Strabomantis biporcatus, Peters, MB. Ak. Wiss. Berlin, 1863, p. 405; Espada, An. Soc. Esp. i. p. 85.

Hylodes biporcatus, Bouleng. Cat. Batr. Sal. p. 220.

Limnophys napæus, Espada, Jorn. Sc. Lisb. iii. p. 60.

Hab. Panama, Veragua (*Peters*).—Ecuador (*Espada*).

Interorbital space deeply concave, the edges of the fronto-parietals prominent, crest-like. Abdomen granular. Vomerine teeth in two arched series behind the choanæ. Head large, broad, with bony ridges. Tympanum a little smaller than the eye. First finger longer than second; fingers with very small, toes without, disks. Two metatarsal tubercles. The foot extends beyond the snout, if the hind limb is brought forwards. Upper parts minutely granular. Dark brown above, with a few black spots edged with yellow; brown below, dotted or marbled with white.

Length of body and limb	75	millim.
„ hind limb	45	„

9. Hylodes euryglossus.

Lithodytes euryglossus, Cope, Proc. Ac. N. Sc. Philad. 1894, p. 196.

Hab. Costa Rica, San José (*Biolley*).

General appearance that of a *Leptodactylus*. No distinct canthus rostralis; end of snout sloping backwards from the edge of the lip. Tympanum a vertical oval, of which the long diameter is two-thirds of that of the eye (sex ?). Vomerine teeth on two transverse ridges, which are on a line with the hind margins of the choanæ. Disks distinct, but small, those of the fingers the smaller. First and second fingers equal; a tarsal fold. Skin smooth everywhere; a discoidal abdominal fold. Brownish-grey above, whitish below; groin reticulated, sides dotted. A black streak above and behind the tympanum; a blackish cross-band between the posterior parts of the orbits; some black longitudinal spots on the dorso-lateral region; lips and limbs cross-banded. Thighs obscurely reticulate behind. (After Cope.)

Length of body	75	millim.
„ hind limb	115	„
„ hind foot	50	„
„ tarsus	20	„

10. Hylodes rhodopis. (Tab. LXVII. figg. C, C'.)

Lithodytes rhodopis, Cope, Proc. Ac. N. Sc. Philad. 1866, p. 323; Proc. Am. Phil. Soc xxxi. p. 337 (1893).

Hylodes rhodopis, part., Bouleng. Batr. Sal. p. 203.

Lithodytes podiciferus, Cope, Journ. Ac. N. Sc. Philad. viii. p. 107, t. 23. fig. 9 (1876); ? Dugès, La Naturaleza, (2) ii. p. 479 (1897).

Hylodes podiciferus, Bouleng. l. c.

Lithodytes muricinus, Cope, Journ. Ac. N. Sc. Philad. viii. p. 108, t. 23. fig. 13 (useless figure) (1876).

Hylodes muricinus, Bouleng. l. c.

Lithodytes habenatus, Cope, Journ. Ac. N. Sc. Philad. viii. p. 109 (1876).

Hab. ? MEXICO, north of Vera Cruz (*Dugès*); COSTA RICA, Pico Blanco and Buenos Aires (*Cope*), La Palma and Cartago (*Underwood*), ? Irazu * (*Rogers*).

Closely allied to *H. sallæi*, *but without granulations on the abdomen.* Form of the body and head as in *H. sallæi*. Tympanum barely one-half the area of the eye (in females). Vomerine teeth and folds of the skin as in *H. sallæi*. Disks of the fingers and toes very small; first and second fingers equal in length. Two small metatarsal tubercles; the subarticular tubercles are not continued on the metatarsals; no tarsal fold. The length of the body is nearly equal to the distance between vent and heel. Coloration variable, some specimens coloured as *H. sallæi*, others more uniformly dark with the ornamental markings indistinct; sometimes uniform brownish-slate-coloured, with a broad white longitudinal band on each side, running from the eye to the groin.

	♀.	♀.	♀.
Length of body	28 millim.	28 millim.	29 millim.
Distance between vent and heel . .	29 „	29 „	32 „

11. Hylodes rugosus.

Hylodes rugosus, Peters, MB. Ak. Wiss. Berlin, 1873, p. 610.

Hab. NICARAGUA, Hacienda Rosa de Jericho (*Rothschuh*); PANAMA, Chiriqui (*Mus. Berol.*).

Habit rather stout, the length of the body equal to the distance between vent and heel; head broad, snout short, with very distinct canthus and subvertical sides. Tympanum half the size of the eye (sex ?). Vomerine teeth in two small groups behind the choanæ, which are equally as wide as the eustachian openings. Upper parts with numerous tubercles and folds; of the latter, a pair of curved)(-like folds on the back, another straight one on each side of the anterior part of the back, and another transverse and undulated across the interorbital space are most conspicuous; tubercles on the upperside of the head and of the lower leg. Abdomen smooth. Disks of the fingers and toes small. First finger a little longer than second. Two metatarsal tubercles. Dark purple above, with a light interorbital cross-band. Upper lip and limbs cross-banded. Hinder side of the thigh marbled with whitish and purplish; throat and chest marbled with brown.

Length of body	16 millim.
Distance between vent and heel	16 „

The young specimen from Nicaragua is the only one I have seen.

* The specimen from this locality, referred to in Cat. Batr. Sal. B. M. p. 203 (*H. podiciferus*), is young and too much deteriorated to admit of identification.

12. **Hylodes venustus,** sp. n. (Tab. LXVIII. fig. C.)

Hab. Mexico, Jalapa (*F. D. G.*).

Habit moderately stout, the length of the body being less than the distance between vent and heel. Head moderately broad, with subacuminate snout; snout with distinct canthus and subvertical sides. Tympanum half the size of the eye (in an adult female), round. Vomerine teeth in two small rounded groups behind the choanæ, well separated from each other. Upper parts smooth, with the exception of a pair of undulated linear folds which run from the superciliary to the end of the transverse processes of the sacral vertebra; also a fold above and round the tympanum. Upper parts of the head smooth, with a few indistinct tubercles on the upper eyelid. Legs and lower parts smooth. Disks of the fingers and toes small; first finger longer than second. Two metatarsal tubercles; subarticular tubercles on the metatarsals. Upper parts light mauve-coloured, back between the linear folds dark purplish; linear folds, a band along the canthus rostralis and above and round the tympanum, an oval spot on each side of the pelvis, and some small irregular spots on the sides of the body, black. No interorbital band. Upper lip and hinder parts of the thighs powdered with grey. Lower parts white, immaculate.

Length of body	35 millim.
Distance between vent and heel	40 ,,

Allied to *H. rugosus*, but differing in the smoothness of the skin, longer hind legs, and coloration.

13. **Hylodes florulentus.**

Lithodytes florulentus, Cope, Proc. Am. Phil. Soc. 1894, p. 336.

Hab. Costa Rica, Boruca (*Cherrie*).

Habit rather stout, the length of the body exceeding the distance between vent and heel. Tympanum vertically oval, about half the width of the eye. *Vomerine teeth in two fascicles close together, behind the choanæ, but well within their inner margins.* Tips of the fingers and toes with very small dilatations; two metatarsal tubercles. Back with a few small tubercles, otherwise smooth. Grey, dusted with blackish; hind limbs black behind, with dark grey cross-bands above, and with the spaces between these bands scarlet. Posterior part of the sides and anterior portion of the abdomen blackish-brown, with crimson and orange spots of various sizes and shapes, the largest in the groin. Fore limbs with cross-bands. (After Cope.)

Length of body	35 millim.
,, hind limb from groin	51 ,,
,, hind foot	24 ,,
,, tarsus	9 ,,

Known from a single specimen, in which the integuments of the head are destroyed.

14. **Hylodes augusti.** (Tab. LXVIII. fig. D.)

Hylodes augusti, Brocchi, Miss. Sc. Mex., Batr. p. 52, t. 16. figg. 1, 1 *a–c*.

Hab. MEXICO, Guanajuato and Tupataro (*Dugès*), Ventanas in Durango, from a mine 1000 feet deep (*Forrer*).

Habit stout, the length of the body being rather more than the distance between vent and heel; snout short, broad, depressed, with the canthus rather obtuse and the loreal region shelving outwards. Tympanum one-half or one-third the size of the eye (in females). Vomerine teeth in two very short oblique groups, close together, and on a level with the hind margin of the choanæ. Back with small scattered tubercles, abdomen smooth. Disks of the fingers and toes exceedingly small. First finger conspicuously longer than second; two metatarsal tubercles; no tarsal fold; subarticular tubercles continued on the metatarsals. Brownish-olive, all the upper parts marbled with dark brown, one or the other of the brown markings assuming the form of cross-bands on the back, and edged with white. Lower parts white.

	♀.	♀.
Length of body	36 millim.	33 millim.
Distance between vent and heel	32 „	29 „

I have long hesitated before identifying our specimens with the species named by Dugès and Brocchi, as the latter figures the vomerine teeth widely distant from each other. However, there is a great agreement between these frogs in all other respects. Unfortunately, the type (as Messrs. Vaillant and Mocquard have kindly informed me) is not preserved in the Paris Museum.

15. **Hylodes laticeps.**

Hylodes laticeps, A. Dum. Ann. Sc. Nat. 3rd ser. Zool. xix. p. 178; Dum. et Bibr. Erpét. Gén. ix. p. 408, t. 99; Brocchi, Miss. Sc. Mex., Batr. p. 49.

Epirhexis laticeps, Cope, Bull. U. S. Nat. Mus. no. 32, p. 17 (name only).

Hab. MEXICO, Yucatan (*Morelet*), Tehuantepec (*Brocchi*).

Head very large, broader than the body; snout broad. Tympanum higher than long; vomerine teeth in two short oblique series behind the choanæ. Skin smooth above and below; a glandular fold above and round the tympanum; another along the side of the back; abdominal disk very distinct. Disks of fingers and toes small; first finger longer than the second. Toes free *. Upper parts greenish-brown, loreal region and tympanum black; whitish below, finely marbled with brown on the throat and limbs.

* Boulenger (Batr. Sal. p. 210) introduces into the diagnosis of this species the words " toes webbed at the base," while A. Duméril describes them as completely free; also the figure quoted represents them without a rudiment of a web.

16. **Hylodes melanostictus.**

Lithodytes melanostictus, Cope, Journ. Ac. N. Sc. Philad. viii. p. 109, t. 23. fig. 10 (1876); Dugès, La Naturaleza, (2) ii. p. 479 (1897).

Hab. MEXICO, north of Vera Cruz (*Dugès*); COSTA RICA, Pico Blanco, alt. 7000 feet (*Cope*).

Habit of body moderate; head flat and wide. Canthus rostralis straight; loreal region and lip oblique. Tympanum a rather narrow, vertical oval, one-fourth of the size of the eye. Vomerine teeth in two short transverse series entirely behind the choanæ, well separated from each other and not extended outwards beyond the level of the inner margins of the choanæ. Skin everywhere smooth. Limbs long, the length of the body being less than the distance between vent and heel. Disks of fingers large, of toes moderate. No web between the toes; first finger shorter than second. Brownish-grey above, dirty white below. Limbs cross-banded rather distantly, the bars extending on the front and back faces of the femur. A white vertebral line from snout to vent, which is bounded on the sides at different points with blackish. A pink band extends from above each tympanum to the end of the ilium, and is broadly bordered with black on the outer side, this colour extending on the sides of the body as oblique black spots. The tympanum is black, and sends a black bar to the cleft of the mouth; two black bars from the orbit to the lip, another along the canthus rostralis. (After Cope.)

Length of body	50 millim.
„ hind limb	96 „ *
„ hind foot	45 „

17. **Hylodes brocchii.** (Tab. LXVIII. figg. A, B.)

Hylodes brocchi (Boulenger), Brocchi, Miss. Sc. Mex., Batr. p. 60, t. 15. figg. 3, 3 *a*.

Hab. GUATEMALA, Vera Paz (*Salvin*); COSTA RICA, La Palma, alt. 1600 metres (*Underwood*).

Abdomen smooth. Habit rather stout; head broad, flattened, with short, broad, rounded snout, the canthus rostralis being distinct and the loreal region sloping. Tympanum two-thirds the size of the eye in a male, one-third in adult females. Vomerine teeth in two short, not very distant, transverse series behind the choanæ. Upper parts nearly smooth, with the exception of a linear fold above and round the tympanum, and another from the eye curving along the side of the back, and disappearing about the middle of the length of the trunk; some small tubercles on the hinder half of the superciliary region, of which one, conical in shape, is very distinct. *Disks of the fingers and toes well developed, truncate,* those of the fingers more so than of the toes. First and second fingers equal in length. One metatarsal

* This measurement seems to have been taken by Cope from the groin, not from the vent as is the case in my descriptions.

tubercle; metatarsals without subarticular tubercles; no tarsal fold. The length of the body is rather less than the distance between vent and heel. Coloration variable: light brownish, with or without large darker spots; upper lip and hind limbs cross-barred; a blackish band-like spot over the tympanum; a dark interorbital band. Sometimes a white vertebral line or a broad vertebral band. The hinder side of the femur either cross-banded or with smaller or larger white spots (fig. B); one specimen with a white line running along the whole length of the hinder side of the posterior leg. Male more uniform dusky, with a white interorbital band.

	♀.	♂.
Length of body	50 millim.	36 millim.
Distance between vent and heel	53 „	40 „

18. **Hylodes nubilus,** sp. n. (Tab. LXIX. fig. A.)

Hab. Costa Rica, Escazu (*Underwood*).

Habit rather slender, the length of the body equal to the distance between the vent and heel. Head flat, moderately wide; snout rather produced, with distinct canthus rostralis and subvertical sides. Tympanum half the area of the eye (in an adult female). Vomerine teeth in two short transverse series, rather approximate to each other, considerably behind the choanæ, but within the level from their inner margins. Choanæ and eustachian tubes equal in width. Skin of the upper parts smooth, only the regions above the shoulder and behind the supraciliary show some small tubercles and short narrow glandular folds; abdomen smooth. Disks of the fingers small, but distinctly developed and larger than those of the toes; first finger longer than second. One metatarsal tubercle, a tarsal fold; metatarsals without subarticular tubercles. Dark purplish, clouded with darker, the darker parts concentrated in four indistinct longitudinal bands, the middle of which reach almost to the snout. Hind limbs and upper lip indistinctly cross-banded; thigh finely marbled behind with brown and whitish. Lower parts whitish, with small blackish-brown marblings, a band along the middle of the throat being pure white.

Length of body	44 millim.
„ hind limb	78 „
Distance between vent and heel	43·5 „

19. **Hylodes alfredi.**

Hylodes alfredi, Boulenger, P. Z. S. 1898, p. 480, t. 39. fig 1.

Hab. Mexico, Atoyac in Vera Cruz (*A. Dugès*).

Habit rather slender, the distance between the vent and heel exceeding the length of the body; head depressed, snout of moderate length, with obtuse canthus rostralis and sloping sides. Tympanum half the size of the orbit (in females). Vomerine teeth in two short oblique series immediately behind the choanæ; choanæ and eustachian openings equally open. Skin smooth above and below. Disks of the fingers large, especially of the two inner ones, much larger than those of the toes; first and second finger subequal in length, third unusually long. A metatarsal tubercle (in the larger of the two specimens a minute outer tubercle); metatarsals without subarticular tubercles. Greyish, finely sprinkled with black and lighter colour. Upper lip and legs without cross-bars; concealed surfaces unspotted; no interorbital band.

Length of body	36 millim.
Distance between vent and heel	39 „
Length of third finger	9 „

20. **Hylodes bransfordi.**

Lithodytes bransfordii, Cope, Proc. Am. Phil. Soc. xxiii. p. 274 (1886).

Hab. NICARAGUA (*Bransford*).

A short-legged species, the heel reaching only to the orbit when the limb is brought forward. Vomerine teeth in two transverse or slightly arched series, close together, well behind the choanæ, and not "extending exterior" to the middle of the latter. Tympanum large, varying in size, equal to, or two-thirds of, the diameter of the eye. Snout short, with "distinct, obtuse" canthus rostralis. Disks of the toes moderate. Two metatarsal tubercles. The skin of the back is thrown into delicate longitudinal parallel folds which are easily lost. Abdomen smooth. In the colour there is much pink on the upper and concealed surfaces; two dark spots on the lip, one below the eye, another behind and above the axilla, others on the parietal region and between the anterior parts of the orbits. Generally a light open chevron, pointing forwards, across the middle of the back, with a dark one in front of it. In a large specimen a pink band extends from the orbit to the ilium. Posterior side of the thigh brown, with light specks. Whitish below, throat sometimes mottled with brown. (After Cope.)

Length of body	25 millim.
„ hind limb from groin	36 „
„ foot	17 „
„ tarsus	7 „

21. **Hylodes diastema.**

Lithodytes diastema, Cope, Journ. Ac. N. Sc. Philad. viii. p. 155 (1875); Proc. Am. Phil. Soc. xxiii. p. 274 (1886).

Hab. NICARAGUA (*Bransford*); PANAMA, Camp Maria Caretta (*Bransford*).

Vomerine teeth in two very short transverse patches behind and within the line of the middle of the choanæ, and separated by an interspace as wide as the length of each. Eustachian openings minute; *tympanum indistinct*, with a diameter less than one-half of that of the eye-slit. Snout narrowly truncate and depressed, with obtuse canthus rostralis. *Digital dilatations large; first finger shorter than the second*; heel reaching beyond the snout, if pressed forwards. Abdomen smooth. Dark brown above, snout paler than interorbital space; pale brown below. (After Cope.)

Length of body	20 millim.
„ hind limb	27 „
„ hind foot	12 „
„ tarsus	6 „

22. Hylodes lanciformis.

Lithodytes lanciformis, Cope, Proc. Am. Phil. Soc. xvii. p. 88 (1877).

Hab. "West Coast of Central America."

Head nearly twice as long as the pelvis. Vomerine teeth in two fasciculi, situated behind the level of the choanæ. Tympanum not narrowed, two-thirds the size of the eye. Snout narrowed, with well-defined canthus rostralis; nostrils nearly terminal. Skin without dermal folds. Disks of fingers small, of toes of medium size. Dark ash-coloured above, with a white vertebral line. A broad black band extends from the end of the snout across the tympanum, where it contracts to a narrow black line extending to near the middle of the side. Hind limbs cross-banded, with the concealed surfaces uniformly dusky. A dark interorbital cross-band, in front of which the snout is very pale. Upper lip and lower parts without spots. (After Cope.)

Length of body	26 millim.
„ hind limb	50 „

23 (24). Hylodes megacephalus.

Lithodytes megacephalus, Cope, Journ. Ac. N. Sc. Philad. viii. p. 110, t. 23. fig. 11 (1875).

Lithodytes pelviculus, Cope, Proc. Am. Phil. Soc. xvii. p. 89 (1877).

Hab. Costa Rica, Las Delicias, Santa Clara, alt. 500 metres (*Underwood, in Mus. Nac. de Costa Rica*), Pico Blanco, alt. 6000 feet (*Gabb*).

Habit rather stout. Head large, broad, *with a raised bony supraciliary ridge, which terminates on the occiput*, rather more swollen in adult than in younger specimens. Snout broad, short, with distinct canthus rostralis, and sloping, concave loreal region. Tympanum vertically oval and two-thirds the size of the eye in a large female; round, nearly the size of the eye in a younger female. Vomerine teeth in two short, approximated, transverse series, behind, but just within the inner margin of, the choanæ. Skin smooth or with minute granules, but with a linear fold above the tympanum, converging towards the middle of the back, not continuous with another similar fold which runs forward from the groin and terminates in the iliac region.

Supraciliary region with a few pointed tubercles. Sides of the body areolated, abdomen smooth. Disks of the fingers and toes very small. Inner finger a little longer than the second. Two metatarsal tubercles, of which the outer is nearly obsolete. No subarticular tubercles. Length of the body rather exceeding the distance between the vent and heel. Upper parts slate- or dull mauve-coloured, with an irregular black mark below the eye, a curved black band above the tympanum, a few small symmetrical black spots on the side of the body, and a light interorbital cross-band. Hind limbs with obscure dark cross-bands; hinder part of the hind limbs, soles, and vent black, marbled with white or greyish. Abdomen white, reticulated with brown.

One specimen with a white vertebral line.

	♀.	♀.
Length of body	66 millim.	32 millim.
Distance between vent and heel	58 „	30 „

This description is taken from two females, both obtained by Mr. Underwood, of which the larger agrees best with Cope's description of *L. megacephalus*, the smaller with his *L. pelviculus*. They are, however, specifically identical, and the slight difference in the formation of the cephalic crests, on which Cope founded his species, is, in my opinion, merely due to age. Moreover, I strongly suspect that the following Copean species is also merely nominal, and ought to be united with *H. megacephalus*.

25. Hylodes gulosus.

Lithodytes gulosus, Cope, Journ. Ac. N. Sc. Philad. viii. p. 112 (1875).

Hab. Costa Rica, Pico Blanco, alt. 6000 feet (*Gabb*).

Said to differ from *H. megacephalus* in the following points:—Head rather smaller; no dermal folds on the back. A black spot across the tympanum, and one under the eye; lower parts brown, irregularly marbled with white.

Known from a single female specimen, body 103 millim. long.

Species the determination of which requires further confirmation.

Hylodes griseus.

Hyla grisea, Hallow. Proc. Ac. N. Sc. Philad. 1860, p. 485.
Hylodes (*Craugastor*) *griseus*, Cope, ibid. 1863, p. 48.
Lithodytes griseus, Cope, Bull. U. S. Nat. Mus. no. 32, p. 16 (1887).

This is a Brazilian frog; its occurrence within the Central-American limits is extremely doubtful, and rests solely on a very short statement by Cope, who enumerates it as a Nicaraguan species.

Hylodes longipes.

Batrachyla longipes, Baird, U. S. & Mex. Bound. Surv. p. 35, t. 37. figg. 1–3 (without description).
Epirhexis longipes, Cope, Journ. Ac. N. Sc. Philad. vi. p. 96 (1866).

The characters of this frog are very imperfectly known; it is said to have been captured forty leagues north of the city of Mexico.

Hylodes lineatus.

Hylodes lineatus, Brocchi (nec Schneid.), Bull. Soc. Philom. (7) iii. p. 22; Miss. Sc. Mex., Batr. p. 59.

Hab. MEXICO * (*Bocourt*).

Head elongate-oval, with the end of the snout truncate; tympanum two-thirds the size of the eye. Tongue oblong, without notch behind. Toes with a rudimentary web; two metatarsal tubercles. Greyish-rose-coloured above, with a very narrow vertebral line.

Hylodes latrans.

Lithodytes latrans, Cope, Amer. Nat. 1878, p. 186; Bull. U. S. Nat. Mus. no. 17, p. 25 (1880).

Size rather large. Skin smooth, that of the abdomen thrown into a disk by a circular fold. Disks of fingers and toes small. Digits long, toes rather short; first finger considerably longer than the second, but shorter than the fourth; heel projecting beyond the snout, when the hind limb is stretched forward. Loreal region oblique. Diameter of the eye exceeding the long diameter of the tympanum by one-half of the latter. Vomerine teeth in two short transverse patches, nearly in a line with the posterior border of the choanæ. Brownish-grey above, with a few large brown spots with pale centres; a brown band connects the orbits round the end of the snout. Limbs cross-banded.—Length of the body 76 millim.; of the hind limb 107 millim.—Cretaceous limestone cliffs of Texas. (After Cope.)

CALYPTOCEPHALUS.

Calyptocephalus, Dum. et Bibr. Erpét. Gén. viii. p. 447.
Phrynocerus, Cope, Proc. Ac. N. Sc. Philad. 1862, p. 157.

Vomerine teeth present. Head covered with a dermal ossification, forming a complete orbit. Tympanum distinct. Toes webbed, without disks. Outer metatarsals incompletely separated. Sternum a cartilaginous plate.

1. Calyptocephalus testudiniceps.

Phrynocerus testudiniceps, Cope, Proc. Ac. N. Sc. Philad. 1862, p. 157.
Calyptocephalus testudiniceps, Cope, Nat. Hist. Rev. 1865, p. 112.

Hab. PANAMA.

Skin with flat tubercles. Yellowish or greenish, marbled with black.

* The exact locality given by Brocchi is "Attillan," which is probably a misprint for Atitlan, a place in the State of Oaxaca.

Fam. BUFONIDÆ.

ENGYSTOMOPS.

Paludicola, Cope, Proc. Ac. N. Sc. Phil. 1864, p. 180 (nec Wagl.).
Engystomops, Espada, An. Soc. Esp. i. p. 86 (1872); Bouleng. Batr. Sal. p. 275.
Microphryne, Peters, MB. Ak. Wiss. Berlin, 1873, p. 616.

Vomerine teeth none. Tympanum distinct or hidden. Toes nearly free, without disks; outer metatarsals united. Parotoids very small and indistinct. Sternum with a bony style.

1. Engystomops pustulosus.

Paludicola pustulosa, Cope, Proc. Ac. N. Sc. Philad. 1864, p. 180.
Microphryne pustulosa, Peters, MB. Ak. Wiss. Berlin, 1873, p. 616; Cope, Journ. Ac. N. Sc. Philad. viii. p. 155 (1875); Proc. Am. Phil. Soc. xviii. p. 267 (1879); Sumichrast, Bull. Soc. Zool. Fr. 1880, p. 189.
Engystomops pustulosus, Bouleng. Batr. Sal. p. 275.
Bufo pustulosus, Brocchi, Miss. Sc. Mex., Batr. p. 86, t. 8. fig. 1.

Hab. MEXICO, Tehuantepec (*Sumichrast*); PANAMA, Chiriqui (*Peters*), Buchio Soldado (*Bransford*).—VENEZUETA (*Peters*).

Tubercular above, smooth below. Tympanum hidden. A conical tubercle on the middle of the inner edge of the tarsus. Greyish above, marbled with darker; throat and breast with a light median stripe. Chiriqui specimens have a white spot in the middle of the back, and a white line along the coccyx.

BUFO.

Bufo, Wagl. Syst. Amph. p. 206.
Otilophus, Cuv. Règne Anim. p. 154.

Auditory organ developed, with the tympanum distinct, more rarely hidden below the integument. Toes more or less webbed, without disks; outer metatarsals united. Parotoids generally well developed. Sternum a cartilaginous plate, sometimes more or less ossified along the middle.

Synopsis of Central-American Species *.

I. Crown of the head without projecting ridges.
A. Toes not more than half-webbed.
1. Metatarsal tubercles not shovel-shaped.
a. Skin smooth or nearly so.
1. *hæmatiticus.*

* No opinion can be formed as regards *Bufo melanogaster,* Hallowell, Proc. Acad. N. Sc. Philad. 1860, p. 486. The name was given to a young Nicaraguan specimen, 13 lines long; it is described in the following

b. Skin of the back tubercular or warty.

2. *punctatus*; 3. *mexicanus*; 4. *monksiæ.*

2. Inner metatarsal tubercle shovel-shaped.

5. *compactilis.*

B. Toes fully webbed.

6. *politus.*

II. A crest round and behind the supraciliary margin; a parietal crest is absent or scarcely indicated *.

7. *marmoreus*; 8. *sternosignatus*; 9. *canaliferus*; 10. *debilis*; 11. *marinus.*

III. Supraciliary crest enormously swollen behind.

12. *occipitalis.*

IV. Supraciliary crest forked behind, with a well-developed parietal prong.

A. Outer fingers moderately long.

1. Tympanum distinct.

a. Crests well or moderately raised.

13. *coccifer*; 14. *coniferus*; 15. *valliceps*; 16. *melanochlorus*; 17. *luetkenii.*

b. Crests enormously developed behind.

18. *typhonius.*

2. Tympanum hidden.

19. *simus*; 20. *intermedius*; 21. *bocourti.*

B. All fingers very slender.

22. *veraguensis*; 23. *auritus.*

1. Bufo hæmatiticus.

Bufo hæmatiticus, Cope, Proc. Ac. N. Sc. Philad. 1862, p. 157; Keferst. Gött. Nachr. 1867, p. 353; Arch. f. Naturg. 1868, p. 295, t. 9. fig. 7; Günth. P. Z. S. 1870, p. 401; Peters, MB. Ak. Wiss. Berlin, 1873, p. 618; Bouleng. Batr. Sal. p. 290; Brocchi, Miss. Sc. Mex., Batr. p. 72.

Hab. NICARAGUA (*Cope*); COSTA RICA, Talamanca (*Cope*), Santo Domingo, Golfe Dulce (*Pittier*), Las Minas, Rio San Juan, Carillo (*Underwood*); PANAMA, Chiriqui (*Peters*), Camp Maria Caretta, Rio Truando (*Cope*).—COLOMBIA.

Crown of the head without projecting ridges, the interorbital space being slightly concave and wider than the supraciliary cover. Snout short, with vertical sides and sharp canthus rostralis, truncated in front. Tympanum distinct, about half the area of the eye. Parotoids large, with an angular ridge which extends forward over the

---

terms:—Supraorbital ridges slightly developed; internal nares of moderate size; tongue elongate, cordiform, rounded posteriorly, free for rather more than half of its length posteriorly; colour grey, with a longitudinal row of subround black spots on either side of the median line; extremities varied with black; underparts yellow, marbled with black.

* It should be borne in mind that in young specimens these cranial crests are much less developed than in mature ones, and entirely absent at a very early age.

tympanum. Upper parts with minute asperities or smooth; lower parts and sides of the head smooth. Fingers and toes with slightly dilated tips: first finger longer than the second; toes slender, with a very short web. Two small soft metatarsal tubercles; tarsal fold indistinct. Dark brown above, generally with a few pinkish spots. A narrow yellowish line runs along the canthus rostralis, the edge of the upper eyelid, and the parotoid ridge, and disappears on the side of the body; a few spots on the sides of the head, and the lateral portion of the parotoid, pink. A pinkish, dark-edged band across the middle of the lower leg. Upper parts of the hands and feet black, reticulated with white. Lower parts brownish, with darker and lighter marbling.

Distance of vent from snout	75	millim.
,, ,, metatarsal joint	78	,,
,, ,, end of fourth toe	111	,,

2. **Bufo punctatus.**

Bufo punctatus, Baird & Girard, Proc. Ac. N. Sc. Philad. 1852, p. 173; U. S. & Mex. Bound. Surv., Batr. p. 25, t. 39. figg. 5–7; Bouleng. Zool. Record for 1882, Rept. p. 23; Ann. & Mag. N. H. 1883, xi. p. 19; Brocchi, Miss. Sc. Mex., Batr. p. 68, t. 7. fig. 2; Cope, Bull. U. S. Nat. Mus. no. 32, p. 10 (1887); Stejneger, N. Amer. Faun. (Dep. Agricult.) no. 3, p. 117 (1890); Dugès, La Natural. (2) ii. p. 482 (1896).

Bufo beldingi, Yarrow, Proc. U. S. Nat. Mus. v. p. 441 (1882).

Hab. North America, Southern United States.—Mexico, Chihuahua, Silao, and Guanajuato (*Dugès*).

Crown of the head without projecting ridges, flat above, very broad behind; with acuminate projecting snout; tympanum distinct, but only about one-third the size of the eye; parotoids small, subtriangular. Upper parts coarsely tubercular, lower parts granular; a fold of the skin on each side of the body from the parotoid towards the loin. First finger longer than the second. Toes one-third webbed; two small metatarsal tubercles, no tarsal folds. Brownish, with red tubercles, surrounded by black dots; scattered black specks on the sides, the thorax, and lower parts of the fore legs.

This species enters the Central-American fauna from the North.

3. **Bufo mexicanus.**

Bufo mexicanus, Brocchi, Bull. Soc. Philom. (7) iii. p. 23; Miss. Sc. Mex., Batr. p. 85, t. 8. fig. 3.

Hab. Mexico, Ciudad in Durango (*Forrer*).

Crown of the head without projecting ridges, the interorbital space being slightly concave and narrower than the broad supraciliary cover. Snout short, with the sides subvertical. Tympanum distinct, three-fifths the width of the eye. Parotoids elliptical. Upper parts somewhat densely covered with rather small tubercles of unequal size. All the lower parts densely granular. First finger very little longer than the second;

toes half-webbed; two metatarsal tubercles, of which the inner is somewhat compressed and pointed, covered with horn; a tarsal fold. Brownish, marbled with darker; lower parts whitish.

Distance of vent from snout	. . . .	68 millim.
,, ,, metatarsal joint	. . . . . .	63 ,,
,, ,, end of fourth toe	. . . . . .	90 ,,

I have seen only one specimen of this form, a female from Ciudad. If it is a distinct species, *B. mexicanus* comes very close to *B. spinulosus* (=*B. halophilus*), in which the size of the tympanum is subject to considerable variation. It is also closely allied to *B. compactilis*, but has the metatarsal tubercle much less dilated and somewhat pointed.

4. Bufo monksiæ.

Bufo monksiæ, Cope, Proc. Am. Phil. Soc. xviii. p. 263 (1879).

Hab. MEXICO, Guanajuato (*Dugès* *).

Crown of the head without projecting ridges; canthus rostralis sharp, loreal region vertical; snout vertically truncated in front. Tympanum concealed; parotoid a wide oval, rather large. Upper parts tubercular, lower parts granular. First and second fingers equal in length. Toes half-webbed; two metatarsal tubercles, small but prominent. Blackish-brown above, with a few small ashy spots; blackish spots on the sides, belly, throat, and limbs. (After Cope.)

Length of body	. . . .	35 millim.
,, hind limb	. . . .	42 ,,
,, hind foot	. . . .	20 ,,

This species is not well established; it seems to have been named from a single young individual, which very probably was only the immature state of one of the other members of the genus, perhaps of one provided with cranial crests in the adult stage, like *B. simus*.

5. Bufo compactilis.

Bufo compactilis, Wiegm. Isis, 1833, p. 661; Peters, MB. Ak. Wiss. Berlin, 1863, p. 80; 1873, p. 624; Bouleng. Batr. Sal. p. 302; Cope, Bull. U. S. Nat. Mus. no. 32, p. 10 (1887); Dugès, Nat. Mex. (2) i. p. 140 (1888); ii. p. 482 (1896).

Anaxyrus melancholicus, Tschudi, Fauna Peruana, Herpet. p. 78, t. 11. fig. 5 (bad; *fide* Peters).

? *Bufo speciosus*, Girard, Proc. Ac. N. Sc. Philad. 1854, p. 86; U. S. & Mex. Bound. Surv. ii. p. 26, t. 40. figg. 5–10 †.

Bufo anomalus, Günth. Batr. Sal. p. 57 (*fide* Peters).

Bufo levifrons, Brocchi, Bull. Soc. Philom. (7) i. p. 187; Miss. Sc. Mex., Batr. p. 70, t. 6. figg. 1–2.

---

* In a list of Mexican toads kindly supplied to me by Dr. Dugès, no mention is made of this species. He complains that Cope did not always give the localities correctly of the species sent to him.

† Considered distinct by Garman, Bull. Essex Inst. xix. 1888, p. 136.

Hab. North America, Texas.—Mexico, Pueblo and Valley of Mexico (*Cope*), Guanajuato and Guadalajara (*Dugès*).

Crown of the head without projecting ridges, the interorbital space being rather flat and narrower than the supraciliary cover. Snout short; tympanum very small and more or less indistinct. Parotoids of moderate size, subquadrangular. Upper parts and sides with small spiny tubercles. First finger longer than the second; toes half-webbed; two metatarsal tubercles, of which *the inner is large, shovel-shaped, with a sharp, horny, black edge,* the outer one tipped with black, but much smaller; tarsus with a series of small rough tubercles along the inner margin. Olive, irregularly spotted with black; abdomen whitish, with a few black spots.

Distance of vent from snout	. . . .	62 millim.
„ „ metatarsal joint	. . . . . .	54 „
„ „ end of fourth toe	. . . . . .	80 „

6. **Bufo politus.**

Bufo politus, Cope, Proc. Ac. N. Sc. Philad. 1862, p. 158; Bouleng. Batr. Sal. p. 292.

Hab. Nicaragua, Greytown (*Cope*).

Crown of the head without projecting ridges; profile of snout sloping towards the front; no canthus rostralis. Tympanum one-fourth the size of the eye; parotoids oval, of moderate size. Skin of the whole body smooth, shining; a series of flat glands on each side of the back; other similar glands on the uppersides of the fore and hind legs, those of the tibia remarkably large. Toes fully webbed, soles smooth, without tubercles, except that formed by the first cuneiform bone, which is very prominent and conical; no tarsal fold. Olive-brown above, glands edged with darker; inner faces of the limbs straw-coloured, with large brown spots. Bright yellow below, reticulated with black. (After Cope.)

7. **Bufo marmoreus.** (Tab. LXX. fig. A.)

Bufo marmoreus, Wiegm. Isis, 1833, p. 661; Peters, MB. Ak. Wiss. Berlin, 1863, p. 81.

? *Bufo sternosignatus,* Keferstein, Gött. Nachr. 1867, p. 352; Wiegm. Arch. f. Naturg. 1868, p. 294; Cope, Journ. Ac. Nat. Sc. Philad. viii. p. 101 (1875); Bull. U. S. Nat. Mus. no. 32, p. 11 (1887).

Bufo argillaceus, Cope, Proc. Ac. N. Sc. Philad. 1868, p. 138; Sumichrast, Bull. Soc. Zool. Fr. 1880, p. 188.

Bufo lateralis, Werner, Zool. Anzeig. 1894, p. 156.

Hab. Mexico, Colima (*Xantus*), Tehuantepec (*Sumichrast, Werner*).

Crown of the head with a slightly raised supraciliary ridge, from which starts a short supratympanic ridge; no parietal branch; interorbital space but slightly concave, not wider than the supraocular cover. Snout short, not pointed, with angular, nearly straight canthus rostralis. Tympanum distinct, circular, nearly four-fifths the area of the eye. Parotoid subtriangular, as large as the supraocular cover. Upper parts with numerous small tubercles, lower parts finely granular; lateral fold very indistinct. First finger longer than the second. Toes two-fifths webbed. Two small metatarsal tubercles; no tarsal fold. Brownish-grey with more or less distinct darker marbling; a whitish vertebral line; sometimes with some whitish spots on the back; whitish markings on the side, more or less confluent into a longitudinal band. A brown interorbital band, concave towards the front. Lower parts uniform whitish.

Distance of vent from snout	. . . .	70 millim.
„ „ metatarsal joint	. . . . . .	64 „
„ „ end of fourth toe	. . . . . .	89 „

The three examples which I have referred to this species are Mexican, and come from one of our editor's collectors; unfortunately, the writing on the label has been effaced, so that collector and locality are unknown. It seems to me probable that the Central-American specimens named *Bufo sternosignatus* by Keferstein and Cope were, in fact, the present species.

I had determined our specimens as *Bufo marmoreus* (Wiegm.) when I received, on loan, the types of *Bufo lateralis*, Werner, and *B. argillaceus*, Cope, by comparison of which the identity of these toads was ascertained. I am greatly indebted to Dr. Franz Werner and the authorities of the National Museum at Washington for their kindness in communicating these examples. Our figure is taken from the type of *B. lateralis*.

8. **Bufo sternosignatus.**

Bufo sternosignatus, Günth. Batr. Sal. p. 68, t. 5. fig. C; ? Keferstein, Gött. Nachr. 1867, p. 352; Wiegm. Arch. f. Naturg. 1868, p. 294; ? Cope, Journ. Ac. N. Sc. Philad. viii. p. 101 (1875); Bull. U. S. Nat. Mus. no. 32, p. 11 (1887); ? Sumichrast, Bull. Soc. Zool. Fr. 1880, p. 188; Bouleng. Batr. Sal. p. 323; Brocchi, Miss. Sc. Mex., Batr. p. 78.

Hab. ? Mexico, Tehuantepec and Coatzacoalcos R. (*Cope*); ? Costa Rica (*Keferstein*).—Colombia; Venezuela.

Crown of the head with a raised supraciliary ridge, which behind swells into a supratympanic ridge; no parietal branch; interorbital space but slightly concave, wider than the supraocular cover. Snout compressed, rather pointed, with sharp, slightly curved canthus rostralis. Tympanum distinct, circular, three-fifths the area of the eye. Parotoid subtriangular, of moderate size. Upper and lower parts coarsely granular; a series of larger tubercles runs from the parotoid towards the groin. First finger longer than the second. Toes two-fifths webbed; two small metatarsal tubercles; no tarsal fold. Greyish-brown, with a whitish vertebral line; on both sides of it symmetrical dark brown spots; a dark interorbital band has the anterior margin but slightly concave; limbs cross-banded. Chest with darker spots, and frequently with a white rectangular cross.

Distance of vent from snout	. . . .	51 millim.
„ „ metatarsal joint	. . . .	47 „
„ „ end of fourth toe	. . . . . .	64 „

I have not seen Central-American examples of this species, and therefore have taken the above notes from the Venezuelan types in the British Museum. I doubt whether the Central-American toads named *B. sternosignatus* by Keferstein and Cope were in reality this species.

Boulenger was right in referring to *B. valliceps* the young specimens which I had regarded as belonging to *B. sternosignatus*.

9. Bufo canaliferus.

Bufo canaliferus, Cope, Proc. Am. Phil. Soc. xvii. p. 85 (1877); Bouleng. Batr. Sal. p. 323; Sumichrast, Bull. Soc. Zool. Fr. 1880, p. 189; Dugès, La Natural. (2) ii. p. 482 (1896).

Hab. MEXICO, W. Tehuantepec (*Sumichrast*), Orizaba, Cerro de Valtierra (Chiapas), and Tabasco (*Dugès*).

A raised ridge above and behind the eye, without parietal branch; interorbital space concave, as wide as the supraocular cover. Snout rather contracted and pointed, with angular curved canthus rostralis. Tympanum distinct, about half the area of the eye. Parotoid elongate, much longer than, but scarcely as wide as, the supraocular cover. Upper parts covered with warts and glands of unequal size; a fold or series of tubercles from the parotoid to the groin. Lower parts finely granular. First toe a little longer than the second. Toes half-webbed; two small metatarsal tubercles, the outer one very small. A tarsal fold is replaced by a series of tubercles. Light greyish or brownish, with a pale vertebral line, dividing two rows of large blackish-brown, more or less ocellated, symmetrical markings: the foremost is the interocular band, the two sides of which form an open, obtuse-angular V.

Distance of vent from snout	. . . .	44 millim.
„ „ metatarsal joint	. . . . . .	44 „
„ „ end of fourth toe	. . . . . .	61 „

10. Bufo debilis.

Bufo debilis, Girard, Proc. Ac. N. Sc. Philad. 1854, p. 87; Bouleng. Batr. Sal. p. 289; Cope, Bull. U. S. Nat. Mus. no. 32, p. 10 (1887); Brocchi, Miss. Sc. Mex., Batr. p. 69.

Bufo insidior, Girard, l. c. p. 88; U. S. & Mex. Bound. Surv., Batr. p. 26, t. 41. figg. 13–18.

Hab. North America, New Mexico and Sonora.—Mexico, Mazatlan (*Cope*), Tamaulipas (*Brocchi*).

Crown of the head broad, flat, rendered slightly concave by the raised supraciliary border; in old specimens sometimes even a slight parietal crest is distinguishable; interorbital space much broader than the supraocular cover. Snout short, contracted, somewhat pointed, with sharp curved canthus rostralis. Tympanum distinct, about half the size of the eye, which itself is rather small. Parotoids rather large, produced behind. Upper parts densely covered with tubercles of subequal size, lower parts granular. First and second fingers equal in length. Toes two-fifths webbed; two small metatarsal tubercles; no tarsal fold. Brownish-ash, uniform or marbled with darker; some of the tubercles black, others white; lower parts white.

	Ad. ♀.
Distance of vent from snout	46 millim.
„ „ metatarsal joint	38 „
„ „ end of fourth toe	55 „

This toad would seem to belong to the fauna of the Southern United States rather than to that of Central America; however, Cope distinctly enumerates Mazatlan among the localities from which he has seen specimens.

11. Bufo marinus.

——, Seba, Loc. Rerum Nat. Thes. i. t. 76. fig. 1; Walbaum, Schr. Ges. naturf. Freund. Berl. v. p. 230 (1784).

Rana marina, Linn. Syst. Nat. i. p. 356.

Bufo marinus, Schneid. Hist. Amphib. p. 219; Girard, U. S. Explor. Exped., Herpet. p. 80, t. 5. figg. 1–3, t. 6. figg. 1–3; Peters, MB. Ak. Wiss. Berlin, 1872, p. 220; 1873, p. 618; Cope, Bull. U. S. Nat. Mus. no. 32, p. 10 (1887); Brocchi, Miss. Sc. Mex., Batr. p. 82; Dugès, La Natural. (2) ii. p. 482 (1896).

Bufo agua, Latr. Rept. ii. p. 13; Günth. Batr. Sal. p. 65; Salvin, P. Z. S. 1860, p. 460; Cope, Proc. Am. Phil. Soc. xxii. p. 379 (1885); Sumichrast, Bull. Soc. Zool. Fr. 1880, p. 188.

Bufo horribilis, Wiegm. Isis, 1833, p. 654; Peters, MB. Ak. Wiss. Berlin, 1863, p. 81.

Hab. Mexico, Presidio de Mazatlan, Ventanas (*Forrer*), Tampico and Colima (*Richardson*), Jalisco, north of the Rio de Santiago (*F. D. G.*), Cordova, Orizaba, Tehuantepec (*Sumichrast*), Vera Cruz (*Cope*), Michoacan, Guadalajara, Cuautla in Morelos, and Tabasco (*Dugès*), Coatzacoalcos (*Street*), Yucatan (*Schott*), Cozumel I. (*Gaumer*); Guatemala, Lanquin and Coban (*Salvin*), Peten (*Cope*); Nicaragua (*Cope*); Costa Rica (*Cope*); Panama (*Cope*), Chiriqui (*Peters*).—Tropical South America; Antilles.

Crown of the head with prominent ridges round the orbit extending forward along the canthus, and emitting a short process towards the parotoids; interorbital space concave, much broader than the supraciliary cover. Snout short. Tympanum rather variable, about one-half or two-thirds the area of the eye. *Parotoids large, sometimes of enormous size.* Upper parts covered with irregular, sometimes (in males?) spinous tubercles. First finger longer than the second; toes half-webbed. Two metatarsal tubercles; a tarsal fold. Brownish, with or without darker spots; lower parts whitish, frequently spotted.

	♀ from Presidio de Mazatlan.
Distance of vent from snout	166 millim.
,, ,, metatarsal joint	124 ,,
,, ,, end of fourth toe	163 ,,

Central-American specimens, at least those of the more temperate districts, do not appear to grow to the enormous size to which the species attains in Brazil and Guiana.

12. **Bufo occipitalis.** (Tab. LXIX. fig. C.)

Bufo occipitalis, Camerano, Att. Accad. Torin. xiv. p. 889; Bouleng. Batr. Sal. p. 310.

Hab. Mexico.

Head deeply concave above, with a straight bony ridge, very prominent and swollen behind. Snout pointed, prominent. Tympanum small and rather indistinct. Parotoid moderate, its lower border forming an angle. Skin covered with small tubercles. First and second fingers equal*; toes almost entirely webbed*; two metatarsal tubercles of moderate size; no tarsal fold.

Distance of vent from snout	45 millim.
,, ,, end of fourth toe	61 ,,

I am indebted to Professor Camerano for communicating to me the sketch taken from the type, and to Count Peracca for several notes on the same specimen.

13. **Bufo coccifer.**

Bufo coccifer, Cope, Proc. Ac. N. Sc. Philad. 1866, p. 130; Journ. Ac. Nat. Sc. Philad. viii. p. 100, t. 23. fig. 5 (1875); Bull. U. S. Nat. Mus. no. 32, p. 11 (1887).

Hab. Mexico, Tehuantepec (*Cope*); Guatemala, Guatemala city (*Cope*); Costa Rica (*Cope*), vicinity of San José, Laguna de la Palma (*Underwood*).

Crown of the head with rather low crests, the parietal and postocular branches forming a right angle; snout short, with sharp, convergent, canthal ridges; interorbital space concave, wider than the supraocular cover. Tympanum distinct, circular, one-

* These characters are not well shown in the figure.

third or one-fourth the area of the eye. *Parotoid projecting, subglobular,* rather longer than the supraocular cover. Upper parts rough from numerous tubercles of unequal size; small tubercles extend forwards even into the interorbital space; lower parts roughly granular. First finger a little longer than the second. Toes one-fourth webbed; two subconical metatarsal tubercles; no tarsal fold. Adult specimens nearly uniform brownish-black, with a yellow vertebral line; lower parts whitish, uniform, or with a few darker spots. Young specimens with brown kidney-shaped spots on the back, and with two brown bands across the interorbital space; lower parts more extensively spotted with brown.

	Ad. ♀.
Distance of vent from snout	65 millim.
„ „ metatarsal joint	55 „
„ „ end of fourth toe	74 „

14. **Bufo coniferus.** (Tab. LXIX. fig. B.)

Bufo coniferus, Cope, Proc. Ac. N. Sc. Philad. 1862, p. 158; Bouleng. P. Z. S. 1898, p. 123; Zool. Record for 1899, Rept. p. 29.

Chilophryne conifera, Cope, l. c. 1863, p. 49.

Bufo ehlersi, Werner, Verh. zool.-bot. Ges. Wien, 1899, p. 481.

Hab. Costa Rica, La Palma and Carillo (*Underwood*); Panama, Turbo (*Schott*).—Colombia.

Crown of the head with distinct ridges, the parietal branch forming a right angle with the postorbital; supraciliary and postorbital ridges forming a curve. Interorbital space and crown very flat, the former wider than the supraocular cover. Snout short, with sharp curved canthus rostralis. Tympanum distinct, subcircular, about two-thirds the area of the eye. *Parotoid very small,* smaller than the supraocular cover. Back of trunk rather smooth or with smaller tubercles, but the sides and the upper parts of the limbs are armed with large pointed tubercles, the largest forming a series from the parotoid to the groin. Lower parts finely granular. Toes and, particularly, fingers slender, but the *first finger is shorter than the second,* sometimes very short. Toes broadly webbed; two flat metatarsal tubercles; no tarsal fold. Dark brown, with a whitish band or series of whitish spots along the lateral series of large tubercles; lower parts light-coloured. Young specimens prettily ornamented with bluish and pink spots.

	Ad. ♀.
Distance of vent from snout	95 millim.
„ „ metatarsal joint	88 „
„ „ end of fourth toe	123 „

15. **Bufo valliceps.**

Bufo valliceps, Wiegm. Isis, 1833, p. 657; Peters, MB. Ak. Wiss. Berlin, 1863, p. 81; Bouleng. Batr. Sal. p. 319; Sumichrast, Bull. Soc. Zool. Fr. 1880, p. 188; Cope, Bull. U. S. Nat. Mus. no. 32, p. 11 (1887); Brocchi, Miss. Sc. Mex., Batr. p. 79.

Bufo cristatus, Wiegm. Isis, 1833, p. 663; Peters, l. c.

Bufo granulosus, Baird & Girard, Proc. Ac. N. Sc. Philad. 1852, p. 173 (nec Spix).

Bufo nebulifer, Girard, Proc. Ac. N. Sc. Philad. 1854, p. 87; U. S. & Mex. Bound. Surv. ii. p. 25, t. 40. figg. 1–4; Salvin, P. Z. S. 1860, p. 460.

Hab. North America, Texas.—Mexico, Presidio de Mazatlan (*Forrer*), Tampico (*Richardson*), Tamaulipas and Puebla (*Cope*), Jalisco (*F. D. G.*), Jalapa (*Höge*), Orizaba, Vera Cruz, Tehuantepec (*Sumichrast*), Atoyac and Teapa (*H. H. Smith*); Guatemala, Dueñas and Vera Paz (*Salvin*); Nicaragua (*Cope*); Costa Rica.

Crown of the head with high projecting crests, the supraciliary ridge forked behind, the parietal branch being very distinct; the highest part of the crests the hindmost part of the supraciliary ridge. Interorbital space deeply concave, broader than the supraocular cover. Snout obtuse. Tympanum distinct, about half the area of the eye. Parotoid triangular, rather small. Upper parts covered with rough tubercles; a series of larger pointed ones from the parotoid along the side of the body. Lower parts granular. Central-American specimens have the back less tubercular than those from Texas, sometimes partly smooth. First finger longer than the second. Toes one-third webbed; two metatarsal tubercles; no tarsal fold. Brownish or yellowish, marbled with darker, sometimes a yellow vertebral band; generally a black interocular cross-bar and a black band on or from the tympanic region; sometimes a few scattered deep black spots on the back. Central-American examples have the lower parts more or less variegated with black.

	Ad. ♀.	Ad. ♀.
Distance of vent from snout	118 millim.	95 millim.
,, ,, metatarsal joint . .	90 ,,	90 ,,
,, ,, end of fourth toe . .	125 ,,	111 ,,

The rough-skinned Northern form passes gradually into the smoother, black-bellied, Central-American race. I should have been inclined to regard the latter as the *Bufo melanochlorus* of Cope, but this author describes that species as having the toes nearly free, which does not apply to any of the specimens examined by me. Also the measurements of the hind leg appear to differ considerably.

16. **Bufo melanochlorus.**

Bufo valliceps, var., Cope, Journ. Ac. N. Sc. Philad. viii. p. 100 (1875).

Bufo melanochlorus, Cope, Proc. Am. Phil. Soc. 1877, p. 85.

Hab. Eastern Costa Rica (*Gabb*).

Crown of the head with projecting ridges, the supraciliary crest being branched behind. Snout obtuse. Tympanum very distinct, half the size of the eye. Parotoids quite small *, subtriangular. *Skin nearly smooth above*, except on the scapular and iliac regions, minutely roughened below; a lateral dermal fold. Fingers and toes long, *nearly free*; no tarsal fold, two metatarsal tubercles; legs short. Green, with deep black spots; throat and thorax black. (After Cope.)

Length of body	47 millim.
„ hind leg	70 „
„ hind foot	30 „

17. Bufo luetkenii.

Bufo luetkenii, Bouleng. Ann. & Mag. N. H. 1891, viii. p. 455.

Hab. Costa Rica, Cartago (*Örsted*).

Crown of the head with low ridges, the parietal branch very low, forming an obtuse angle with the postorbital. Interorbital space concave, as wide as the broad supra-ocular cover. Snout short, rather compressed, with sharp, nearly straight canthus. Tympanum very distinct, deeper than long, about half the area of the eye. *Parotoid very small*, smaller than the area of the supraocular cover. Upper parts with few very small tubercles, a lateral series of somewhat larger ones from the parotoid towards the groin. Lower parts finely granular. *First finger longer than second.* Toes fully half-webbed. Two small metatarsal tubercles; no tarsal fold, but a series of small asperities. Brownish, with darker and some whitish spots; a light vertebral band.

	Ad. ♀.
Distance of vent from snout	78 millim.
„ „ metatarsal joint	65 „
„ „ end of fourth toe	87 „

18. Bufo typhonius.

Rana typhonia, Linn. Syst. Nat. p. 356.

Bufo typhonius, Schneid. Hist. Amphib. i. p. 207; Peters, MB. Ak. Wiss. Berlin, 1872, p. 224; 1873, pp. 618, 624; Bouleng. Batr. Sal. p. 317; Zool. Record for 1884, Rept. p. 17.

Rana margaritifera, Laur. Syn. Rept. p. 30.

Bufo pleuropterus, Schmidt, Denkschr. Ak. Wiss. Wien, 1858, p. 252, t. 2. figg. 17–19; Cope, Journ. Ac. Sc. Philad. viii. p. 155 (1875).

? *Bufo ocellatus*, Günth. Batr. Sal. p. 64.

Bufo alatus, Thominot, Bull. Soc. Philom. viii. p. 151 (1884).

Hab. Panama, Chiriqui (*Peters*), Soldado (*Cope*), Veragua (*Schmidt*).—Tropical South America.

Supraciliary ridge raised into a high crest, continued to above the tympanum, and

* First described as "large" a few lines before, but this seems to have been a *lapsus calami.*

passing into a lateral series of conical tubercles; the crest is highest above the tympanum, much less developed in the young than in mature specimens. Snout pointed, compressed, with sharp canthus. Tympanum distinct, round, two-thirds the size of the eye. Parotoid smaller. Upper parts with small, flat tubercles, lower roughly granular. First and second fingers equal. Toes not quite half-webbed; two small metatarsal tubercles; no tarsal fold. Light brownish, with some irregular ashy or whitish spots; lower parts whitish.

	Ad. ♀.
Distance of vent from snout	84 millim.
,, ,, metatarsal joint	78 ,,
,, ,, end of fourth toe	103 ,,

19. **Bufo simus.**

Bufo simus, Schmidt, Denkschr. Ak. Wiss. Wien, 1858, p. 254.

Bufo intermedius, Cope, Proc. Am. Phil. Soc. xxii. p. 379 (1885); Bull. U. S. Nat. Mus. no. 32, p. 10 (1887) (nec Günth.); Dugès, La Nat. Mex. (2) ii. p. 482 (1896).

Hab. MEXICO, Presidio, Milpas in Durango (*Forrer*), Guanajuato, Huazteca Potosina, and Silao (*Dugès*), Sierra de Bolaños and Jalisco (*Buller*), Omilteme and Amula (*H. H. Smith*), Vera Cruz (*Cope*), Orizaba (*Sumichrast*); PANAMA, Chiriqui (*Schmidt*).

The projecting ridges of the head are rather feebly developed, even in the adult; the supraorbital converge in front, and terminate behind in a fork, the prongs of which are short; young examples without ridges. The interorbital space is concave (in the adult) and broader than the supraocular cover. Snout rather short, with the sides slightly sloping. *Tympanum quite hidden under tubercular skin.* Parotoids of moderate size, prominent, elliptical, immediately behind the hind margin of the orbit. Upper parts rather densely covered with small, sometimes spiny, tubercles; lower parts coarsely granular. First finger longer than second; toes two-thirds webbed; two metatarsal tubercles, of which the inner is somewhat compressed; a series of pointed tubercles takes the place of a tarsal fold. Greyish or brownish above, marbled with darker and sometimes with pink blotches; lower parts uniform whitish, sometimes, and in the young always, spotted with black.

	♀.	♀.
Distance of vent from snout	87 millim.	75 millim.
,, ,, metatarsal joint . . .	71 ,,	70 ,,
,, ,, end of fourth toe . .	102 ,,	98 ,,

This species was originally described from young examples, the largest of which was only 26 millim. long. At that age, and even in specimens up to 50 millim. in length, the cephalic crests are not developed, the crown being as flat as in the flat-headed toads generally. Yet I should have hesitated to refer our Mexican specimens to *B. simus* of Schmidt, as this author describes his species with an abdomen "*fere glabrum*," whilst our examples have it distinctly granulated. However, fortunately, the Natural History Museum has received, through Dr. Werner, one of the type-specimens of Schmidt's *Bufo simus*; it is only 17 millim. long, but has a distinctly granulated abdomen.

Bufo simus is closely related to *B. intermedius* (Günth.), from which it differs only in having the tympanum hidden under a tubercular skin. The specimens from Omilteme and Amula have a particularly rough skin on the upper parts of the body, as well as on the lower.

20. **Bufo intermedius.**

Bufo intermedius, Günth. Batr. Sal. p. 140, t. 9. fig. A; Boulcng. Batr. Sal. p. 307.

Hab. Mexico, Jalisco (*Godman*).—Andes of Ecuador.

The projecting ridges of the head are rather feebly developed; the supraciliary, convergent in front, terminate behind in a fork with short prongs. Interorbital space concave, and, in the middle of its length, rather broader than the supraocular cover. Snout short, with distinct, short, convergent canthi, and with the sides slightly sloping. *Tympanum not covered by tubercular skin*, though rather small, one-third or one-half the area of the eye. Parotoids of moderate size, prominent, elongate elliptical, longer than, and as wide as, the supraocular cover, immediately behind the hind margin of the orbit. Upper parts with small, flat, not densely packed, tubercles, which do not extend on to the interorbital space; a lateral series of somewhat more conical tubercles; lower parts finely granulated. First finger longer than second; toes two-thirds webbed; two metatarsal tubercles, of which the inner is somewhat compressed; a very fine serrature takes the place of a tarsal fold. Greyish or greenish, marbled with darker; lower parts uniform whitish.

	Spec. Jalisco.
Distance of vent from snout	65 millim.
" " metatarsal joint	48 "
" " end of fourth toe	86 "

This is the only Central-American specimen which I have found to agree with *B. intermedius* in the distinctness of the tympanum and the comparatively smooth texture of the skin. The form *B. simus* is much more common and more widely spread; and I therefore conclude that the specimens named *B. intermedius* by Cope and Dugès belong to *B. simus* rather than to the present form.

21. **Bufo bocourti.**

Bufo bocourti, Brocchi, Bull. Soc. Philom. (7) i. p. 186; Miss. Sc. Mex., Batr. p. 84, t. 7. fig. 1; Müller, Verh. Ges. Basel, vii. p. 672 (1885).

Hab. Guatemala, Totonicapam (*Bocourt*).

Crown of the head with projecting ridges, branched behind, the postorbital branch forming a right angle with the main (supraorbital) ridge, and the parietal branch being rather swollen, short, bent inwards. Snout short, vertically truncate. *Tympanum hidden*; parotoid narrow, oblong. Upper parts with large tubercles, lower coarsely granular. First and second fingers equal in length; toes half-webbed; two metatarsal tubercles; a short tarsal fold. Sides marbled with black; upper lip with a white margin below the eye; a yellowish vertebral line. (After Brocchi.)

22. **Bufo veraguensis.**

Bufo veraguensis, Schmidt, Denkschr. Ak. Wiss. Wien, 1858, p. 253, t. 2. fig. 20, t. 3. fig. 21; Peters, MB. Ak. Wiss. Berlin, 1873, p. 618.

Hab. PANAMA, Chiriqui (*Peters*), Veraguas (*Schmidt*).

Crown of the head with a projecting ridge round the eye, from which a short parietal crest starts obliquely inwards; interorbital space concave. *Tympanum hidden.* Parotoids of moderate size, prominently oblong. Upper parts tubercular, *a series of larger tubercles on each side of the back.* Fingers very slender, *first shorter than second.* Toes entirely webbed, but the web is deeply notched. No tarsal fold. Greyish above, marbled with blackish-brown; lighter below. (After Schmidt.)

This apparently well-characterized species does not seem to have been recognized since the time of its discovery.

23. **Bufo auritus.**

Bufo auritus, Cope, Journ. Ac. N. Sc. Philad. viii. p. 99, t. 23. fig. 4 (1875).

Hab. COSTA RICA, Talamanca (*Cope*).

Crown of the head flat, bounded by a vertical supraciliary ridge, forked behind. Snout narrow, produced, the nostrils being nearer to each other than to the orbit. Anterior half of the tympanum distinct. Parotoids quite small, surmounted by several dermal spines. Tubercles of the skin small, spinulose, most numerous on the sides, absent below. *Fingers very slender*, the two inner subequal. Toes half-webbed. No tarsal fold; solar and plantar tubercles insignificant. Brown; a square patch on the crown, an oblique band from the occiput towards the side, and a spot on each side of the sacrum, black; abdomen marbled with black. (After Cope.)

Length of body	31 millim.
„ fore foot	10 „
„ hind limb	38 „
„ hind foot	21 „

NANNOPHRYNE.

Nannophryne, Günth. P. Z. S. 1870, p. 401; Cope, Proc. U. S. Nat. Mus. xii. p. 142 (1889).
Cranopsis, Crepidius, and *Ollotis,* Cope, Journ. Ac. N. Sc. Philad. viii. pp. 96, 97, 98 (1875).

Differs from *Bufo* by the absence of ostia pharyngea and tympanum. Palmation of toes weak. Parotoids present.

1. Nannophryne fastidiosa.

Cranopsis fastidiosus, Cope, Journ. Ac. N. Sc. Philad. viii. p. 96, t. 23. fig. 1 (1875).

Hab. Costa Rica, Pico Blanco 2500 feet (*Gabb*).

Cranial integument entirely ossified, rugose. The supraciliary crest is elevated and continued to the parotoid; it is swollen, particularly behind, forming a rugose osseous plate behind the orbit. Upper part of the snout and interorbital space concave. Parotoid short, subtriangular. Upper parts studded with round warts; a series of larger tubercles from the parotoid to the groin. Toes half-webbed; metatarsal tubercles very obscure; no tarsal fold. Yellowish-brown above, dirty brown below, sides with a blackish band. (After Cope.)

Length of body	58 millim.
,, hind limb	60 ,,
,, hind foot	35 ,,

2. Nannophryne epiotica.

Crepidius epioticus, Cope, Journ. Ac. N. Sc. Philad. viii. p. 97, t. 23. fig. 2 (1875).

Hab. Costa Rica, Pico Blanco 5000 feet (*Gabb*).

Crown of the head without ossification; of crests are present a canthal, supraciliary, and supratympanic; the supraciliary sends inwards a short tuberosity at the posterior third of the orbit; supratympanic very protuberant. Parotoid very small, rounded; skin everywhere rugose with minute tubercles, which are sparse on the dorsal, dense on the ventral regions. Digits of all the feet enclosed in the skin, leaving the longest toe projecting; inner digits of both feet rudimentary. Femur half enclosed in the inguinal integument. Soles smooth, without tubercles or folds. Brownish-black, sometimes with a brown median dorsal band. (After Cope.)

Length of body	35 millim.
,, hind limb (free portion)	30 ,,
,, hind foot	18 ,,

3. Nannophryne cœrulescens.

Ollotis cœrulescens, Cope, Journ. Ac. N. Sc. Philad. viii. p. 98, t. 23. fig. 3 (1875).

Hab. Costa Rica, Pico Blanco 3000 to 5000 feet (*Gabb*).

Crown of the head without ossifications, but the supraciliary ridges are slightly prominent, and continued behind in straight lines by two angles of the parietal bone. Parotoids present; the skin of the back and sides is studded with numerous spaced tubercles; lower parts nearly smooth. Toes half-webbed; metatarsal tubercles obsolete; no tarsal fold. Blackish above, tubercles brown; black below, thickly marked with light-blue spots. (After Cope.)

Length of body	25 millim.
„ hind limb	23 „
„ hind foot	13 „

Fam. RHINOPHRYNIDÆ.

RHINOPHRYNUS.

Rhinophrynus, Dum. et Bibr. Erpét. Gén. viii. p. 757.

Form of the body as in *Engystoma.* Snout pointed, but with the front truncated, terminating in a flat circular disk. Auditory apparatus imperfectly developed; no tympanum or cavum tympani or eustachian tubes. Skin smooth; a large parotoid with smooth surface, hidden below the skin. Sternum rudimentary. Toes webbed.

1. Rhinophrynus dorsalis.

Rhinophrynus dorsalis, Dum. et Bibr. Erpét. Gén. viii. p. 758, t. 91. fig. 2; Günth. P. Z. S. 1858, p. 339; Batr. Sal. p. 127, t. 12; Bouleng. Batr. Sal. p. 329; Cope, Bull. U. S. Nat. Mus. no. 32, p. 11 (1887); Rovirosa, La Nat. Mex. (2) i. pp. 156–160, t. 15. figg. A–C (1888); Dugès, La Nat. Mex. (2) ii. p. 98, t. 5.

Rhinophrynus rostratus, Brocchi, Bull. Soc. Philom. (7) i. p. 196; Miss. Sc. Mex., Batr. p. 98, t. 9. fig. 1.

Hab. MEXICO, Vera Cruz (*Sallé*), Tuxpan (*Cope*), Chiapas, Tabasco (*Dugès*), Tehuantepec (*Sumichrast*), Campeche (*Morelet*); GUATEMALA, Peten (*Berendt*).

Metatarsus with a large shovel-shaped tubercle. Bluish-grey, with yellowish spots on the sides and along the vertebral line, the latter forming a more or less complete band.

This Batrachian shares in Tabasco, with *Phrynosoma*, the popular name of "Camaléon."

Fam. SCAPHIOPODIDÆ.

SCAPHIOPUS.

Scaphiopus, Holbr. N. Amer. Herpetol. i. p. 85.

Scaphiopus and *Spea,* Cope, Journ. Ac. N. Sc. Philad. vi. p. 81 (1866).

Habit stout, toad-like; pupil vertical. Upper jaw and vomer toothed; tongue entire or nearly so. Tympanum present, but generally hidden. Sacral vertebra with strongly dilated diapophyses, anchylosed with coccyx; manubrium cartilaginous. Skin tubercular; a generally small parotoid. Toes webbed; outer metatarsals separated by a web; a shovel-shaped metatarsal tubercle (fossorial).

1. Scaphiopus couchii.

Scaphiopus couchii, Baird, Proc. Ac. N. Sc. Philad. vii. p. 62 (1855); U. S. & Mex. Bound. Surv. ii., Rept. p. 28, t. 35. figg. 1–6 (1859); Cope, Proc. Ac. Nat. Sc. Philad. 1863, p. 52; Brocchi, Miss. Sc. Mex., Batr. p. 26; Bouleng. Batr. Sal. p. 434; Cope, Bull. U. S. Nat. Mus. no. 32, p. 12 (1887); Mocquard, Nouv. Arch. Mus. i. p. 342, t. 13. fig. 6 (fore limb) (1899).

Scaphiopus varius, Cope, Proc. Ac. N. Sc. Philad. 1863, p. 52.

Hab. North America, Lower California, Texas.—Mexico, Presidio (*Forrer*), Tampico (*Godman*), Tamaulipas (*Smithson. Instit.*).

The skin of the crown of the head is very thin, or actually merged into bone; no pectoral gland. Openings of the eustachian tubes distinct, but much smaller than choanæ. Tongue slightly notched behind. No canthus rostralis, sides of the head sloping. Skin of the back with small flat tubercles; lower parts smooth. Greenish marbled with grey; lateral tubercles white.

Length of body . . . . 53 millim.

2. Scaphiopus rectifrenis.

Scaphiopus rectifrenis, Cope, Proc. Ac. N. Sc. Philad. 1863, p. 53; Bouleng. Batr. Sal. p. 435.

Hab. Mexico, Tamaulipas and Coahuila (*Smithson. Instit.*).

Upper surface of head bony, rough; no pectoral gland. Canthus rostralis strong; loreal region subvertical. Tympanum hidden. Skin of head and body sparsely tubercular; a weak antepectoral fold. Above brown, tubercles tipped with brown; a brown line extends from each eye to the sacrum, converging posteriorly. Below light yellow or brownish. (After Cope and Boulenger.)

Length of body . . . . 2 in. 8 lines.

3. Scaphiopus hammondii.

Scaphiopus hammondii, Baird, U. S. Pacif. R. R. Reports, xi., Rept. p. 12, t. 28. fig. 2 (1859); Cope, Proc. Ac. N. Sc. Philad. 1863, p. 53; Bull. U. S. Nat. Mus. no. 32, p. 12 (1887).

Spea hammondii, Cope, Journ. Ac. N. Sc. Philad. vi. p. 81 (1866).

Scaphiopus bombifrons, Cope, Proc. Ac. N. Sc. Philad. 1863, p. 53.

Scaphiopus stagnalis, Cope, in Wheeler's Report of Geogr. & Geol. Exped. West of 100th Merid. v. p. 525, t. 25. figg. 6–8 (young).

Scaphiopus dugesii, Brocchi, Bull. Soc. Philom. iii. p. 23 (1879); Miss. Sci. Mex., Batr. p. 24, t. 9. fig. 4; Bouleng. Batr. Sal. p. 436.

Hab. North America, Missouri, New Mexico, Texas, California.—Mexico, Chihuahua 8000 feet (*Brimley*), Guanajuato (*Dugès*).

Skin of the crown of the head rather thick. Openings of the eustachian tubes distinct, but smaller than choanæ; tongue without posterior notch. Snout very short, without canthus, its sides subvertical. Skin of the back, and sometimes of the lower thigh, densely studded with tubercles; abdomen with smaller, more scattered, flat tubercles. Upper parts greyish, uniform or with brownish marbling; lower parts white.

Length of body . . . . 53 millim.

4. Scaphiopus multiplicatus.

Scaphiopus multiplicatus, Cope, Proc. Ac. N. Sc. Philad. 1863, p. 51.
Spea multiplicata, Cope, Journ. Ac. N. Sc. Philad. vi. p. 81 (1866).

Hab. MEXICO, valley of Mexico (*Potta*).

Crown of the head covered with skin. A strong fold across the gular region; sides with longitudinal folds. Parotoid very large, extending beyond the scapula. Upper surfaces coarsely tubercular, skin of the limbs nearly smooth. A large gland occupies most of the length of the tibia. Abdomen minutely, pubic region coarsely, rugose. Choanæ smaller than the ostia pharyngea. Tongue entire. Above ashy-brown, below yellowish. (After Cope.)

Length of body . . . . 5 in. 1 line.

Fam. HYLIDÆ.

HYLA.

Hyla sp., Laur. Syst. Amphib. p. 32.
Hyla sp. and *Trachycephalus,* Dum. et Bibr. Erpét. Gén. viii.; Günth. Batr. Sal.
Hyla, Calamites, Hypsiboas, Trachycephalus, Ranoidea, Centrotelma, Scytopis, Osteocephalus, Acrodytes, Smilisca, Dryomelictes, Cinclidium, Cincloscopus, Cope, Journ. Ac. Nat. Sc. Philad. vi. (1866–67), and Ll. al.
Hyla, Bouleng. Batr. Sal. p. 337.
Hyla, Plectrohyla, and *Cauphias,* Brocchi, Bull. Soc. Philom. i. pp. 92, 129 (1877).
Hyliola *, Mocquard, Nouv. Arch. Mus. Hist. Nat. i. p. 337 (1899).

Pupil horizontal. Tongue entire or but slightly emarginate behind. Vomerine teeth †. Fingers and toes dilated into terminal disks, the latter always webbed.

* This genus is based on the ossification of a cartilage intercalated between the terminal and penultimate phalanges of the fourth toe; in a non-ossified and rudimental condition it seems to be present in the majority of, if not in all, the species. Insignificant as the taxonomic value of this character appears to be, I should have gladly availed myself of it for division of the numerous species of this genus. But at present the condition of this part of the skeleton is unknown in a large proportion of the species, nor have I the material for supplying the deficiency.

† Abnormally in some of the small species vomerine teeth may be wanting: see *H. prosoblepon, H. nana.*

Auditory organ developed; tympanum present or hidden under the skin. Outer metatarsals united or slightly separated. Females without dorsal pouch.

Synopsis of Central-American Species.

I. Head covered with soft skin.
A. Vomerine teeth arranged in two fascicles or short transverse series.
1. Vomerine teeth placed between the choanæ.
a. Tympanum more or less distinct.
α. Eyes lateral or mainly so.
aa. Fingers free.
αα. Upper parts smooth.
* Sides of the head or body ornamented with longitudinal colour-bands.
1. *eximia*; 2. *bocourti*; 3. *nana*; 4. *cherrii*; 5. *microcephala*.
** Sides without longitudinal bands †.
6. *elæochroa*; 7. *bistincta*.
ββ. Upper parts with tubercles.
8. *splendens*; 9. *copii*; 10. *boulengeri*.
bb. Fingers narrowly (less than one-third) webbed.
11. *mocquardi*; 12. *quinquevittata*; 13. *phæota*; 14. *tæniopus*; 15. *puma*.
cc. Fingers more or less broadly (one-third or more) webbed.
αα. Tympanum one-half or one-third the area of the eye.
16. *baudini*; 17. *venulosa*; 18. *sordida*; 19. *gabbi*; 20. *pseudopuma*; 21. *godmani*; 22. *uranochroa*.
ββ. Tympanum less than one-third the area of the eye.
23. *leucophyllata*; 24. *underwoodi*; 25. *nigripes*; 26. *molitor*; 27. *miotympanum*.
β. Eyes directed forward.
28. *prosoblepon*.
b. Tympanum hidden under the skin.
29. *pulverata*; 30. *glandulosa*; 31. *guatemalensis*; 32. *crassa*.
2. Vomerine teeth placed behind the level of the hind margin of the choanæ.
33. *punctariola*; 34. *spilomma*; 35. *miliaria*.
B. Vomerine teeth in two curved series forming together an arch.
36. *crepitans*; 37. *albomarginata*.
II. Skin of the head coalesced with the bone, rough.
38. *rudis*.

1. Hyla eximia.

Hyla eximia, Baird, Proc. Ac. N. Sc. Philad. 1851, p. 61; U. S. Mex. Bound. Surv., Rept. p. 29, t. 38. figs. 8–10; Peters, MB. Ak. Wiss. Berlin, 1869, p. 880; Bouleng. Batr. Sal. p. 378;

† *Hyla plicata*, Brocchi, Miss. Sc. Mex., Batr. p. 35, t. 12. fig. 1, would appear to belong to this division. It is imperfectly known from a single specimen in the Paris Museum, preserved for a long time and said to be from Mexico. Its most striking character seems to be in the small size of the disks, which are not half the size of the tympanum, which itself is described as small, with a diameter only one-third of that of the eye.

Brocchi, Miss. Sc. Mex., Batr. p. 32, t. 13. figs. 3–4; Cope, Bull. U. S. Nat. Mus. no. 32, p. 14 (1887).

Hyla euphorbiacea, Günth. Batr. Sal. p. 109, t. 10. fig. C.

Hyla gracilipes, Cope, Proc. Ac. N. Sc. Philad. 1865, p. 194 *.

Hyla staufferi, Cope, Proc. Ac. N. Sc. Philad. 1865, p. 195.

Hyla eximia, subsp. *staufferii*, Cope, Bull. U. S. Nat. Mus. no. 32, p. 14 (1887).

Hab. MEXICO, Chihuahua, alt. 7000–8000 feet (*Brimley*), Presidio and Ciudad (*Forrer*), city of Mexico (*Doorman*, *H. H. Smith*), Sierra de Cuyatlan, Guadalajara, alt. 3800 met. (*Buller*), Guanajuato (*Dugès*), Amula (*H. H. Smith*), Jalisco and South Mexico (*F. D. G.*).

Vomerine teeth between the choanæ, in two transverse groups or slightly oblique series; choanæ rather more open than eustachian ostia. Snout of moderate length, rather depressed, with obtuse canthus and sloping sides. Distance of the nostril from the eye nearly as long as the diameter of the latter, and more than the distance of the nostrils from each other. Interorbital space wider than the upper eyelid. Tympanum distinct, one-half or one-third the area of the eye. Fingers free, with disks as large as those of the toes, equal to or somewhat exceeding the area of the tympanum. Toes one-third webbed or even less; a tarsal fold. Upper parts smooth, very rarely tubercular †; lower parts granular; a fold across the chest; no other folds. Back greenish, bluish, or brownish, uniform or with blackish, sometimes light-edged, oblong spots; a blackish line along the canthus rostralis continued as a broader band from behind the eye, across the tympanum, and along the side of the body; towards the back it is sharply defined, generally with a white edge, but towards the abdomen it gradually passes into the light coloration of the lower parts; it is also white-edged below, between the eye and the shoulder. Upper jaw with a semitransparent horn-coloured border; limbs cross-banded; thighs with the concealed parts not pigmented. Male with an external subgular vocal sac.

	♂ (Mexico).	♀ (Amula).
Greatest width of the head	12 millim.	11·5 millim.
Distance of vent from snout	33 „	30 „
„ „ heel	30 „	30 „
„ „ metatarsal joint	38 „	37·5 „
„ „ end of fourth toe	53 „	53 „

* I cannot help thinking that the words "eyes little prominent, one-half tympanic disk," in Cope's description of *H. gracilipes*, are a *lapsus calami*. No Central-American *Hyla* has so small an eye. Also two specimens in the Natural History Museum, named *H. gracilipes*, do not differ in this respect from others undoubtedly *H. eximia*.

† Tubercular skin observed once only, in a female from Presidio.

2. **Hyla bocourti.**

Hyla regilla, Brocchi, Miss. Sc. Mex., Batr. p. 31, t. 13. fig. 2 (not B. & G.).
Hyliola bocourti, Mocquard, Nouv. Arch. Mus. i. p. 341 (1899); Bull. Soc. Philom. (9) i. p. 164 (1899).

Hab. Mexico, Tepic (*Diguet*); Guatemala, Alta Vera Paz (*Bocourt*).

Closely allied to *H. eximia*, but with the posterior face of the thigh ornamentally coloured.

Vomerine teeth between the choanæ, in two transverse groups; choanæ and eustachian ostia nearly equal. Snout of moderate length, rather depressed, with obtuse canthus and sloping sides. Distance of the nostril from the eye nearly as long as the diameter of the latter, and more than the distance of the nostrils from each other. Interorbital space wider than the upper eyelid. Tympanum distinct, one-half the area of the eye. Fingers free, with disks somewhat larger than those of the toes, and somewhat smaller than the tympanum. Toes barely one-third webbed; (tarsal fold, if present, quite indistinct). Upper parts smooth, lower granular. Upper parts bluish, with large blackish longitudinal spots, which commence between the eyes. Side of the snout blackish, which colour is continued as a band across the tympanum and along the side of the body; this band is sharply defined towards the back, and finely edged with white, but towards the abdomen it gradually passes in an irregular outline into the light coloration of the lower parts; it has also an incomplete whitish lower edge between the eye and the tympanum. Upper jaw without a horn-coloured border; limbs indistinctly cross-banded, and the posterior face of the thigh is finely marbled with black and white on a greyish ground.

	♀.
Greatest width of head	15 millim.
Distance of vent from snout	37 „
„ „ heel	33 „
„ „ metatarsal joint	43 „
„ „ end of fourth toe	59 „

This description is taken from one of the specimens collected by Bocourt, and kindly communicated to me by Professor Vaillant; I have deposited it in the Natural History Museum.

3. **Hyla nana,** sp. n. (Tab. LXXIII. fig. A.)

Hab. Mexico, Cuernavaca (*H. H. Smith*).

Vomerine teeth (sometimes indistinct or even entirely absent) between the choanæ, in two fascicles; choanæ wider open than eustachian ostia. Snout of moderate length, rather pointed, not depressed, with angular canthus and subvertical sides. Eye of moderate size, its diameter equal to its distance from the nostril, and more than the distance between the nostrils. Interorbital space wider than the upper eyelid. Tympanum distinct, one-third the area of the eye. Fingers free; toes two-thirds webbed, the web not reaching the disks of the third and fifth. Disks rather smaller than, or equal to, tympanum. No tarsal fold. Upper parts smooth, lower granular. A fold across the chest. Back uniform olive-coloured; a brown band along the canthus rostralis is continued along the sides of the head and body, fading behind, and passing below gradually into the whitish of the lower parts; its anterior or supratympanic portion is faintly edged with whitish above, but there is no white margin on any portion of the lower margin; no horn-coloured border to the upper jaw; limbs without cross-bands and with the concealed parts pigmentless. Male with a large external subgular vocal sac.

	♂.	♀.
Width of head	7·5 millim.	10 millim.
Distance of vent from snout	22 "	25 "
" " heel	19·5 "	22·5 "
" " metatarsal joint	26 "	29 "
" " end of fourth toe	35·5 "	39 "

This is one of the smallest of the Central-American *Hylæ*, the male, of which the measurements are given, being fully mature with a large vocal sac. Several specimens show vomerine teeth only indistinctly, and in one female specimen they have disappeared entirely. Enlarged figures of a toothed and a toothless palate are given on Tab. LXXIII.

4. **Hyla cherrii.**

Hyla cherrei, Cope, Proc. Ac. N. Sc. Philad. 1894, p. 195.

Hab. Costa Rica, Alajuela, Rio Alfaro (*Mus. Nac. de Costa Rica*, *San José*).

Vomerine teeth between the very large choanæ, at the apex of a ridge which proceeds from the anterior margin of the latter. Snout not prominent, with vertical sides, and with straight angular canthus; distance of the nostrils from the eye a little less than the diameter of the latter. Tongue longer than wide. Tympanum one-half the area of the eye. Fingers almost webless; toes fully webbed. Skin smooth above. The yellowish coloration of the back is abruptly separated from the colour of the sides by a narrow white stripe which runs from the eye to the sacrum; humerus and femur not pigmented. (After Cope.)

Width of head	8 millim.
Length of body	26 "
" hind limb	45 "
" hind foot	20 "

5. **Hyla microcephala.**

Hyla microcephala, Cope, Proc. Am. Phil. Soc. 1886, p. 281.

Hab. PANAMA, Chiriqui (*U. S. Nat. Mus.*).

Vomerine teeth in two fascicles between the choanæ. Snout short, rather deep, not prominent; eyes little prominent. Tympanum round, its diameter one-third of that of the eye. Fingers free; toes webbed nearly to the disks of the third and fifth; metatarsal tubercles indistinct; disks smaller than tympanum. Skin smooth above. Cream-coloured: a pale brown band with a narrow yellow upper edge from the end of the snout to the groin; another narrow brown band on each side of the back, extending forwards to the orbit. (After Cope.)

Width of head	8 millim.
Length of body	27·5 „
„ hind limb (from groin)	41·5 „
„ hind foot	19 „

6. **Hyla elæochroa.**

Hyla elæochroa, Cope, Journ. Ac. N. Sc. Philad. viii. p. 105, t. 26. fig. 3 (very bad) (1876); Bull. U. S. Nat. Mus. no. 32, p. 14 (1887).

Hab. COSTA RICA, Sipurio (*Gabb*); PANAMA, Pacific side (*Bransford*).

Vomerine teeth between the choanæ, in two short oblique series, more or less convergent behind. Snout elongate; tongue longer than wide. Distance of the nostrils from each other about half of their distance from the eye. Eyes large, four or five times the area of the tympanum. *Fingers entirely free*; toes half-webbed. Upper parts smooth, abdomen finely areolate. Uniform olivaceous, without spots. (After Cope.)

Width of head	9 millim.
Length of body	26 „
„ hind limb	44 „
„ hind foot	19 „

7. **Hyla bistincta.**

Hyla bistincta, Cope, Proc. Am. Phil. Soc. 1877, p. 87; Bull. U. S. Nat. Mus. no. 32, p. 14 (1887).

Hab. MEXICO, Vera Cruz and Zacualtipan (*Cope*).

Vomerine teeth in small fasciculi between the choanæ. Head, particularly the interorbital region, broad and flat; canthus rostralis distinct. Tympanum exceedingly small, one-fifth the diameter of the eye, overhung by a thick dermal fold. Fingers free; toes three-fourths webbed; foot shorter than tibia and femur. Upper parts smooth, greyish-blue, lower yellowish; sides marbled with these two colours; posterior face of femur pale brown, with a few yellow specks along the superior border. Lip not white-bordered. (After Cope.)

Length of body	45 millim.
Width of head	15 ,,
Length of hind limb	67 ,,
,, hind foot	31 ,,

8. **Hyla splendens.**

Hyla splendens, Schmidt, Denkschr. Akad. Wiss. Wien, 1858, p. 244, t. 1. fig. 7.

Hab. Panama, Chiriqui, near Boca del Toro.

Vomerine teeth between the choanæ, in a transverse series interrupted in the middle. Snout rounded, its upperside gently sloping downwards, with distinct canthus and slightly concave sides. Tympanum of moderate extent; eustachian ostia very small, much narrower than choanæ. *Fingers nearly free,* only the two outer ones connected by a rudimentary web; no external rudiment of pollex. Toes rather long, *half-webbed, the web not extending to the disks of the fourth and fifth toes,* unless it be as a narrow fringe. *Upper parts with flat tubercles*; abdomen, chest, and throat granular; *a deep fold across the chest,* and another fold over the tympanum. Yellowish-green, shining; sides of the thigh, hinder face of the upper arm, and the upperside of the foot marbled with black; upperside of the arm and lower leg cross-banded. (After Schmidt.)

Length of body	48 millim.
,, hind leg	80 ,,

9. **Hyla copii.** (Tab. LXX. fig. C.)

? *Hyla affinis,* Baird, Proc. Ac. N. Sc. Philad. 1854, p. 61; U. S. Mex. Bound. Survey, Rept. p. 29, t. 38. figs. 4–7 (nec Spix).

? *Hyla arenicolor,* Cope, Journ. Ac. N. Sc. Philad. vi. p. 84 (1866); Bull. U. S. Nat. Mus. no. 32, p. 14 (1887); Stejneger, N. Amer. Fauna, no. 3, p. 117 (1890).

Hyla copii, Bouleng. Ann. & Mag. N. H. 1887, xx. p. 53; 1888, i. p. 189.

? *Hyliola digueti,* Mocquard, Bull. Soc. Philom. (9) i. p. 165, t. 1. fig. 4 (1899).

Hab. North America, Texas.—Mexico, Colonia Garcia, Chihuahua (*Brimley*), Guanajuato (*Dugès*), Jalisco (*F. D. G., Diguet*), Rancho Santa Cruz, Real de la Tesca (*Buller*).

Vomerine teeth in two groups between the rather narrow choanæ. Snout of moderate length, depressed, with indistinct canthus and sloping sides. Distance of the nostril from the eye equal to diameter of the latter. Interorbital space as wide as the upper eyelid. Tympanum distinct, one-third the area of the eye. *Fingers free,* with disks a little larger than those of the toes, and nearly the size of the tympanum. No external rudiment of pollex. Toes short, the fourth, measured from metatarsal joint, being two-fifths the length of the body; *toes broadly webbed*, the web extending to the disks of the fourth and fifth; a tarsal fold. *Upper parts tubercular*, the tubercles extending forwards beyond the interorbital space; abdomen coarsely, throat finely granular; a smooth space with *a very distinct fold across* the chest. No other folds or appendages. Upper parts brownish-olive with greyish spots; upper lip and limbs cross-barred, posterior face of hind limb mottled. Lower parts white; throat brownish or with black specks. Male with an external gular vocal sac.

	♀ (Jalisco).
Greatest width of head	15 millim.
Distance of vent from snout	40 ,,
,, ,, heel	39 ,,
,, ,, metatarsal joint	47 ,,
,, ,, end of fourth toe	65 ,,

Boulenger has already pointed out the discrepancies in the accounts of this species given by Baird and himself, the Editor of the 'American Naturalist' (1888, p. 80) having identified *H. affinis* with *H. copii*. Whilst it is still uncertain whether Baird had before him a distinct species, or whether his description and figure are inaccurate, I adopt the name given by Boulenger.

10. Hyla boulengeri.

Scytopsis boulengeri, Cope, Bull. U. S. Nat. Mus. no. 32, p. 12 (1887).

Hab. NICARAGUA (*MacNeil*).

Vomerine teeth between the choanæ, in two transverse fasciculi, together forming a *series which is convex backwards*. Snout subacuminate, with canthus and sides concave; diameter of the eye equal to its distance from the nostril, and twice that of the tympanum. Choanæ double the size of the pharyngeal ostia. Fingers free; web between the toes short, extending to the antepenultimate phalanx of the fourth toe; none between first and second. Skin nearly smooth, but with numerous small warts on the top of the head and shoulder; abdomen areolate, throat smooth. (A glandular patch above the tympanum *.) Light grey, with indistinct darker cross-bands on the back and limbs. Concealed faces of femur and tibia light-coloured, with purplish-black cross-bands; several spots of the same colour in the groin. (After Cope.)

* Cope does not mention this in his description, but as he refers this frog to his genus *Scytopis*, which is characterized by an accumulation of sebaceous glands above the tympanum, it may be supposed that the species resembles *Hyla venulosa* in this respect.

Length of head	37 millim.
„ hind limb	63 „
„ hind foot.	25 „

11. **Hyla mocquardi,** sp. n. (Tab. LXXIII. fig. B.)

Hab. Guatemala or Panama.

Vomerine teeth in two groups, between the choanæ. Snout not depressed, as long as the eye *, rather pointed, with angular canthi which would meet at an acute angle, and with subvertical sides. The eye is rather longer than its distance from the nostril, and much longer than the distance between the two nostrils; interorbital space not much wider than the upper eyelid. Tympanum two-fifths of the area of the eye. The three outer fingers one-fourth, toes half-webbed; all the disks small, much smaller than the tympanum. No tarsal fold. Upper parts smooth, lower granular. When carried forward, the heel reaches the middle of the eye. Upper parts light-coloured, with irregular brownish blotches having a tendency to form longitudinal bands. A brown band starts from the loreal region, and extends to the middle of the side of the body, where it becomes paler; in its anterior portion it is sharply edged with white above and below, the lower edge terminating behind the tympanum, the upper extending farther backwards. In the larger specimen the posterior face of the thigh is crossed by five very short black cross-bars, the smaller specimen having this part pigmentless.

Width of head	7·5 millim.
Distance of vent from snout	25 „
„ „ heel	22 „
„ „ metatarsal joint	29 „
„ „ end of fourth toe	36 „

The two specimens before me were collected either in Panama or in Guatemala.

12. **Hyla quinquevittata.**

Hyla quinque-vittata, Cope, Proc. Am. Phil. Soc. xxiii. p. 273 (1886).

Hab. Nicaragua (*Bransford*).

Vomerine teeth between the choanæ, in two rounded fasciculi, close together. Snout rather acuminate, projecting beyond the mouth, with distinct and concave canthus; tongue rather longer than broad. Tympanum two-fifths the diameter of the eye. Fingers with a slight rudiment of web at their bases; web of the toes reaching only the middle of the penultimate phalanges of the third and fifth. Upper parts smooth.

* Measured along the longitudinal axis.

Light grey, with five darker longitudinal bands, the median of which expands anteriorly so as to form a large subtriangular spot between the eyes; femur with one, tibia with three cross-bands; concealed surfaces unspotted. (After Cope.)

Length of body	29 millim.
„ hind limb	43 „
„ hind foot	19 „

13. **Hyla phæota.**

Hyla phæota, Cope, Proc. Ac. N. Sc. Philad. 1862, p. 358; Werner, SB. Bayer. Akad. Wiss. xxvii. p. 97 (1897).

Hyla labialis, Peters, MB. Ak. Wiss. Berlin, 1863, p. 463; 1873, p. 618.

Hab. PANAMA (*Werner*), Chiriqui (*Peters*).—COLOMBIA, Turbo (*Cope*), Bogota (*Peters*).

Vomerine teeth in two transverse series between the large choanæ. Head broad, depressed, with rounded canthus rostralis. Tympanum more than half the extent of the eye. The three outer fingers slightly webbed; nearly three phalanges of the fourth toe free, the web, however, margining the outer side. The tibio-tarsal articulation reaches well beyond the end of the snout, when the hind limb is carried forwards. Upper parts smooth; a strong fold from the eye over the tympanum. Brownish-olive above, with a longitudinal band or series of spots on the back, bifurcating anteriorly; a spot on the coccyx and near the vent brown. Upper jaw, postorbital region, and the margins of the upper eyelid and of the limbs white. A dark band along the canthus rostralis is continued over the tympanum to the scapular region; femur with narrow cross-bands and with the hinder side reticulate or uniformly coloured; tibiæ more broadly cross-banded; sides with brown vermiculations; lower parts whitish. Male with an unpaired vocal sac. (After Cope.)

Length	22 lines.

The descriptions of other authors who believe that they have recognized Cope's species differ in several noteworthy particulars. Thus, Peters states that "the tympanum is scarcely visible, its diameter being two-fifths of the diameter of the eye"; the web between the toes is deeply excised, but "otherwise nearly complete, only the last phalanx of the fourth toe being free." On the other hand, Werner describes the tympanum as "equal to $\frac{3}{4}$ of the diameter of the eye. Fingers with $\frac{1}{3}$, toes with nearly complete webs"; he also ascribes to this species a tarsal fold.

14. **Hyla tæniopus,** sp. n. (Tab. LXXII. fig. F.)

Hab. MEXICO, Jalapa (*Trujillo*).

Vomerine teeth in two transverse series, between the choanæ. Head broad, depressed, broader than the body, with the eyes obliquely directed forwards; snout short, not longer than the eye, with rounded canthus and slightly sloping sides; the diameter of the eye equals its distance from the nostril, and is much more than the distance between the nostrils. Interorbital space much wider than upper eyelid. Tympanum one-third the area of the eye. The three outer fingers one-fourth webbed; toes nearly fully webbed, the web reaching the disks of the third and fifth fingers, but leaving the last phalanx of the fourth free. Disks of the fingers rather larger than of the toes, and nearly the size of the tympanum. *A tarsal fold.* When carried forward, the heel reaches the eye. Upper parts smooth, lower granular; no fold across the chest. Light purplish, with or without some darker mottling; lower parts white, with scattered black specks on the throat and anterior part of the abdomen; sides marbled with black and yellow; hind limbs with darker cross-bands, three on the femur, three on the tibia, two on the tarsus, and two on the metatarsus; concealed parts not pigmented; the cross-bands are darkest on the edges, lighter in the centre.

Width of head	11 millim.
Distance of vent from snout	30 „
„ „ heel	25 „
„ „ metatarsal joint	31 „
„ „ end of fourth toe	42 „

15. **Hyla puma.**

Hyla puma, Cope, Proc. Am. Phil. Soc. 1884, p. 183.

Hab. NICARAGUA (*Cope*).

Vomerine teeth between the choanæ, in two transverse patches. *Snout short, the nostril being separated from the orbit by a space equal to the long diameter of the tympanum*; canthus rostralis well marked, concave. Tympanum an oblique oval, its long diameter two-thirds of the eye-fissure. Pharyngeal ostia much narrower than choanæ. Fingers with a rudimentary web; toes half-webbed. Digital disks half the size of tympanum. Smooth above; abdomen areolate. Uniform dark brown above, uniform light brown below. (After Cope.)

Width of head	18 millim.
Length of body	45 „
„ hind limb	71 „
„ hind foot	31 „

16. **Hyla baudini.** (Tab. LXXI.)

Hyla baudinii, Dum. et Bibr. Erpét. Gén. viii. p. 564; Cope, Proc. Ac. N. Sc. Philad. 1863, p. 48; Günth. Batr. Sal. p. 105; Bouleng. Batr. Sal. p. 371; P. Z. S. 1898, p. 124; Brocchi, Miss. Sc. Mex., Batr. p. 29, t. 14. fig. 4.

Smilisca baudinii (*daulinia*), Cope, Proc. Ac. N. Sc. Philad. 1865, p. 194; 1866, p. 127; Journ. Ac. N. Sc. Philad. vi. p. 85 (1866); Bull. U. S. Nat. Mus. no. 32, p. 13 (1887).

Hyla vanvloeti, Baird, Proc. Ac. N. Sc. Philad. 1854, p. 61; U. S. & Mex. Bound. Surv., Rept. t. 38. figg. 1–3.

Hyla vociferans, Baird, U. S. & Mex. Bound. Surv., Rept. t. 38. figg. 11–13.

Hyla muricolor, Cope, Proc. Ac. N. Sc. Philad. 1862, p. 359.

? *Hyla pansosana*, Brocchi, Bull. Soc. Philom. (7), i. p. 125 (1899); Miss. Sc. Mex., Batr. p. 34, t. 12. fig. 2.

Hab. North America, Texas.—Mexico, Presidio and Ventanas (*Forrer*), Tampico (*Richardson*), Guadalajara (*Major*), Mascota in Jalisco (*Buller*), Orizaba (*F. D. G.*), Mirador (*Sartorius*), Tehuantepec and Cordova (*Sumichrast*), Rio Papagaio in Guerrero, Teapa and Frontera in Tabasco (*H. H. Smith*), Cozumel I. (*Gaumer*), Yucatan (*Schott*); Guatemala, Vera Paz (*Salvin*), Peten (*Berendt*); Nicaragua, Hacienda Rosa de Jericho, alt. 3250 feet, and Matagalpa (*Rothschuh*), Chontales Mines (*Rix*); Costa Rica, Cariblanco de Sarapiqui and La Hondura (*Underwood*), Valley of San José (*Van Patten*).—Colombia, Buenaventura (*Rosenberg*); Northern Ecuador (*Rosenberg*).

Vomerine teeth in two transverse series, between the wide choanæ. Snout broad, not depressed, with distinct canthus, and sloping concave sides. Distance of the nostril from the eye equal to the diameter of the latter. Interorbital space broader than the upper eyelid. Tympanum distinct, two-thirds the area of the eye. The three outer fingers are one-third webbed, and provided with disks which are equal to those of the toes, but smaller than the tympanum. Toes fully webbed, with subarticular tubercles; *a very distinct tarsal fold*; no external rudiment of pollex. Upper parts smooth; a slight fold over the tympanum; no fold above the vent; no appendage at the heel; a fold across the chest. Abdomen granular; throat and chest smooth. Brownish, with darker subsymmetrical markings on the back and crown of the head, upper lip and limbs with dark cross-bands; a subtriangular lighter (in young specimens white) spot below the eye, and a narrow blackish stripe along the fold behind the tympanum. A very common variety is green or greyish-olive above, sometimes with rounded black spots; a deep black band starts from the eye, covers the tympanum, and is lost behind the shoulder; it is bordered below by a white line which fringes the upper lip; the limbs are cross-barred in this as in the other variety, but not the upper lip. Lower parts whitish, very rarely with black spots. Male with an external vocal sac on each side of the throat.

	♀ (Costa Rica).
Greatest width of head	25 millim.
Distance of vent from snout	75 „
„ „ heel	77 „
„ „ metatarsal joint	100 „
„ „ end of fourth toe	126 „

Rothschuh (*in litt.*) describes the colours of a living specimen from the Hacienda

Rosa de Jericho as green on the back, and golden yellow with black spots on the abdomen ; a black tympanic band ; limbs brownish, cross-barred.

17. Hyla venulosa.

Rana venulosa, Laur. Syn. Rept. p. 31.

Hyla venulosa, Daud. Rainettes, p. 35, t. 13; Dum. et Bibr. Erpét. Gén. viii. p. 560; Peters, MB. Ak. Wiss. Berlin, 1872, p. 219; 1873, p. 618; Günth. Batr. Sal. p. 104; Bouleng. Batr. Sal. p. 364; Ann. & Mag. N. H. 1882, x. p. 327.

Scytopis venulosus, Cope, Journ. Ac. N. Sc. Philad. vi. p. 85 (1866) ; Bull. U. S. Nat. Mus. no. 32, p. 12 (1887).

Hyla zonata, Spix, Spec. Nov. p. 41, t. 12. fig. 1.

Hyla bufonia, Spix, l. c. p. 42, t. 12. fig. 2.

Hyla tibiatrix, Laur. Syn. Rept. p. 34.

Hyla vermiculata, Dum. et Bibr. Erpét. Gén. viii. p. 563.

Hyla lichenosa, Günth. Batr. Sal. p. 102, t. 8. fig. C; Brocchi, Miss. Sc. Mex., Batr. p. 33, t. 14. fig. 2.

Scytopis hebes, Cope, Proc. Ac. N. Sc. Philad. 1862, p. 354.

Hyla nigropunctata, Bouleng. Batr. Sal. p. 366.

Hyla pænulata, Brocchi, Miss. Sc. Mex., Batr. p. 45, t. 14. fig. 1 (not good).

Hab. Mexico, Presidio (*Forrer*), Tampico (*Richardson*), Jalapa (*Höge*), Vera Cruz and Cordova (*Sallé*), Misantla (*H. H. Smith*), Frontera, Tabasco (*H. H. Smith*); Guatemala (*Cope*, *Boucard*); Nicaragua (*Cope*); Costa Rica, Bebedero (*Underwood*); Panama, Chiriqui (*Peters*).—Tropical America.

Vomerine teeth between the wide choanæ, nearer to their hind margin; they form generally a transverse series slightly interrupted in the middle, or each of the two halves may be slightly arched. Snout broad, short, obtuse, not depressed, with indistinct canthus, and sloping concave sides. Distance of the nostril from the eye equal to the diameter of the latter. Interorbital space wider than the upper eyelid. Tympanum distinct, about one-half the area of the eye, but smaller in specimens with a glandular skin, and sometimes larger in such as have the glands but little developed. The three outer fingers one-half or one-third webbed; their disks equal those of the toes, and are generally of the same size as the tympanum. No external rudiment of a pollex. Toes three-fourths webbed, with strong subarticular tubercles; no tarsal fold. Upper parts generally tubercular and glandular; a fold above the tympanum often assumes the appearance of a parotoid, suppressing portion of the tympanum; no fold on the side or above the vent; folds across the chest or throat; no appendage at the heel. Lower parts, including the throat, coarsely granular. In some specimens the skin of the upper parts is nearly smooth. Back brownish, generally with large, subsymmetrical, dark patches, more or less confluent into a pair of longitudinal bands, sharply defined against a lighter lateral band; hind limbs with broad dark cross-bands. Lower parts uniform whitish. Male with a non-retractile vocal sac behind each angle of the mouth.

	♀ (Tampico).
Greatest width of head	25 millim.
Distance of vent from snout	80 „
„ „ heel	66 „
„ „ metatarsal joint	84 „
„ „ end of fourth toe	111 „

This species shows a somewhat perplexing amount of variation in the shape of the vomerine series, size of the tympanum, development of cuticular glands, and coloration. In many Mexican specimens the coloration during immaturity is peculiar: the light lateral band is pure white, and many of the tubercles on the sides of the body and on the legs are deep black (*H. lichenosa*). Boulenger has described a living specimen:— "When handled, it exuded a great quantity of poisonous fluid . . . . . ; this fluid, of a milky appearance, coagulates instantaneously, sticking to the fingers in a very disagreeable manner; it has a strong odour, resembling that of peaches, and affects very disagreeably the mucous membrane of the nostrils, causing a strong itching."

18. Hyla sordida.

Hyla sordida, Peters, MB. Ak. Wiss. Berlin, 1863, p. 460.

Hab. PANAMA, Veraguas.

Vomerine teeth in two short transverse series, between the choanæ, which are a little larger than the eustachian ostia. Snout with rounded canthus, obtuse, the diameter of the eye being equal to its distance from the end of the snout; distance between the nostrils a little less than their distance from the eye. Diameter of the tympanum one-third that of the eye. The three outer fingers are almost entirely webbed, and even the two inner fingers are connected by a rudimentary membrane. The web between the toes is much excised, but reaches to the disks, leaving free only one-half of the last phalanx of the fourth toe. Disks of toes and fingers nearly equal. *When the hind limb is carried forwards, three-fourths of the tarsus reach beyond the end of the snout.* Upper parts smooth, lower granular. Bluish-grey, sides of the body and posterior face of the thighs finely marbled with white and brown; limbs with cross-bands. Lower parts white. Male with a subgular vocal sac. (After Peters.)

Width of head	12 millim.
Length of body	34 „
„ hind limb	60 „
„ foot	27 „

To judge from the description, this frog must be most closely allied to *Hyla gabbi*, and may be even specifically the same. The only noteworthy difference seems to be the comparative length of the hind limb. The two specimens on which Peters based this species were both males.

19. **Hyla gabbi.** (Tab. LXX. fig. B.)

Hyla gabbii, Cope, Journ. Ac. N. Sc. Philad. viii. p. 103 (1877); Bouleng. P. Z. S. 1898, p. 478 *.
Hyla salvini, Bouleng. Batr. Sal. p. 372; Werner, Verh. zool.-bot. Ges. Wien, 1896, p. 344.

Hab. Costa Rica, Sipurio (*Cope*), La Palma, Monte Rotundo, alt. 1600 metres, and Laguna de la Palma (*Underwood*), Cartago and San José (*Mus. Brit.; Mus. Nac. de Costa Rica, San José*).

Vomerine teeth in a transverse series, interrupted in the middle, between the choanæ, which are moderately open. Snout broad, short, obtuse, rather depressed, with rather obtuse canthus and with sloping concave sides. Distance of the nostril from the eye equal to the diameter of the latter. Interorbital space wider than the supraocular cover. Tympanum distinct, one-half the area of the eye. *When the hind limb is carried forwards, the heel reaches the end of the snout.* The three outer fingers are three-fourths webbed, and also the two inner fingers are connected by a narrow web; their disks as large as those of the toes and as the tympanum. No external rudiment of pollex. Toes almost entirely webbed; a tarsal fold. Upper parts smooth or nearly so; abdomen and throat rather coarsely granular; a short fold above the tympanum, and sometimes a transverse one across the chest. No fold or appendage above the vent or on the heel; forearm with a narrow tubercular fringe. Upper parts greyish, sides and anterior and posterior faces of the hind limbs reticulated; *no spots on the lips.*

	♀.
Greatest width of head	18 millim.
Distance of vent from snout	55 „
„ „ heel	56 „
„ „ metatarsal joint	72 „
„ „ end of fourth toe	92 „

The figure is taken from one of the typical specimens of *H. salvini.*

20. **Hyla pseudopuma,** sp. n. (Tab. LXXII. figg. A, B.)

Hyla puma, Bouleng. P. Z. S. 1898, p. 478 (not Cope).

Hab. Costa Rica, La Palma, alt. 1600 metres (*Mus. Nac. de Costa Rica, San José*).

Vomerine teeth in two transverse patches, slightly separated from each other, between the posterior part of the choanæ. Snout short, rounded, depressed, with obtuse and concave canthus and with sloping concave sides. The diameter of the eye

* The specimen in the Natural History Museum, identified with Cope's *Hyla gabbii*, has the fingers as broadly webbed as in typical *H. salvini*; but it has the upper lip margined with whitish, as mentioned by Cope in his description of *H. gabbii.* I do not consider this to be sufficient for the distinction of two species.

equals its distance from the nostril, and is rather more than the distance of the nostrils from each other. Tympanum circular, about one-half or one-third the area of the eye, *its diameter being much less than that of the eye.* Choanæ rather narrow, but wider than the eustachian ostia. Tongue broader than long, slightly notched behind. Interorbital space much wider than the upper eyelid. *The three outer fingers one-third webbed*, with disks of the size of the tympanum. *Toes three-fourths webbed*, the web reaching the disks of the third and fourth toes, and leaving free the last phalanx of the fourth; disks rather smaller than those of the fingers; metatarsal tubercle small; *no tarsal fold.* When carried forwards, the heel reaches the front margin of the orbit. Upper parts, throat, and chest generally smooth; abdomen areolated; a slight fold above the tympanum; no other fold or appendage. Upper parts plum-coloured, with a very obscure darker hourglass-shaped patch on the back, beginning between the eyes; cross-bands on the hind limbs; concealed surfaces of femur uniform dark, like back, but without spots. Lower parts white, throat speckled with brown. The groin where the two colours meet shows some marbling or spotting. Another variety has a greyish or white back; a broad deep brown fascia on the side of the body starts from the eye and tympanum, and is variegated with more or less numerous yellow spots along its lower margin. Male with an internal vocal sac, and with the skin of the pollex thickened and black.

	♂.	♀.
Width of head	12·5 millim.	14 millim.
Distance of vent from snout	40 „	41 „
„ „ heel	38 „	37 „
„ „ metatarsal joint	47 „	48 „
„ „ end of fourth toe	65 „	63 „

21. **Hyla godmani**, sp. n. (Tab. LXXII. fig. E.)

Hab. Mexico, Misantla and Jalapa (*F. D. G.*); ? Costa Rica, La Palma (*Mus. Nac. de Costa Rica*, *San José*).

Vomerine teeth between the choanæ, in two slightly oblique series convergent behind; tongue as broad as long. Head broad, depressed; snout * as long as the eye; diameter of the eye a little more than its distance from the nostril, and much more than the distance between the nostrils; canthi rostrales obtuse and rather concave; sides of the snout sloping; interorbital space wider than upper eyelid. Tympanum one-third (or a little less) of the area of the eye. Fingers half-webbed, the web nearly reaching the disks of the fourth and second fingers, and leaving free the last phalanx of the third; a rudimentary web connects the first and second fingers at the base; the disks equal nearly the tympanum, and are larger than those of the toes. Toes four-fifths webbed; a distinct tarsal fold. When carried forward, the heel reaches the end of the snout or a little beyond. Upper parts smooth; abdomen finely granular. Purplish or reddish

* Always measured along the longitudinal axis.

olive, with fine but obscure marbling; sides sometimes more distinctly marbled with brown, black, and white; sides of the head not spotted; concealed surfaces of the limbs not spotted. Lower parts whitish.

	♀.	♀.
Width of head	14 millim.	13·5 millim.
Distance of vent from snout	37 ,,	37 ,,
,, ,, heel	38 ,,	36 ,,
,, ,, metatarsal joint . .	51 ,,	46 ,,
,, ,, end of fourth toe . .	64 ,,	61 ,,

Two specimens from La Palma, Costa Rica, seem to me to belong to the same species, although they differ in two of the technical characters. The tympanum is conspicuously larger, about one-half the area of the eye, and the web between the fingers a little broader. Otherwise the agreement as regards general habit and proportion and coloration is so striking that I must hesitate to describe this as a distinct species.

22. **Hyla uranochroa.** (Tab. LXX. fig. D.)

Hyla uranochroa, Cope, Journ. Ac. N. Sc. Philad. viii. p. 103, t. 27. fig. 4 (bad) (1878); Bouleng. Batr. Sal. p. 377.

Hab. COSTA RICA, Rio Sucio (*Rogers*), La Palma, alt. 1600 metres (*Tristan and Underwood*), Sipurio (*Cope*).

Vomerine teeth in two transverse groups, between the rather narrow choanæ. Snout short, depressed, with canthus and rather sloping sides. Distance of nostril from the eye equal to the diameter of the latter. Interorbital space wider than the upper eyelid. Tympanum distinct, one-half the area of the eye. Fingers one-third webbed, with disks which are rather larger than those of the toes, but smaller than the tympanum; no external rudiment of pollex. Toes three-fourths webbed; no tarsal fold. Upper parts smooth, abdomen granular; no folds or appendages. Uniform purplish-blue above (? green in life), white below; a narrow, sharply defined, yellowish line runs round the edge of the upper lip, slightly dilated below, and touching, the eye; the line is continued above the shoulder along the side, where it forms a sharp boundary between the colours of the upper and lower parts. Metatarsus and outer toe with a yellow edge; a cross-line above, and tubercles round, the vent yellow. Concealed surfaces of thigh uncoloured.

	♀.
Greatest width of head	10 millim.
Distance of vent from snout	32 ,,
,, ,, heel	30·5 ,,
,, ,, metatarsal joint	40 ,,
,, ,, end of fourth toe	50 ,,

This species has the peculiar coloration of *Hyla carolinensis*, which it represents in the southern parts of Central America. However, it differs by a broader head and by the absence of a tarsal fold. The position of the vomerine teeth is the same in both species. One of the Costa Rica specimens has a brown ground-colour of the upper parts, marbled with darker.

23. **Hyla leucophyllata.** (Tab. LXXII. figs. C & D.)

Hyla leucophyllata, Beiris, Schrift. Berl. Ges. naturf. Freund. iv. p. 178, t. 41. fig. 4; Dum. et Bibr. Erpét. Gén. viii. p. 607; Bouleng. Batr. Sal. p. 387.
Hyla elegans, Wied, Beitr. Naturg. Bras. i. p. 529; Abbild. tab. —.
Hyla triangulum, Günth. P. Z. S. 1868, p. 489, t. 38. fig. 4.
Hyla ebraccata, Cope, Proc. Ac. Nat. Sc. Philad. 1874, p. 68.

Hab. Nicaragua (*Bransford*); Costa Rica (*Mus. Nac. de Costa Rica, San José*); Panama (*Mus. Brit.*).—Tropical America.

Vomerine teeth in two very small groups between the choanæ. Snout short, broad, not depressed, with the canthus rather obtuse, and with the sides convex. The distance between the nostrils equals their distance from the eye, and is less than the diameter of the latter. Tympanum small, rather indistinct, about one-fourth the area of the eye. Interorbital space wider than the upper eyelid. The three outer fingers are half-webbed, with disks about as large as the tympanum; rudiment of pollex scarcely projecting. Toes almost entirely webbed; no tarsal fold. Upper parts smooth, abdomen granular; a fold across the chest. Back white or yellowish; generally with a deep brown hourglass-shaped spot, enclosing behind the coccygeal region; a brown band on the side of the head and body; limbs brown, with various large white spots. Femur always colourless.

The coloration varies greatly, the white and brown ground-colours forming various patterns. Of the two Panama specimens which I have seen, one (fig. C) has the back uniform white, but the lateral brown band is present, the upper lip having a white margin; also the limbs are almost entirely white, only the upper arm and the fingers being brown. The other example (fig. D) is ornamented with the dorsal hourglass-shaped figure; but this figure, like the other brown spots, encloses a number of round, small, white dots; the upper lip has a white margin, as in the other specimen. An individual from Nicaragua described by Cope seems to have had a very similar coloration.

Greatest width of head	12	millim.
Distance of vent from snout	38	„
„ „ heel	36	„
„ „ metatarsal joint	45	„
„ „ end of fourth toe	64	„

24. **Hyla underwoodi.**

Hyla microcephala, Bouleng. P. Z. S. 1898, p. 481, t. 39. fig. 3 (nec Cope).
Hyla underwoodi, Bouleng. Ann. & Mag. N. H. 1899, iii. p. 277.

Hab. Costa Rica, Bebedero (*Underwood*).

Vomerine teeth in two small groups well separated from each other, between the choanæ. Head rather small, a little longer than broad; snout not depressed, with angular canthi, which convergent form a somewhat acute angle in front; the distance between the nostrils is much less than their distance from the orbit, or than the diameter of the eye. Tympanum distinct, one-fourth the area of the eye. Fingers one-third, toes three-fourths webbed; disks of both subequal, and about the size of the tympanum. *No tarsal fold.* Upper parts smooth; abdomen granular; no folds. Pinkish, powdered with brown; a brown band, edged with white above, runs along the side of the snout, and through the eye to the side of the body, gradually fading behind; the white edge marks distinctly the canthus and margin of upper eyelid, and is continued above the tympanum across the shoulder. Thighs not pigmented. Lower parts white.

Greatest width of head	8·5 millim.
Distance of vent from snout	28·5 ,,
,, ,, heel	28 ,,
,, ,, metatarsal joint	34 ,,
,, ,, end of fourth toe	46 ,,

25. **Hyla nigripes.**

Hyla nigripes, Cope, Journ. Ac. N. Sc. Philad. viii. p. 104, t. 23. fig. 7 (1876).

Hab. Costa Rica (*Cope*).

Vomerine teeth between the choanæ, in two short transverse series. *Shape of head oval, tongue longer than broad*; canthus rostralis a little concave; sides of the snout oblique. Tympanum one-fourth or one-fifth the area of the eye; eustachian ostia and choanæ nearly equal. Fingers with the web extending to about one-half the length of the longest; web of toes not extending to the end of the penultimate phalanx; disks of moderate size. Smooth above, finely areolate below. Dark brown, limbs and feet blackish; groin finely white and blackish marbled; no lateral border or band Posterior face of femur black; no spots on any other region. (After Cope.)

Width of head	13 millim.
Length of body	39 ,,
,, hind limb	65 ,,
,, hind foot	16 ,,

26. **Hyla molitor.**

Hyla molitor, Schmidt, Denkschr. Ak. Wiss. Wien, xiv. p. 425, t. 1. figg. 8, 9 (1858); Peters, MB. Ak. Wiss. Berlin, 1873, p. 618.

Hab. PANAMA, Chiriqui Lagoon, near Boca del Toro.

Vomerine teeth between the choanæ, in two groups well separated from each other. Snout with distinct strongly convergent canthi and with sloping sides. Tympanum small, obliquely elliptical; eustachian ostia nearly as wide as the choanæ. The three outer fingers with a rudimentary web; toes fully webbed to the disks. Upper parts smooth; no fold across the chest; an indistinct one above the tympanum. Upper parts greyish blue, uniform or marbled with black; a series of black dots along the canthus rostralis; sides and thighs generally marbled with blackish. (After Schmidt.)

Length of body	33 millim.	38 millim.
„ hind limb	53* „	60 „

27. **Hyla miotympanum.**

Hyla miotympanum, Cope, Proc. Ac. N. Sc. Philad. 1863, p. 47; Bull. U. S. Nat. Mus. no. 32, p. 14 (1887); Peters, MB. Ak. Wiss. Berlin, 1871, p. 651.

Hyla microtis, Peters, l. c. 1869, p. 880.

Hab. MEXICO, Jalapa and Zacualtipan (*Cope*), Puebla (*Berkenbusch*), Mirador (*Sartorius*), Vera Cruz and Tehuantepec (*Sumichrast*).

Vomerine teeth in two slightly oblique series converging backwards between the choanæ, which are much wider than the eustachian ostia. Snout broadly rounded, without distinct canthus, and with sloping sides; nostrils as far distant from each other as from the eyes. *Tympanum extremely small, scarcely visible.* Fingers one-half webbed or less, with disks which are larger than the tympanum. Toes broadly webbed. Upper parts smooth, lower granular; no folds or appendages. Purplish above, with or without a few lighter specks and indistinct darker reticulations; a yellow band along the margin of the upper jaw, continued on the side of the body, this portion sometimes not well separated from the abdominal shade. Limbs not barred; concealed portions of the thigh not coloured. (After Cope and Peters.)

Width of head	13 millim.
Length of body	37 „
„ hind limb	60 „

Cope describes the lateral band as yellow. Peters states: "On the boundary between the dark coloration of the back and the lighter of the abdomen there is a narrow line or a punctated series of black colour"; he does not mention the lateral band.

* 33 in Schmidt's description is obviously a misprint.

28. **Hyla prosoblepon.**

Hyla prosoblepon, Böttger, Katal. Batr. Senckenb. Mus. p. 45 (1892).
Hylella puncticrus, Bouleng. Ann. & Mag. N. H. 1896, xviii. p. 341.

Hab. Costa Rica, Limon (*Fleischmann*), La Palma (*Underwood*); Panama, Bugaba (*Champion*).

Vomerine teeth (if present) in two transverse series, between the choanæ. Head broad, depressed, with very short snout, and with *large eyes, directed forwards*; canthus obtuse, concave, sides of the snout sloping. Diameter of the eye a little more than its distance from the nostril, and much more than the distance between the nostrils; upper eyelid narrow, narrower than the interorbital space. Tympanum very small, deeply sunk, about one-sixth the area of the eye. The two outer fingers one-third webbed, but the third and second are connected only by a rudiment of a web; toes four-fifths webbed. The disks of the fingers are larger than those of the toes and than the tympanum. When carried forwards, the heel reaches beyond the end of the snout. Upper parts smooth, abdomen coarsely areolated. Light purplish above with scattered round dark purple or blackish dots; lower parts whitish; the concealed parts of the limbs not pigmented, the remainder coloured like the back. Humerus of the male with a projecting, curved, compressed osseous process.

	♀.
Width of head	9 millim.
Distance of vent from snout	25 „
„ „ heel	27 „
„ „ metatarsal joint	35 „
„ „ end of fourth toe	42 „

A small species, in which the vomerine teeth appear to be as frequently absent as present; specimens with and without them occur in the same locality.

29. **Hyla pulverata.**

Hyla pulverata, Peters, MB. Ak. Wiss. Berlin, 1873, p. 614.

Hab. Panama, Chiriqui.

Vomerine teeth in two groups between the choanæ, which are much wider than the narrow eustachian ostia. Snout as long as the eye, with obtuse canthus. *Tympanum very small and covered with pigmented skin.* The three outer fingers broadly webbed, the web extending to the disks of the second and fourth fingers, but not to that of the third; toes broadly webbed. Upper parts smooth, abdomen granulated; no fold above the tympanum. Colourless, with scattered white points. (After Peters.)

Length of body	24 millim.
Width of head	8 „
Length of hind limb	45 „
„ foot	20 „

30. **Hyla glandulosa.**

Hyla glandulosa, Bouleng. Ann. & Mag. N. H. 1883, xii. p. 164.

Hab. Guatemala (*F. D. G.*).

Vomerine teeth between the choanæ, in two small groups. Snout of moderate length, not depressed, with angular canthus and slightly sloping sides; diameter of the eye rather longer than its distance from the nostril or than the distance between the nostrils; *tympanum hidden* under thickened pigmented skin; choanæ narrower than the eustachian ostia. Fingers long, one-fourth webbed, with rather large disks; toes two-thirds webbed; no tarsal fold. Upper parts studded with small flat tubercles, lower parts granular; a fold above the tympanum. Uniform greenish above, whitish below; a black line from the nostril along the canthus and supratympanic fold; a whitish cross-band above the vent; femur not pigmented, except on a narrow strip of its upper surface; tarsus with white margin.

	♀.
Distance of vent from snout	35 millim.
„ „ heel	32 „
„ „ metatarsal joint	42 „
„ „ end of fourth toe	57 „

31. **Hyla guatemalensis.**

Plectrohyla guatemalensis, Brocchi, Bull. Soc. Philom. (7) i. p. 92 (1877).
Cauphias guatemalensis, Brocchi, l. c. p. 129; Miss. Sc. Mex., Batr. p. 62, t. 12. fig. 3.
Hyla guatemalensis, Bouleng. Batr. Sal. p. 396.

Hab. Guatemala, Patzcicia * (*Bocourt*).

Vomerine teeth between the choanæ, in two small groups. Snout very short, broad, rounded, with high concave sides, and with distinct canthi which converge at a very obtuse angle. *Tympanum completely hidden*; eustachian ostia and choanæ equally open. The three outer fingers with a rudimentary web; a spur-like projection of the pollex (male?); toes entirely webbed; a tarsal fold; disks well developed. General appearance toad-like. Head, sacral region, sides, and calves with tubercles; abdomen strongly granulated. A fold across the chest. Greyish, marbled with greenish; lower parts whitish. (After Brocchi.)

Length of body	59 millim.

32. **Hyla crassa.**

Cauphias crassum, Brocchi, Bull. Soc. Philom. (7) i. p. 131 (1877); Miss. Sc. Mex., Batr. p. 64, t. 12. fig. 4.
Hyla crassa, Bouleng. Batr. Sal. p. 396.

* The locality, which is situate in the arid central region of Guatemala, is given as "Pacicilla" by Brocchi.

Hab. MEXICO (*Brocchi*).

Extremely similar to *H. guatemalensis,* from which it is said to differ in having the fingers quite destitute of a web, and without a projecting pollex. Skin of the back finely granular, shagreened. A curved fold over the tympanic region. Male with a gular vocal sac.

Length of body	66 millim.

33. **Hyla punctariola.**

Hyla punctariola, Peters, MB. Ak. Wiss. Berlin, 1863, p. 462.
? *Hyla punctariola,* Cope, Journ. Ac. Nat. Sc. Philad. viii. p. 106 (1876).

Hab. ? COSTA RICA; PANAMA, Veraguas.

Vomerine teeth behind the level of the choanæ, in two small, rather approximated groups. Snout scarcely longer than the eye; distance between the nostrils a little less than their distance from the eye, the diameter of which is thrice that of the tympanum; eustachian ostia not smaller than the choanæ. Fingers free, with moderate disks; toes broadly webbed, the web leaving free the last two phalanges of the fourth toe. When carried forwards, the heel does not reach the end of the snout. Upper parts smooth. Brownish; a cross-band between the eyes, a large spot on the snout, and transverse spots on the upper jaw whitish; limbs, finger, and toes alternately crossed by brown and white; lower parts whitish, throat brown, punctulated with white. (After Peters.)

	♀.
Width of head	8·5 millim.
Length of body.	22·5 ,,
,, hind limb	34 ,,
,, hind foot	15 ,,

Cope himself suspects that the Costa Rica specimens, which he has referred to this species, may be distinct; they seem to have somewhat longer limbs, and have a very different coloration. He distinguishes, besides, three colour-varieties, which he names *pictipes, mœsta,* and *monticola.*

34. **Hyla spilomma.**

Hyla spilomma, Cope, Proc. Am. Phil. Soc. xvii. p. 86 (1877).

Hab. MEXICO, Vera Cruz, Cosamaloapam (*Sumichrast*).

Vomerine teeth behind the level of the choanæ, in two transverse series. Head broad; snout very short, with obsolete canthus. Tympanum two-fifths the area of the eye; choanæ small. Fingers free, with disks as large as the tympanum; toes short, about half-webbed. Skin of the back thickened, and studded with large obtuse warts. When carried forwards, the heel does not reach the end of the snout. Light purplish-brown above, sometimes with a blackish band on each side of the back; eye spotted with yellow. (After Cope.)

Width of head	14	millim.
Length of body	38	,,
,, hind limb	52	,,
,, hind foot	23	,,

35. Hyla miliaria.

Hypsiboas miliarius, Cope, Proc. Am. Phil. Soc. xxiii. p. 272 (1886).

Hab. NICARAGUA (*Bransford*).

Vomerine teeth behind the level of the choanæ, in two transverse series. Head flat and depressed, wider than long; snout broadly rounded, as long as the eye, almost without canthus. Tympanum three-fifths the diameter of the eye; eustachian ostia smaller than the choanæ. *Pollex free for most of its length, terminating in a flattened cone.* Both fingers and toes broadly webbed. When carried forwards, the heel reaches the end of the snout. Upper parts thickly covered with small tubercles, which are largest and most prominent on the top of the head; fore leg and hind foot with an outer tubercular fringe. Upper parts dark plum-coloured, lower yellowish with spots of the colour of the back; fringes, a spot on the heel and below the vent, and also behind the eye, yellowish. (After Cope.)

Width of head	25	millim.
Length of body	62	,,
,, thigh (from groin)	25	,,
,, tibia	32	,,
,, tarsus	19	,,
,, foot	25	,,

36. Hyla crepitans.

Hyla pardalis, part., Spix, Spec. Nov. p. 34 (see Peters, MB. Ak. Wiss. Berlin, 1872, p. 212).

Hyla pardalis, Günth. Batr. Sal. p. 99.

Hyla crepitans, Wied, Beitr. Naturg. Bras. p. 525; Abbild. tab. —; Bouleng. Batr. Sal. p. 352.

Hypsiboas crepitans, Wagl. Syst. Amph. p. 200; Cope, Bull. U. S. Nat. Mus. no. 32, p. 13 (1887).

Hyla palmata, part., Dum. et Bibr. Erpét. Gén. viii. p. 545.

Hyla doumercii, Dum. et Bibr. l. c. p. 551 (see Peters, l. c. 1873, p. 612).

Hyla levaillantii, Dum. et Bibr. l. c. p. 550 (see Peters, l. c.).

Hyla pugnax, Schmidt, Denkschr. Ak. Wiss. Wien, 1858, p. 243, t. 1. figg. 5, 6; Reinh. & Lütk. Vidensk. Meddel. 1866, p. 193.

Hyla xerophylla, Bouleng. Batr. Sal. p. 358 (an = *H. xerophylla,* Dum. et Bibr., Peters ?).

Hab. PANAMA, Chiriqui (*Schmidt*).—TROPICAL SOUTH AMERICA.

Vomerine teeth between, and extending to behind, the wide choanæ, in two oblique, arched series: ⌈ ⌉. Snout broad and depressed, with rather indistinct canthus and sloping sides. Distance of the nostril from the eye equal to the diameter of the latter, but more than the distance between the nostrils. Interorbital space wider than the supraocular cover. Tympanum three-fourths the area of the eye. The three outer fingers one-third, more rarely one-fourth webbed, their disks equal to those of the toes, but much smaller than the tympanum. A conspicuously projecting rudiment of the pollex. Toes three-fourths webbed, with moderate subarticular tubercles; large specimens with a slight tarsal fold. Upper parts smooth; a transverse fold or series of tubercles above the vent. Abdomen with the proximal part of the throat granular. Upper parts marbled with brown or uniform whitish or pinkish; sides with narrow vertical blackish bars; hinder part of the thigh with similar bars, which extend across towards the front, though much fainter. Male with a subgular vocal sac.

	♀.	♂.
Greatest width of head	24 millim.	20 millim.
Distance of vent from snout	70 „	60 „
„ „ heel	75 „	64 „
„ „ metatarsal joint . . .	93 „	83 „
„ „ end of fourth toe . .	115 „	105 „

37. **Hyla albomarginata.**

Hyla albomarginata, Spix, Spec. Nov. p. 33, t. 8. fig. 1; Dum. et Bibr. Erpét. Gén. viii. p. 555; Günth. Batr. Sal. p. 98; Peters, MB. Ak. Wiss. Berlin, 1872, p. 207; Bouleng. Batr. Sal. p. 356.

Hypsiboas albomarginata, Wagl. Syst. Amphib. p. 201; Cope, Proc. Am. Phil. Soc. 1886, p. 273.

Hyla infulata, Wied, Beitr. Naturg. Bras. i. p. 533; Burmeist. Erläut. Faun. Bras. p. 97, t. 30. figg. 1–6.

Hyla punctata, Wied, Abbild. tab. —.

Hab. NICARAGUA (*Bransford*); COSTA RICA, San Carlos (*Underwood*).—TROPICAL AMERICA.

Vomerine teeth between the wide choanæ, in two rather long oblique series, generally arched: ⌈ ⌉. Snout broad and depressed, with somewhat pointed nose, rather indistinct canthus, and sloping sides. Distance of the nostril from the eye a little more than the diameter of the latter. Interorbital space much wider than the upper eyelid. Tympanum distinct, one-third the area of the eye. The three outer fingers are more than half-webbed, and provided with disks at least as large as the tympanum, and larger than the disks of the toes. Toes almost entirely webbed, with moderate subarticular tubercles; no tarsal fold. Upper parts smooth; generally a very distinct fold, which starts from the eye and descends towards the middle of the side; a cross-fold above the vent; a short cutaneous appendage at the heel. The rudimentary pollex scarcely projects beyond the skin. Abdomen granular, throat and chest smooth. Fold across the chest incomplete. Yellowish or pinkish, uniform or minutely speckled with purple or with some irregular white spots; concealed parts of the thighs not coloured. Lower parts whitish. The fold over the vent and the edge of the hind foot from the heel milky-white. Male with a subgular vocal sac.

	Brazil.	Costa Rica.
Greatest width of head	19 millim.	16 millim.
Distance of vent from snout	51 „	46 „
„ „ heel	53 „	48 „
„ „ metatarsal joint	68 „	64 „
„ „ end of fourth toe	89 „	81 „

The single Central-American specimen which I have examined differs from those of Brazil and Guiana in having the lateral fold of the skin only indicated for a very short distance, scarcely extending beyond the tympanum.

38. **Hyla rudis.**

Hyla rudis, Mocquard, Bull. Soc. Philom. (9) i. p. 163, t. 1. fig. 3 (1899).

Hab. Mexico, Guadalajara (*Diguet*).

The skin is adherent to the bone and rough on the upper surface of the head and in the loreal region.

Vomerine teeth in two small groups, between the choanæ. Head depressed, broader than long; snout conspicuously longer than the eye, with rugose canthus; distance of the nostril from the eye equal to the diameter of the latter, which is twice the length of the tympanum. Interorbital space much broader than the supraocular cover. Fingers free, toes one-third webbed; disks smaller than tympanum. No tarsal fold. When carried forward, the heel reaches the eye. Upper parts finely granulated; a fold above the tympanum and across the thorax. Brownish-olive, with some darker spots; lower parts whitish. (After Mocquard.)

Length of body	35 millim.

HYLELLA.

Hylella, Reinh. & Lütk. Vidensk. Meddel. 1861, p. 199; Bouleng. Batr. Sal. p. 419.
Exerodonta, Brocchi, Bull. Soc. Philom. (7) iii. p. 19 (1879).

Pupil horizontal. Tongue entire or but slightly emarginate behind. Vomerine teeth none. Fingers and toes dilated into terminal disks, the latter always webbed. Auditory organ developed; tympanum present or hidden under the skin. Outer metatarsals united.

This genus comprises dwarf species of *Hyla*, in which the vomerine teeth have disappeared; its separation from *Hyla* is merely artificial, and not more justified than that of *Ixalus* from *Polypedates*, or of *Liohyla* from *Hylodes*. The species composing it also belong to similarly divergent types; and of the following Central-American forms, *Hylella fleischmanni* and *Hylella picta* are more nearly allied respectively to *Hyla prosoblepon* and *Hyla eximia* than they are to each other. That vomerine teeth may abnormally be absent in some species of *Hyla* and *Agalychnis* has been mentioned elsewhere. On the other hand, I think that species in which the tympanum with the tympanic cavity has been lost should be kept separate (*Cophomantis*, Peters).

1. Hylella sumichrasti.

Exerodonta sumichrasti, Brocchi, Bull. Soc. Philom. (7) iii. p. 20 (1879); Miss. Sc. Mex., Batr. p. 48, t. 15. fig. 2.
Hylella platycephala, Cope, Am. Phil. Soc. xviii. p. 267 (1879).
Hylella sumichrasti, Bouleng. Ann. & Mag. N. H. 1891, viii. p. 456.
Hab. MEXICO, Santa Efigenia and Tapana (*Sumichrast*).

Head short, broad, and flat; snout short, with distinct convergent canthi; tympanum distinct, its diameter about one-fourth of that of the eye. Fingers with a very short web; toes two-thirds webbed. When carried forwards, the heel reaches the eye. Upper parts smooth, lower areolate. Light ash-coloured above, yellowish below.

Width of head	11 millim.
Length of body	33 „
„ hind limb	45 „
„ hind foot	19 „

2. Hylella picta, sp. n. (Tab. LXXIII. fig. C.)

Hab. MEXICO, Jalapa (*Trujillo*).

Head moderately broad, not depressed; snout moderately long, rather pointed, with subvertical sides, and with angular straight canthi convergent into a point, forming an acute angle; distance between the nostrils much less than the diameter of the eye, which equals its distance from the nostril; interorbital space not much broader than upper eyelid. Tympanum distinct, about one-fifth the area of the eye. Fingers with a rudimentary basal web, toes three-fourths webbed; no tarsal fold. Disks as large as, or larger than, tympanum. When carried forwards, the heel reaches the end of the snout. Upper parts smooth, lower granular. Back light purplish, with scattered, round, dark purple, lighter-edged spots of the size of the tympanum; on each side of the back a whitish line running along the canthus rostralis, above the tympanum, and becoming fainter in the groin; sides below this line darker than back; lower parts whitish; limbs mottled, concealed parts pigmentless.

Width of head	7·5 millim.
Distance of vent from snout	21 „
„ „ heel	21 „
„ „ metatarsal joint	27 „
„ „ end of fourth toe	37 „

This species seems to come nearest to *H. sumichrasti*, but it is differently coloured and has a much more pointed snout.

The figure is twice the natural size.

3. **Hylella fleischmanni.** (Tab. LXXIII. fig. D.)

Hylella fleischmanni, Böttger, Bericht Senckenb. Ges. 1893, p. 251; Bouleng. Zool. Record for 1894, Rept. p. 41.

Hylella chrysops, Cope, Proc. Ac. N. Sc. Philad. 1894, p. 196.

Hab. Costa Rica, San José (*Fleischmann & Biolley*), La Palma (*Underwood*), Alajuela, Rio Alfaro (*Mus. Nac. de Costa Rica*), San José.

Head broad and depressed; snout very short, without canthus; eyes of moderate size, projecting forwards, with narrow upper eyelids, much narrower than the interorbital space. Tympanum small, entirely hidden below the general integument; pharyngeal ostia open, but narrower than the choanæ. The third and fourth fingers half-webbed, the second and third connected by a rudimentary web, the first free. Toes three-fourths webbed. All the disks moderately developed. When carried forwards, the heel reaches beyond the snout. Upper parts smooth, lower granular. Apparently uniform pinkish white above, but slightly magnified the skin of the back is found to be covered with minute black specks, between which pure white points are more sparsely scattered. Lower parts whitish; concealed parts of the limbs pigmentless. Male with an external subgular vocal sac.

Width of head	9 millim.
Distance of vent from snout	22 „
„ „ heel	22 „
„ „ metatarsal joint	28·5 „
„ „ end of fourth toe	38 „

The figure is twice the natural size.

NOTOTREMA.

Notodelphys, Weinland, MB. Ak. Wiss. Berlin, 1854, p. 372.

Nototrema and *Opisthodelphys*, Günth. Batr. Sal. pp. 115, 117.

Nototrema, Bouleng. Batr. Sal. p. 416.

Pupil horizontal. Tongue entire or but slightly emarginate behind. Vomerine teeth. Fingers and toes dilated into terminal disks; the latter webbed. Outer metatarsals united. Females with a dorsal pouch for the reception of the ova.

1. **Nototrema oviferum.** (Tab. LXXIV. fig. A.)

Notodelphys ovifera, Weinland, MB. Ak. Wiss. Berlin, 1854, p. 372; Arch. f. Anat. Physiol. 1854, p. 449, tabb. 17–19.

Opisthodelphys ovifera, Günth. Batr. Sal. p. 117.

Nototrema oviferum, Bouleng. Batr. Sal. p. 418.

Hab. MEXICO, Cordova (*Sallé*); COSTA RICA, Carrillo (*Underwood*).—VENEZUELA.

In the adult the upper surface of the head is entirely osseous and margined by series of bony spines, largest along the edges of the occipital and parietal regions. Vomerine teeth in two fascicles between the hind margins of the choanæ. Tympanum large, not much smaller than the eye. Fingers with a rudimentary web, toes one-third webbed. Disks large, but smaller than tympanum. A tarsal fold. Greenish with black markings; limbs with black light-edged cross-bars; abdomen lighter coloured, with subsymmetrical whitish markings. Male without vocal sac.

Length of body	60 millim.
„ hind limb	95 „

It may appear hazardous to refer a very young frog from Costa Rica, which has not yet completed its metamorphosis, to a species and genus of the development of which we know so little as of *Nototrema*, especially when that specimen does not show some of the features which are considered characteristic of *N. oviferum*. Yet the most essential points in which our example does differ, such as the soft integument of the upperside of the head or the absence of vomerine teeth, or the less-developed digital web, are those which we should expect to find in a less specialized condition in the very young of that species. This frog may belong to a new species of *Nototrema*; but there is an ydult specimen of *N. oviferum* in the British Museum, preserved for many years and much discoloured, in which, on nearer inspection, certain marks are still faintly visible which significantly agree with the pattern of the young example: such are the curious loop-like green figure round the humeral joint on each side of the chest and the black cross-bands of the lower leg, which are almost linear, much narrower than those of the thigh.

The body of the specimen (Tab. LXXIV. fig. A) is 28 millim. long; the tail has entirely disappeared, but the mouth is not yet cleft to its angle, an internal fold of the mucous membrane tying the proximal portion of the mandible to the upper jaw *. Vomerine teeth absent; choanæ closed by the mucosa, although visible as a depression. Tongue slightly emarginate behind. Outlines of the large tympanum clearly distinct, but the membrane merged into the deep black of the temporal region. Fingers without web, toes half-webbed. Disks of moderate size.

* This ligament is preserved on the right side only; I have been obliged to tear it on the left side to obtain a view of the tongue and inside of the mouth.

Back greenish-olive, with a few small, irregularly scattered, black specks. Throat and abdomen black, with a loop-like green band on each side of the chest. A black vertical stripe on each side of the extremity of the snout, continued as a black band along the canthus; widening behind the orbit it covers the temporal region and extends to behind the humerus; sides of the body green, with a large black spot about the middle and another in the groin. Forearms nearly unspotted above, but black-banded below, one band running round the wrist. Femur black-banded, rather irregularly; lower leg with black bands, which are broad below, but stretch as mere lines across its upper face. Tarsus and toes with similar cross-bands. All these markings have a more or less distinct, very narrow, white edge.

AGALYCHNIS.

Agalychnis, Cope, Proc. Ac. Nat. Sc. Philad. 1865, p. 194.
Agalychnis and *Phyllomedusa* sp., Bouleng. Batr. Sal. pp. 422, 424.

Pupil erect. Tongue slightly notched behind. Vomerine teeth generally present. Fingers and toes dilated into terminal disks; the latter webbed. Outer metatarsals united. Females without dorsal pouch.

1. **Agalychnis moreletii.**

Hyla moreletii, A. Dum. Ann. Sc. Nat. xix. p. 169 (1853); Keferstein, Götting. Nachr. 1867, p. 356; Arch. f. Naturg. 1868, p. 297; Brocchi, Miss. Sc. Mex., Batr. p. 31, t. 13. fig. 1.

Agalychnis moreletii, Cope, Journ. Ac. Nat. Sc. Philad. viii. p. 107 (1877); Bouleng. Batr. Sal. p. 422; Werner, Verh. zool.-bot. Ges. Wien, 1896, p. 351.

Hyla holochlora, Salvin, P. Z. S. 1860, p. 460, t. 32. fig. 2.

Hab. Guatemala, Vera Paz (*Morelet, Salvin*); Nicaragua, Matagalpa (*Rothschuh*); Costa Rica, Cartago and Irazu (*Mus. Brit.*), Escazu (*Underwood*), San José (*van Patten; Mus. Nac. de Costa Rica*).

Head broad, snout rather longer than the eye, with sloping sides and obtuse canthus. Size of tympanum variable, two-thirds or one-half the length of the eye, rarely less. Fingers broadly, at least one-half webbed, a web between the first and second being constantly present; disks very large. Toes two-thirds or three-fourths webbed. When carried forward, the heel reaches the nostril or the end of the snout. Green above, very rarely with some round yellow spots; concealed parts of the limbs and hands and feet yellowish, with the exception of the outer finger and toe, which are coloured like the back; a white line along the outer edge of the forearm, lower hind limb, and tarsus.

Width of head		25	millim.
Distance of vent from snout		75	„
„	„ heel	72	„
„	„ metatarsal joint	92	„
„	„ end of fourth toe	120	„

The descriptive diagnosis, as given above, may comprise several races; at any rate, specimens from Guatemala may be readily recognized by their broader digital web, which reaches the disks of the two outer fingers, and by their larger tympanum. Two of the Costa Rica specimens show, besides, some of those round yellow spots by which a resemblance to a dead leaf is attained, and which are present also in some other Tree-Frogs of Tropical America.

2. Agalychnis callidryas.

Hyla callidryas, Cope, Proc. Ac. N. Sc. Philad. 1862, p. 359.
Agalychnis callidryas, Cope, Bull. U. S. Nat. Mus. no. 32, p. 15 (1887).

Hab. Mexico, near Cordova (*Sumichrast*); Panama (*Field*).

Head elongate; sides of snout subvertical, with rounded canthus. Tympanum nearly as large as the eye, obliquely elliptical. Vomerine teeth in two oblique series. Fingers one-fourth webbed, with very large disks; toes half-webbed. When carried forward, the heel reaches the end of the snout. Bluish-purple above, lower parts and concealed parts of the limbs uniform saffron-yellow; numerous oblique bands of a lighter yellow on the sides. (After Cope.)

Length of body . . . . 45 millim.

3. Agalychnis helenæ.

Agalychnis helenæ, Cope, Proc. Am. Phil. Soc. 1884, p. 182; 1886, p. 274; Peracca, Boll. Mus. Zool. Torino, xi. no. 253, p. 12 (1896).

Hab. Mexico, Teapa, Tabasco (*H. H. Smith*); Nicaragua, Hacienda Rosa de Jericho, alt. 3250 feet (*Rothschuh*); Costa Rica, San Carlos (*Underwood*); Panama, Darien, Rio Sabana (*Festa*).

Head broad; snout as long as the eye, with subvertical sides and distinct canthus. Tympanum two-thirds the length of the eye, coloured and rather indistinct. The two outer fingers are nearly two-thirds webbed, the web not quite reaching the disk; the web between the second and third fingers much narrower, only about one-fourth, and that between first and second still more reduced; disks about size of tympanum. Toes two-thirds webbed. When carried forward, the heel reaches the end of the snout. Green above; sides yellow, with large black spots which are separated by vertical lines or narrow bands of the yellow ground-colour; concealed parts of the limbs dark purplish ; hands and feet yellow, with the exception of the outer finger and toe, which are coloured like the back. Lower parts yellowish-white.

Width of head		22	millim.
Distance of vent from snout		65	„
„ „	heel	65	„
„ „	metatarsal joint	85	„
„ „	end of fourth toe	110	„

In this species the vomerine teeth are sometimes inconspicuous and may be entirely absent.

Rothschuh (*in litt.*) describes the colours of several Nicaraguan specimens from life. One, adult, had a green back, the inner face of the thighs sky-blue, and the feet brick-red. Another, also adult, had all the upper parts sky-blue and the narrow lateral stripes yellow. A smaller specimen was green on the back, with the concealed portions of the limbs violet ; feet red ; iris dark red.

4. **Agalychnis dacnicolor.**

Phyllomedusa dacnicolor, Cope, Proc. Ac. N. Sc. Philad. 1864, p. 161 ; Bouleng. Batr. Sal. p. 426; Ann. & Mag. N. H. 1882, x. p. 328.
Agalychnis dacnicolor, Cope, Bull. U. S. Nat. Mus. no. 32, p. 15 (1887).
Hab. Mexico, Colima (*Xantus*), Mazatlan (*Bischoff*), Presidio (*Forrer*).

Head broad, snout longer than the eye, which is rather small; sides of snout sloping, canthus obtuse. Tympanum two-thirds the diameter of the eye. Vomerine teeth in two short, straight, transverse series, between the anterior margins of the choanæ. Fingers one-fourth webbed, with disks which are larger than those of the toes, but smaller than the tympanum. The three outer toes one-third webbed, the three inner ones barely one-fourth. When carried forward, the heel reaches the eye. Uniform bluish above, rarely with a few scattered round red spots; sides marbled with reddish-brown and with small round white spots; forearm and foot margined with white.

Width of head		30	millim.
Distance of vent from snout		84	„
„ „	heel	69	„
„ „	metatarsal joint	91	„
„ „	end of fourth toe	113	„

5. **Agalychnis lemur.**

Phyllomedusa lemur, Bouleng. Batr. Sal. p. 425.
Agalychnis lemur, Cope, Bull. U. S. Nat. Mus. no. 32, p. 15 (1887).
Hab. Costa Rica.

Snout moderately broad, as long as the eye, with subvertical sides and distinct canthus. Tympanum one-third the diameter of the eye. Vomerine teeth none. Fingers free, disks rather larger than tympanum, and conspicuously larger than those of the toes; toes webbed at the base only. When carried forward, the heel reaches the end of the snout. Purplish or cream-coloured, uniform or with scattered red specks. Concealed parts of the limbs not pigmented.

Width of head	13	millim.
Distance of vent from snout	42	,,
,, ,, heel	40	,,
,, ,, metatarsal joint	55	,,
,, ,, end of fourth toe	68	,,

PTERNOHYLA.

Pternohyla, Bouleng. Ann. & Mag. N. H. 1882, x. p. 326.

Pupil transversely oval. Tongue without notch behind. Vomerine, no parasphenoid teeth. Fingers and toes with disks, the latter webbed. Outer metatarsals united; metatarsal tubercle enlarged, compressed. Females without dorsal pouch.

This genus is intermediate between the rough-headed *Hylæ* and *Triprion.*

1. **Pternohyla fodiens.** (Tab. LXXIV. fig. B.)

Pternohyla fodiens, Bouleng. Ann. & Mag. N. H. 1882, x. p. 326.

Hab. MEXICO, Presidio (*Forrer*).

Head osseous, rough, margin of the upper jaw reverted; canthus rostralis distinct; sides of the snout sloping, concave, dilated; nostrils separated by a flat space; interorbital space concave, as wide as the orbit. Tympanum half the area of the eye. Upper parts granular. Fingers free, toes one-third webbed. Disks small. Brownish, largely marbled with darker, each spot with a darker edge. Thighs and groin, in life, sulphur-yellow, marbled with dark brown.

Width of head	23	millim.
Distance of occiput from snout	17	,,
,, vent from snout	61	,,
,, ,, heel	45	,,
,, ,, metatarsal joint	55	,,
,, ,, end of fourth toe	80	,,

Boulenger, *l. c.*, says that "it is slow in its movements, and not a good climber. Its habits are more burrowing than arboreal. It is not able to climb up a glass, but burrows itself deeply in the moss by means of its metatarsal shovels, the movements executed in this proceeding being exactly those of *Pelobates.*"

TRIPRION.

Pharyngodon, Cope, Proc. Ac. N. Sc. Philad. 1865, p. 193.
Triprion, Cope, ibid. 1866, p. 127.
Diaglena, Cope, Bull. U. S. Nat. Mus. no. 32, p. 12 (1887) ; Bouleng. Ann. & Mag. N. H. 1891, viii. p. 456.
Tetraprion, Stejneger, Proc. U. S. Nat. Mus. xiv. p. 167 (1891).

Pupil vertical or transversely oval. Tongue entire behind. Vomerine and parasphenoid teeth. Integuments of the head merged into the bones. Fingers and toes with disks, the latter webbed. Outer metatarsals united.

In some of the specimens palatine teeth have been observed, the presence of which has led to generic separation (*Tetraprion*). These teeth are in such a rudimentary condition, and consequently so unreliable a character, that this distinction is scarcely justified. The appearance of these odontoids goes morphologically hand in hand with, and is of no greater taxonomic value than, the ossification of the external integument of the head. The three species of *Triprion* are evidently so nearly allied that, in this instance, I cannot admit even the difference of the shape of the pupil as of generic value.

1. **Triprion petasatus.**

Pharyngodon petasatus, Cope, Proc. Ac. N. Sc. Philad. 1865, p. 193.
Triprion petasatus, Cope, l. c. 1866, p. 127; Journ. Ac. N. Sc. Philad. vi. t. 25. figg. 7 & 8; Bouleng. Batr. Sal. p. 431.

Hab. Mexico, Yucatan.

Canthus rostralis forming a prominent ridge, which terminates in a strong projection at the eye, the space between the projections being deeply concave; occipital ridge raised; nostrils separated from each other by a very narrow osseous bridge. Sphenoid teeth in a simple series. Tympanum half the area of eye. Fingers one-fourth, toes two-thirds webbed. Disks much smaller than tympanum. Upper parts brownish, with darker spots; head with some lighter spots.

Width of head	21 millim.
Distance of occiput from snout	24 „
„ vent from snout	70 „
„ „ heel	53 „
„ „ metatarsal joint	68 „
„ „ end of fourth toe	84 „

2. **Triprion spatulatus.** (Tab. LXXIV. fig. C.)

Triprion spatulatus, Günth. Ann. & Mag. N. H. 1882, viii. p. 279.
Diaglena spatulata, Cope, Bull. U. S. Nat. Mus. no. 32, p. 12 (1887); Bouleng. Ann. & Mag. N. H. 1891, viii. p. 456.

Hab. Mexico, Presidio (*Forrer*).

Canthus rostralis not very prominent, terminating in an obtuse tuberosity at the eye; space between the eyes rather concave; occipital ridge scarcely raised; nostrils separated by a broad arched bridge. Sphenoid teeth in a narrow band. Tympanum half the area of eye. Fingers free; toes half-webbed. Disks smaller than tympanum. Uniform light olive.

Width of head	25 millim.
Distance of occiput from snout	30 „
„ vent from snout	81 „
„ „ heel	61 „
„ „ metatarsal joint	76 „
„ „ end of fourth toe	103 „

The pupil is transversely oval when expanded, and subquadrangular when more contracted, but never vertical. It can be shut entirely.

The palatine ridges show some odontoid roughnesses, but they are not equally developed in the three examples examined by me. In one specimen they are visible only on the outer part of the ridge of the left palatine, in another they are confined to the outer portions of the ridges of both palatines, and in the third the ridges are more conspicuous and beset with odontoids in nearly their whole length.

Division BATRACHIA CAUDATA.

Fam. SALAMANDRIDÆ.

MOLGE.

Triton, Laur. Syn. Rept. p. 37 (nec Linn.).
Molge, Merr. Tent. p. 185.
Diemyctelus (et *Notophthalmus*) (Rafinesque), Cope, Bull. U. S. Nat. Mus. no. 34, p. 202.

Palatine teeth in two longitudinal straight or slightly convex series. Parasphenoid toothless. Tongue free on both sides. Toes five. Tail compressed.

1. Molge meridionalis.

Diemyctelus miniatus, subsp. *meridionalis,* Cope, Bull. U. S. Nat. Mus. no. 17, p. 30 (1880).
Molge meridionalis, Bouleng. Ann. & Mag. N. H. 1888, i. p. 24.
Diemyctelus viridescens meridionalis, Cope, Bull. U. S. Nat. Mus. no. 34, p. 211 (1889).

Hab. North America, Texas, San Diego (*Taylor*).—Mexico, Matamoros, Tamaulipas (*Couch*).

Head depressed; back without crest. Palatine teeth in two series, convergent and confluent in front. Gular fold distinct. Three pits on each side of the head, behind the eye, are more developed in the male than in the female, in which they may be entirely absent. Fingers and toes depressed, free; inner finger and inner and outer toes very short, outer finger at least half as long as the adjoining. Olive above, marbled with lighter and with blackish specks; yellowish or orange-coloured below, with larger round black spots.

	♂.	♀.
Total length	104 millim.	112 millim.
Length of tail	54 „	58 „

AMBLYSTOMA.

Amblystoma, Tschudi, Batr. p. 92; Bouleng. Batr. Grad. p. 38.
Siredon, Wagl. Syst. Amph. p. 209 (larval form).
Axolotes, Owen, Ann. & Mag. N. H. 1844, xiv. p. 23 (larval form).
Amblystoma, Chondrotus, Linguælapsus, Cope, Amer. Nat. 1887, p. 88.

Palatine teeth in a long transverse series, continuous or slightly interrupted in the middle; parasphenoid toothless. Tongue free on the sides, slightly free in front, with radiating folds. Toes five. Tail more or less compressed.

This genus, which is well represented in North America, ranges into Mexico. Some of the southern forms grow to full size and attain to sexual maturity whilst they retain their larval characteristics. These forms are generally known by their Mexican name of Axolotls. Some remain in a larval condition through the whole of their life, others complete their metamorphosis whilst still being sexually immature or after having reproduced their species. However, the observations made in Europe on specimens living in aquaria seem to show that the metamorphosis of Axolotls into perfect *Amblystoma* is exceptional in the artificial conditions under which they were bred.

1. **Amblystoma tigrinum.**

Salamandra tigrina, Green, Journ. Ac. Nat. Sc. Philad. v. p. 116.
Triton tigrinus, Holbr. N. Amer. Herpet. v. p. 79, t. 26.
Ambystoma tigrinum, Gray, Batr. Grad. p. 35.
Amblystoma tigrinum, Bouleng. Batr. Grad. p. 43.
Ambyostoma californiense, Gray, P. Z. S. 1853, p. 11, t. 7.
Ambystoma weismanni, Wiedersheim, Zeitschr. wiss. Zool. xxxii. p. 216, tabb. 11, 12.
Amblystoma mexicanum (Shaw), Brocchi, Miss. Sc. Mex., Batr. p. 104, t. 17. fig. 1, t. 17 bis. fig. 6.
(For complete synonymy, see Boulenger, l. c.)

Hab. North America.—Mexico, Ciudad in Durango (*Forrer*), Volcan de Orizaba (*Sallé*), Colonia Briguela and La Cumbre de los Arrastrados in Jalisco 8500 feet (*Buller*).

Palatine teeth extending from the posterior outer angle of the choana to that on the other side, either in a straight line or curving forwards, continuous or slightly interrupted in the middle. A pair of carpal and tarsal tubercles small, but distinct.

Twelve costal grooves. Brown or blackish, uniform, or with more or less numerous yellow spots of variable size and shape.

Total length 200 millim., of which the tail takes about one-half.

Axolotl.

Axolotl, Cuv. in Humboldt's Obs. Zool. tabb. 12, 14; Home, Philos. Trans. xxi. p. 419, tabb. 21–23; Bouleng. Batr. Grad. p. 44.
Gyrinus mexicanus, Shaw, Nat. Misc. 1800, p. 343.
Siren pisciformis, Shaw, Zool. iii. p. 612, t. 140. fig. 12.
Siredon axolotl, Wagl. Ic. Amphib. t. 20.
Axolotes guttatus, Owen, Ann. & Mag. N. H. 1844, xiv. p. 23.
Axolotes maculatus, Owen, l. c.
Siredon mexicanus, Baird, Journ. Ac. N. Sc. Philad. i. p. 292 (1849).
Amblystoma mexicanum, Cope, Proc. Ac. N. Sc. Philad. 1867, p. 184; Bull. U. S. Nat. Mus. no. 32, p. 7 (1887); Brocchi, Miss. Sc. Mex., Batr. p. 104, t. 17. fig. 2, t. 17 bis. figg. 4, 5.
Siredon humboldtii, Dum. et Bibr. Erpét. Gén. viii. p. 177.
Siredon harlanii, Dum. et Bibr. l. c. p. 181, t. 95.
Siredon dumerilii, Dugès, Ann. Sc. Nat. xv. p. 2, t. 10; La Nat. Mex. i. p. 241, t. 5 (1870).
Siredon tigrina, Velasco, La Nat. Mex. iv. pp. 209–233, tabb. 7–9.

On metamorphosis, cfr. A. Dum. Compt. Rend. 1865, lx. p. 765, lxi. p. 775; Arch. Mus. ii. p. 265; Marsh, Am. Journ. xlvi. p. 364; Tegetmeier, P. Z. S. 1870, p. 160; Weismann, Zeitschr. wiss. Zool. xxv. p. 297; Chauvin, ibid. xxvii. p. 522; Joly, Rev. Sc. Nat. Montpell. 1872, p. 7; Vaill. Bull. Soc. Philom. 1876; Osborn, Amer. Nat. 1900, p. 551.

In the large series of Mexican Axolotls which are available for my examination, three forms may be distinguished, which, however, differ only in coloration :—

1. The entire body covered with small, round, black spots. Albinos are not rare among wild specimens as well as among those bred in captivity. They attain a length of 260 millim., of which the tail measures one-half. To this form belong the Axolotls found in the vicinity of the city of Mexico (*Doorman*), in Lake Chalco (*R. James*), in Lake Texcoco (*F. D. G.*); also the majority of examples reared in England may be referred to this form.

2. The sides of the body ornamented with irregular yellow spots, sometimes confluent and assuming the shape of a band; the upper parts generally uniformly blackish or with round black spots. These Axolotls are, on the whole, sometimes smaller than the preceding; I have seen them from the Colonia Briguela and La Cumbre de los Arrastrados, elevation 8500 feet, in Jalisco, and from La Laguna, Juanacatlan, elevation 6500 feet (*Buller*).

3. Upper parts marbled with brown and yellow, both colours being distributed in about equal proportions. Of this form I have seen only one specimen, which is the type of Owen's *A. maculatus*; it came from the Sierra Madre, Chihuahua, and does not differ structurally or in its general form from the other Axolotls.

The types of *S. dumerilii*, Dugès, were from the Laguna de Patzcuaró, and those of *S. tigrina*, Velasco, from the Lago de Santa Isabel, near Guadalupe Hidalgo, in the valley of Mexico.

2. **Amblystoma altamirani.**

Amblystoma altamirani, Dugès, La Nat. Mex, 1896, p. 459, t. 29 *.

Hab. Mexico, Monte de Las Cruces (*Dugès*).

Palatine teeth in two straight oblique lines, commencing near the hind margins of the choanæ, and converging in a forward direction, not continuous in front. A pair of tarsal and carpal tubercles, small, but distinct. Tips of the fingers and toes hardened. Tail strongly compressed. Twelve costal grooves. Olive above, densely marbled with darker; uniform light-coloured below.

	Dugès.	N. H. Mus.
Total length	247 millim.	160 millim.
Length of tail	145 „	91 „

Dr. Dugès states that this species is confined to considerable altitudes. The specimen in the Natural History Museum has completed its metamorphosis.

SPELERPES.

Spelerpes, Baird, Journ. Ac. N. Sc. Philad. i. p. 283 (1847); Bouleng. Batr. Grad. p. 60.
Geotriton, Bonap. Faun. Ital.
Geotriton, Cylindrosoma, Œdipus, Pseudotriton, Tschudi, Batr. pp. 93, 94.
Bolitoglossa, Dum. et Bibr. Erpét. Gén. viii. p. 88.
Œdipina, Keferstein, Arch. f. Naturg. 1868, p. 299.
Ophiobatrachus, Gray, Ann. & Mag. N. H. 1868, ii. p. 297.

Palatine teeth in two straight or oblique transverse series; parasphenoid teeth present. Tongue free all round, supported by a central pedicel. Toes five. Tail cylindrical or slightly compressed. Nostrils small.

Much uncertainty exists in the distinction of the Central-American species. The technical characters on which they have been established, such as the extent of the interdigital web, the length of the tail, the extent of the palatine series of teeth, the degree of separation of the parasphenoid groups, the coloration, &c., must be subject to some variation; but without a much larger series of well-preserved examples than exists in any museum, or, at any rate, is at my disposal, it is impossible to ascertain the degree of variation within the limits of a species. At present, I am inclined to think that the number of species has been unduly multiplied.

* I have not been able to verify this reference, no copy of the volume of the work being in any of the London libraries; my information is based upon MS. notes kindly communicated by Dr. Dugès, and upon a specimen sent by him to the Natural History Museum.

Synopsis of the Central-American Species.

I. Body moderately elongate, with well-developed limbs.

A. Digits free or with a short web.

1. *chiropterus*; 2. *gibbicaudus*; 3. *cephalicus*; 4. *bellii*; 5. *leprosus*.

B. Digits half-webbed.

6. *bocourti*; 7. *subpalmatus*.

C. Digits broadly webbed, only the last phalanx of the middle digits projecting.

8. *morio*; 9. *robustus*.

D. Digits completely enclosed in the web.

10. *variegatus*; 11. *lignicolor*; 12. *rufescens*; 13. *yucatanus*.

II. Body and tail worm-shaped, with very weak limbs.

14. *lineolus*; 15. *uniformis*.

1. Spelerpes chiropterus.

Spelerpes chiropterus, Cope, Proc. Ac. N. Sc. Philad. 1863, p. 54; 1869, pp. 105, 106; Bouleng. Batr. Grad. p. 67.

Spelerpes orculus, Cope, Proc. Ac. N. Sc. Philad. 1865, p. 196; 1866, p. 132.

Hab. MEXICO, Cuernavaca (*Dugès*), Mirador near Vera Cruz (*Sartorius*), Orizaba, alpine zone (*Sumichrast*), Jalapa (*Godman*).

Body moderately elongate, with 11 (or 12) costal grooves *, which extend across the back and abdomen; limbs moderately developed, with short free digits; the inner finger and toe rudimentary, dilated into a short lobe. Tail cylindrical, slightly longer than trunk. Series of palatine teeth arched, not extending outwards beyond the choanæ, separated from the parasphenoid groups which are not confluent. Brown above, with or without a yellowish band; lighter below; sometimes uniform black.

Total length 80 millim., the tail measuring 48 millim.

2. Spelerpes gibbicaudus.

Spelerpes gibbicaudus, Blatchley, Proc. U. S. Nat. Mus. xvi. p. 38 (1893).

Hab. MEXICO, Orizaba (*Sumichrast*).

Body rather slender, with 12 costal grooves, measuring over three and a half times the distance from snout to gular fold. Head narrow. Tail cylindrical. Digits short, free. Series of palatine teeth straight, extending outwards beyond the choanæ; parasphenoid teeth in two bands diverging but slightly posteriorly. Uniform brown, dusted with yellow. (After Blatchley.)

Total length	85 millim.
Head to gular fold	10 „
Length of tail	39 „

* 13 according to Cope's original description, in which the inguinal fold seems to have been included in the number; but in his later notes the number is stated to be 11.

3. Spelerpes cephalicus.

Spelerpes cephalicus, Cope, Proc. Ac. N. Sc. Philad. 1865, p. 196; 1869, p. 106.

Hab. N.E. MEXICO (*Cope*).

A short stout species, with 11 costal grooves; hind limb stout, fingers and toes flat, depressed, slightly webbed, margined, the inner very short. Tail but little compressed, rather shorter than body. Series of palatine teeth nearly straight, not in contact, externally extending beyond choanæ, separated from the parasphenoid groups, which are not confluent. Black, paler on the sides. Total length 65 millim. (After Cope.)

4. Spelerpes bellii.

Spelerpes bellii, Gray, Batr. Grad. p. 46; Cope, Proc. Ac. N. Sc. Philad. 1860, p. 372; 1869, p. 105; Bull. U. S. Nat. Mus. no. 32, p. 8; Bouleng. Batr. Grad. p. 68; Brocchi, Miss. Sc. Mex., Batr. p. 110, t. 20 bis; Blatchley, Proc. U. S. Nat. Mus. xvi. p. 37 (1893).

Œdipus platydactylus, Baird, Journ. Ac. N. Sc. Philad. i. p. 282 (1847).

Bolitoglossa mexicana, Dum. et Bibr. Erpét. Gén. ix. p. 93, t. 104. fig. 2.

Hab. MEXICO, city of Mexico (*Doorman*), Putla (*Boucard*), Zacualtipan (*Bernad*), La Cumbre de los Arrastrados, Talpa, Mascota, Jalisco (*Buller*), Guanajuato (*Dugès*), Jalapa (*Höge*), Orizaba 8000–14,000 feet (*Sumichrast, Blatchley*), Omilteme (*H. H. Smith*).

Body moderately elongate, with 12 costal grooves; head as broad as long, snout very short; limbs moderately developed; fingers and toes very short, depressed, united by a short web. Tail cylindrical. Series of palatine teeth slightly arched, extending outwards beyond the choanæ, separated from the parasphenoid groups, which are divided by a longitudinal groove and divergent behind. Black, back with two series of large, symmetrical, yellow spots; the spots of the two series may be confluent into chevrons or may be absent altogether.

	♂.	♀.
Total length	220 millim.	256 millim.
Length of tail	125 „	136 „

5. Spelerpes leprosus.

Spelerpes leprosus, Cope, Proc. Ac. N. Sc. Philad. 1869, p. 105; Bouleng. Batr. Grad. p. 68; Brocchi, Miss. Sc. Mex., Batr. p. 109, t. 19. figg. 1–2.

Hab. MEXICO, alpine region of Vera Cruz (*Sallé, Sumichrast, Dugès*), Jalapa (*Höge, Godman*), Popocatepetl, 9000 feet, in pine-forest (*Godman*).

Body moderately elongate, with 12 or 13 costal grooves; head longer than wide; limbs rather weak; fingers and toes short, depressed, with a short web, the inner very short. Tail cylindrical, about half the total length. Series of palatine teeth slightly arched, extending outwards beyond the choanæ, separated from the parasphenoid groups, which are more or less separated from each other and divergent behind. Black, sometimes with powdery yellowish patches, rarely with a reddish dorsal band.

	♂.	♀.
Total length	90 millim.	105 millim.
Length of tail	45 „	52 „

Spelerpes laticeps, Brocchi, Miss. Sc. Mex., Batr. p. 110, t. 18. fig. 1, from Vera Cruz (*Sallé*), does not seem to me distinct from *S. leprosus*. It has been distinguished on account of its parasphenoid teeth, which are in contact with each other; but this condition is closely approached by specimens which I have examined, and which cannot be held distinct from such as have the two groups of teeth somewhat more apart.

Not less problematical appear to me three other species, two of which are named by Brocchi in the same work, viz. *Spelerpes rostratum* from pine-forests of Totonicapam in Guatemala (p. 112) and *Spelerpes sulcatum* from Mexico (p. 112, t. 20. fig. 2). The third is *Spelerpes orizabensis*, Blatchley, Proc. U. S. Nat. Mus. xvi. p. 38, taken on Mt. Orizaba at an altitude of 11,000 feet.

6. **Spelerpes bocourti.** (Tab. LXXVI. fig. A.)

? *Spelerpes bocourti*, Brocchi, Miss. Sc. Mex., Batr. p. 111, t. 18. fig. 2.

Hab. ? Guatemala, pine-forests of Totonicapam (*Bocourt*); Costa Rica, La Palma (1600 metres) and Cariblanco (*Biolley & Underwood*).

Body rather stout, with 12 costal grooves; the length of the head to the gular fold is contained three and a third or three and a half times in the distance of the latter from the vent. Head broad, not much longer than wide. Tail rather compressed. Limbs moderately developed; digits half-webbed. The series of palatine teeth extends outwards beyond the choanæ; parasphenoid teeth confluent into a single group. Black, a whitish ring round the root of the tail; end of the tail and sometimes the extremity of the snout and toes whitish.

	Ad. ♀.	Ad. ♀.	Hgr.
Total length	190 millim.	176 millim.	94 millim.
Length of tail	90 „	96 „	43 „

7. **Spelerpes subpalmatus.**

Spelerpes subpalmatus, Bouleng. Ann. & Mag. N. H. 1896, xviii. p. 341.

Hab. Costa Rica, La Palma (*Underwood*).

Body moderately stout, with 12 costal grooves; the length of the head to the gular fold is one-third of the distance between the latter and the vent. Tail cylindrical, not compressed. Limbs moderately developed; digits half-webbed. The series of palatine teeth does not extend outwards beyond the choanæ; parasphenoid teeth confluent into a single group. Greyish; back pinkish, marbled with grey.

Total length	119 millim.
Length of tail	70 ,,

These characters are taken from the largest of the specimens for which the species has been established, as I do not feel sure that all the young individuals associated with the larger one should be referred to it.

8. Spelerpes morio.

Geotriton carbonarius, Cope, Proc. Ac. N. Sc. Philad. 1868, p. 313.

Œdipus morio, Cope, l. c. 1869, p. 103; Journ. Ac. N. Sc. Philad. viii. p. 96 (1876); Proc. Am. Phil. Soc. xviii. p. 271 (1879).

Spelerpes morio, Bouleng. Batr. Grad. p. 70; Brocchi, Miss. Sc. Mex., Batr. p. 113.

Hab. MEXICO, Jalapa (*Höge*), Tlalpam (*Dugès*); GUATEMALA (*Van Patten*); COSTA RICA, plateau of Cartago (*Zeledon*).

Body moderately elongate, with 12 or 13 costal grooves; head longer than wide; limbs rather weak; digits very short, connected by a broad membrane which leaves free the last phalanx only. Tail cylindrical. Series of palatine teeth slightly arched, extending outwards beyond the choanæ, separated from the parasphenoid groups, which are in contact with each other for the greater part of their length. Black, lower parts and sides with indistinct whitish specks.

Total length	97 millim.
Length of tail	47 ,,

9. Spelerpes robustus.

Œdipus robustus, Cope, Proc. Ac. N. Sc. Philad. 1894, p. 194.

Hab. COSTA RICA, slopes of the Volcan de Irazu (*Cooper*).

Body robust, with 12 costal folds; length of the head to gular fold two-ninths of that of the head and body. Tail cylindrical, the extremity moderately compressed. Digits broadly webbed, the last phalanx, however, being free, except that of the inner digits which is entirely enclosed. Series of palatine teeth straight, extending outwards beyond the choanæ; parasphenoid group not divided anteriorly. Uniform plumbeous. (After Cope.)

Total length	206 millim.
Head and body	98 ,,
Length to gular fold	22 ,,
,, of tail	108 ,,

10. **Spelerpes variegatus.** (Tab. LXXV.)

Salamandra variegata, Gray, in Griffith's Anim. Kingd. ix. p. 107.
Œdipus variegatus, Gray, Batr. Grad. p. 48; Cope, Bull. U. S. Nat. Mus. no. 32, p. 8.
Spelerpes variegatus, Strauch, Salam. p. 84; Bouleng. Batr. Grad. p. 73.
Bolitoglossa mexicana, var., Dum. et Bibr. Erpét. Gén. t. 104. fig. 1.
Spelerpes mexicanum, Brocchi, Miss. Sc. Mex., Batr. p. 113, t. 18 bis. figg. 1–4.
Geotriton carbonarius, Cope, Proc. Ac. N. Sc. Philad. 1860, p. 373.
Œdipus carbonarius, Cope, l. c. 1869, p. 103; Proc. Am. Phil. Soc. xviii. p. 267 (1879).
Œdipus salvinii, Gray, Ann. & Mag. N. H. 1868, ii. p. 297.
Spelerpes salvini, Brocchi, l. c. p. 117, t. 18. figg. 3–4.
Spelerpes sp., Müller, Verh. naturf. Ges. Basel, 1878, p. 645, t. 3. figg. C & D; 1883, p. 298 (*S. salvini*).
Spelerpes mülleri, Brocchi, l. c. p. 116, t. 20. figg. 3–5.
Spelerpes punctatum, Brocchi, l. c. p. 115.
Spelerpes attitlanensis, Brocchi, l. c. p. 115, t. 19. figg. 3–4.

Hab. MEXICO, city of Mexico (*Doorman*), Jalapa (*Höge, Trujillo*), Orizaba and Tehuantepec (*Sumichrast*); BRITISH HONDURAS (*Wickham*); GUATEMALA, Pacific coast (*Salvin*), Atitlan and Alta Vera Paz (*Bocourt*); COSTA RICA, San Carlos * (*Underwood*).

Body rather slender, with 12 or 13 costal grooves; the length of the head to the gular fold is two-sevenths or less of the distance of the latter from the vent; head longer than wide; limbs weak; digits very short, enveloped in a broad web; tail cylindrical. Series of palatine teeth extending outwards beyond the choanæ, separated from the parasphenoid teeth, which form one undivided group. Black below and on the sides; back buff-coloured, the two colours sharply defined; an elongate triangular or subtriangular black spot occupies almost invariably the crown of the head and nape of the neck, and is more or less produced behind. The buff-coloured dorsal portion is either uniform or, more commonly, black-spotted, the spots being irregular as regards number, shape, size, or distribution: they may be large and few in number (as in var. *salvinii*, fig. D), or confluent and covering nearly the whole back (Pacific coast, fig. B), or confluent into two longitudinal bands (British Honduras, fig. A), or confluent into a median band with lateral transverse branches (fig. C). In the latter variety the black lateral and lower parts are finely marbled with whitish.

The variations of colour are extremely diversified, and do not seem to be connected with special localities. The illustrations given here represent only some of the most striking aberrations from the typical form.

11. **Spelerpes lignicolor.**

Spelerpes (*Œdipus*) *lignicolor*, Peters, MB. Akad. Wiss. Berlin, 1873, p. 16.

Hab. PANAMA, Chiriqui (*Champion*).

* A specimen from this locality, named *Spelerpes morio*, is in the Nat. Hist. Mus.

Body moderately stout, with 12 or 13 costal grooves; the length of the head to the gular fold is one-third of the distance of the latter from the vent. Head longer than wide. Limbs well developed, digits entirely connected by membrane. Tail cylindrical, longer than the body. The series of palatine teeth nearly meet in the middle, and do not extend outwards beyond the choanæ; parasphenoid teeth confluent into a single group. Black; upper parts light rufous, sharply defined from the black sides, sprinkled with black, the black forming indistinct longitudinal bands; throat and chest with minute whitish specks.

Total length	125 millim.
Length of tail	70 „

12. Spelerpes rufescens.

Œdipus rufescens, Cope, Proc. Ac. N. Sc. Philad. 1869, p. 104; Proc. Am. Phil. Soc. xviii. p. 267 (1879); Bull. U. S. Nat. Mus. no. 32, p. 8.

Spelerpes rufescens, Bouleng. Batr. Grad. p. 71.

Hab. Mexico, Vera Cruz, Orizaba, Cordova, Tehuantepec (*Sumichrast*), Tabasco and Chiapas (*Dugès*).

Body stout, with 12 costal grooves; head longer than wide. Limbs weak, digits entirely connected by membrane. Tail cylindrical, shorter than the body. The series of palatine teeth meet nearly in the middle and do not extend outwards beyond the choanæ; parasphenoid teeth in two approximate patches, narrowing in front and but little divided behind. Black, upper parts with a rufous cast; brown below, with white specks. (After Cope.)

Perhaps not distinct from *S. lignicolor*.

13. Spelerpes yucatanus.

Spelerpus (*Œdipus*) *yucatanus*, Peters, SB. Ges. nat. Fr. Berlin, 1882, p. 137.

Spelerpes yucatanicus, Bouleng. Batr. Grad. p. 72; Zoolog. Record for 1882, Rept. p. 26.

Hab. Mexico, Yucatan (*Mus. Brit. et Mus. Berol.*).

Habit rather slender; body with 12 costal grooves; head longer than wide; limbs rather weak, digits entirely united by a broad web; tail cylindrical. The series of palatine teeth sweep outwards in a curve, but not beyond the choanæ, and nearly meet in the middle; they are separated from the parasphenoid group, which is not medially divided. Upper parts reddish-grey, powdered with purple; lower parts and sides blackish, with whitish specks irregularly distributed.

Total length	107 millim.	125 millim.
Length of tail	51 „	52 „

Known from two specimens; very closely allied to *S. variegatus*.

14. **Spelerpes lineolus.**

Spelerpes lineolus, Cope, Proc. Ac. N. Sc. Philad. 1865, p. 197; Bouleng. Batr. Grad. p. 74.
Opheobatrachus lineolus, Cope, l. c. 1869, p. 101.

Hab. MEXICO, Orizaba (*Sumichrast*).

Habit very slender, the cylindrical tail being about twice as long as the body; body with 13 or 14 costal grooves; head small, longer than wide. Limbs very weak, with the digits reduced to rudiments. Parasphenoid teeth in two long groups confluent in front, divergent behind. Black.

Total length	120 millim.
Length of tail	85 „

15. **Spelerpes uniformis.**

Œdipina uniformis, Keferstein, Arch. f. Naturg. 1868, p. 299; Nachricht. Ges. Wiss. Götting. 1868, p. 331; Günth. Zool. Record for 1868, p. 131.
Ophiobatrachus vermicularis, Gray, Ann. & Mag. N. H. 1868, ii. p. 298.
Opheobatrachus vermicularis, Cope, Proc. Ac. N. Sc. Philad. 1869, p. 102; Journ. Ac. N. Sc. Philad. viii. p. 96 (1876).
Spelerpes uniformis, Bouleng. Batr. Grad. p. 75.
Spelerpes vermicularis, Brocchi, Miss. Sc. Mex., Batr. p. 118, t. 20. fig. 1.

Hab. MEXICO, Vera Cruz (*Dugès*); COSTA RICA, La Palma, alt. 1600 met. (*Biolley*), Bebedero and Irazu (*Underwood*).

Habit extremely slender, worm-shaped, the cylindrical tail being more than twice as long as the body. Body with 18 or 19 costal grooves; head much longer than wide. Limbs very weak, with the digits reduced to rudiments. Parasphenoid teeth forming an undivided elongate group. Black, lighter below.

		♀.
Total length	178 millim.	173 millim.
Length of tail	124 „	115 „

THORIUS.

Thorius, Cope, Am. Nat. 1869, p. 222.

Palatine teeth in a transverse series; parasphenoid teeth present. Tongue free all round, supported by a central pedicel. Toes five. Tail cylindrical, slightly compressed towards its end. Nostrils very large.

1. **Thorius pennatulus.**

Thorius pennatulus, Cope, Am. Nat. 1869, p. 222; Proc. Ac. N. Sc. Philad. 1869, p. 111; Bouleng. Batr. Grad. p. 79, t. 3. fig. 2 (head).

Hab. MEXICO, Orizaba (*Sumichrast*).

Body rather elongate, with 13 costal grooves, its length being nearly four times the distance of the snout from the gular fold. Limbs weak, digits rudimentary. Parasphenoid teeth confluent into a single group. Brown, with a dark band along the side, edged with whitish above.

Total length	48 millim.
Length of tail	23 „

Division BATRACHIA APODA.

DERMOPHIS.

Dermophis, Peters, MB. Akad. Wiss. Berlin, 1879, p. 937.

Eyes distinct; tentacle in a circular groove, in front and below the level of the eye; rudimentary scales imbedded in the skin. Mandibulary teeth behind the main series indistinct, if present at all.

1. Dermophis mexicanus.

Siphonops mexicanus, Dum. et Bibr. Erpét. Gén. viii. p. 284; A. Dum. Mém. Soc. Sc. Nat. Cherbourg, ix. p. 23, fig. 10 (1863); Brocchi, Miss. Sc. Mex., Batr. p. 120, t. 21. fig. 2; Sumichrast, Bull. Soc. Zool. Fr. 1880, p. 190.

Dermophis mexicanus, Peters, MB. Akad. Wiss. Berlin, 1879, p. 937, fig. 6; Bouleng. Batr. Grad. p. 98, t. 8. fig. 2.

Hab. MEXICO, Tehuantepec (*Sumichrast*), Teapa in Tabasco (*H. H. Smith*); GUATEMALA, Pacific coast (*Salvin*); PANAMA (*Dow*).

Habit stout, the circumference of the body * being about one-seventh or two-thirteenths of its length. Snout rounded, moderately projecting beyond the mouth, its length equal to the distance between the eyes. 150–170 circular folds, of which the (about) 50 anterior and 20 posterior are complete, the intermediate being alternately complete or interrupted on the abdomen. Brownish or plumbeous above, whitish below; folds brownish-black.

	Mexico.	Guatemala.	Panama.	Locality?
Total length	407 millim.	451 millim.	410 millim.	240 millim.
Circumference . . .	60 „	62 „	60 „	38 „

Of this species I have before me a sufficient number (ten) of specimens to ascertain that there are no inner mandibulary teeth; one specimen only (from Panama) shows

* In this and the following descriptions I use the circumference of the body for comparative measurement in preference to the diameter, as the latter is frequently altered, and may be even misleading in specimens in which the shape of the body is distorted.

some roughnesses of the bone which may be taken as indicating a tendency towards tooth-formation.

2. **Dermophis gracilior,** sp. n. (Tab. LXXVI. fig. B.)

Hab. PANAMA, Chiriqui (*Champion*).

Habit moderate, the circumference of the body being nearly one-sixth of its length. Snout with rounded margin, projecting, shorter than the distance between the eyes. Tentacle very little nearer to the eye than to the nostril. No teeth within the mandibulary main series, which is composed of 26 teeth. 162 circular folds, of which the 40 anterior and 22 posterior are complete; from about the fortieth anterior annulus a secondary series of incomplete folds interrupted on the abdomen makes its appearance, the folds gradually approaching the median line of the abdomen until, towards the end of the body, they become complete and indistinguishable from the folds of the main series. Plumbeous above, paler or whitish below, the annuli not marked with black.

Total length	334 millim.
Circumference of body	38 „

CŒCILIA.

Cœcilia (L.), Wagl. Syst. Amph. p. 198; Peters, MB. Akad. Wiss. Berlin, 1879, p. 934.

Eyes distinct or hidden under the skin; tentacle valvular, in a horseshoe-shaped pit, below the nostril; rudimentary scales imbedded in the skin; a few teeth in the mandible behind the main series.

1. **Cœcilia isthmica.**

Cœcilia rostrata, Günth. P. Z. S. 1859, p. 417 (nec Cuv.).
Cœcilia isthmica, Cope, Proc. Am. Phil. Soc. xvii. p. 91 (1877); Bouleng. Batr. Grad. p. 94, t. 6. fig. 1.
Cœcilia guentheri, Peters, MB. Akad. Wiss. Berlin, 1879, p. 936.

Hab. PANAMA, Isthmus of Darien, eastern side (*Cope*).—W. ECUADOR.

Habit moderately stout, the circumference of the body being about one-tenth of its length. Snout subtruncated, very prominent, shorter than the distance between the eyes. Eyes distinct. 124–142 circular folds, interrupted on the back and belly, the last 16 or 18 being complete. Two teeth behind the main series of the mandible, near the symphysis. Plumbeous, paler below.

Total length	610 millim.*
Circumference of body	60 „

* 660 millim., in Cat. Batr. Grad. (1882) p. 94, seems to be a misprint.

HERPELE.

Herpele, Peters, MB. Akad. Wiss. Berlin, 1879, p. 939.

Eyes hidden below the bones; tentacle globular, in a circular pit, below and somewhat behind the nostril. Rudimentary scales imbedded in the skin near the folds. Mandibulary teeth in two series.

1. Herpele ochrocephala.

Cœcilia ochrocephala, Cope, Proc. Ac. N. Sc. Philad. 1866, p. 132; Brocchi, Miss. Sc. Mex., Batr. p. 119, t. 21. fig. 1.

Herpele ochrocephala, Cope, Proc. Am. Phil. Soc. xxii. p. 279 (1885); Bouleng. P. Z. S. 1895, p. 409.

Cœcilia polyzona, Fischer, in Peters, MB. Akad. Wiss. Berlin, 1879, p. 937; Arch. für Naturg. 1880, p. 215, t. 8. figg. 1–4; Bouleng. P. Z. S. 1895, p. 407.

Hab. PANAMA (*Cope, Mus. Brit.*).—COLOMBIA.

Habit slender, the circumference of the body being contained from 15 to 19 times in its length. Snout narrow, projecting. 200–209 circular folds, nearly all complete. The inner mandibulary series formed by 10–12 teeth. Tentacle at the lower side of the snout. Yellowish-plumbeous, with the folds darker, and with the head of a lighter colour.

	Fischer No. 1.	Fischer No. 2.	Cope.	Fischer in B.M.	Panama in B.M.
Total length . . .	675 millim.	650 millim.	325 millim.	565 millim.	450 millim.
Circumference . .	37 „	33 „	...	32 „	30 „
Diameter	11 „	11 „	6·4 „	10 „	9 „

The identification of Fischer's *Cœcilia polyzona* with Cope's *Herpele ochrocephala* requires explanation. There are two specimens in the Natural History Museum, one of which is determined as Cope's species, the description by that author well agreeing with it. The other, although a little more slender, clearly belongs to the same species, but it has been received from Fischer's collection, a portion of which was purchased in 1887, after his death, under the name of *C. polyzona*. There can be no doubt that it was so named by him, although it was not one of the two typical specimens from Caceres.

Fischer himself says that his species appeared to be closely allied to Cope's *H. ochrocephala*; but he was misled by the latter's statement that the proportions of this *C. ochrocephala* were near to those of *Siphonops mexicanus*, which, in fact, is one of the most robust species of the order. It is more surprising that Peters should have failed to recognize the pertinence of *C. polyzona* to his genus *Herpele*. But on comparing the accounts given by Peters and Fischer (1879 and 1880) I come to the conclusion that Peters had received from Fischer only the diagnosis, but not the specimens, when he read his paper before the Berlin Academy, and that he did not examine the position of the eye when he received the latter at a somewhat later period.

GYMNOPIS.

Gymnopis, Peters, MB. Akad. Wiss. Berlin, 1874, p. 216; 1879, p. 938; Bouleng. P. Z. S. 1895, p. 409.

Eyes hidden below the bones; tentacle globular, in a circular pit, nearer the commissure of the jaws than to the nostril. Rudimentary scales imbedded in the skin. Mandibulary teeth in two series.

1. Gymnopis multiplicata.

Gymnopis multiplicata, Peters, MB. Akad. Wiss. Berlin, 1874, p. 616, t. 1. fig. 1; 1879, p. 939, fig. 7; Bouleng. P. Z. S. 1895, p. 410.
Siphonops simus, Cope, Proc. Am. Phil. Soc. xvii. p. 91 (1877).
Gymnopis sima, Cope, Proc. Am. Phil. Soc. xxii. p. 171 (1885).

Hab. COSTA RICA, Cariblanco (*Underwood & Biolley*); PANAMA, Veragua.—COLOMBIA, Antioquia.

Head moderate, the circumference of the body being one-ninth of its length; snout rounded, prominent. 228–250 circular folds, of which the (about) 20 anterior and 30 posterior are complete, the intermediate being alternately complete and interrupted on the abdomen. Plumbeous, lighter below.

Total length	450 millim.
Circumference of body	50 „

2. Gymnopis proxima.

Siphonops mexicanus, Cope, Journ. Ac. Nat. Sc. Philad. viii. p. 96 (1875).
Siphonops proximus, Cope, Proc. Am. Phil. Soc. xvii. p. 90 (1877).
Gymnopis proxima, Cope, l. c. p. 171 (1885).

Hab. COSTA RICA, eastern coast (*Gabb*).

Habit moderate, the circumference of the body being about one-eighth of its length; snout flattened, narrow, produced; (tentacular fossa close to eye*); 129 complete annuli. Plumbeous, paler below. (After Cope.)

Total length	425 millim.
Diameter of body	17 „

* This expression, used by Cope before he was acquainted with the real position of the eye, must now appear inappropriate. What Cope took for an eye is the tentacular canal, which is apparent through the skin as a whitish patch (well shown by Peters in fig. 7 of his paper of the year 1879), and which occupies a position in which one would expect the eye to be placed. Peters himself was mistaken as to the nature and position of the eye when he established the genus *Gymnopis* in 1874.

3. **Gymnopis oligozona.**

Siphonops oligozonus, Cope, Proc. Am. Phil. Soc. xvii. p. 91 (1877).
Gymnopis oligozona, Cope, l. c. xxii. p. 171 (1885); Bouleng. P. Z. S. 1895, p. 410.

Hab. GUATEMALA (*Mus. Brit.*).

Habit slender, the circumference of the body being about one-seventeenth of its length; snout narrow, projecting; (tentacular fossa close to the eye *, which is invisible); total number of circular folds 166, "the principal series containing 119, of which the 14 anterior and 42 posterior are complete; of the second series (none in anterior 72 annuli) 34 incomplete and 13 complete." (After Cope.)

Total length	263 millim.
Diameter of body	5 „

The single specimen which I have seen of this species is not in a state of good preservation.

* See footnote, p. 308.

INDEX.

[Names in small capitals refer to Families, &c.; those in roman type to the chief reference to each species included in the work; those in italics to species incidentally mentioned, synonyms, &c.]

NOTE

FOR PURPOSES OF THIS REPRODUCTION THE FOLLOWING PLATES HAVE BEEN REDUCED TO 90% OF ORIGINAL SIZE

PRINTED BY TAYLOR AND FRANCIS, RED LION COURT, FLEET STREET.

EMYS ORNATA. JUV. & PULL.
(nat. size.)

. Mintern del et lith. Mintern Bros. imp.

EMYS SALVINI.

P. Mintern del et lith. Mintern Bros. imp.

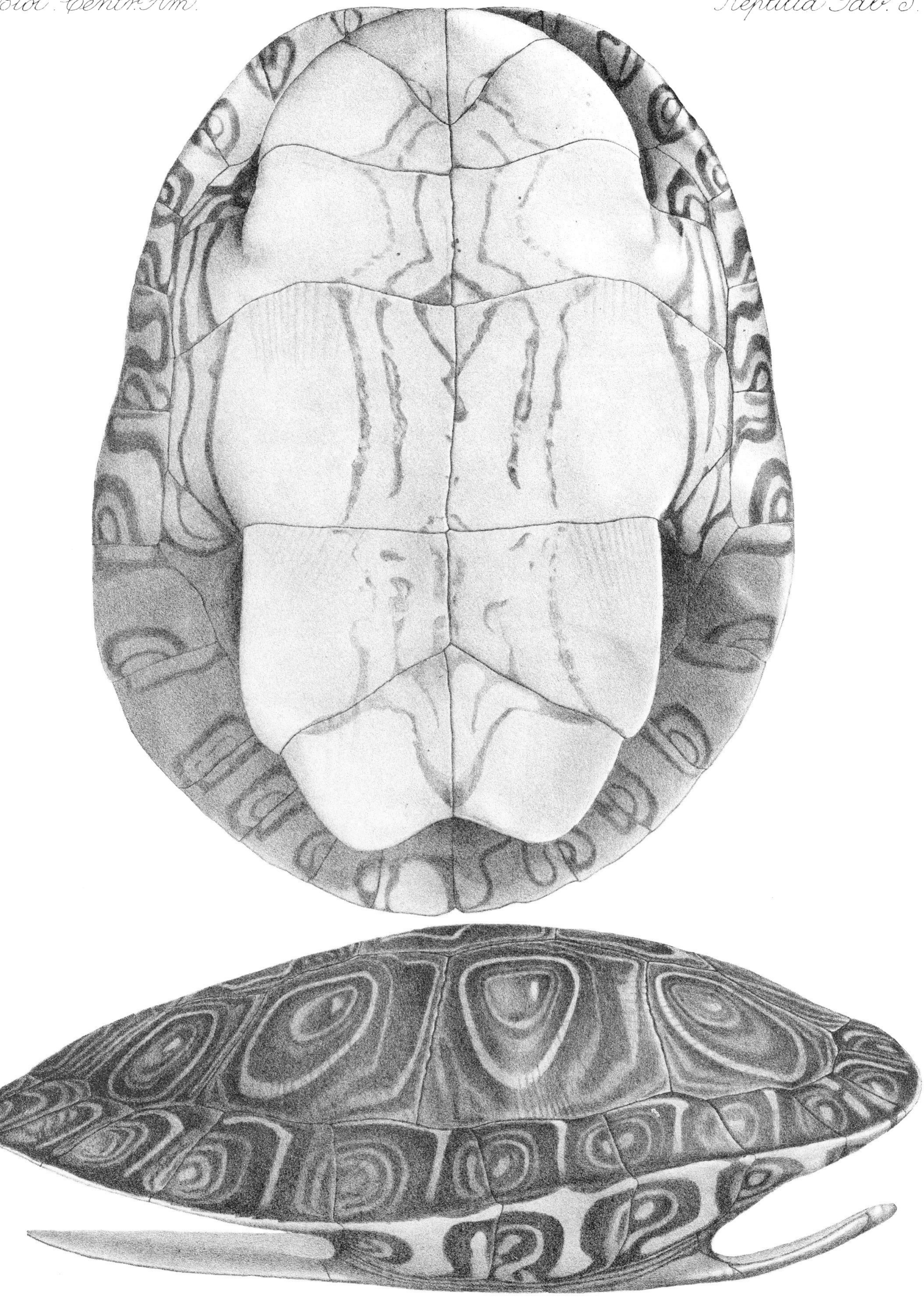

EMYS SALVINI.

Mintern del et lith. Mintern Bros. imp.

EMYS UMBRA.

R. Mintern del et lith. Mintern Bros. imp.

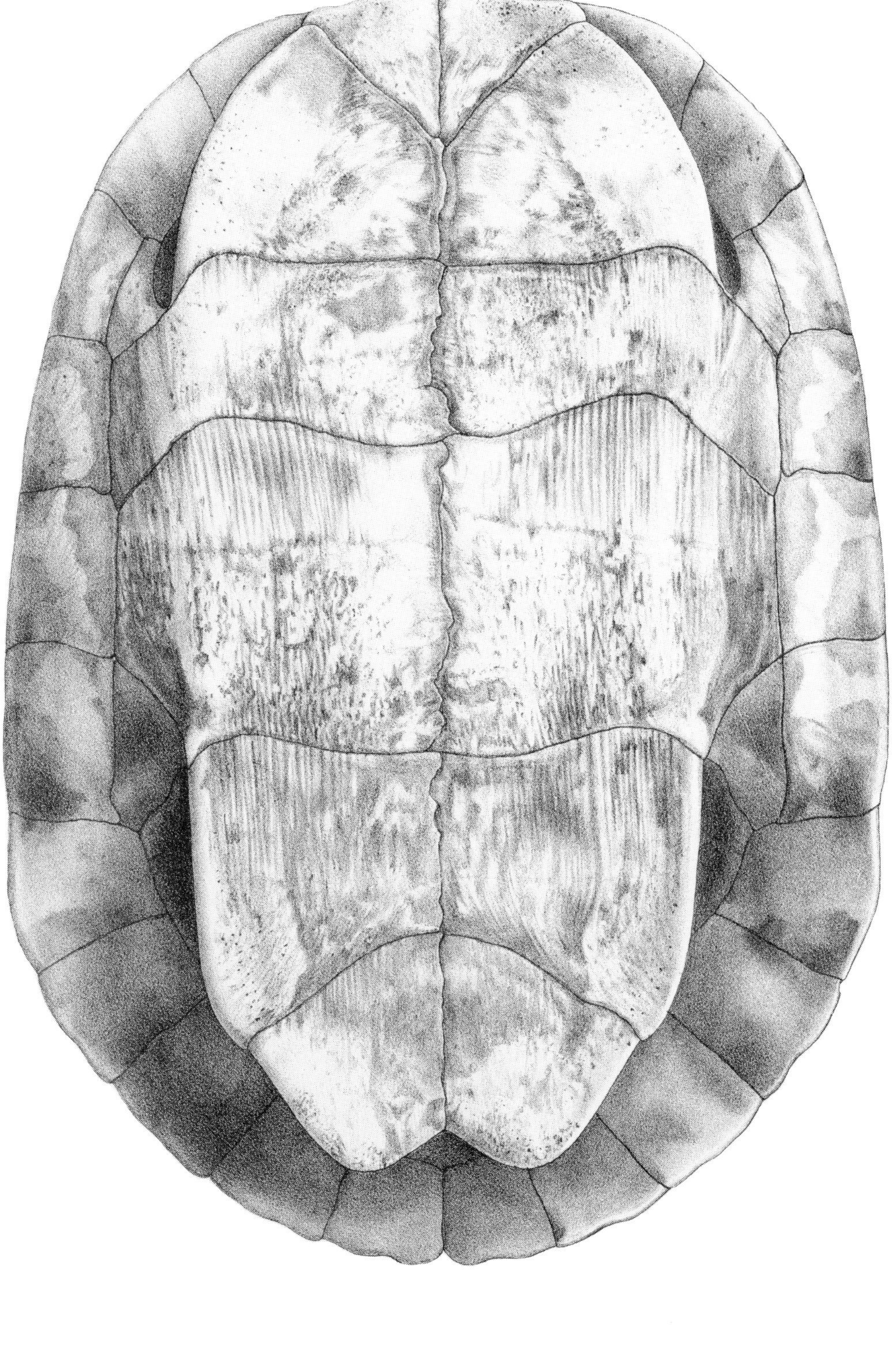

EMYS UMBRA.
(2/3 nat. size.)

R. Mintern del et lith. Mintern Bros. imp.

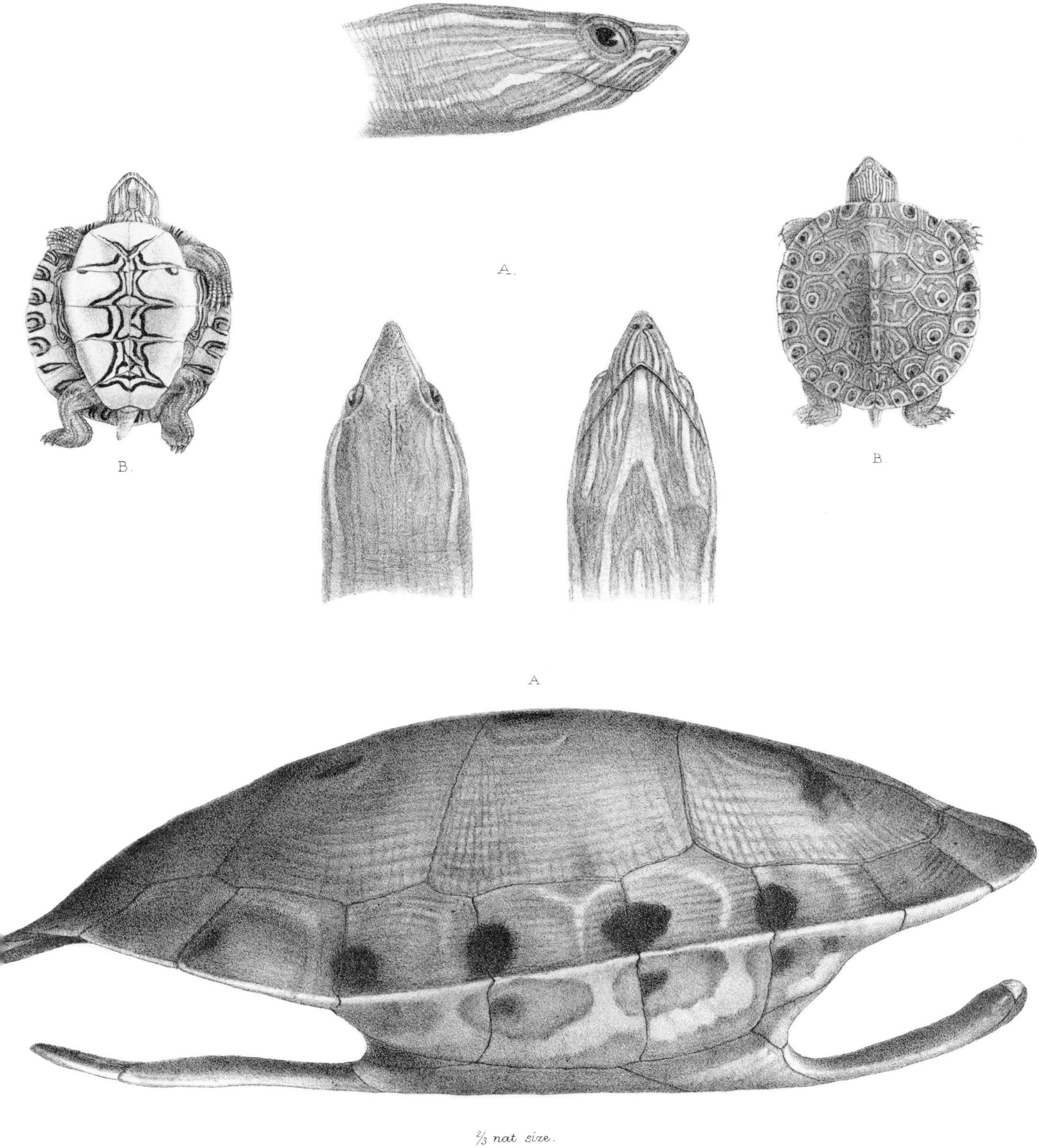

2/3 nat. size.

A. EMYS UMBRA. B. EMYS CATASPILA.

R. Mintern del et lith. Mintern Bros. imp.

EMYS PULCHERRIMA.

Mintern del et lith. Mintern Bros. imp.

A. EMYS PULCHERRIMA. B. EMYS AREOLATA.
C. C'. C''. EMYS RUBIDA.

Mintern del et lith. Mintern Bros. imp.

CINOSTERNUM CRUENTATUM.

R. Mintern del et lith. Mintern Bros. imp.

♀

♂

R. Mintern del et lith.

CINOSTERNUM CRUENTATUM.

Mintern Bros. imp.

♀ ♂

CINOSTERNUM CRUENTATUM.

R. Mintern del et lith. Mintern Bros. imp.

CINOSTERNUM HIRTIPES.

A ♀ TRES MARIAS. B. ♀. MAZATLAN. C. ♂ MAZATLAN.

R. Mintern del et lith. Mintern Bros. imp.

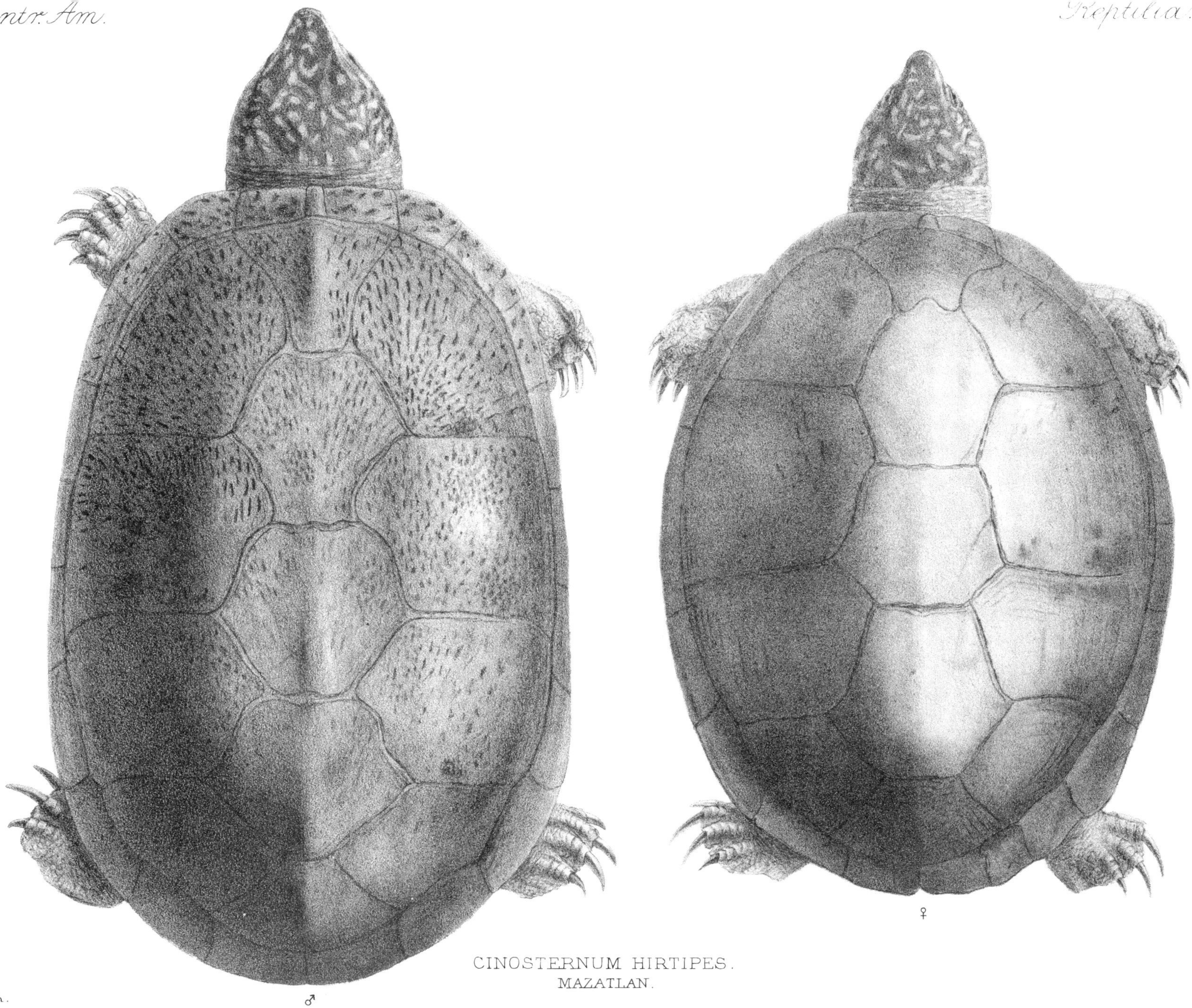

♂ ♀

CINOSTERNUM HIRTIPES.
MAZATLAN.

R. Mintern del et lith. Mintern Bros. imp.

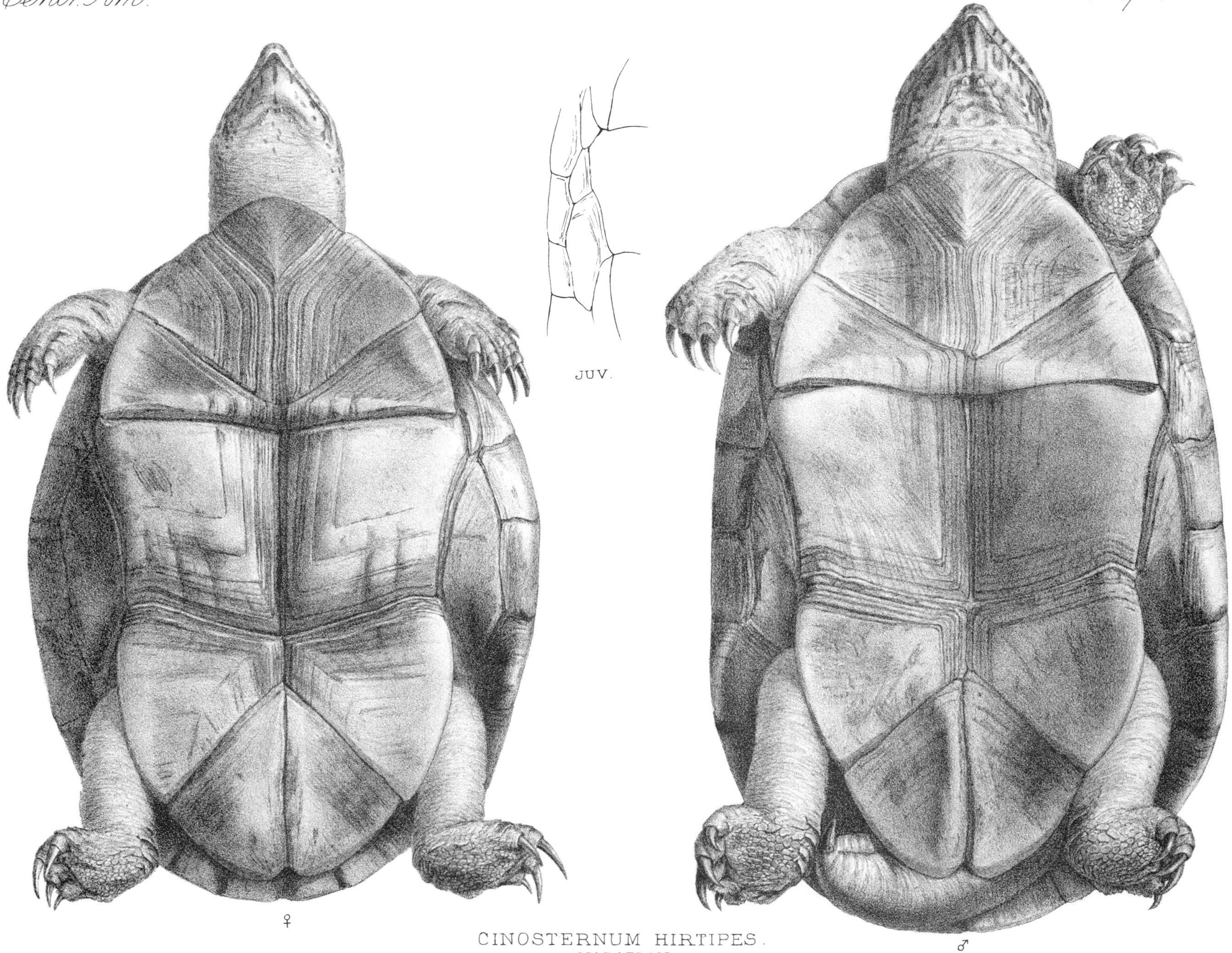

CINOSTERNUM HIRTIPES.
MAZATLAN.

R. Mintern del et lith.

Mintern Bros. imp.

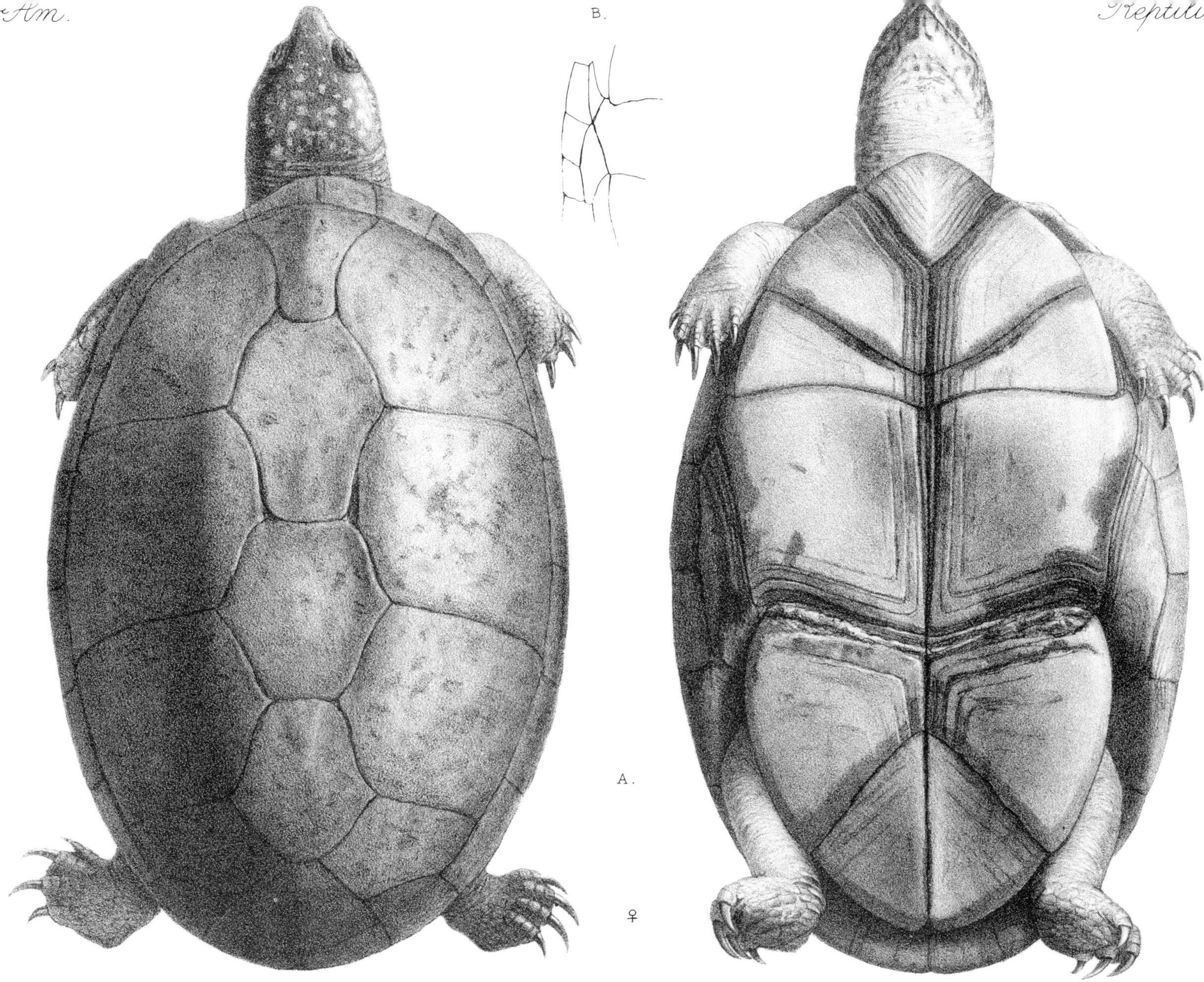

B. A. ♀

CINOSTERNUM HIRTIPES.
TRES MARIAS.
A. ADULT. ♀. B. HALF GROWN. ♀.

R. Mintern del et lith. Mintern Bros. imp.

♂

CINOSTERNUM LEUCOSTOMUM

Smit del et lith. Mintern Bros. imp.

♀

CINOSTERNUM LEUCOSTOMUM

Smit del et lith. Mintern Bros. imp.

A. CINOSTERNUM BREVIGULARE. B. CINOSTERNUM COBANUM.

Mintern del et lith. Mintern Bros. imp.

AMEIVA SURINAMENSIS, ♂.
(COSTA RICA.)

R. Mintern del et lith.

Mintern Bros. imp.

CNEMIDOPHORUS MARIARUM.
(AD. ♂ AND YOUNG.)
TRES MARIAS ISLANDS.

Mintern del et lith. Mintern Bros. imp.

CNEMIDOPHORUS GUTTATUS.
AD. AND YOUNG.

. Mintern del et lith. Mintern Bros. imp.

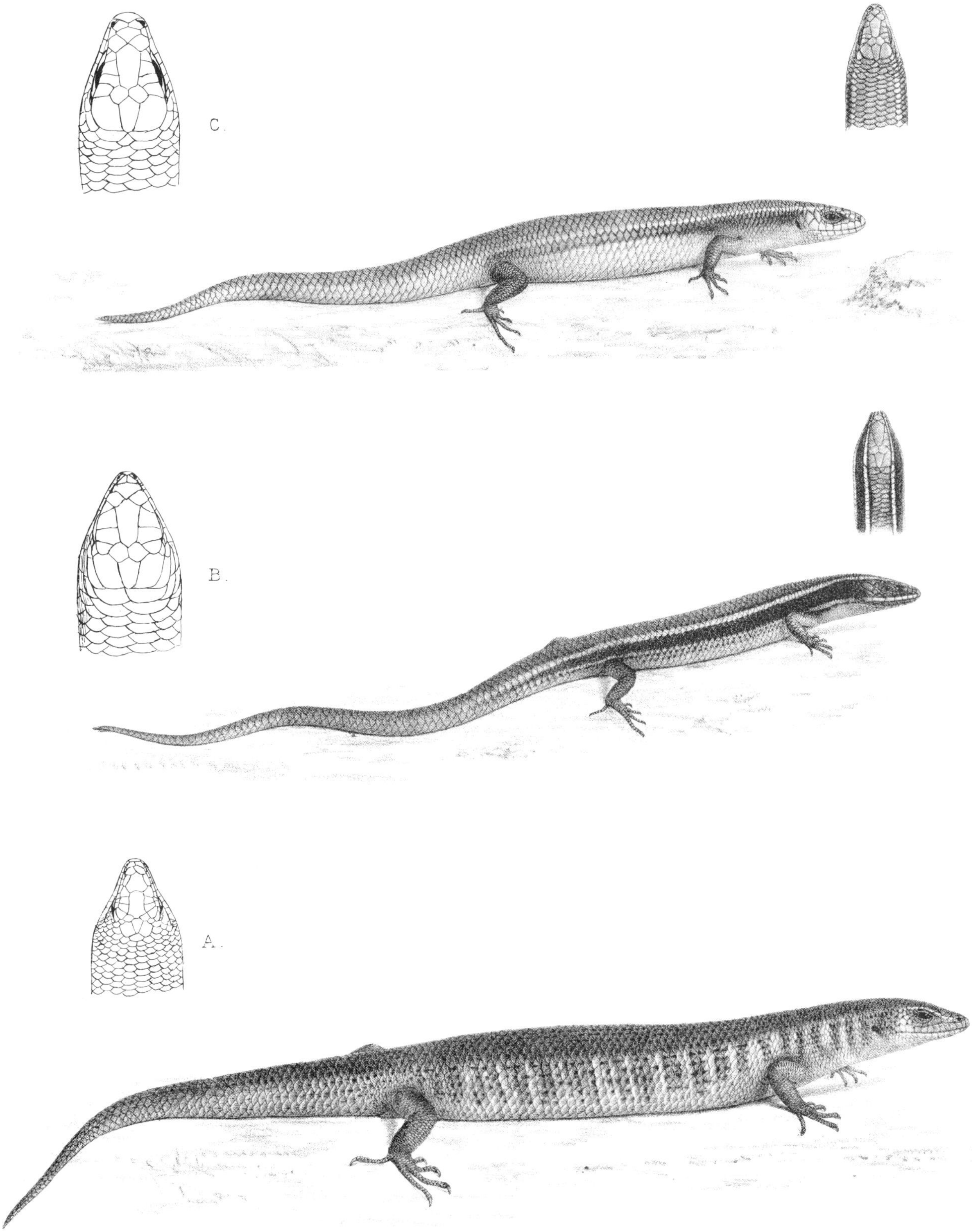

A. DIPLOGLOSSUS STEINDACHNERI. B. EUMECES BREVIROSTRIS. C. EUMECES BOCOURTI.

Mintern del et lith. Mintern Bros. imp.

R. Mintern del et lith. DIPLOGLOSSUS MONOTROPIS. Mintern Bros. imp.

A. GERRHONOTUS OAXACÆ, AD. & YG. B. GERRHONOTUS GRAMINEUS. C. GERRHONOTUS MORELETII.

R. Mintern del et lith. Mintern Bros. imp.

A–D GERRHONOTUS COERULEUS. E. GERRHONOTUS OBSCURUS.

R. Mintern del et lith. Mintern Bros. imp.

HELODERMA HORRIDUM.
(YOUNG.)
(TEHUANTEPEC.)

R. Mintern del et lith. Mintern Bros. imp.

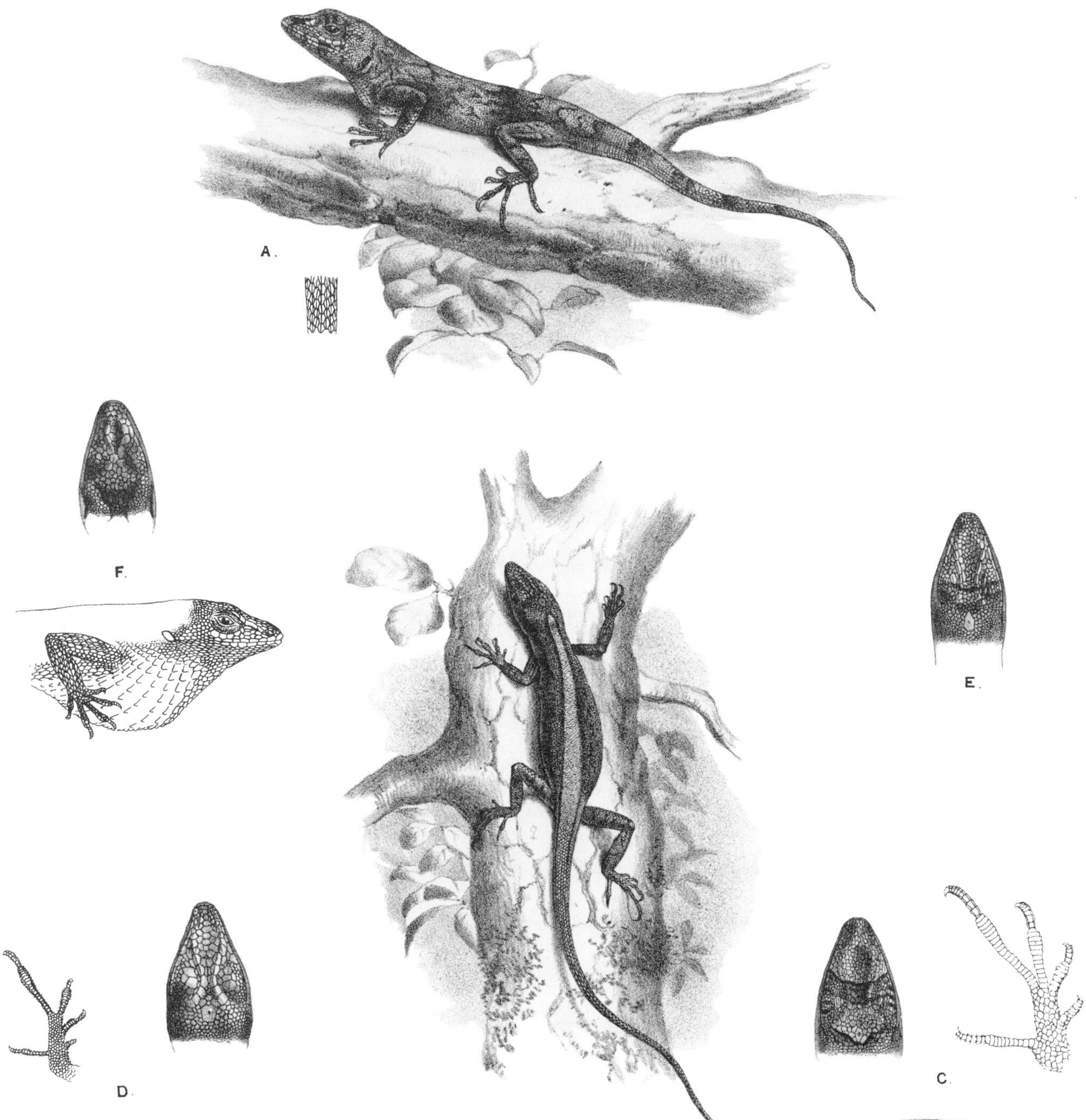

ANOLIS

A. A. SALVINI. B. A. SALLÆI. C. A. TROPIDOLEPIS. D. A. USTUS. E. A. GODMANI. F. A. CRASSULUS.

Peter Smit del. et lith. Mintern Bros. imp.

R. Mintern del. et lith. BASILISCUS VITTATUS, ♂ & ♀. Mintern Bros. imp.

CTENOSAURA COMPLETA, AD. ♂ (½ NAT. SIZE.)
(BONACCA ISL.)

R. Mintern del. et lith. Mintern Bros. imp.

R. Mintern del. et lith. Mintern Bros. imp.

CTENOSAURA COMPLETA, AD ♀ (NAT. S.)
(CHIAPAM)

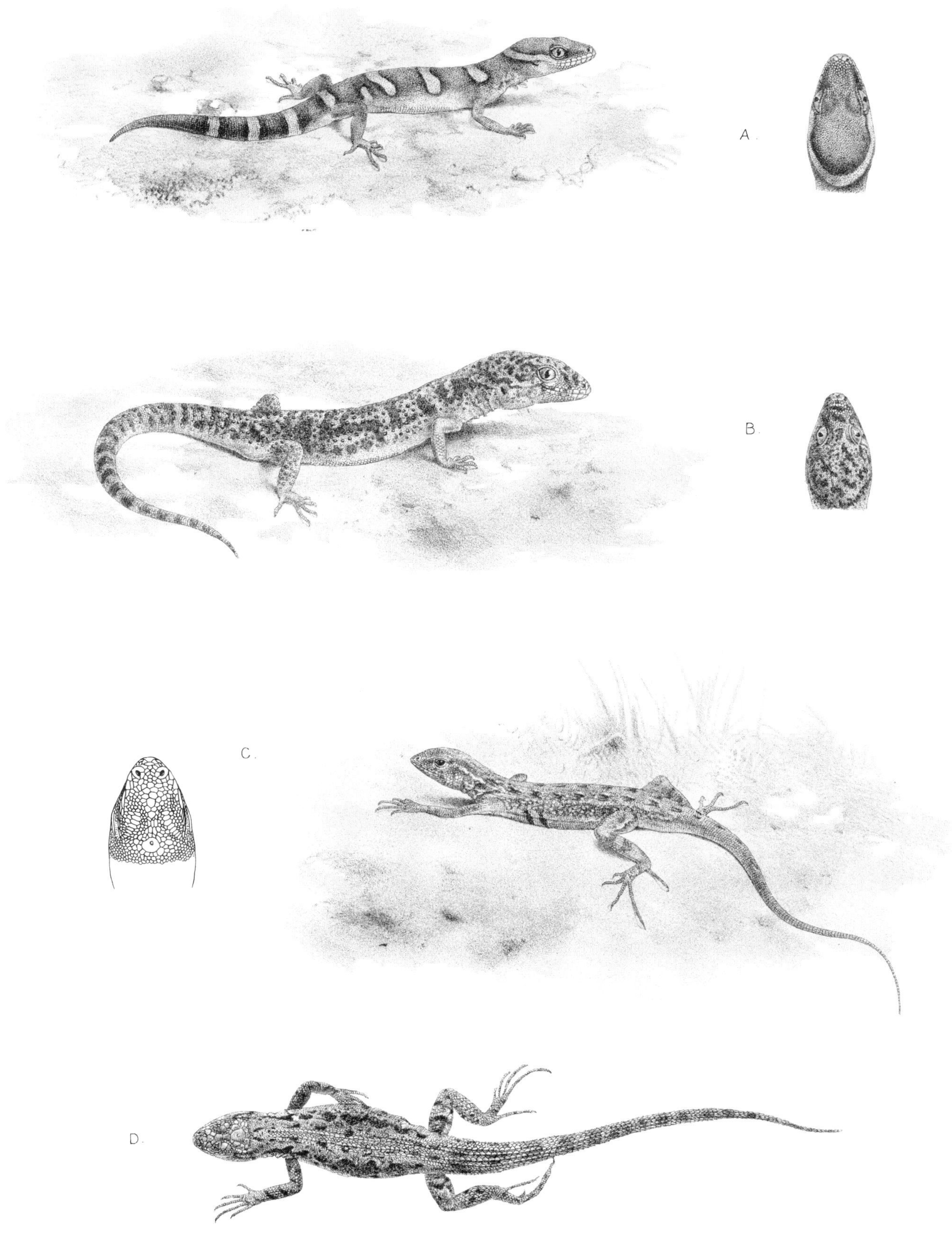

A. EUBLEPHARIS FASCIATUS. B. EUBLEPHARIS DOVII. C. HOLBROOKIA PROPINQUA. D. UTA LATERALIS.

ter Smit del. et lith. Mintern Bros. imp.

A. SCELOPORUS OMILTEMANUS. B. SCELOPORUS PLEUROLEPIS. C. SCELOPORUS RUBRIVENTRIS.

ter Smit del. et lith. Mintern Bros. imp.

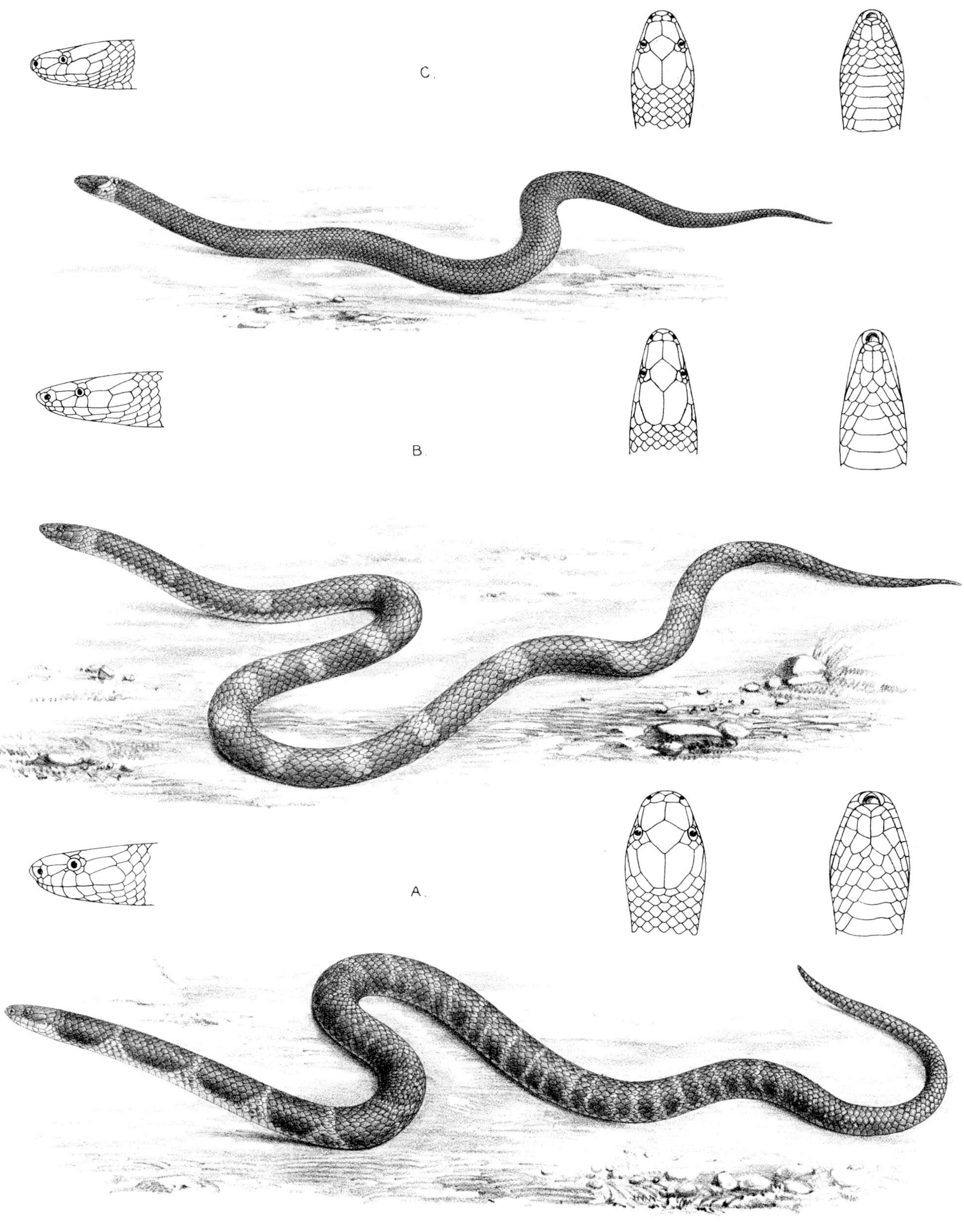

GEOPHIS.
A. G. OMILTEMANA. B. G. CHALYBÆA VAR. QUADRANGULARIS. C. G. MOESTA.

Mintern & J. Green del. et lith. Mintern Bros. imp.

A. GEOPHIS FASCIATA. B. CONOPSIS NASUS.

Mintern & J. Green del. et lith. Mintern Bros. imp.

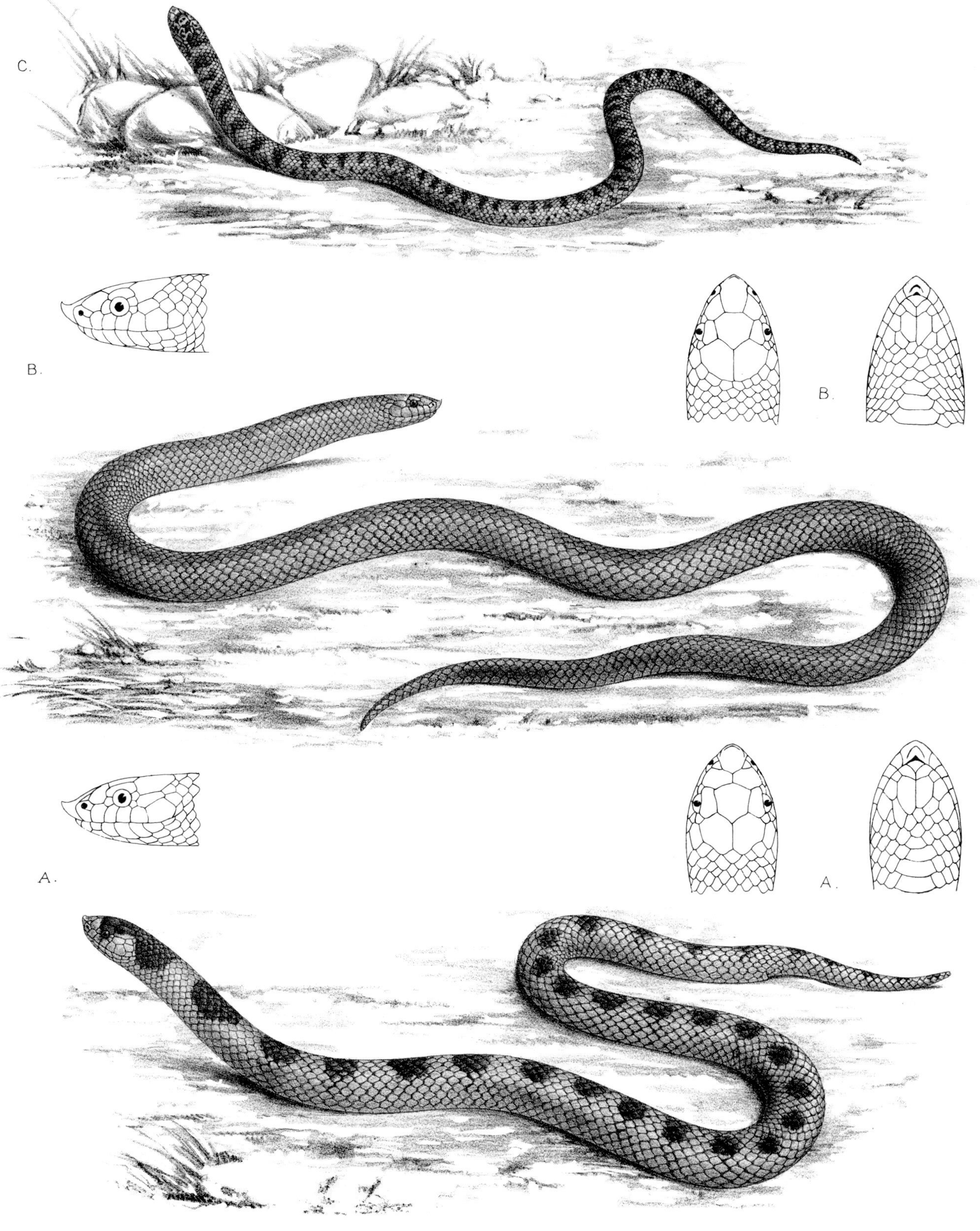

A. FICIMIA QUADRANGULARIS. B, C. FICIMIA OLIVACEA.

Mintern & J. Green del. et lith. Mintern Bros. imp.

R. Mintern & J. Green del. et lith. Mintern Bros. Chromo.

A. ELAPOCHRUS ÆQUALIS. B & C. HOMALOCRANIUM MICHOACANENSE.

A. ABLABES DECIPIENS. B. LIOPHIS EPINEPHELUS.

..ntern. & J. Green del. et lith. Mintern Bros. imp.

VARIETIES OF CORONELLA ANNULATA.
A. YUCATAN. B. MEXICO. C. TEHUANTEPEC.

R. Mintern & J. Green del. et lith. Mintern Bros. Chromo.

R. Mintern & J. Green del. et lith.

Mintern Bros. imp.

A. CORONELLA LEONIS.   B. CORONELLA GODMANI.

A. DROMICUS LAUREATUS. B. DROMICUS OMILTEMANUS.

tern & J. Green del. et lith. Mintern Bros. imp.

R. Mintern & J. Green del. et lith. COLUBER CHLOROSOMA. Mintern Bros. imp.

$\frac{2}{3}$

SPILOTES SALVINI.

Mintern & J. Green del. et lith. Mintern Bros. Chromo.

$\frac{3}{4}$

SPILOTES PŒCILONOTUS.

Mintern & J. Green del. et lith. Mintern Bros. imp.

SPILOTES ARGUS.

Mintern & J. Green del. et lith. Mintern Bros. imp.

ZAMENIS RHOMBIFER AD & YG.

Mintern & J. Green del. et lith. Mintern Bros. imp

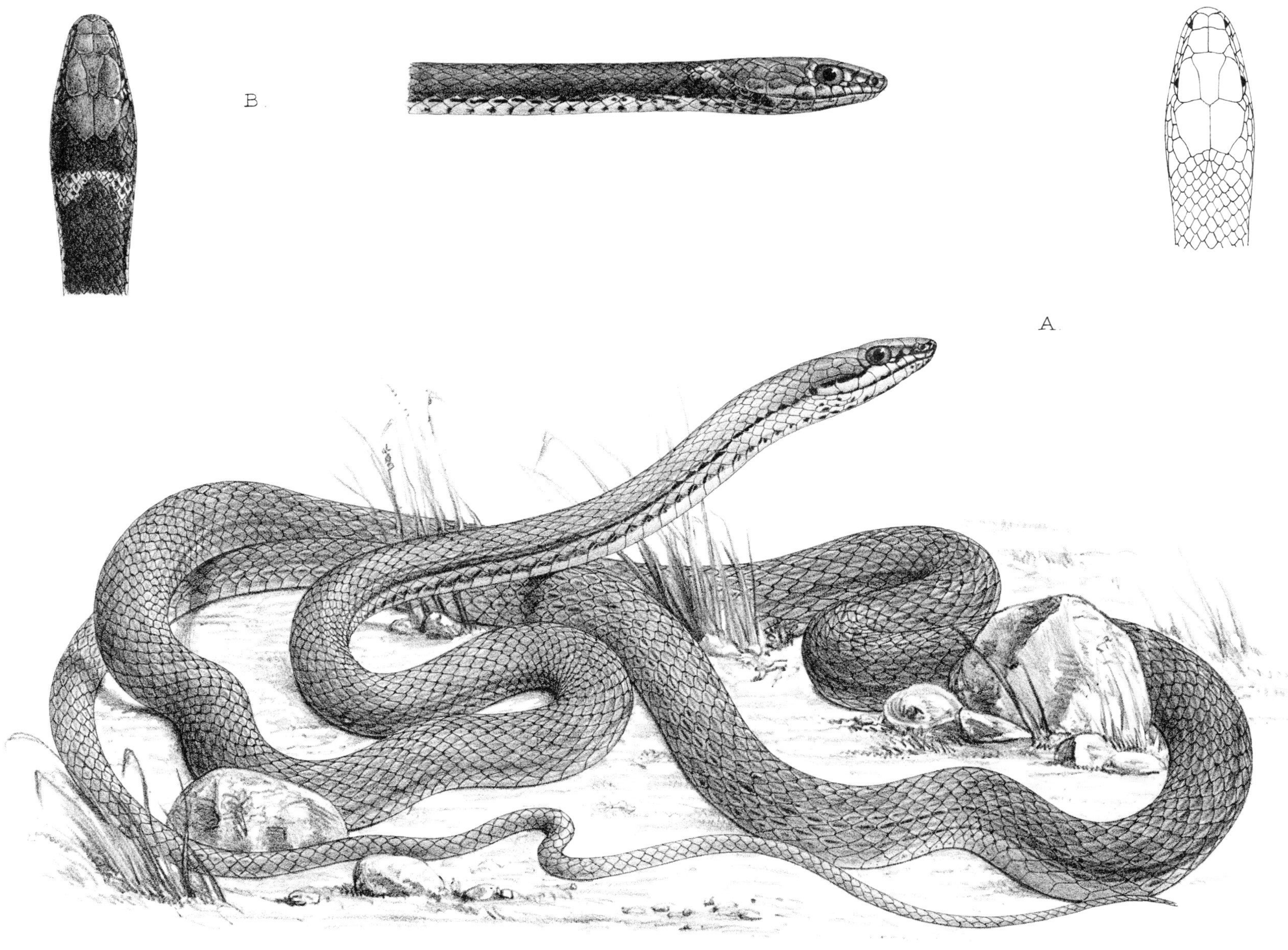

A. ZAMENIS SEMILINEATUS. B. ZAMENIS ORNATUS

R. Mintern & J. Green del. et lith. Mintern Bros. imp.

3/4

PITUOPHIS LINEATICOLLIS.

Mintern & J. Green del. et lith. Mintern Bros. imp.

LEPTOPHIS MODESTUS.

Mintern & J. Green del et lith. Mintern Bros. imp.

R. Mintern & J. Green del. et lith. Mintern Bros. imp.

A. HYPSIGLENA DISCOLOR. B. HYPSIGLENA LATIFASCIATA. C. LEPTOGNATHUS ANNULATUS.

R. Mintern & J. Green del. et lith. Mintern Bros Chromo.

A. TROPIDODIPSAS DUMERILI. B. TROPIDODIPSAS FASCIATA.

A, B. MESOPELTIS DIMIDIATUS. C. HENICOGNATHUS VENUSTISSIMUS.

J. Green & R. E. Mintern del et lith. Mintern Bros. Chromo lith.

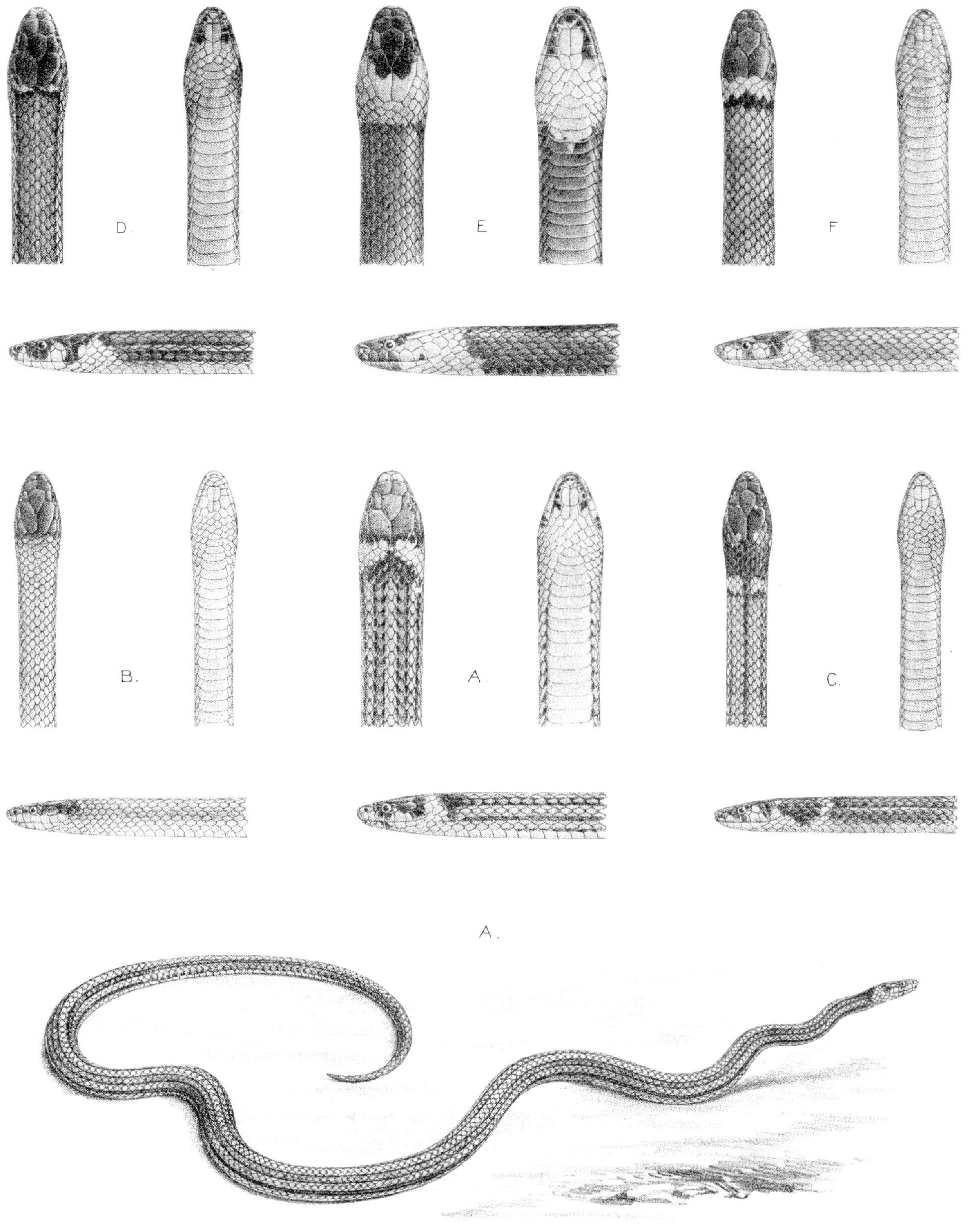

HOMALOCRANIUM
A. VIRGATUM. B. ATRICEPS. C. ARMILLATUM. D. JANI.
E. MOESTUM. F. BOULENGERI.
(Figures of the heads twice the natural size.)

J. Green del. et lith. Mintern Bros. imp.

A. TACHYMENIS DECIPIENS. B. LEPTODIRA SPLENDIDA.

Mintern & J. Green del et lith. Mintern Bros. imp.

C.

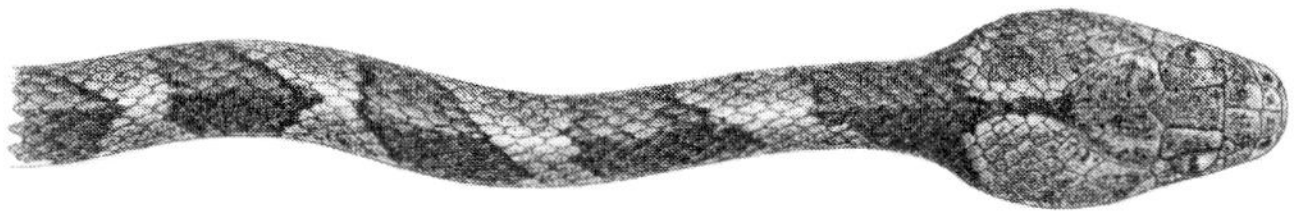

B.

A.

A. & B. LEPTODIRA PERSONATA. C. LEPTODIRA RHOMBIFERA.

Mintern & J. Green del et lith. Mintern Bros. imp.

B.

A.

R. Mintern & J. Green del. et lith. Mintern Bros. imp.

A. LEPTODIRA POLYSTICTA. B. LEPTODIRA OCELLATA.

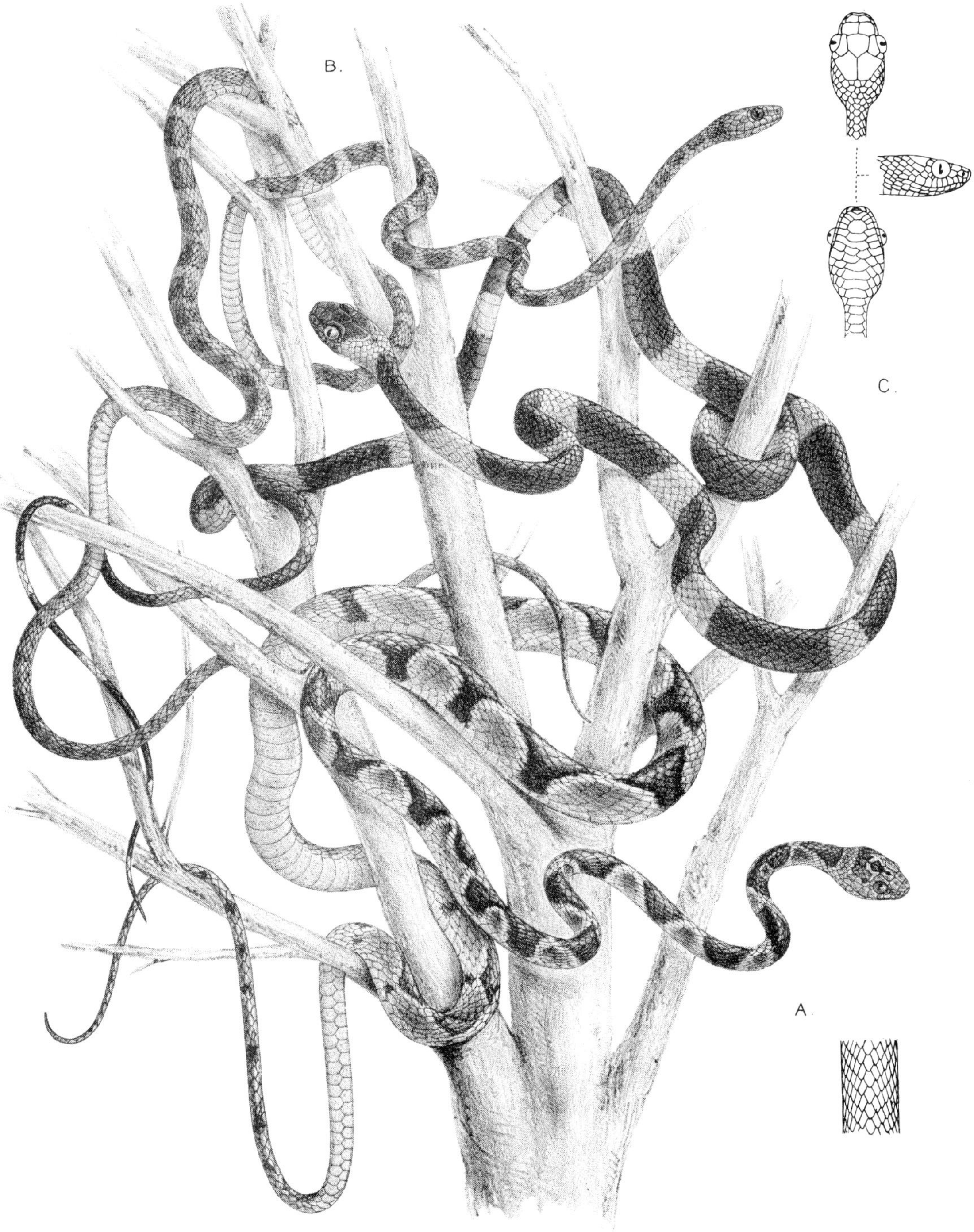

A. DIPSAS SPLENDIDA. B. DIPSAS GRACILLIMA. C. NEOPAREAS BICOLOR.

R. Mintern & J. Green del. et lith. Mintern Bros. imp.

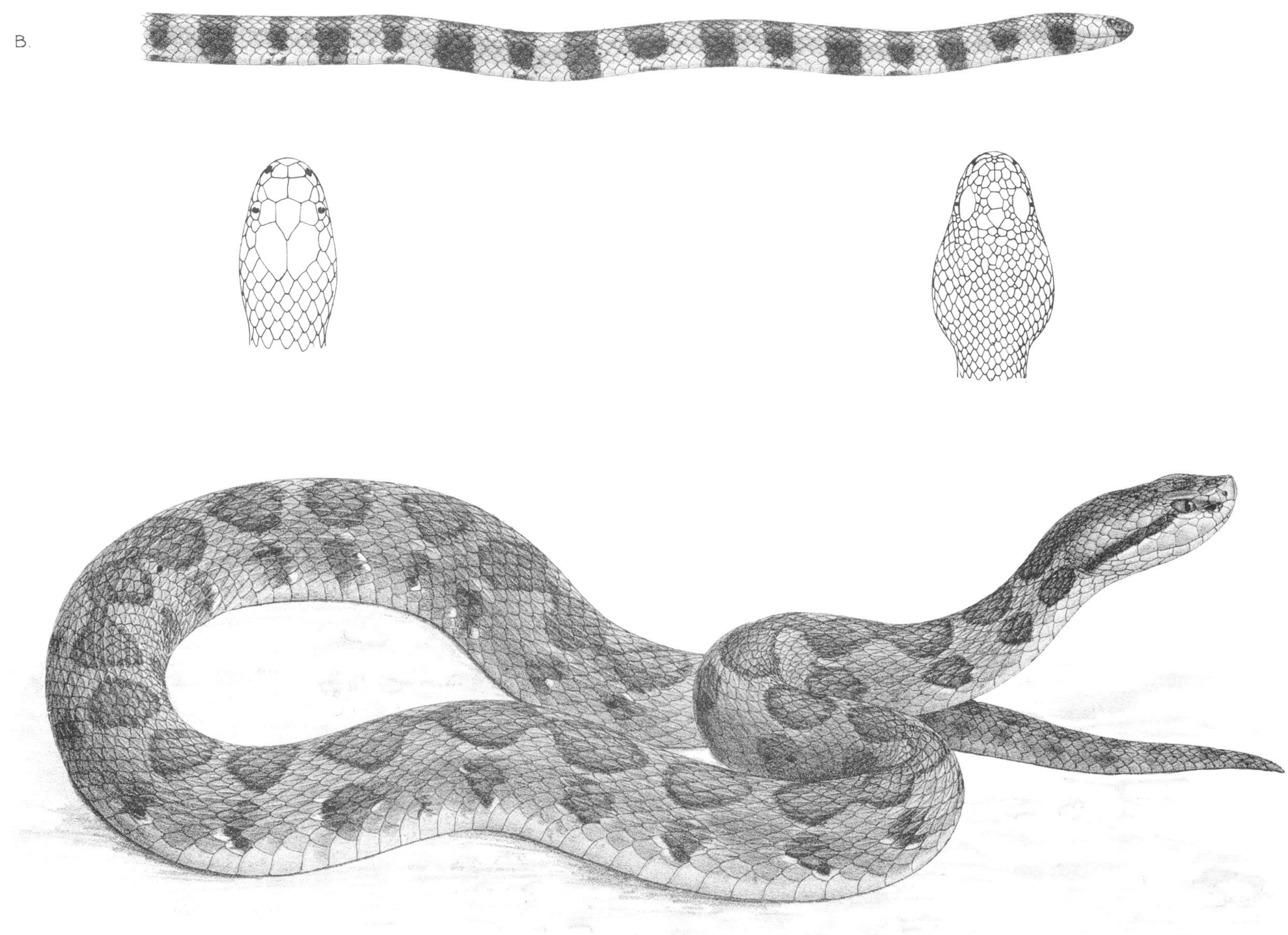

R. Mintern & J. Green del. et lith. Mintern Bros. imp.

A. BOTHRIECHIS GODMANI. B. ELAPS RUATANUS.

R. Mintern & J. Green del. et lith.

A. & B. ANCISTRODON BILINEATUS. C. CROTALUS OMILTEMANUS.

Mintern Bros. imp.

R. Mintern & J. Green del. et lith. Mintern Bros. imp.

A. CROTALUS SALVINI. B. CROTALUS PALLIDUS.

A.

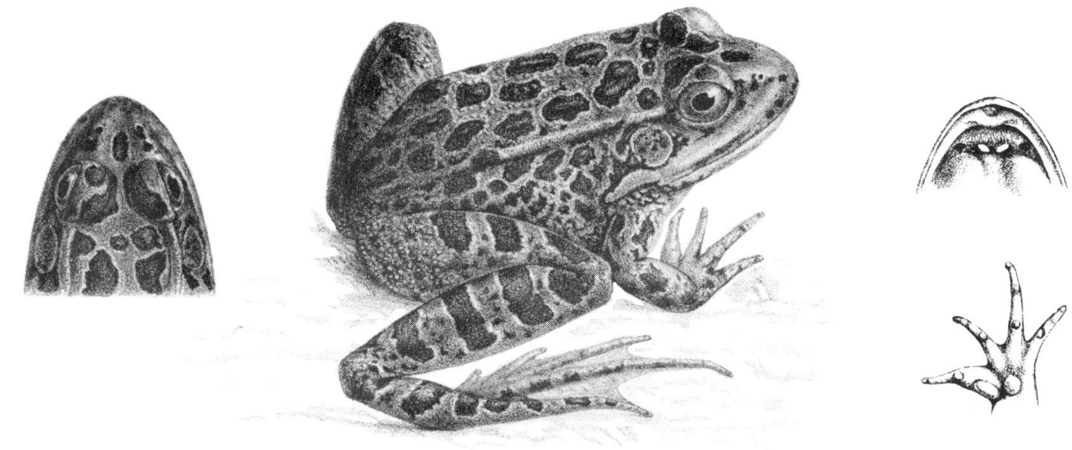

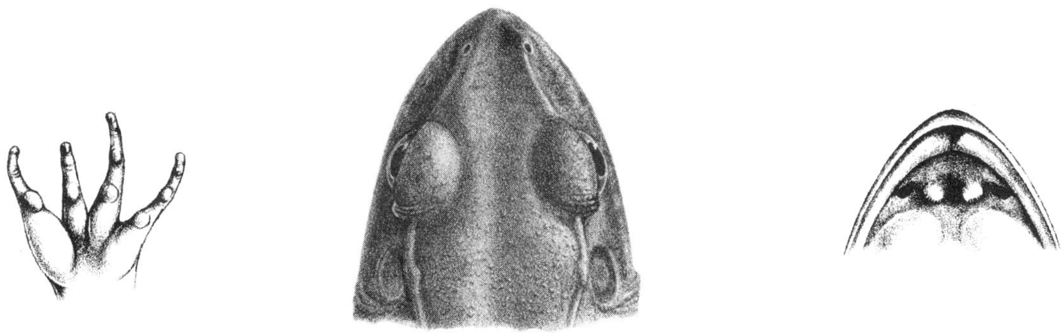

B.

A. RANA FORRERI. B. RANA BONACCANA.

J. Green & R. Mintern del et lith. Mintern Bros. imp.

A.

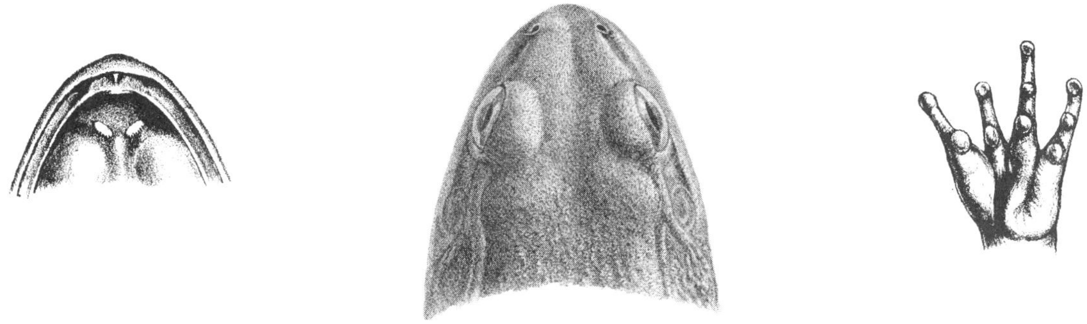

B.

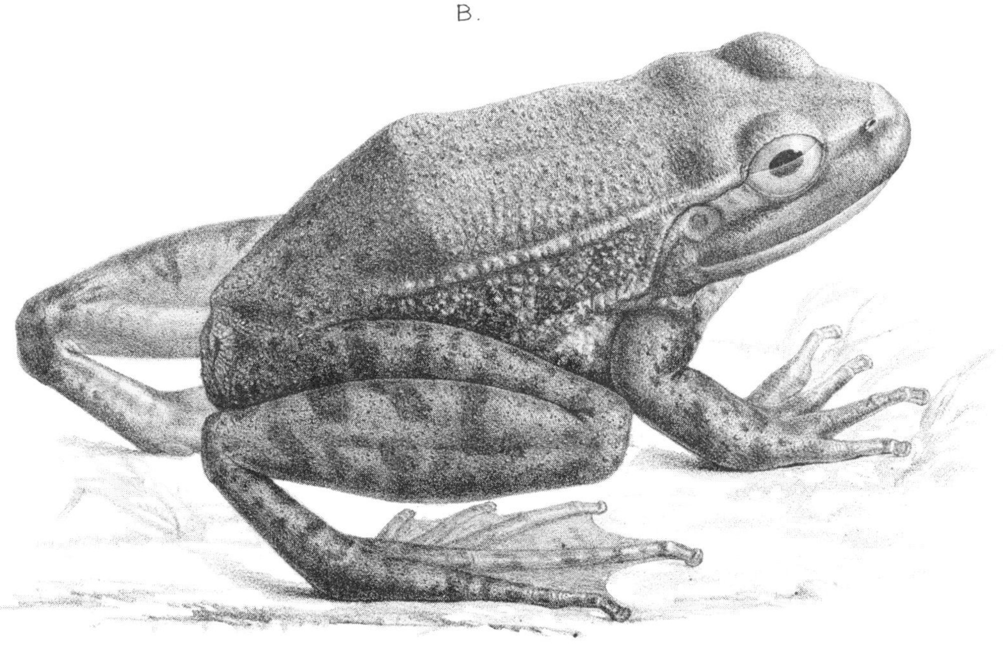

A. RANA OMILTEMANA. B. RANA PUSTULOSA.

J. Green & R. Mintern del. et lith. Mintern Bros. imp.

A. B. C. ATELOPUS VARIUS.   D. ENGYSTOMA ELEGANS.

J. Green del. et lith.   Mintern Bros. Chromo.

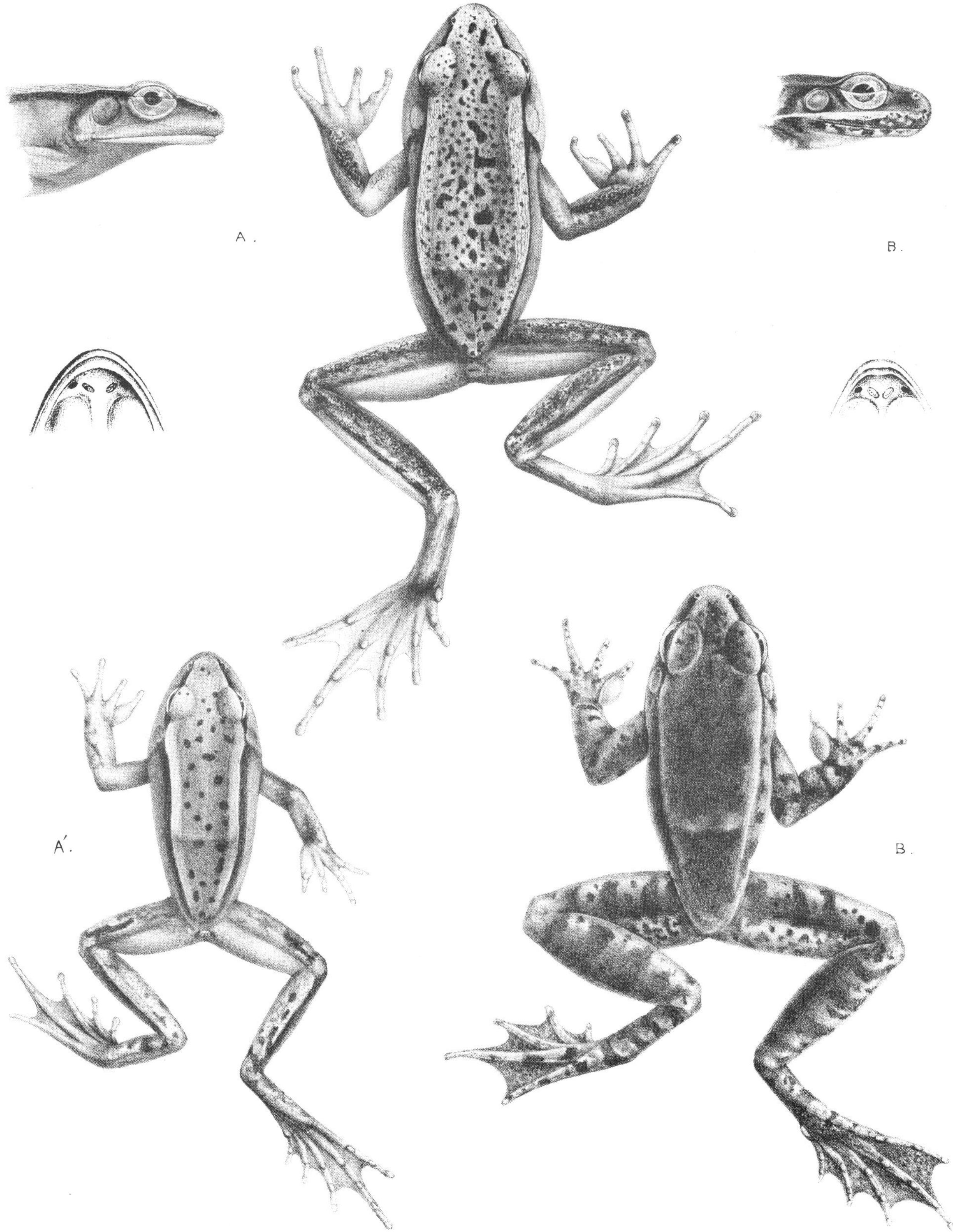

A. A'. RANA GODMANI B. RANA MELANOSOMA.

J. Green del. et lith. Mintern Bros. imp.

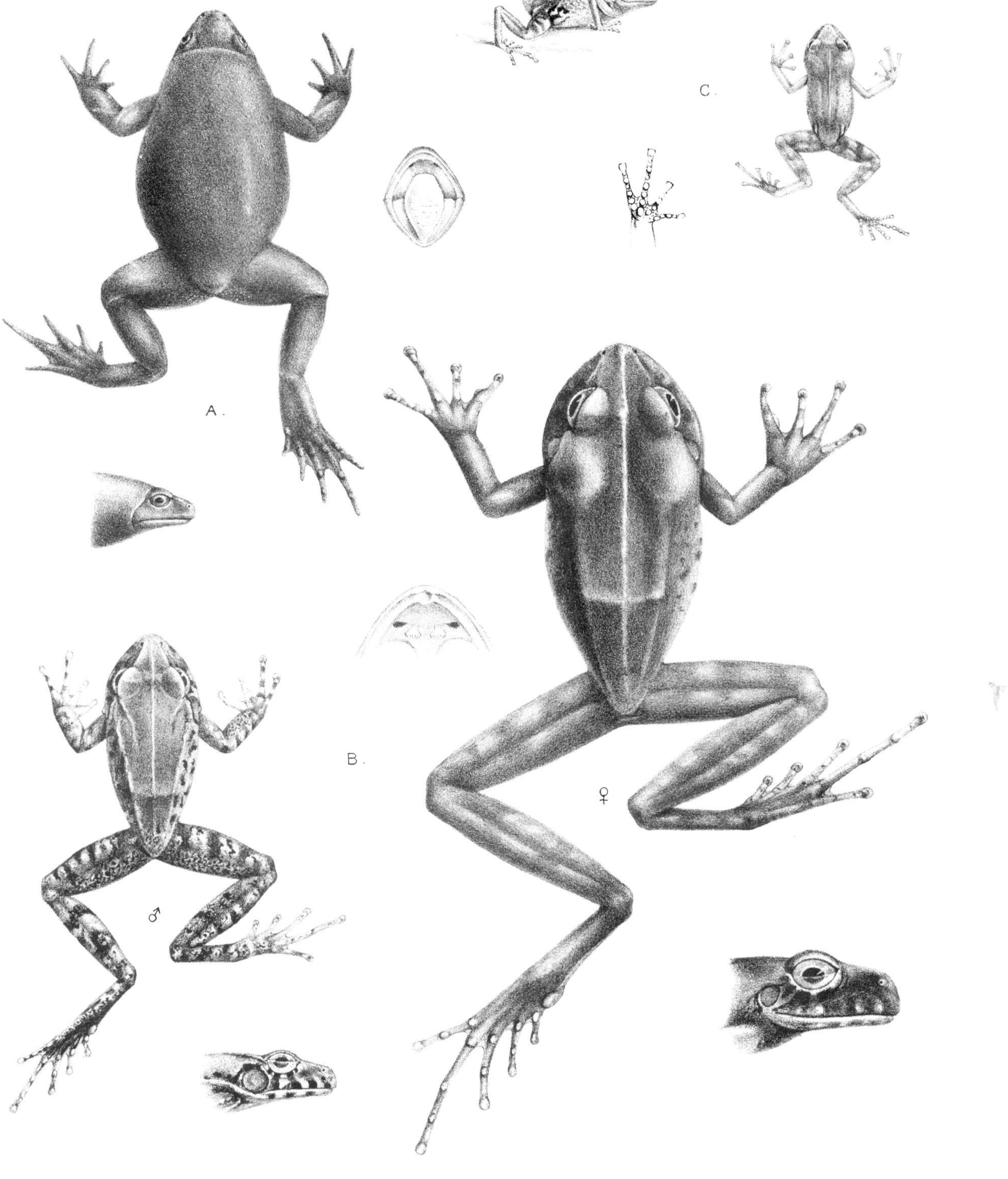

A. GLOSSOSTOMA ATERRIMUM. B. LIOHYLA FLEISCHMANNI.
C. TOMODACTYLUS AMULÆ.

J. Green del. et lith. Mintern Bros. imp.

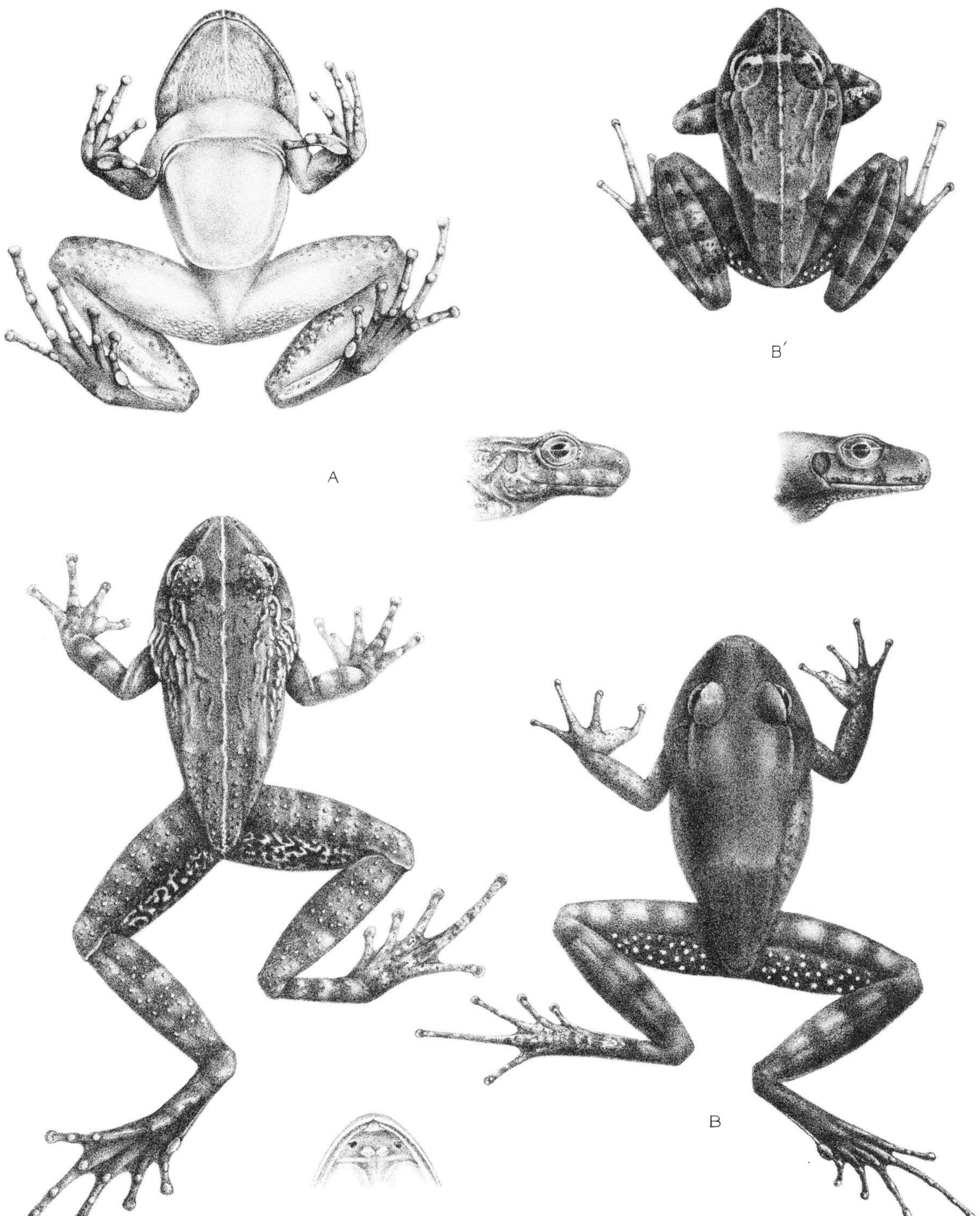

A. LIOHYLA ENGYTYMPANUM. B. B′. LIOHYLA PITTIERI.

J. Green del. et lith. Mintern Bros. imp.

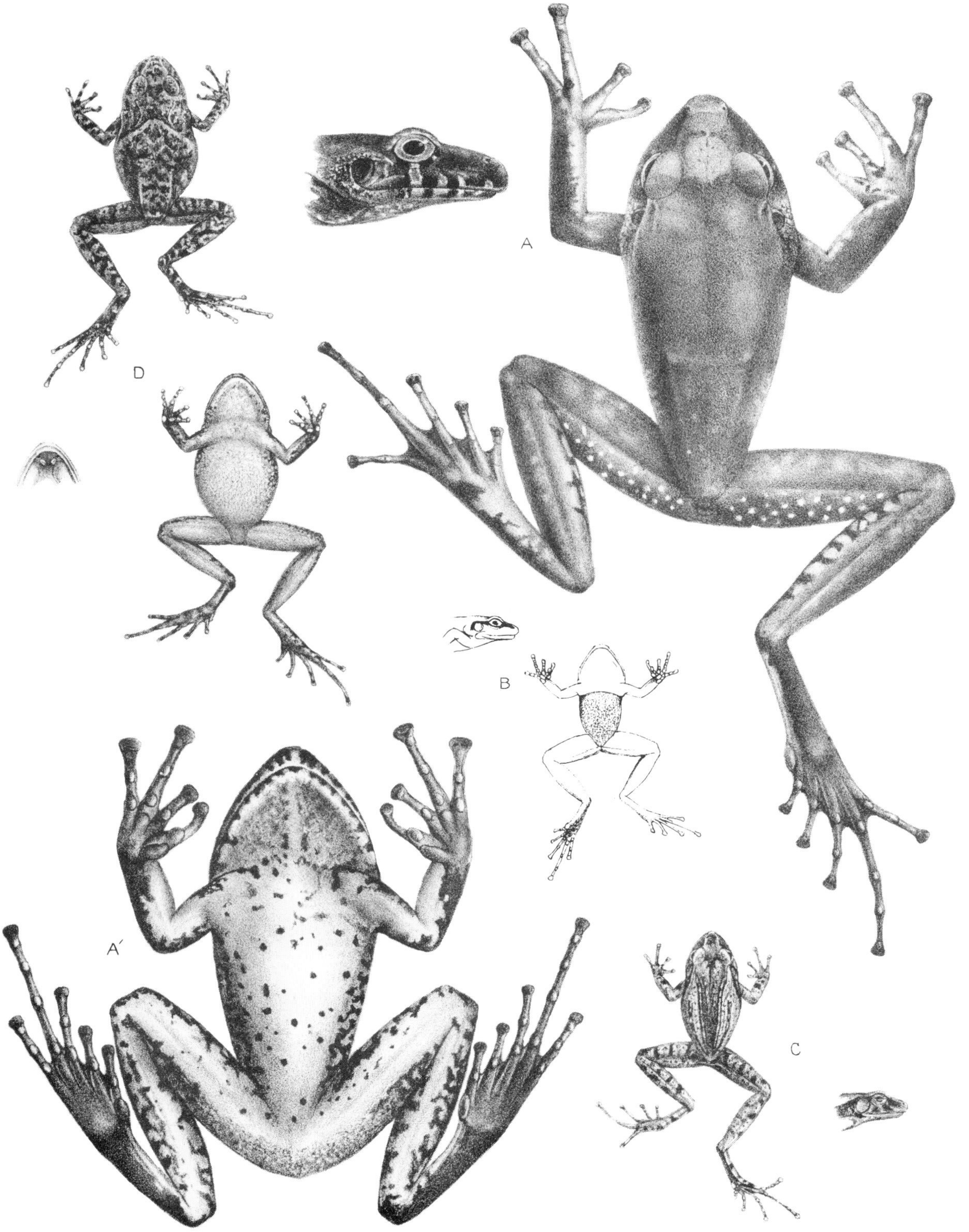

A. A′. LIOHYLA GUENTHERI. B. HYLODES PLICATUS. C. HYLODES POLYPTYCHUS.
D. HYLODES UNDERWOODI.

J. Green del. et lith. Mintern Bros. imp.

HYLODES.
A. H. PLATYRHYNCHUS. B. H. CALCITRANS. C.C′. H. RHODOPIS.

J. Green del. et lith. Mintern Bros. imp.

HYLODES.

A. H. BROCCHII. B. H. BROCCHII, VAR. C. H. VENUSTUS. D. H. AUGUSTI.

J. Green del. et lith. MinternBros. imp.

A. HYLODES NUBILUS. B. BUFO CONIFERUS. C. BUFO OCCIPITALIS.

J. Green del. et lith. Mintern Bros. imp.

A. BUFO MARMOREUS. B. HYLA GABBI. C. HYLA COPII.
D. HYLA URANOCHROA.

J. Green del. et lith. Mintern Bros. imp.

HYLA BAUDINI.

A & B. COSTA RICA. C. JALISCO. D. NICARAGUA.

J. Green del. et lith. Mintern Bros. Chromo.

HYLA.

A, B. H. PSEUDOPUMA. C, D. H. LEUCOPHYLLATA. E. H. GODMANI. F. H. TÆNIOPUS.

J. Green del. et lith. Mintern Bros. imp.

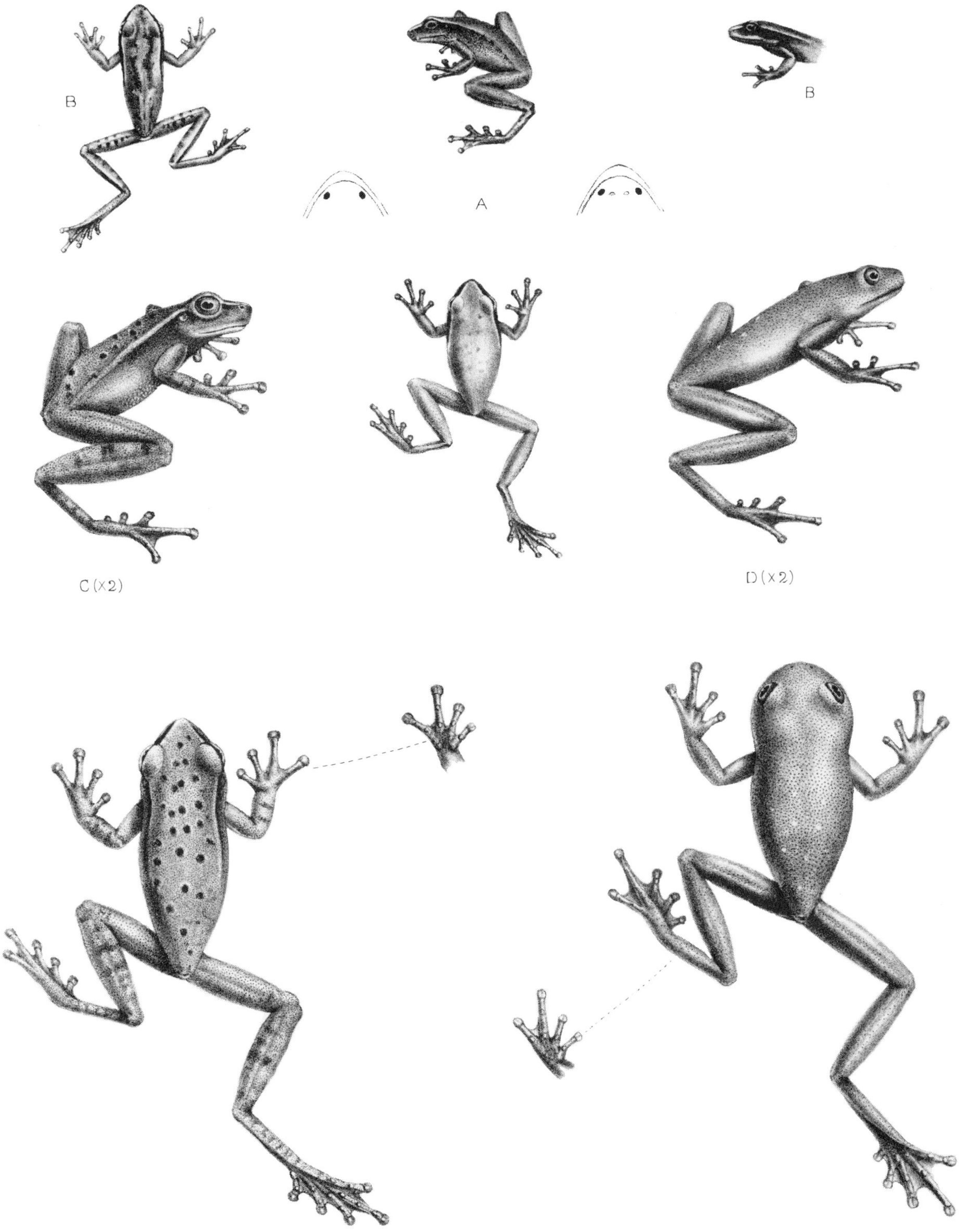

A. HYLA NANA. B. HYLA MOCQUARDI. C. HYLELLA PICTA.
D. HYLELLA FLEISCHMANNI.

J. Green del et lith. Mintern Bros imp.

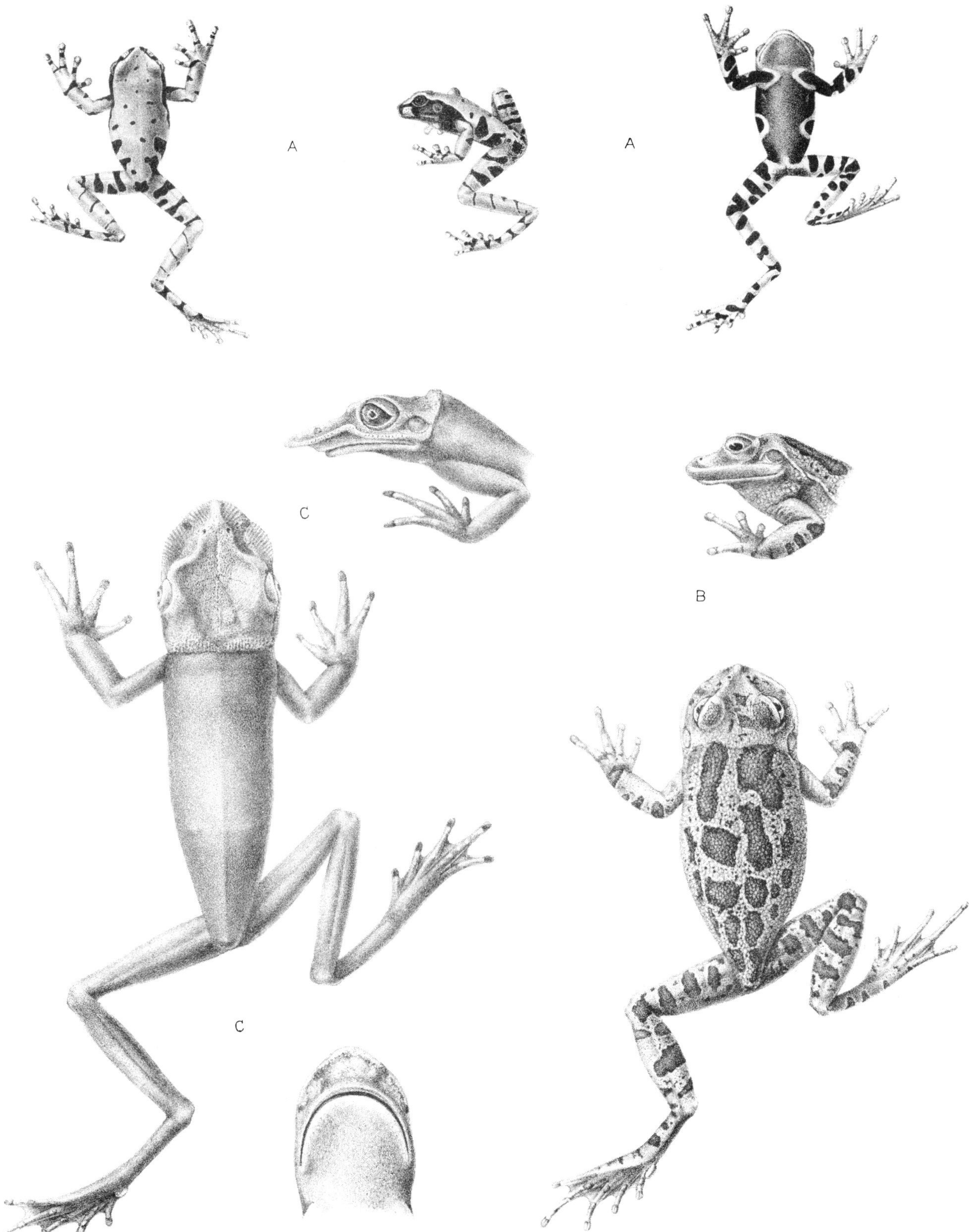

A. NOTOTREMA OVIFERUM, juv. B. PTERNOHYLA FODIENS. C. TRIPRION SPATULATUS.

J Green del. et lith. Mintern Bros. imp.

SPELERPES VARIEGATUS. VARIETIES.

J. Green del. et lith. Mintern Bros. imp.

A. SPELERPES BOCOURTI. B. DERMOPHIS GRACILIOR.

J. Green del. et lith. Mintern Bros. imp.

PUBLICATIONS OF THE
SOCIETY FOR THE STUDY OF AMPHIBIANS AND REPTILES

PUBLICATIONS of the Society may be purchased from the Publications Secretary, Douglas H. Taylor, Department of Zoology, Miami University, Oxford, Ohio 45056, U.S.A. Prices are effective as of March 1987 and subject to revision. Please make checks payable to "SSAR." Prices include shipping charges (book rate) within the USA; all overseas orders will be billed only for the *additional* postage charges in excess of domestic rates. Publications sent at customer's risk; however, packages can be insured at customer's cost. Overseas customers must make payment in USA funds, by International Money Order, or by MasterCard or VISA (in which case account number and expiration date must be provided). Items marked "out-of-print" are no longer available.

CONTRIBUTIONS TO HERPETOLOGY

Book-length monographs, comprising taxonomic revisions, results of symposia, and other major works. Issued irregularly by subscription.

No. 1. *Reproductive Biology and Diseases of Captive Reptiles,* by James B. Murphy and Joseph T. Collins (eds.). 1980. Results of a Society-sponsored symposium, including papers by 37 authors. 287 p., illus. Paperbound $25.00.

No. 2. *The Turtles of Venezuela,* by Peter C. H. Pritchard and Pedro Trebbau. 1984. An exhaustive natural history, covering half of the turtle species of South America. 414 p., 48 color plates (25 watercolor portraits and 165 photographs of turtles and habitats), 16 maps. Regular edition, clothbound $45.00; patron's edition, two leatherbound volumes in slipcase, signed and numbered by authors and artist $300.00; set of 25 watercolor prints of turtles, on heavy paper stock and in protective wrapper $30.00.

No. 3. *Introduction to the Herpetofauna of Costa Rica / Introducción a la Herpetofauna de Costa Rica,* by Jay M. Savage and Jaime Villa R. 1986. Bilingual edition, in English and Spanish, with distribution checklist, bibliographies, and extensive keys. 220 p., illus. (one color plate). Clothbound $30.00.

FACSIMILE REPRINTS IN HERPETOLOGY

Photolithographic reprints of classic and important books and papers. Most titles typically have extensive new introductions by leading authorities. Issued irregularly by subscription.

ANDERSON, J. 1896. *Contribution to the Herpetology of Arabia.* Introduction and new checklist of Arabian amphibians and reptiles by Alan E. Leviton and Michele L. Aldrich. 160 p., illus. (one plate in color). Clothbound $24.00.

BELL, T. 1842-1843. *Herpetology of the "Beagle."* Part 5 of Darwin's classic, "Zoology of the Voyage of H.M.S. Beagle," containing descriptions of amphibians and reptiles collected on the expedition. Introduction by Roberto Donoso-Barros. 100 p., 20 plates. Paperbound $12.00, clothbound $17.00.

BOJANUS, L. H. 1819-1821. *Anatome Testudinis Europaeae.* The standard atlas of turtle anatomy. Introduction by Alfred Sherwood Romer. 200 p., 40 foldout plates. Paperbound $25.00.

BOULENGER, G. A. 1877-1920. *Contributions to American Herpetology.* A series of collected papers from various journals; complete in 18 parts totalling 900 p., numerous illustrations. Introduction by James C. Battersby. Paperbound.

Table of Contents: Parts 1-10, 1877-1897, 20 p., $2.00.

Part 1: 1877-1881, 6 papers, 32 p., $2.00.
2: 1882-1883, 12 papers, 48 p., $2.00 (available only as part of a complete set).
3: 1884-1885, 10 papers, 44 p., $2.00.
4: 1886-1887, 10 papers, 56 p., $2.00.
5: 1888, 13 papers, 32 p., $2.00.
6: 1889-1890, 10 papers, 48 p., $2.00.
7: 1891-1893, 11 papers, 44 p., $2.00.
8: 1894, 5 papers, 40 p., $2.00.
9: 1895-1896, 11 papers, 52 p., $2.00.
10: 1897, 2 papers, 56 p., $2.00.

Table of Contents: Parts 11-18, 1898-1920, 12 p. $2.00.

Part 11: 1898, 4 papers, 52 p., $2.00.
12: 1899-1900, 9 papers, 48 p., $2.00.
13: 1901-1902, 7 papers, 34 p., $2.00.

14: 1903-1904, 8 papers, 28 p., $2.00.
15: 1905-1911, 12 papers, 42 p., $2.00.
16: 1912-1913, 6 papers, 46 p., $2.00.
17: 1914-1919, 8 papers, 38 p., $2.00.
18: 1920, 3 papers, 64 p., $2.00.

Index, 44 p., $2.00.

Complete set: 18 parts plus Index and Tables of Contents, in parts as issued, $35.00.

BULLETIN OF THE ANTIVENIN INSTITUTE OF AMERICA. Volumes 1-5, 1927-1932. Complete set of an important journal devoted to venomology and to the taxonomy, biology and distribution of snakes, lizards, salamanders and crocodilians. Introduction by Sherman A. Minton. 575 p., 163 photographs, maps, index. Paperbound $25.00, clothbound $28.00.

CAMP, C. L. 1923. *Classification of the Lizards*. The foundation of modern lizard systematics. New preface by the author and an introduction by Garth Underwood. 220 p., 112 figures, index. Out-of-print.

CHANG, M. L. Y. 1936. *Amphibiens Urodèles de la Chine*. The standard work on Chinese salamanders, with a new checklist by Arden H. Brame. 168 p., illus., 5 plates. Paperbound $8.00.

COPE, E. D. 1864. *Papers on the Higher Classification of Frogs*. Reprinted from Proceedings of the Academy of Natural Sciences of Philadelphia and Natural History Review. 32 p. Paperbound $2.00.

COPE, E. D. 1871. *Catalogue of Batrachia and Reptilia Obtained by McNiel in Nicaragua; Catalogue of Reptilia and Batrachia Obtained by Maynard in Florida*. 8 p. Paperbound $2.00.

COPE, E. D. 1892. *The Osteology of the Lacertilia*. An important contribution to lizard anatomy, reprinted from Proceedings of the American Philosophical Society. 44 p., 6 plates. Paperbound $4.00.

COWLES, R. B. and C. M. BOGERT. 1944. *A Preliminary Study of the Thermal Requirements of Desert Reptiles*. The foundation of thermoregulation biology, with extensive review of recent studies by F. Harvey Pough. Reprinted from Bulletin of American Museum of Natural History. 52 p., 11 plates. Paperbound $5.00.

DUNN, E. R. 1926. *Salamanders of the Family Plethodontidae*. A recognized classic treatment of the plethodontid salamanders, including tropical and European species. Introductions by David B. Wake and Arden H. Brame. 480 p., illus., 3 plates, 86 maps, index. Paperbound $16.00.

ESCHSCHOLTZ, F. 1829-1833. *Zoologischer Atlas* (herpetological sections). Descriptions of new reptiles and amphibians from California and the Pacific. Introduction by Kraig Adler. 32 p., 4 plates. Paperbound $3.00.

ESPADA, M. J. DE LA. 1875. *Vertebrados del Viaje al Pacifico: Batracios*. Espada's major work on South American frogs. Introduction by Jay M. Savage. 208 p., 6 plates. Clothbound $18.00.

FITZINGER, L. J. 1843. *Systema Reptilium*. An important nomenclatural landmark for herpetology, including Amphibia as well as reptiles; world-wide in scope. Introduction by Robert Mertens. 128 p., index. Paperbound $12.00, clothbound $16.00.

GLOYD, H. K. 1940. *The Rattlesnakes, Genera* Sistrurus *and* Crotalus. Introduction and new checklist by Hobart M. Smith and Herbert M. Harris. 300 p., plus 31 plates of photographs, index. Clothbound $25.00.

GRAY, J. E. 1825. *A Synopsis of the Genera of Reptiles and Amphibia*. Reprinted from Annals of Philosophy. 32 p. Paperbound $3.00.

GRAY, J. E. 1831-1844. *Zoological Miscellany*. A privately printed journal, devoted mostly to descriptions of amphibians, reptiles and birds from throughout the world. Introduction by Arnold G. Kluge. 86 p., 4 plates. Paperbound $6.00, clothbound $10.00.

GÜNTHER, A. 1885-1902. *Biologia Centrali-Americana. Reptilia and Batrachia*. The standard work on Middle America, with 76 full-page plates (12 in full color). Introductions by Hobart M. Smith, A. E. Gunther, and Kraig Adler. 575 p. clothbound $50; separate set of 12 color plates, in protective wrapper $18.

HOLBROOK, J. E. 1842. *North American Herpetology*. Five volumes bound in one. The classic work by the father of American herpetology. This is an exact facsimile of the second edition, including all 147 plates (20 reproduced in full color). Introduction and checklists by Richard and Patricia Worthington and by Kraig Adler. 1032 p. Regular edition, clothbound $60.00; patron's edition, out-of-print.

JUNIOR SOCIETY OF NATURAL SCIENCES (CINCINNATI, OHIO). 1930-1932. Herpetological papers from the society's Proceedings, with articles by Weller, Walker, Dury and others. 56 p. Paperbound $2.00.

KIRTLAND, J. P. 1838. *Zoology of Ohio* (herpetological portion). 8 p. Paperbound $1.00.

LECONTE, J. E. 1824-1828. *Three Papers on Amphibians*, from the Annals of the Lyceum of Natural History, New York. 16 p. Paperbound $1.00.

LINNAEUS, C. 1766-1771. *Systema Naturae* (ed. 12) and *Mantissa Plantarum* (herpetological portions from both). 56 p. Introduction by Kraig Adler. Out-of-print.

LOVERIDGE, A. 1946. *Reptiles [and Amphibians] of the Pacific*. The standard review of the herpetofauna of the Pacific region including Australia and extending from Indonesia to Hawaii and the Galápagos Islands. 271 p., 7 plates, 1 double-page map, index. Paperbound $17.00.

MCILHENNY, E. A. 1935. *The Alligator's Life History*. The most complete natural history of the American alligator. Introduction by Archie Carr and a review of recent literature by Jeffrey W. Lang. 125 p., 18 photographs and a portrait. Paperbound $18.00, clothbound $22.00.

MCLAIN, R. B. 1899. *Contributions to North American Herpetology*. Three privately printed papers containing important distributional records and the description of a new form. 28 p., index. Paperbound $2.00.

ORBIGNY, A. D' [and G. BIBRON]. 1847. *Voyage dans l'Amérique Méridionale*. This extract comprises the complete section on reptiles and amphibians. 14 p., 9 plates. Out-of-print.

RAFINESQUE, C. S. 1820. *Annals of Nature* (herpetological and ichthyological sections), 4 p. Paperbound $0.50.

RAFINESQUE, C. S. 1822. *On Two New Salamanders of Kentucky*. 2 p. Paperbound $1.00.

RAFINESQUE, C. S. 1832-1833. *Five Herpetological Papers from the Atlantic Journal*. 4 p. Paperbound $1.00.

SOWERBY, J. DEC., E. LEAR and J. E. GRAY. 1872. *Tortoises, Terrapins, and Turtles Drawn From Life*. The finest atlas of turtle illustrations ever produced. Introduction by Ernest E. Williams. 26 p., 61 full-page plates. Clothbound $20.00.

SPIX, J. B. VON and J. G. WAGLER. 1824-1825. *Herpetology of Brazil*. The most comprehensive and important early survey of Brazilian herpetology. Introduction by P. E. Vanzolini. 400 p., 98 plates (one in color). Clothbound $36.00.

TROSCHEL, F. H. 1850 (1852). Cophosaurus texanus, *neue Eidechsengattung aus Texas*. 8 p. Paperbound $1.00.

TSCHUDI, J. J. VON. 1838. *Classification der Batrachier*. A major work in systematic herpetology, with introduction by Robert Mertens. 118 p., 6 plates. Paperbound $10.00.

TSCHUDI, J. J. VON. 1845. *Reptilium Conspectus*. Descriptions of new reptiles and amphibians from Peru. 24 p. Paperbound $2.00.

VANDENBURGH, J. 1895-1896. *Herpetology of Lower California*. A review of the herpetology of Baja California (collected papers). 101 p., 11 plates, index. Paperbound $6.00.

WIEGMANN, A. F. A. 1834. *Herpetologia Mexicana*. Introduction by Edward H. Taylor. 66 p., 10 plates (folio, measuring 10 x 14 inches). Out-of-print.

WILCOX, E. V. 1891. *Notes on Ohio Batrachians*. 3 p. Paperbound $1.00.

WILLISTON, S. W. 1925. *Osteology of the Reptiles*. Covers living and extinct forms, with introduction by Claude W. Hibbard. 304 p., 191 text figures, index. Paperbound $12.00, clothbound $18.00.

WRIGHT, A. H. and A. A. WRIGHT. 1962. *Handbook of Snakes of the United States and Canada, Vol. 3, Bibliography*. Out-of-print since about 1969, this bibliography is a necessary companion to Volumes 1 and 2. 187 p. Clothbound $18.00.

HERPETOLOGICAL REVIEW

The Society's official newsletter, international in coverage. Most issues contain extensive listings of current literature, a section that from 1983 became a separate serial, *Recent Herpetological Literature*. Issued quarterly as part of Society membership. All numbers paperbound as issued.

Volume 1, number 1 (1967), $3.00.
number 2 (1968), $3.00.
number 3 (1968), $3.00.
number 4 (1968), $3.00.
number 5 (1968), $3.00.
number 6 (1969), $3.00.
number 7 (1969), out-of-print.
number 8 (1969), $3.00.
number 9 (1969), $3.00.
Volume 2, number 1 (1970), $3.00.
number 2 (1970), $3.00.
number 3 (1970), $3.00.
number 4 (1970), $3.00.
Volume 3, number 1 (1971), $3.00.
number 2 (1971), out-of-print.
number 3 (1971), $3.00.
number 4 (1971), $3.00.
number 5 (1971), $3.00.
number 6 (1971), $3.00.
Volume 4, number 1 (1972), out-of-print.
number 2 (1972), $3.00.
number 3 (1972), $3.00.
number 4 (1972), $3.00.
number 5 (1972), $3.00.
number 6 (1972), $3.00.
Volume 5, number 1 (1974), $3.00.
number 2 (1974), out-of-print.
number 3 (1974), $3.00.
number 4 (1974), $3.00.
Volume 6, number 1 (1975), $3.00.
number 2 (1975), $3.00.
number 3 (1975), $3.00.
number 4 (1975), out-of-print.
Volume 7, number 1 (1976), $3.00.
number 2 (1976), $3.00.
number 3 (1976), $3.00.
number 4 (1976), $3.00.
Cumulative Index for Volumes 1-7 (1967-1976), $5.00.
Volume 8, number 1 (1977), $3.00.
number 2 (1977), $3.00.
number 3 (1977), $3.00.
number 4 (1977), $3.00.
Volume 9, number 1 (1978), $3.00.
number 2 (1978), $3.00.
number 3 (1978), $3.00.
number 4 (1978), $3.00.
Volume 10, number 1 (1979), $3.00.
number 2 (1979), $3.00.
number 3 (1979), $3.00.
number 4 (1979), $3.00.
Volume 11, number 1 (1980), $3.00.
number 2 (1980), $3.00.
number 3 (1980), $3.00.
number 4 (1980), $3.00.
Volume 12, number 1 (1981), $3.00.
number 2 (1981), $3.00.
number 3 (1981), $3.00.
number 4 (1981), $3.00.
Volume 13, number 1 (1982), $3.00.
number 2 (1982), $3.00.
number 3 (1982), $3.00.
number 4 (1982), $3.00.

Volume 14, number 1 (1983), $3.00.
number 2 (1983), $3.00.
number 3 (1983), $3.00.
number 4 (1983), $3.00.
Volume 15, number 1 (1984), $3.00.
number 2 (1984), $3.00.
number 3 (1984), $3.00.
number 4 (1984), $3.00.
Volume 16, number 1 (1985), $3.00.
number 2 (1985), $3.00.
number 3 (1985), $3.00.
number 4 (1985), $3.00.
Volume 17, number 1 (1986), $3.00.
number 2 (1986), $3.00.
number 3 (1986), $3.00.
number 4 (1986), $3.00.

HERPETOLOGICAL CIRCULARS

Miscellaneous publications of general interest to the herpetological community. Issued irregularly and by subscription. All numbers paperbound, as issued.

No. 1. *A Guide to Preservation Techniques for Amphibians and Reptiles* by George R. Pisani. 1973. 22 p., illus. $2.00.

No. 2. *Guía de Técnicas de Preservación de Anfibios y Reptiles* por George R. Pisani y Jaime Villa. 1974. 28 p., illus. $1.00.

No. 3. *Collections of Preserved Amphibians and Reptiles in the United States* compiled by David B. Wake (chair) and the Committee on Resources in Herpetology. 1975. 22 p. Out-of-print.

No. 4. *A Brief Outline of Suggested Treatments for Diseases of Captive Reptiles* by James B. Murphy. 1975. 13 p. $2.00.

No. 5. *Endangered and Threatened Amphibians and Reptiles in the United States* compiled by Ray E. Ashton, Jr. (chair) and the 1973-74 SSAR Regional Herpetological Societies Liaison Committee. 1976. 65 p. Out-of-print.

No. 6. *Longevity of Reptiles and Amphibians in North American Collections* by J. Kevin Bowler. 1977. 32 p. $2.00.

No. 7. *Standard Common and Current Scientific Names for North American Amphibians and Reptiles* (1st ed.) by Joseph T. Collins, James E. Huheey, James L. Knight and Hobart M. Smith. 1978. 36 p. $3.00.

No. 8. *A Brief History of Herpetology in North America Before 1900* by Kraig Adler. 1979. 40 p., 24 photographs, 1 map. $2.00.

No. 9. *A Review of Marking Techniques for Amphibians and Reptiles* by John W. Ferner. 1979. 42 p., illus. $3.00.

No. 10. *Vernacular Names of South American Turtles* by Russell A. Mittermeier, Federico Medem and Anders G. J. Rhodin. 1980. 44 p. $2.50.

No. 11. *Recent Instances of Albinism in North American Amphibians and Reptiles* by Stanley Dyrkacz. 1981. 36 p. $3.00.

No. 12. *Standard Common and Current Scientific Names for North American Amphibians and Reptiles* (2nd ed.) by Joseph T. Collins, Roger Conant, James E. Huheey, James L. Knight, Eric M. Rundquist, and Hobart M. Smith. 1982. 32 p. $3.00.

No. 13. *Silver Anniversary Membership Directory*, including addresses of all SSAR members, addresses and publications of the herpetological societies of the world, and a brief history of the Society. 1983. 56 p., 4 photographs. $2.00.

No. 14. *Checklist of the Turtles of the World with English Common Names* by John Iverson. 1985. 14 p. $3.00.

No. 15. *Cannibalism in Reptiles: A World-wide Review* by Joseph C. Mitchell. 1986. 37 p. $4.00.

RECENT HERPETOLOGICAL LITERATURE

A comprehensive listing of titles and citations, with authors' addresses. Prior to 1983 these lists were published in *Herpetological Review*. Issued irregularly as part of Society membership. Paperbound, as issued. Series now suspended.

Number 1 (1983), 66 p., $3.00.
2 (1984), 44 p., $3.00.

CATALOGUE OF AMERICAN AMPHIBIANS AND REPTILES

Loose-leaf accounts of taxa prepared by specialists, including synonymy, definition, description, distribution map and comprehensive list of literature for each taxon. Covers amphibians and reptiles of the entire Western Hemisphere. Issued by subscription, although individual accounts or groups of accounts may be purchased.

THE CATALOGUE: Imprinted binder, taxonomic tabs, and accounts 1-150, $60.00.

IMPRINTED POST BINDER: $20.00.

SYSTEMATIC TABS (Ten tabs to fit binder: "Class Amphibia," "Order Caudata," etc.), $4.00.

CATALOGUE ACCOUNTS: Numbers 1-25, 26-50, 51-75, 76-100, 101-125, 126-150, 151-175, 176-190, 191-205, 206-220, 221-245, 246-270, 271-295, 296-320, 321-345, 346-370, 371-390, and 391-410. Price for each group of numbered accounts is $10.00.

INDIVIDUAL ACCOUNTS: $0.50 each (please order by account number). In a number of instances only xerox copies of *individual* accounts can be supplied. However, complete sets and the groups of numbered accounts listed above are all originals.

JOURNAL OF HERPETOLOGY

The Society's official scientific journal, international in scope. Issued quarterly as part of Society membership. All numbers paperbound as issued. Table of Contents are supplied *gratis* with orders for complete volumes.

Volume 1, numbers 1-4 (1968), out-of-print.
Volume 2, numbers 1-2 (1968), out-of-print.
numbers 3-4 (1968), out-of-print.
Volume 3, numbers 1-2 (1969), out-of-print.
numbers 3-4 (1969), $5.00.
Volume 4, numbers 1-2 (1970), out-of-print.
numbers 3-4 (1970), out-of-print.
Volume 5, numbers 1-2 (1971), $5.00.
numbers 3-4 (1971), out-of-print.
Volume 6, number 1 (1972), out-of-print.
number 2 (1972), out-of-print.
numbers 3-4 (1972), out-of-print.
Volume 7, number 1 (1973), out-of-print.
number 2 (1973), $5.00.
number 3 (1973), $5.00.
number 4 (1973), $5.00.
Volume 8, number 1 (1974), out-of-print.
number 2 (1974), $5.00.
number 3 (1974), $5.00.
number 4 (1974), $5.00.
Volume 9, number 1 (1975), out-of-print.
number 2 (1975), out-of-print.
number 3 (1975), $5.00.
number 4 (1975), $5.00.
Volume 10, number 1 (1976), $5.00.
number 2 (1976), $5.00.
number 3 (1976), $5.00.
number 4 (1976), $5.00.
Cumulative Index for Volumes 1-10 (1968-1976), $7.00.
Volume 11, number 1 (1977), $5.00.
number 2 (1977), $5.00.
number 3 (1977), $5.00.
number 4 (1977), out-of-print.
Volume 12, number 1 (1978), out-of-print.
number 2 (1978), $5.00.
number 3 (1978), $5.00.
number 4 (1978), $5.00.
Volume 13, number 1 (1979), $5.00.
number 2 (1979), $5.00.
number 3 (1979), $5.00.
number 4 (1979), $5.00.
Volume 14, number 1 (1980), $5.00.
number 2 (1980), $5.00.
number 3 (1980), $5.00.
number 4 (1980), $5.00.
Volume 15, number 1 (1981), $5.00.
number 2 (1981), $5.00.
number 3 (1981), $5.00.
number 4 (1981), $5.00.
Volume 16, number 1 (1982), $5.00.
number 2 (1982), $5.00.
number 3 (1982), $5.00.
number 4 (1982), $5.00.
Volume 17, number 1 (1983), $5.00.
number 2 (1983), $5.00.
number 3 (1983), $5.00.
number 4 (1983), $5.00.
Volume 18, number 1 (1984), $5.00.
number 2 (1984), $5.00.
number 3 (1984), $5.00.
number 4 (1984), $5.00.
Volume 19, number 1 (1985), $5.00.
number 2 (1985), $5.00.
number 3 (1985), $5.00.
number 4 (1985), $5.00.
Volume 20, number 1 (1986), $5.00.
number 2 (1986), $5.00.
number 3 (1986), $5.00.
number 4 (1986), $5.00.

PUBLICATIONS OF THE OHIO HERPETOLOGICAL SOCIETY

OHS was the predecessor to the Society for the Study of Amphibians and Reptiles. All publications international in scope. Paperbound as issued.

Volume 1, numbers 1-4, plus Special Publications 1-2 (all 1958), facsimile reprint, $3.00.
Volume 2, number 1 (1959), $3.00.
number 2 (1959), out-of-print.
number 3 (1960), out-of-print.
number 4 (1960), out-of-print.
Volume 3, number 1 (1961), out-of-print.
number 2 (1961), $0.50.
number 3 (1962), $0.50.
number 4 (1962), $0.50.
Volume 4, numbers 1-2 (1963), $2.00.
number 3 (1964), $1.00.
number 4 (1964), $1.00.
Volume 5, number 1 (1965), $1.00.
number 2 (1965), $1.00.
number 3 (1966), $1.00.
number 4 (1966), $1.00.
Special Publication 3 (1961), out-of-print.
Special Publication 4 (1962), $1.00.

OTHER MATERIALS AVAILABLE FROM THE SOCIETY

Color Prints

FIJI IGUANAS. Two full-color prints (7 x 10 inches) of *Brachylophus fasciatus* and *B. vitiensis*, as published in *Journal of Herpetology*, vol. 15, no. 3, July 1981. Special printing limited to 50 sets, individually signed by the artist David M. Dennis. Set of two, $15.00.

SILVER ANNIVERSARY COMMEMORATIVE PRINT. A full-color print (11 1/2 x 15 1/4 inches) of a Gila monster (*Heloderma suspectum*) on natural background, from a watercolor by David M. Dennis. Issued as part of Society's 25th Anniversary in 1982. Edition limited to 1000. $8.00 each or $5.00 in quantities of 10 or more.

PRODUCTION SPECIFICATIONS

TYPESETTING AND LAYOUT OF NEW INTRODUCTIONS: Keyboarding by Cornell University, Ithaca, New York, U.S.A. (Carol Abarbanell, Stacey Coil) and University of Colorado, Boulder, Colorado, U.S.A. (Phyllis O'Connell). Type was set in Times face on a Macintosh computer using Macwrite and Microsoft Word programs.

PRINTING: Cushing-Malloy, Inc., Ann Arbor, Michigan, U.S.A. (Thomas F. Weber). The text is printed on 50-pound Glatfelter's natural stock and the black-and-white plates on 70-pound Mead's Moistrite matte stock. Both papers are acid- and groundwood-free stocks.

COLOR PLATES: Brodock Press, Inc., Utica, New York, U.S.A. (Craig S. Brodock, Robert Hausle, Donald W. Weagley). The color plates were laser scanned and are printed on 70-pound acid-free Lustro dull text stock. A separate set of the color plates, with special title page, was also issued (400 sets).

BINDING: John H. Dekker and Sons, Grand Rapids, Michigan, U.S.A. The spine is covered in leather-simulated Type II Tara cloth (Whitman Products Ltd., West Warwick, Rhode Island, U.S.A.) and the covers and endpapers in Multicolor Antique papers (Process Materials Corp., Carlstadt, New Jersey, U.S.A.).

DATE OF PUBLICATION: 15 September 1987.

PLACE OF PUBLICATION: Athens, Ohio, U.S.A.

NUMBER OF COPIES: 800.